C.H.Dowding

D1236675

Third International Conference On
STABILITY IN SURFACE MINING

June 1, 2, 3 - 1981
Vancouver, British Columbia Canada

C. O. Brawner, Editor

Advisory Organizations

Society of Mining Engineers of AIME
B.C. and Yukon Chamber of Mines
B. C. Mining Association
B.C. Dept. of Mines and Petroleum Resources

Published by the
Society of Mining Engineers
of the
American Institute of Mining, Metallurgical, and Petroleum Engineers, Inc.

New York, New York • 1982

Copyright © 1982 by
The American Institute of Mining, Metallurgical,
and Petroleum Engineers, Inc.

Printed in the United States of America
By Edwards Brothers Inc., Ann Arbor, Michigan

All rights reserved. This book, or parts thereof, may not be
reproduced in any form without permission of the publisher.

Library of Congress Catalog Card Number 81-70690
ISBN 0-89520-292-1

Conference Organizing Committee

Symposium Chairman

C. O. Brawner, P. Eng.

Mineral Engineering Department
University of British Columbia
Vancouver, BC

Advisory Committee

K. Barron

Coal Mining Research
  Center
Edmonton, Alberta

A. Brown

Golder Associates Inc.
Denver, Colorado

K. Broadbent

Kennecott Mining Company
Engineering Center
Salt Lake City, Utah

R. Call

R.D. Call Inc.
Tucson, Arizona

W. Hustralid

Colorado School of
  Mines
Golden, Colorado

Advisory Organizations

Society of Mining Engineers of AIME
B.C. and Yukon Chamber of Mines
B.C. Mining Association
B.C. Dept. of Mines and Petroleum Resources

iii

C. O. BRAWNER, EDITOR

C.O. Brawner has spent 29 years in geotechnical engineering specializing in soil mechanics, landslides, earth and tailings dams and rock mechanics relating to transportation and mining engineering.

He is a graduate of the University of Manitoba and Nova Scotia Technical College.

Mr. Brawner spent 10 years with the British Columbia Department of Highways, the last six as Senior Materials and Testing Engineer. From 1963 to 1978 he was a principal and ultimately president of Golder Brawner and Associates, Consulting Geotechnical Engineers. Since 1978 he has been on the staff of the Mining Department of the University of British Columbia teaching Geomechanics in Mining as well as providing specialist review consulting to the Mining and Transportation industries. In this capacity he has served on numerous geotechnical review boards including Syncrude Canada Ltd., the North East Coal Program in B.C. and the State Electric Commission of Australia.

Mr Brawner has been a member of the N.R.C. Associate Committee for Geotechnical Research, Vice Chairman of the Canadian Advisory Committee on Rock Mechanics, Chairman of the CACRM Committee on Mine Waste Embankments, Specialist Consultant to the Canadian Department of Mines for the development of the "Design Guide for Mine Waste Enbankments in Canada", Specialist Technical Advisor to the World Health Organization, Special Technical Advisor to the B.C. Uranium Commission for Uranium Waste, Chairman of CANCOLD Committee on Industrial and Tailings Dams and Program Chairman of the First, Second and Third Conferences on Surface Mine Stability, First International Conference on Stability in Coal Mining, First International Conference on Mine Drainage and First International Conference on Uranium Mine Waste Disposal.

He is author of over 65 technical papers, Supplementary Reporter to the Third International Rock Mechanics Conference, Penrose Lecturer for the Geological Society of America and invited lecturer to over 70 conferences, universities and engineering institutes internationally, including United States, Britain, Japan, Germany, China and Russia.

Brawner was awarded the Presidents Medal by the Canadian Association of Good Roads, Award of Merit of the Association of Consulting Engineers of Canada, the Meritorious Service Award of the Association of Professional Engineers of British Columbia and a Publication Award by the Society of Mining Engineers of AIME.

FOREWORD

Ten years have passed since the Second International Conference on Stability in Open Pit Mining. In that time many advances have been made in technology relating to Structural Geologic interpretation techniques, instrumentation to monitor ground water pressures and permeability, blasting and its influence on stability, methods to monitor movement and interpret the mechanics of movement, and analysis methods to evaluate stability quantitatively.

Important developments in Investigation, Research and Design include the recognition of the influence of a "Key Block" in failure, the portential application of Rock Classification for surface rock design, stability analysis in heavily jointed rock, estimation of ground water drawdown and the development of a single core orientation technique.

Improvement in stabilization technology includes more effective techniques to drain slopes, particularly the addition of vacuum drainage, significant improvement in blasting techniques to reduce damage to the final slopes, slope reinforcement with tensioned anchor cables or untensioned vertical reinforcement near the toe, blasting to break up thin, weak layers and reduce pore water pressures, and major improvements in electronic distance measuring devices to monitor movements.

The proceedings of this conference will provide an extremely useful upgrading of the state of knowledge in surface rock mechanics and stability.

The program was developed in three parts:

1) State of the Art of Rock Slope Stability

2) Investigation, Research and Design for Stability in Surface Mining

3) Case examples of Stability in Surface Mining

Many internationally recognized experts contributed to make the program a great success.

The willingness of Mining Companies to present case examples was particularly gratifying. Only with such reviews can we realistically assess the validity and application of theoretical concepts, testing and analysis to the solution of practical problems of design and stabilization.

It is hoped that these proceedings will generate extended interest in Rock Mechanics related to Surface Mining.

A total of 37 papers were presented by speakers from seven countries. Twenty three countries were represented.

C.O. Brawner
Editor

v

Table of Contents

1. State of the Art — Rock Slope Stability

2. Investigation, Research and Design

3. Case Examples

1
State of the Art
Rock Slope Stability

Chapter 1

THE ROLE OF SLOPE STABILITY
IN THE ECONOMICS, DESIGN AND OPERATION
OF OPEN PIT MINES - AN UPDATE

Michael Richings

Associate
Golder Associates
Denver, Colorado

ABSTRACT

Although the role of slope stability has not changed, there have been changes in the mining industry which affect the geotechnician engaged in slope stability studies. The deposits currently being mined or evaluated are technically difficult, capital intensive, and economically marginal. In addition, many mining companies are now either controlled by oil companies or are financing projects through bank loans. The mining investor or decision maker is, therefore, often more sophisticated financially, but less sophisticated in mining terms. The impact of these changes will be to force the geotechnician and mining engineer to attempt to quantify risk and financial implications of a slope design.

In conclusion, this paper outlines several questions which mining engineers should be asking those engaged in geotechnical studies and indicates the necessity for the mining engineer and geotechnician to work together and quantify the financial risk associated with a geotechnical design.

INTRODUCTION

This paper is written from the viewpoint of a mining engineer who is responsible for determining the viability of new deposits and ensuring the continued profitability of existing ones. This requires the preparation of designs or plans that utilize, among other data, information supplied by the geotechnician engaged in pit wall and

1

waste embankment stability studies. It is my intent to show that, although the role of slope stability (Stewart, Kennedy 1977) in open pit mining economics, design, and operation has not substantially changed, there have been changes in the mining industry and in the deposits being mined or evaluated that have had a significant impact on the geotechnician and mining engineer.

CHANGES

The changes described below have developed for various reasons, but they have a common component. That component is the increasing need for the engineer to express his answers in financial terms.

The first change is that open pit projects that were in their infancy 10 years ago are now mature, and some are experiencing major instability problems. So, for the first time, the geotechnician is in the position of living with his past mistakes and is now having to recommend solutions that will have an immediate impact on the economic position of the mine.

Second, there has been a significant change in the nature of ore-bodies being mined or evaluated. As a general rule, they are of lower grade, more remotely located, and more technically difficult to develop. It should be noted at this point that not all of the difficulties that we now encounter are inherent in the orebody. Some have been generated by the imposition of new mining regulations. Typical of the technical difficulties encountered in mines currently being evaluated are:

o Very high pit walls (2,000 feet)
o 1,000 foot high waste dumps
o Million ton per day mining rates
o Extreme climates
o Unusual material types such as leached cappings, laterites, and bentonitic clays

These and other factors have resulted in much higher initial investment in relation to productive capacity and, overall, in more marginal economics. For example, large projects being developed in the late 1960's and early 1970's required initial investments in the $40 to $150 million range; a project like Bougainville, at $400 million, is an extreme exception. Today, even projects in developed areas of the United States will cost in excess of $300 million. More remote projects involve investments of up to $1 billion, with $2 to $3 billion being contemplated in some areas.

The third change has been in the nature of the financing for these projects. The need for capital in the mining industry has outstripped most mining companies' ability to provide this financing internally. Mining companies have, therefore, been prime targets for oil companies anxious to diversify, or, alternatively, they have been required to turn to bank loans to finance projects. Some of these are business loans that are secured by assets and revenue from sources not associated with the project. Others, however, are financed by using the project or mining venture itself as the security. In this case, the viability of the project must be clearly demonstrated prior to any financial commitment. These changes are evidenced by the fact that mining companies' debt ratios (Ballmer 1981) have increased from 10% in the 1960's to approximately 35% today. In many major mines now in operation or being developed, the individual making the final decision is not a mining man, but either an oil company executive or a banker.

IMPACTS OF CHANGES

As a result of these changes, I can identify two related impacts on the geotechnician/mining engineer engaged in slope stability studies. The first one is readily apparent. The second is only just appearing, but can be expected to enter more into future studies.

First, slope stability, waste embankment, and other geotechnical studies have become a necessity. The potential to minimize early expenditures using good geotechnical designs, the economic sensitivity of existing operation to failures, and the imposition of new regulations have resulted in undreamed-of amounts of money, compared with 10 years ago, being spent on geotechnical studies.

The second impact is the need to present the results in a manner that can be used by the mining engineer to assess the risk and then to translate this risk into financial terms. Because the individual evaluating the investment alternatives will be heavily financially oriented, it is important that the mining engineer responsible for the development or operation bridge the gap and quantify the mining risks in understandable terms.

It is this aspect that creates difficulties for miners. One approach (Pentz 1981) is to calculate the reliability (or uncertainty) for a given slope design. Using this method, the potential financial benefits of assuming various degrees of risk can be made. To provide the manager with all the information he requires to make a decision, a final step is necessary. That step is to undertake an analysis which compares this risk and benefit with other risks and financial implications of accepted business parameters such as prime forecasts, budgeted costs, and productivities. Such an analysis will provide him with a basis for a decision. The manager can then determine the reliability of the required slope design, or in the case of a decreasing reliability associated with steeper slope design, he can determine how much risk he is willing to assume.

CONCLUSION

I have therefore summarized questions that should be asked and have attempted to illustrate the manner in which they should be answered.

o How much money should be spent on gathering data or performing slope stability and other geotechnical studies?

The program should obtain sufficient data to enable designs to be made, so what when they are part of the overall development, they will not hazard the successful performance of the operation and hence its financial return. It is normally expected that about 10 to 20 percent of the total engineering hours must be spent on the design of mining plant and facilities before an estimate of sufficient reliability can be generated as the basis for a final feasibility study. This is not a magic figure, but one reached after a number of years of experience. Slope stability studies have now been carried out on a number of projects that have progressed from feasibility study to mature mine, and some of them have, of course, experienced failures. Therefore, I believe there should now be sufficient information available to at least make a start on determining what program should be carried out to minimize or prevent geotechnical surprises in relation to the overall project viability.

o What is the probability of a failure at a give slope angle?

I believe the mining engineer over the last few years has relinquished his responsibilities for making the judgment on pit slope angles and leaned heavily on the geotechnician to provide him with a single answer. There is, of course, no single answer. The mining engineer who is in a position to see all of the picture should make the decision on what slope design he can live with. To do this, he must be supplied with the probability of failure associated with the slope design angle and this should be expressed in some useful form.

o What are the consequences of a failure?

In order to assess the implications of the failure, the mining engineer must know whether the failure will be large or small, and whether it will be fast or slow. With that information, he may be able to say whether or not he can live with the slide, and he will at least be able to begin to quantify its economic implications. The answer will be very different for an interim in-pit slope and one which contains a permanent haul road or defines a pit limit bordering on a metallurgical plant or major dump.

o Where does time fit into the equation?

This is a broad area of discussion. Some factors that have a direct bearing on the mine economics are:

- For a given section of pit wall (dump face), is the risk of failure independent of time?

- During the development of the mine, are there periods when the risk of failure is greater than at other times?

- In the areas most likely to fail, what will the rate of failure be?

I would not like to leave you with the impression that the geotechnician is going to have to do all the work. Because of those changes I described earlier, we mining engineers are going to have to look very closely at our planning and esti-mating procedures, as they will undoubtedly come under more stringent review.

Finally, I believe the role of slope stability is the same today as it was 10 years ago. However, the impact of being wrong is far greater now with the more marginal, more difficult projects of today. Furthermore, the investor is at the same time more sophisti-cated financially, but less sophisiticated in mining terms.

It is, therefore, necessary for the mining engineer and geotech-nician to work together to obtain an assessment of the risk and financial impact of a slope or geotechnical design.

REFERENCES

Ballmer, R.W., "The Moving Targets of Mine Development", 1981
    Jackling Lecture, A.I.M.E. annual meeting, Chicago, Illinois,
    February, 1981.

Pentz, D.L., "Slope Stability Analysis Techniques Incorporating
    Uncertainty in Critical Parameters", 3rd International Conference
    of Stability in Surface Mining, Vancouver, B.C., S.M.E. of
    A.I.M.E., June, 1981.

Stewart, R.M. and Kennedy, B.A., "The Role of Slope Stability in the
    Economics, Design and Operation of Open Pit Mines", Stability
    in Open Pit Mining, S.M.E. of A.I.M.E., 1977.

Question

What do you consider to be a useful form of the answer to the
question of what is the probability of failure of a slope.

Answer

The most useful form of answer is one which allows the risks
associated with a particular course of action to be expressed in
financial terms. This, of course, cannot be done by the geotechnician alone but must be done in conjunction with the mining engineer
to evaluate the costs and benefits of a strategy.

Question

There are so many variables, and even mechanisms, of failure still
developing -- how do you justify applying probability theory to
predict risk failure.

Answer

The very large investments required for many of today's technically
difficult and economically marginal deposits require that the
best efforts possible be utilized to quantify the risk. If the use
of probability theory will permit the calculation of the risk
associated with a given course of action, this can be compared with
the other internal business risks. This process will permit the
investor to make reasonable decisions and thereby justify then the
use of probability theory.

Chapter 2

INFLUENCE OF ROCK STRUCTURE ON STABILITY

Carl D. Broadbent and Zavis M. Zavodni

Manager, Mining Engineering Department
Kennecott Minerals Company
Salt Lake City, Utah

Field Mining Engineer
Kennecott Minerals Company
Salt Lake City, Utah

ABSTRACT

During the past 15 years, rock mass structural properties have become acknowledged as the focal point for rock slope design. Research into the methods for measuring and incorporating structure into the design effort has taken many directions and has produced a variety of useful techniques. Among the findings at Kennecott has been the realization that structure-controlled instability can be categorized as regressive or progressive depending on the relative geometries of slope and structure and on the strength characteristics of the structures. It was further identified that a failure condition can initiate as regressive and in time become progressive should external stimuli increase or structure characteristics be altered due to extensive displacements. The transition point has been shown to be a useful indicator of the expectable collapse time.

This paper defines three fundamental failure types as dictated by structure and offers discussion on the practical value of such a classification to slope design and failure control. Several prominent examples of failures typifying the categories of structure-controlled instability are discussed. An application of failure prediction based on the regressive-to-progressive transition point is described.

INTRODUCTION

Major geologic structural controls have been considered in the rudimentary design of cut rock slopes ever since man has been inclined to construct these facilities. But as recently as 20 years ago, some of the world's foremost civil engineers were still attempting to utilize the mechanics of soils for rock slope design in situations where the structure was not prominent.

7

It is interesting that during the early 1960's several groups independently began research on the methods for measuring and incorporating geologic structure in their slope designs. We are all aware of the important contributions from the Royal School of Mines (London) and from the University of the Witwatersrand. Kennecott was also involved with its Kimbley Pit research which has been often reported. The primary result of the various investigations during the 60's was that geologic structure, regardless of how elusive its characteristics, must always be considered in rock slope designs.

At Kennecott, an understanding of the role structural systems play in slope stability has evolved along with technological applications for slope design and failure control at each of its open pit mines. Our files contain numerous examples of successful designs and of rock failures--both of our own and of other situations that have come to our attention. These case histories are invaluable for leading to better understandings of reality in this imprecise science.

One result of the ongoing work at Kennecott was a realization of the predictable differences in the propensity of various failing slope blocks to displace over time. The differences being principally a function of the slope vs structure geometry. It is the principal objective of this paper to underscore the geometric and other mechanical differences between three fundamental failure types and to show how and why these differences are identified and used.

The Concept of Progressive vs Regressive

During our analyses of many rock slope failures, it became apparent that all existing or potential failures must be either regressive or progressive depending on the tendency of the condition to become more stable or more unstable. A regressive failure in our definition is one that shows short term decelerating displacement cycles if disturbing events external to the rock are removed from the slope environment. A progressive failure, on the other hand, is one that will displace at an accelerating rate, usually an algebraically-predictable rate, to the point of collapse unless active and effective control measures are taken. The regressive failure condition has two basic structural configurations depending on the geometric relationship of slope and structure.

Figure 1 defines the conditions that must exist for each type of large scale failure (greater than 100,000 tons). In the figure, column 1 shows the relative geometry of slope and structure, where the structure line represents either a dip or plunge of a single or combination of several discontinuities; column 2 relates the geometries and the effective mechanical properties semi-quantitatively; and columns 3 and 4 describe some well documented failures that typify these conditions. For this purpose, a simple structural control (col. 3) regards a condition where one or several continuous or near-continuous features form the primary mechanical system.

STRUCTURE ATTITUDE	PRIMARY STRUCTURE	
	SIMPLE CONTROL — One or Two Surfaces	COMPLEX CONTROL — Multiple Structures
α = Mean Structure Dip or Plunge β = Slope Angle ϕ = Angle of Friction		
Type I — Regressive 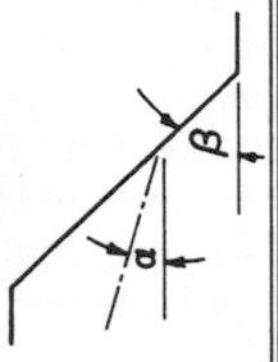 $\alpha < \beta$ $\phi < \beta$	**MINERAL CREEK** Structural Control — Single Fault with Gouge Weight — 11 Million T Height — 50 M Mean Structure Dip Approx. 10° Slope (β) Variable External Stimuli — Blasting, Water Unique Feature — Predictable & Regular Response to Blasting	Uncommon Type
Type II — Progressive 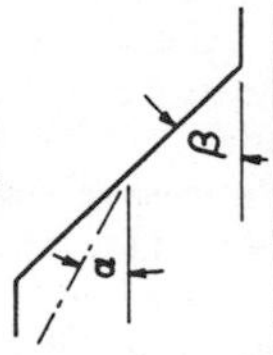 $\alpha < \beta$ $\phi > \beta$	**LIBERTY PIT** Structural Control — 2 Intersecting Faults Weight — 7 Million T Height — 175 M Overall Slope (β) — 33° Mean Plunge (α) — 19° External Stimuli — ? Unique Feature — Regularity of Fault Surfaces & Attitudes	**KIMBLEY PIT** Structural Control — Single Joint System Weight — 1.1 Million T Height — 160 M Overall Slope (β) — 58° (Initial) Mean Structure Dip (α) — 65° External Stimuli — None Unique Feature — Induce Failure by Explosive Undercut, (β) — 69° (Final)
Type III — Regressive / Progressive 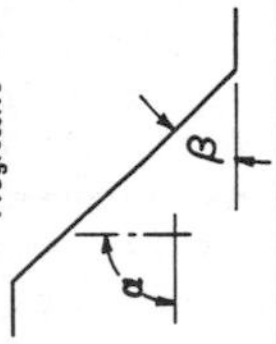 $\alpha \geq \beta$ $\phi \vee$	Uncommon Type	**CHUQUICAMATA** Structural Control — Multiple Fault Joints Weight — ? (Est. 1 Million T) Height — 151 M Overall Slope (β) — 42° — Various Mean Structure Dip (α) — Various External Stimuli — Heavy Pit Blasting, Frequent Earthquakes Unique Feature — Rubble Confined Above Pit Bottom

Figure 1.   Rock Failure Types Based on Structure/Slope Characteristics

Complex structural control, on the other hand, regards a mechanical system comprised of numerous discontinuous surfaces of one or more structural systems. The foregoing classification covers all possible relationships of structure, slope and strength, where strength refers to the rock discontinuities as opposed to the rock substance.

The value of differentiating between these three failure types and of properly identifying them in the field should be apparent. With this information:

(a) an optimum slope design can be achieved, or
(b) an effective correction program can be designed if a failure exists.

The optimum design for a mine slope is not one that assures absolute freedom from failure under all foreseeable conditions as it would be for many civil structures. A mine design may be acceptable if the potential failure condition would be regressive and controlable. An inherent progressive condition, on the other hand, must usually be designed against at all reasonable costs.

Once a failure situation is recognized, the determination of its regressive or progressive status will enable the proper and most effective correction efforts to be initiated. A regressive condition would be corrected through removal or minimizing of the external cause. A progressive failure, on the other hand, requires immediate planning for the likely collapse. In either case, monitoring systems and data interpretation must be oriented toward assessing the effectiveness of control programs and identifying a reversal of status, i.e., a regressive system becoming progressive—as often happens—or a progressive system reverting to a regressive mode—an unlikely event.

The following sections report general displacement characteristics of these several types of inherent conditions of instability.

Displacement Characteristics of the Structural Types

Type I Regressive Condition – A common condition of instability is represented by the Type I (regressive) structural system in the "simple control" mode. There are numerous examples of Type I instability and their displacement characteristics are typified by curve A in Figure 2. On the curve as shown, cycle initiation points 1, 2 and 3 plot as linear displacement assuming the discontinuous data. In reality, these points describe either an accelerating or decelerating displacement trend as revealed from continuous monitoring programs. The characteristic that qualifies this curve as regressive is the deceleration of each cycle between external stimuli at points 1, 2 and 3.

Broadbent and Ko (1971) show that each displacement cycle (e.g. 2 to 3) is predictable in a relatively uniform environment; and that generally, the points (1, 2, 3) will describe acceleration.

For a Type I failure to be initiated, an external disturbance(s) to the rock and structure is (are) required.

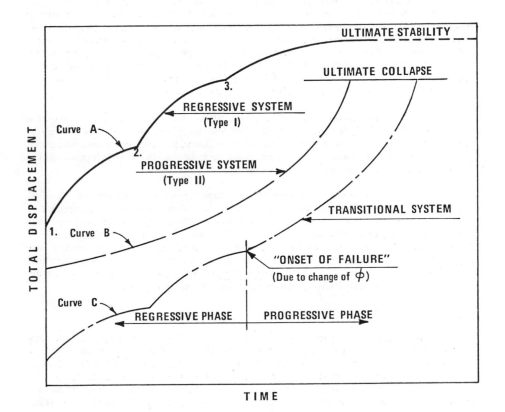

Figure 2.   Typical Displacement Curves

Type II Progressive Condition – Structural systems that initiate as Type II or eventually become progressive as a result of an effective strength reduction are often reported because the end result can be spectacular. The qualifying characteristic is a structure dip greater than the effective structure strength which manifests itself in positive exponential displacement as shown in Figure 2 (curve B). Decelerating cycles may be present but would be subtle and nearly indistinguishable from the long term trend.

The time period over which progressive displacement of a large scale failure takes place is usually short, 4–45 days. It appears to be somewhat related to the mass of the failure, although efforts to quantify a relationship (Zavodni and Broadbent, 1980) have not been successful.

Transitional Condition — Data are available to support the contention that many, if not most, economically significant failures began life as regressive and because of varied elements became progressive. This situation, which is described by displacement curve C of Figure 2, led to the formula for failure prediction reported by Zavodni and Broadbent (1980). The prediction relies upon characteristics and relationships of the regressive phase and the "onset of failure" point in the time history of the failure.

The transitional condition failures are significant because Type I structural conditions are often controlled, or the highly unstable Type II condition fails immediately during construction; but, transitional conditions (curve C) require extensive engineering, control, protection and reconstruction.

Type III Regressive/Progressive Condition — Toppling or "wedge induced" (Calder and Blackwell, 1980) failures would be characteristic of Type III structural conditions. These types are unique and have highly varied characteristics. They can be regressive, particularly in the case of wedge induced instability, or progressive as typified by domino-type falls referred to as toppling. We might say Type III conditions are <u>uncharacteristic</u> and demand intensive study and analysis if sensible improvement is to be achieved.

Examples of Structural Types

The example chosen to illustrate a Type I structure is the Mineral Creek failure which occurred in 1971. The kinematics of this failure were reported by Broadbent and Ko (1971) at which time it was shown that displacement occurred in cycles roughly related to peaks of pit blasting activity. Figure 3 is a cross section of this failure. The mean structure dip is less than the slope angle and less than the effective strength of the structure. Because $\alpha < \beta$ and $\alpha < \phi$, this Type I failure would not displace unless one or more external stimuli were present.

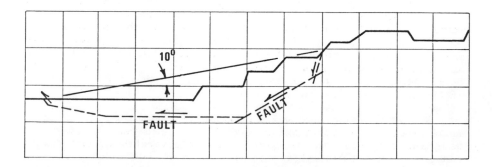

Figure 3. Mineral Creek Failure Cross-Section

Having been identified as regressive (this term and a formalized concept had not been conceived at this time), mine blasting in the immediate vicinity and surface water infiltration were reduced with the result that displacement ceased. The structural system was eventually mined out at a more expeditious time.

Type II Structure (simple control) – One of the most often reported structural controlled mine slope failures was the Liberty Pit failure in Nevada. The data suggest this failure initiated as a Type I but became a progressive (Type II) failure; probably as gouge was created or lubricated, and as asperities were sheared off. The characteristics of the structural system are given in Figure 1. The time-displacement rate history of the failure is presented in Figure 4. This figure, which was first published by Zavodni and Broadbent (1980), shows the basis for the technique of predicting failure collapse times. The following several paragraphs are taken from that paper.

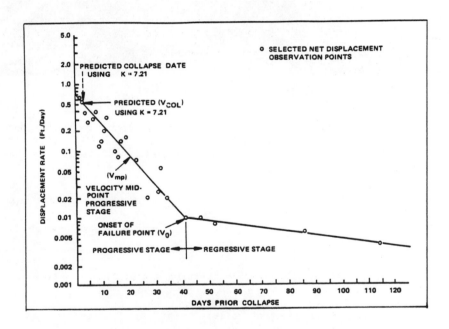

Figure 4. Liberty Pit Transitional System Displacement Rate Curve and Failure Collapse Prediction

Figure 5.   Liberty Pit Failure

A close examination of nine major transitional condition rock slides progressing to total collapse revealed a semi-quantitative empirical relationship for failure collapse prediction. It was observed that:

$$\frac{V_{mp}}{V_o} = K \qquad\qquad\qquad\qquad (1)$$

where,

V_{mp} =   velocity at mid-point in the progressive failure stage (Figure 4)

V_o =   velocity at onset of failure point

K =   constant (ave. = 7.21, σ = 2.11, range 4.6 - 10.4)

Knowing that the general equation for a semi-log straight line fit has the form

$$V = Ce^{St} \tag{2}$$

where,

V = velocity (ft/day)

S = slope of line (days^{-1})

C = constant

t = time (days)

e = base of natural logarithm

and assuming $t = 0$ at the collapse onset point, equation 2 takes the following form for the progressive failure stage:

$$V = Voe^{St} \tag{3}$$

From this equation and the empirical relationship of equation 1, one can determine the velocity at the collapse point (V_{col}) as

$$V_{col} = K^2 Vo \tag{4}$$

Equation 4, in conjunction with a semi-log plot such as in Figure 4, enables one to estimate the number of days until total collapse, once the failure onset point is reached, and the progressive stage failure displacement rate pattern is established from the monitoring record. This appears to be a useful empirical relationship.

Type II Structure (complex control) – The Kennecott slope stability research program at the Kimbley Pit produced a Type II progressive failure by explosive undercutting at the slope toe. The undercut increased the slope angle from the original designed and mined 56° slope to an effective 69° slope. After undercutting, the slope failed to a residual angle of about 58°. Figure 6 shows the Schmidt plot of that structural condition. Data regarding this failure were reported by Broadbent (1972).

Type III Structure (simple control) – Kennecott has noted wedge failures (110 m. high) bounded by two near planar continuous fault systems whose line of intersection exceeds the overall slope angle by 4-5° and is not daylighted. These relatively shallow slides (30-60 m. deep) occur typically on a regressive one cycle basis. Mining levels are offset some 0.7-3 m. with little or no rubblizing or debris flow.

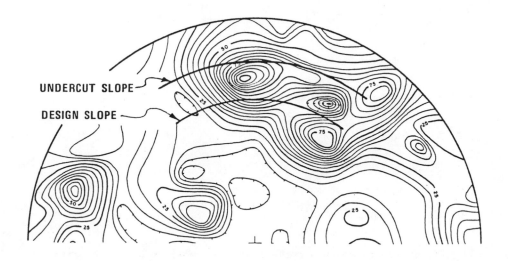

Figure 6.   Kimbley Pit Contour Plot

Type III Structure (complex control) – One of the more interesting structure controlled failures in recent years was the large
high slope failure at the Chuquicamata pit in Chile.

The Chuquicamata failure is of technical interest for several
reasons.   The structural geology and time displacement history
(Kennedy and Niermeyer, 1970) are well documented. Excellent before
and after photographs are available. It is a graphic example of a
Type III transitional failure of a high slope, and the external disturbances causing ultimate collapse are in large part probably
related to earthquake seismicity. The failure occurred in the high,
straight north wall of the Chuquicamata pit. Rock in the vicinity of
the failure was hard and competent. Ground water pressure was apparently absent. The structure consisted of continuous, well developed
steeply dipping fault systems and steeply dipping discontinuous
jointing. The significant faults and joints and the plunges of faults
and joint intersections were steeper than the overall slope of the
north wall, thus putting this failure in the Type III category.
Figure 7 shows the typical spacial attitudes of these structures.

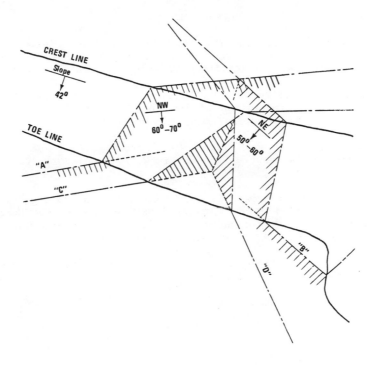

Figure 7. Chuquicamata Failure Structural Control

The before and after photographs reveal that it was only the upper benches that actually collapsed and that the bottom benches were barely covered with the talus. Essentially no rubble covered the pit bottom, and had the pit been several benches deeper, the rubble might not have reached the bottom of the pit.

Closing Remarks

This paper classifies fundamental large scale rock failure types as dictated by structure attitude and discontinuity strength characteristics. The value of differentiating between the failure types and of properly identifying them is of paramount importance to optimum slope design and an effective correction program.

REFERENCES

Broadbent, C. D., 1971, "The Practical Side of Mining Research at Kennecott Copper Corporation," Proceedings of the Second International Conference on Stability in Open Pit Mining, Vancouver.

Broadbent, C. D., and Ko, K. C., 1971, "Rheologic Aspects of Rock Slope Failures," Proceedings of the 13th U. S. Symposium on Rock Mechanics, University of Illinois, Urbana.

Calder, P. N. and Blackwell, G., 1980, "Investigation of a Complex Rock Slope Displacement at Brenda Mines," CIM Bulletin, Vol. 73, No. 820.

Kennedy, B. A., and Niermeyer, K. E., 1970, "Slope Monitoring Systems Used in the Prediction of a Major Slope Failure at the Chuquicamata Mine, Chile," Proceedings of Symposium on Planning Open Pit Mines, Johannesburg; A. A. Balkema, Amsterdam.

Zavodni, Z. M., and Broadbent, C. D., 1980, "Slope Failure Kinematics," CIM Bulletin, Vol. 73, No. 816.

Chapter 3

THE INFLUENCE AND CONTROL OF
GROUNDWATER IN LARGE SLOPES

Adrian Brown

Principal
Golder Associates
Denver, Colorado

INTRODUCTION

The primary tool which is available to improve the stability of an open pit mine at a given slope angle is control of groundwater pressure. This paper sets out the methods by which water pressure control can be achieved in large mines in different materials.

The emphasis of this paper is on large mine slopes because these slopes exhibit the greatest need for groundwater control, the greatest economic benefit from groundwater control, and the greatest technical challenge in achieving groundwater control.

STABILIZATION USING WATER PRESSURE CONTROL

Groundwater pressures reduce the stability of slopes in two fundamental ways, as shown schematically in Figure 1. In the example, a horizontal failure surface intersects a vertical joint or weak zone. Water pressure acts:

o On the base plane to reduce the normal force on the sliding surface and hence reduce frictional resistance to sliding, and

o On the vertical back plane, to create a driving force for the system.

19

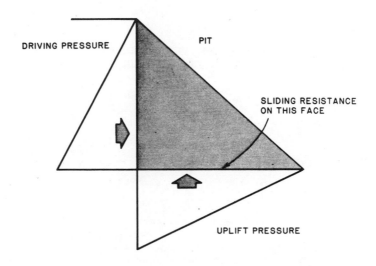

Fig. 1.   Example of Effect of Groundwater
           Pressure on Slopes.

This particular slope stability example was chosen to have water as the only driving force, so that water pressure reduction clearly improves the stability of the system dramatically. It is of interest to note that in this example, about 70% of this increased stability comes from reduction of the driving pressure, and about 30% comes from increase of the frictional resistance on the base plane.

A powerful way to illustrate the need for groundwater pressure reduction is to contrast the extreme cases of fully saturated slopes and completely depressurized slopes. The relationship between slope angle and the primary factors which determine slope angle (height, friction, cohesion, and material unit weight) has been elegantly presented for five groundwater scenarios by Hoek and Bray (1977). Subtracting the slope angle for a completely depressurized slope from that for a completely saturated slope gives the increased slope angle available through water pressure control. The resulting diagram is presented as Figure 2.

The significance of the savings available for mine operators becomes clear when it is recognized that an increase of 10° in slope angle in most large mines halves the required stripping.

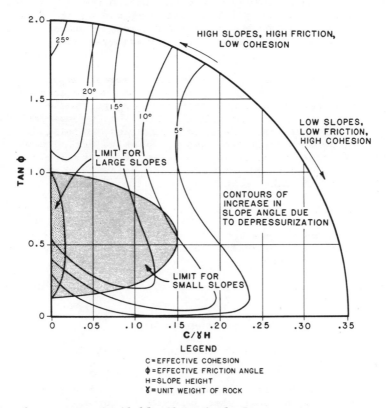

Fig. 2.   Maximum Available Slope Angle Increase as
a Result of Groundwater Pressure Control.

THE NATURE OF LARGE SLOPE FAILURES

The nature of most large slope "failures" (using the term in the
geomechanical sense that some materials in the slope have yielded)
is usually more complex than the simple example above, or the model
used by Hoek and Bray. Such failures are rarely rapid, except over
a bench or two, and bench scale failures, while an operational and
safety concern, are beyond the scope of the present discussion.

Large pit wall failures usually begin as slow movement of the
surface of most of the wall. At first, the rates are in the order
of a few millimeters per day. If mining continues, these movements
gradually accelerate over a period of months or even years. Surface
disturbance inside the pit become more marked, and bench access and

integrity becomes impossible to maintain. Cracking at or outside
the crest is common. Movements are much more marked in wet periods
(rain, snowmelt) and immediately following mining at or near the
toe. If nothing is done about these failures, the movements become
more and more rapid until finally a major translational failure of
the wall occurs.

If the conditions within the slope are observed, it is seen that
these failure mechanisms are deep seated. The movement zone often
extends into most of the region beneath the slope to the total pit
depth (i.e., the shaded zone in Figure 1). What is happening during
failure is that the interlocking rock particles are being disturbed
in this entire zone, and the initial cohesive strength is being
reduced. At some point in this process, the strength of a continuous daylighting failure surface drops below that required for
stability, and the wall fails in a translational manner.

Most big pit failures do not reach this stage. Water pressure
control or slope flattening or both are usually used to reduce
movements. However, some big pit slopes have failed without translational failure because the continual movements have become
operationally intolerable.

For the purposes of this paper, the following points about
failure are important:

o Large pit failures are deep seated

o Water pressure is an important driving force when it is
   present

o Slope control can, in general, only be achieved by slope
   flattening or deep water pressure control

The remainder of the paper evaluates the ways in which deep water
pressure control can be obtained.

PARAMETERS CONTROLLING WATER PRESSURE IN SLOPES

In slope stabilization using water pressure control, the focus is
on reducing groundwater pressure. Typically, any stabilization
involves installing pressure control devices at a number of locations and waiting for the water pressure to drop in the material
around and between those devices. The process focuses on obtaining
a major water pressure reduction in an acceptable length of time.
There are two disciplines which have traditionally dealt with
matters of transient groundwater pressure changes: groundwater
hydrology and soil mechanics.

Groundwater Hydrology Approach

Groundwater hydrology has its foundation in the provision of water supply from aquifers. Ever since the classic paper by Theis in 1935, groundwater hydrologists have used two major parameters to describe the characteristics of aquifers:

Transmissivity (T)--which is defined as the quantity of water which flows through a unit width of an aquifer under a unit hydraulic gradient. It has the units $[L^2/T]$.

Storage Coefficient (S)--which is defined as the quantity of water produced by a unit area of the aquifer when the aquifer is subjected to a unit head reduction. It has no units.

Both these parameters implicitly include the thickness of the aquifer, and so for mine hydrology work it has been usual to divide them by the material thickness to produce three corresponding parameters:

Hydraulic Conductivity (k)--which is defined as the quantity of water which flows through a unit area of a material under a unit head gradient. It has the units $[L/T]$.

Specific Storage (S_s)--which is defined as the volume of water which is produced from a unit volume of saturated material when it is subjected to a unit head reduction. It has the units $[L^{-1}]$.

Drainable Porosity (n_d)--which is defined as the volume of water produced per unit volume of material as the water table moves through the material. It has no units.

The basic general differential equation for heads in a groundwater system is as follows for a planar one-dimensional case:

$$\frac{\partial h}{\partial t} = \frac{T}{S} \frac{\partial^2 h}{\partial x^2} = \frac{k}{S_s} \frac{\partial^2 h}{\partial x^2}$$

where

$\qquad$ h = head
$\qquad$ t = time
$\qquad$ x = distance

Soil Mechanics Approach

By comparison, the soil mechanics approach to this particular problem comes from the study of the settlement of buildings, and is embodied in consolidation theory. The classic statement of this theory is in a book by Terzaghi and Peck, first published in 1948, and repeated in later editions (Terzaghi and Peck, 1967). There are three parameters which characterize this approach, as follows:

Hydraulic Conductivity (k)--as before.

Coefficient of Volume Compressibility (m_v)--which is defined as the change of pore volume per unit total volume per unit of pressure increase. It has the units inverse pressure [LT^2/M].

Coefficient of Consolidation (c_v)--which is defined by the following equation:

$$c_v = \frac{k}{\rho_w g m_v}$$

where ρ_w = density of water

   g = acceleration due to gravity

The standard equation for consolidation in a one-dimensional case is:

$$\frac{\partial u}{\partial t} = c_v \frac{\partial^2 h}{\partial x^2}$$

where $u = \rho_w g h$ = the pore-water pressure

Thus, by substitution, the differential equation can be converted to:

$$\frac{\partial h}{\partial t} = c_v \frac{\partial^2 h}{\partial x^2}$$

Comparison Between Parameters

Clearly, from the two main differential equations,

$$\frac{T}{S} = \frac{k}{S_s} = c_v$$

and c_v, the coefficient of consolidation, controls the time rate of change of head at any point in a transient groundwater pressure system. This parameter will be used extensively in this paper.

WATER PRESSURE CONTROL METHODS

There are four major water pressure control methods which appear economic today in the context of most large mines. They are:

o Unaided drainage
o Horizontal drainholes
o Wells
o Drainage adits

They are listed in order of increasing cost, increasing effectiveness, and increasing operational convenience.

In addition, for very low permeability materials, the physical unloading associated with mining can also lower pore water pressures enough to create a useful stabilizing effect.

Each of these five strategies is evaluated below in the context of their effectiveness in stabilizing slopes in large open pit mines.

Unaided Drainage

Method. Unaided drainage uses the excavation itself as a collector drain. Water flows into the pit and is pumped out from there.

Theory. The effectiveness of the method as a function of coefficient of consolidation of the host rock is shown in Figure 3 for a typical mining case. Conditions selected were:

    Pit depth = 500 meters
    Time for material to depressurize = 1 year
    Depth of critical zone = 500 meters

The 1 year depressurization time is appropriate as the critical areas are adjacent to the pit floor, which is usually continually deepening. Thus, relatively rapid pressure control in this area is important.

If we assume that 80% reduction in water pressure is required, then this method of pressure control will be effective in long pits with coefficients of consolidation of 2×10^{-2} meters squared per second or greater, and in radial pits with coefficients of consolidation of about 2 meters squared per second or greater.

Practice. These values are available in most relatively free draining rocks, with average horizontal hydraulic conductivities greater than 10^{-6} meters per second. (Note that average specific storage of most rock materials is about 10^{-6} meters^{-1}.) However, even a few meters of clay in (say) faults

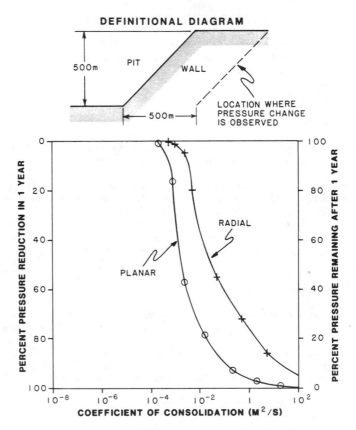

Fig. 3.   Effectiveness of Unaided Drainage as a
          Function of Pit Shape and C_v.

parallel to the face will so reduce the effective hydraulic conductivity as to render this method of stabilization ineffective.   In
flat-bedded sedimentary systems it can work well, provided the
egress of water at the face is not sufficiently concentrated to
cause piping of any poorly cemented materials.

Operational Aspects.   Operationally, the major disadvantage of
unaided drainage is the uncontrolled flow into the pit, most of it
coming to the pit floor and lower slope areas.   This flow (often in
excess of 0.1 cubic meters per second because of the relatively high
hydraulic conductivity) makes operations difficult, and also must
usually be treated before discharge, as it is degraded in quality
when it flows onto the pit floor.   Thus, most mines which could
stabilize the walls by this method choose not to do so for other
reasons.

Horizontal Drainholes

Method. Horizontal drains are drilled into the wall from the pit floor as the pit deepens and provide a pressure relief conduit for the wall material. Typically, they penetrate the wall in a grid, with spacing both vertically and horizontally in the 25 to 100 meter range. Holes are usually in the 200 to 300 millimeter diameter range and are cased with slotted PVC or steel liner pipe.

Theory. The effectiveness of horizontal drains in parts of the wall which they penetrate is shown on Figure 4 as a function of the average coefficient of consolidation of the wall material. As can

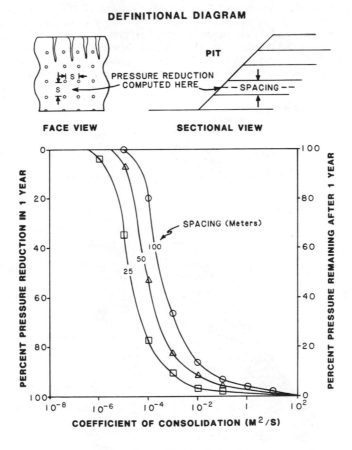

Fig. 4. Effectiveness of Horizontal Drains as a
         Function of Spacing and C_v.

be seen, 80% pressure reduction after 1 year can be achieved with
coefficients of consolidation down to 10^{-3} meters squared per
second with a drain spacing of 25 meters or more. In rock mines
this indicates that this strategy can be effective in wall material
penetrated by drains down to an average hydraulic conductivity of
10^{-9} meters per second.

Practice. The practical advantage of horizontal drains is that they
penetrate structures in the wall and provide pressure relief behind
them. As many large mines exhibit steeply dipping structures filled
with low permeability material, this method of pressure relief is
extraordinarily effective.

The primary disadvantage of horizontal drains is that they are
very difficult to advance beyond 100 to 150 meters. Their
effectiveness beyond this reach is typically modest, so that they
are frequently inadequate for the total pressure control of deep
mines. This effect is shown on Figure 5. However, they can be and
often are used as a part of a total pressure relief strategy.

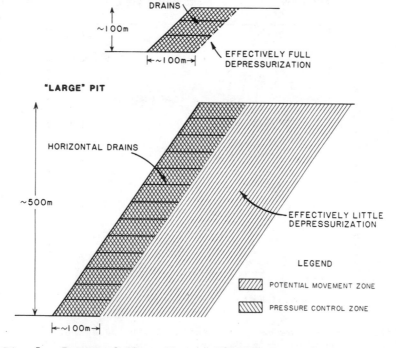

Fig. 5.   Impact of Slope Size on Effectiveness of
          Horizontal Drains.

Operational Aspects. Horizontal drainhole systems are difficult to maintain in an operating mine. Ideally, water should be collected from each hole and piped to a point from which it can be pumped from the pit. This avoids resaturating the slope immediately below the drain, partially nullifying the effect of lower drains. The drains must also be maintained with annual cleaning and occasional replacement. Both of these aspects require maintaining access to the benches from which the holes were drilled, which is usually difficult and makes steep slopes difficult to achieve.

Finally, the drains are usually lost on each push back, and the network of holes, pipes, sumps, and pumps needed in an operating system creates obstacles for normal pit operations.

There are, however, many successfully operating horizontal drain systems in mines around the world, and they provide a low capital water pressure control strategy in relatively low permeability materials.

Wells

Method. The classical dewatering strategy for mines is pumped wells, and there are many systems in use today. The usual installation involves wells around the periphery of the mine, augmented with some in pit wells in larger mines. Wells are typically screened over their entire length with an inexpensive screen, and gravel packed when installed in poorly cemented materials. Pumping is usually by submersible electric or shaft driven turbine pumps. Unlike water supply wells, dewatering wells are pumped until the drawdown approaches the bottom of the well in order to maximize the pressure reduction effect. Water from wells is typically piped directly to discharge without becoming contaminated. Treatment of discharge is occasionally needed.

Theory. In essence, most well systems seek to cut off water entry into the mine slope area by creating a low pressure line along the mine crest. The effectiveness of an infinite line of wells is shown in Figure 6, for various well spacings and coefficients of consolidation. Practical well spacings for large mines are in the 200 to 500 meter range, which indicates that they will be effective in depressurizing a mine slope when the average horizontal coefficient of consolidation exceeds about 1 meter squared per second. Note that this value is quite strongly dependent upon well spacing.

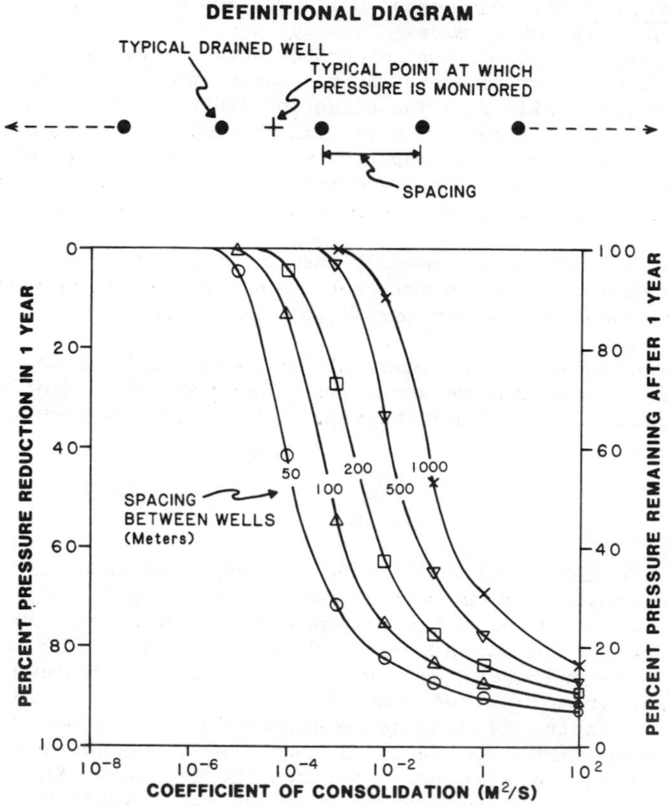

DEFINITIONAL DIAGRAM

Fig. 6.   Effectiveness of Vertical Wells as a
Function of Spacing and C_v.

Practice. The practical limit to the ability of well dewatering
systems is usually the nature of the wall materials between the line
of wells and the toe of the slope. For large slopes, the wells may
be some 1,000 meters from the toe, and any major, low permeability
feature in this distance will reduce the effect of the system. For
this reason, it is common for wells to be used in conjunction with
horizontal drains. Also for this reason, wells do best in flat-
bedded sedimentary materials.

The location of the wells is important. Wells are very sensitive
to slope movement and should therefore be located outside the
potential movement zone. If they are not, then they tend to fail at
the very moment when they are most needed.

Operational Aspects. From an operating point of view, well systems have some good and some bad aspects.

On the positive side, the systems can usually be placed outside the active operating area, and are not an obstacle to mining. On the negative side, wells are an alien technology to most mine operations personnel, and require constant attention. This can cause difficulties in the overall operation of the mine.

Drainage Adits

Method. The drainage adit or gallery has been a standard (if expensive) depressurization option for at least a century. The basic design of the gallery is shown on Figure 7. The adit is

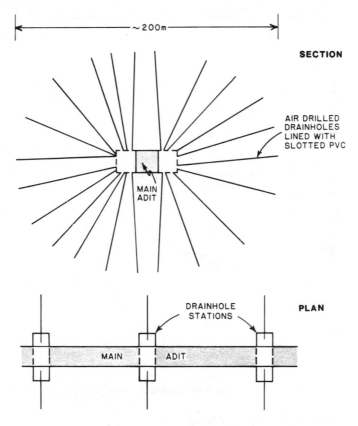

Fig. 7. Typical Drainage Adit Layout (Schematic).

usually advanced by conventional means, with stations slashed out every 50 to 100 meters for fan drilling of drainholes. Typical drainhole length is also 50 to 100 meters, at about 50 millimeter diameter. Slotted PVC pipe is usually inserted into the drains to keep them open.

Theory. The effectiveness of adits was investigated by first evaluating the effectiveness of a single adit, and then multiple adits for large mine slopes. The results of analysis for a single adit are shown on Figure 8. This is computed for an effective adit radius of 100 meters and indicates that for a mine 500 meters deep, a single adit will be effective for average coefficients of consolidation in excess of about 0.1 meters squared per second.

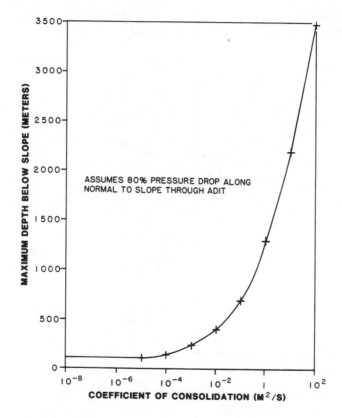

Fig. 8.   Maximum Depth of Cover for Effective
          Adit Performance.

For materials with lower values of coefficient of consolidation, several adits may be needed. The analysis of adit spacing as a function of average coefficient of consolidation and depth of cover over the adit is given in Figure 9. The cutoff in adit effectiveness occurs when radial drains are no longer effective to drain the material between each drain.

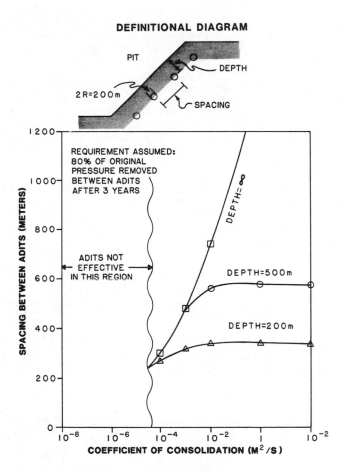

Fig. 9. Required Spacing Between Adits as a
Function of Depth and C_v.

The conclusions of the study of the theory of adits is summarized below for pits in the 500 meter deep range.

$C_v(m^2/sec)$	Adit Requirement
Less than 10^{-4}	Adits ineffective
10^{-4} to 10^{-3}	Multiple, close spaced adits close to pit surface
10^{-3} to 10^{-2}	Two adits, deeply buried one at toe and one mid-height
Greater than 10^{-2}	One adit located beneath the crest, at about toe level. Expect high initial flows.

Practice. The practical application of drainage adits rests also on geologic conditions. While, in theory, one adit can depressurize very large volumes of rock, provided the coefficient of consolidation is reasonably high, in actual operation the effect may only spread as far as the nearest thick, gouge-filled discontinuity. Adits work best in relatively isotropic materials for similar reasons.

One of the major benefits of adits is that they can be installed well ahead of the mine development, to give the maximum time for pressure reduction to occur. They can also, of course, be established outside the movement zone of the mine for operational safety. In any case, they are not particularly sensitive to modest movements.

Operational Aspects. Drainage adits offer several operational advantages. First, they are outside pit operations and need not influence them. Second, they collect all drainage water in one location for easy handling and discharge.

The primary operational disadvantage of drainage adits is their inflexibility. If the pit is expanded or the slope location changed, the adit system is difficult and expensive to modify.

Unloading

Method. Continual unloading of slopes is a normal procedure in most types of open pit mining. It occurs whenever a wall is pushed back.

Theory. When a load is removed from a column of material, it expands. If the material is porous and saturated with water, the load removal causes an instantaneous drop in the pressure of the pore fluid equal to the stress reduction due to the load removal. In time, the pressure returns to a new equilibrium as water is drawn in from the boundaries. For materials with low coefficients of consolidation, this equilibrium may take many years; in these materials, unloading is the only practical method of groundwater control.

The theory has been approached in two ways: first, by looking at the pressure effect of unloading at shallow depths, and second, by looking at the entire unloading history of a large mine.

The shallow effects have been simulated by unloading a single 10 meter cut and checking the remaining head after 3 years at various depths. These results are shown on Figure 10. From this figure, it

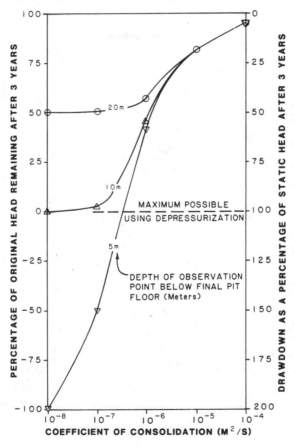

Fig. 10. Effectiveness of Unloading as a Strategy for Stabilization--Shallow Case Showing Effect of 10 Meter Cut After 3 Years.

can be seen that for coefficients of consolidation less than about 10^{-6} meters squared per second, the surface of the mine would stay effectively depressurized.

The deep effects have been simulated by unloading a column of material sequentially as mining progresses, and evaluating the distance below the slope in which the groundwater pressure has been reduced below zero. This data is presented in Figure 11. As can be seen, it would appear that a mine in a material with a coefficient of consolidation less than about 10^{-4} meters squared per second will develop useful unloading water pressure reduction in its life. It will, however, be necessary to accept considerably flatter angles for the slopes in the early part of the life of the mine.

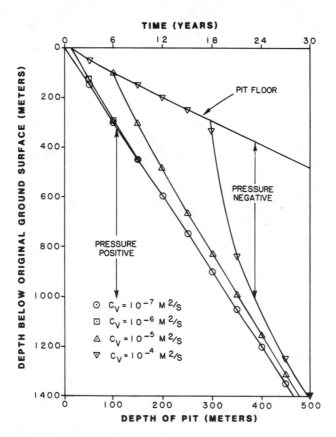

Fig. 11.   Depth of Negative Pressure Zone as Pit Develops as a Function of C_v.

In summary, the unloading phenomenon can be described as follows:

$C_v(m^2/sec)$	Unloading Effect
More than 10^{-4}	No useful depressurization
10^{-4} to 10^{-6}	Depressurization effect increases through life of mine
Less than 10^{-6}	Depressurization effective even at early times

Practice. The phenomenon described above has actually been observed in a number of man-made cuts, most notably the Panama Canal (Banks, 1972) and railway cuttings in London Clay (Skempton, 1977). In this latter material, the coefficient of consolidation would appear to be in the order of 10^{-8} meters squared per second, based on its liquid limit. The author is not aware of any mine where there has been a positive identification of pore water pressure being depressed by this method; however, it is suspected that a considerable number of mines are presently stable for this reason.

Clearly, the deep geology is of critical importance. There must be no major conduits into the mine area from depth to allow rapid pore pressure equilibration.

Operational Aspects. If mining is to take place in materials which exhibit this phenomenon, then the mining method should be modified to maximize the effect. Specifically, each part of the mine should be excavated at least annually, so that material is removed from over the entire mine area. Permanent slopes should be avoided until the end of the project. Upon cessation of operations, the pit would gradually begin to fail, with the failures continuing slowly until pore water pressure equilibrium was established. In the case of London Clay and the Panama Canal cuts, this process has taken more than 50 years.

Summary of Groundwater Pressure Control Options

Table 1 identifies the primary results of the above evaluation. The ranges of values of coefficient of consolidation for which each of the strategies outlined is valid are shown on Figure 12. In addition, an approximate material characterization has been presented to guide users in material types which might be expected to exhibit the nominated coefficients of consolidation.

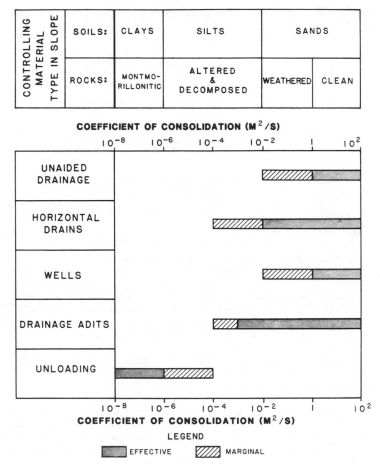

Fig. 12.   Summary of Groundwater Pressure Control
           Strategies for Large Mine Slopes.

Several aspects become immediately obvious.

o   Materials exhibiting an effective coefficient of
    consolidation in excess of 1 meter squared per second can be
    readily depressurized. Flow control is more of an issue
    than pressure control in these materials.

o   Materials exhibiting an effective coefficient of
    consolidation of between 1 meter squared per second and
    10^{-3} meters squared per second can be depressurized
    using combinations of horizontal drains and wells, or by
    adits.

o   Materials exhibiting an effective coefficient of consolidation below 10^{-6} meters squared per second will generally depressurize themselves if mining is designed to take advantage of the unloading phenomenon.

o   Materials exhibiting an effective coefficient of consolidation of between 10^{-3} meters squared per second and 10^{-6} meters squared per second are virtually impossible to economically depressurize. These materials include fine silts and highly altered rocks. Electro-osmosis may be the only available method, and this is prohibitively expensive on a large pit scale.

CONCLUSIONS

This paper has sought to identify the benefits to be gained by groundwater control in large open pit mines, and the ways in which these benefits can be obtained in commonly encountered mining materials. The conclusions are as follows:

o   Up to 20° increase in slope angle is available as a result of groundwater pressure removal.

o   The parameter which allows a determination of the feasibility of groundwater pressure removal is the coefficient of consolidation (c_v).

$$c_v = \frac{T}{S} = \frac{k}{S_s}$$

o   Materials with a c_v in excess of 1 meter squared per second can be readily depressurized using unaided drainage, horizontal drains, wells, or adits. These materials include sands and clean fractured rock.

o   Materials with a c_v between 10^{-3} and 1 meter squared per second can be depressurized with more difficulty, using an adit system or a combination of wells and horizontal drains. These materials include fine sands, silts, and weathered, gouge-filled rocks.

o   Materials with an effective c_v lower than 10^{-6} meters squared per second will depressurize themselves due to the effects of unloading. These materials include massive clays and montmorillonitic rocks.

o   Materials with a c_v between 10^{-6} and 10^{-3} meters squared per second are exceedingly difficult to depressurize economically on a mining scale. These materials include fine silts and highly altered or decomposed rocks.

REFERENCES

Banks, D.C., 1972, "Study of Clay Shale Slopes," 13th Symposium on
    Rock Mechanics, University of Illinois.

Hoek, E., and Bray, J.W., 1977, Rock Slope Engineering, Revised
    Second Edition, Insititution of Mining and Metallurgy, London.

Morgensten, N.R., 1971, "The Influence of Groundwater on Stability,"
    Stability in Open Pit Mining, C.O. Brawner and V. Milligan,
    (eds), Society of Mining Engineers, New York.

Skempton, A.W., 1977, "Slope Stability of Cuttings in Brown London
    Clay," Special Lecture, Proceedings Ninth International
    Conference on Soil Mechanics and Foundation Engineering, Tokyo,
    pp. 261-270.

Theis, C.V., 1935, "The Relation Between the Lowering of the
    Piezometric Surface and the Rate and Duration of Discharge of a
    Well Using Groundwater Storage," Trans. Amer. Geophysical
    Union, Vol. 16, pp. 519-524.

Terzaghi, K., and Peck, R.B., 1967, Soil Mechanics in Engineering
    Practice, Wiley, New York.

Question

Would you prefer groundwater level monitoring around the pit
perimeter before selecting a suitable pit dewatering scheme.

Answer

I prefer groundwater pressure monitoring. The thing that pushes
slopes down is not simply the presence of water, but the presence of
pressurized water. Therefore I tend to prefer peizometers (usually
of standpipe type, even in low hydraulic conductivity material) to
measure water pressure. Open-hole water levels must be treated with
extreme caution in geomechanics.

Question

Vertical sand drains have been used for decades to accelerate the
dissipation of excess pore water pressure below road embankments
etc. constructed over low permeability soils. Many people now
question whether drains were as necessary as first thought, because
sand partings within the material have acted as drainage paths.
Perhaps some vertical and horizontal drains are not necessary because
joints and other features act as sand partings. How then does one
identify where drains will be effective or not before you pay for
putting them in.

Answer

The question of what the <u>effective</u> coefficient of consolidation is in a given slope rock mass depends on the purpose for which the data is to be used. For example, the effective value for a predominantly clean rock with a one meter clay-filled zone between the water pressure control feature and the location where the pressure control is needed is very similar to the value for the clay. While the paper does not address the question of methods of measurement, it is a critical and difficult step in the design of slope depressurization systems.

Question

Would the deepening of the pit change or affect the scheme needed for depressurizing the walls.

Answer

Absolutely. I did not cover the impact of size in the paper, but several things tend to happen as a pit gets deeper. First, the strategy needed for stabilization gets more expensive and more complex, as per the following list:
- unaided drainage
- horizontal drains
- wells plus horizontal drains
- drainage adits

Particularly the step from horizontal drains to augmented systems is a major undertaking, and should take place at a pit slope height in the order of the length of production horizontal drains.

Chapter 4

INFLUENCE OF BLASTING ON SLOPE STABILITY;
STATE-OF-THE-ART

L. L. Oriard

Consultant
Huntington Beach, California

ABSTRACT

   In order to predict the influence of blasting on slopes, one must
first understand the action of explosives, the manner in which rock
is broken or displaced, and how seismic waves are transmitted, and the
nature of these seismic waves. In this paper, the author describes
the action of high explosives on rock masses, beginning at the source
of the explosion, extending through the zone of rupture to a distance
where only elastic waves of low energy persist. The paper describes
the relationships between explosives charge quantities and distances,
including the character and intensity of the seismic waves in various
types of terrain.

   In looking at the analysis of slopes subjected to seismic waves,
it has been common practice to assume a simplified model in which the
anticipated acceleration is applied horizontally to the slope as an
equivalent static force, or a model in which there is a rigid base
subjected to mechanical shaking. Such models do not accurately portray blasting activity taking place within or near the slope in
question. In dealing with blasting effects, it is important to understand seismic wave types, wave lengths, attenuation and transmission
paths, as well as vibration intensity.

   In most cases, those slopes that are the least stable under static
loading will also be the least stable under dynamic loading, although
there may be rare exceptions. Thus, the more that is known about a
slope statically, the better will be the assessment of it dynamically.

UNDERSTANDING BLASTING PHENOMENA

   One cannot proceed very far with a discussion of blasting effects
on slopes without recognizing the need for an understanding of the
full range of effects, beginning at the source and extending to such

distances that only low seismic levels remain. Without an understanding of the various phenomena involved, it is easy to be drawn into inappropriate assumptions about blasting effects (whether ground rupture or shaking) and/or the use of inappropriate methods of stability analysis.

At the second conference in this series, the writer presented a discussion of blasting effects (Oriard, 1971). The writer will not repeat that discussion in detail. However, a brief overview of a few of the previous comments will provide the background for a better understanding of the present paper.

THE BLASTING SOURCE - THE INELASTIC ZONE

It seems useful to discuss the near-source phenomena for at least two reasons. One is for the purpose of controlling breakage. The other is to point out the distinction between vibration effects and those non-vibration, inelastic effects near the source, such as rock rupture, block motion and gas venting. The latter effects have an important bearing on the stability of bench faces and the near-surface portions of excavated slopes.

When an explosive charge is detonated in a borehole in rock, a high-pressure shock wave is transmitted to the rock, followed by a longer-acting pressure under the action of the expanding gases. The shock wave develops a very high pressure, capable of crushing the rock for a distance of the order of 1 to 3 charge radii, but is dissipated very quickly. The propagating pressure pulse develops radial cracks around the borehole, and these are further advanced by the continuing expansion of the explosives gases. Although a larger number of cracks may begin at the perimeter of the crushed zone, it is common for a group of some 8 to 12 cracks to become more prominent and extend to greater distances than the others. According to the principles of fracture mechanics, less energy is consumed in extending the more prominent, existing cracks than to develop new ones. Under the action of the initial stress waves, additional cracks also develop at the locations of flaws within the rock mass, because these flaws provide points of stress concentration.

There was a considerable amount of research done on various aspects of explosives action on rock during the 1950's and 1960's. A limited amount of such research continues at certain universities which have special interests in the subject, e.g. U. of Maryland (Fourney and Barker, 1979), others. In general, this on-going research tends to confirm the basic concepts developed during earlier investigations and construction experience, with certain refinements being advanced as research continues.

The amount of rock directly ruptured, fractured or displaced by the blast is a function of a number of different variables involved in the blast design. Some of the more important, (in addition to the characteristics of the rock mass itself) are:

1. Charge size.
2. Charge concentration or charge spatial distribution.
3. Type of explosive.
4. Depth of burial (distance to any free surface).
5. Coupling to the rock.
6. Sequence and timing intervals of detonation of multiple charges.

Trends in Blasting Technology

In recent years, the principal directions of expanding blasting technology have been found at opposite ends of the spectrum. On the one hand, there has been an increase in very large blasts and large-scale operations, including explosive excavation and cratering technology. Several countries have shown an interest in ejecta dam feasibility (casting ejecta across a valley by blasting) (Oriard, 1976). On the other hand, delicate excavations are more common, also, including refinements of conventional perimeter blasting technology and new developments in fracture-control blasting (Oriard, 1981).

It seems likely that the earliest concepts of blasting were developed around cratering principles, then modified for bench blasting as more sophisticated drilling equipment evolved. Every few years or so there seems to be an interest in reviving the cratering concepts, or expanding on them as they might apply to bench blasting or other types of blasting. In their simplest forms, the cratering principles apply more directly to the case of heavy, concentrated charges (ideally, the point-charge concept), whereas bench-blasting concepts were developed around the use of long cylindrical charges whose lengths are very great compared to their diameters. The debate is somewhat academic because of the great need to tailor any concept very precisely to the specific conditions encountered at the particular site involved. It makes little difference what name is given to a method. It is always important to place the right amount of explosive at the right location to accomplish the work.

The main attraction of a "method" is to simplify the procedures of blast design. Although this is understandably attractive, it has a tendency to develop wrong practices. It is a deception to consider that the geological world is uniform and that formulated approaches are better than site-specific designs. The explosives engineer would develop better skills if he were constantly honing them to the specific conditions of the site. Formulated approaches should be used chiefly to prevent the first test of an inexperienced person from being disastrously designed.

Cratering and Bench Blasting

Single, concentrated charges placed below a horizontal ground surface will form a crater if detonated sufficiently close to the sur-

face. A flat, shallow crater is formed if there is a shallow depth
of burial. The crater becomes larger in volume as the depth of the
charge burial increases, but beyond the optimum depth of burial, a
decreasingly smaller crater is broken, and the rock is merely fractured, not ejected. At great depth of burial, the surface remains
undisturbed, although there is a zone of crushing and fracturing a-
round the charge location. In order to take charge size into account
in such cratering experiments, it is customary to scale the depth of
burial, and the crater volume, by the cube root of the charge weight,
based on empirical data that tend to show such a correlation. For
example, if we wish to double the radius of a true crater developed
by the detonation of a point charge, we will have to increase the
charge weight by the cube of 2, or 8 times. The apparent crater,that is
left open after the blast, does not precisely fit this cube root law
because of ejected material that falls back into the crater through
gravity action. According to the above, then, the "scaled depth of
burial" of a point charge $(dob)_{pt}$ is:

$$(dob)_{pt} = \frac{DOB}{W^{1/3}} \quad \text{(by definition)}.$$

Cratering principles can be applied to explosives excavation (in
which material is purposely ejected or excavated directly by the
explosives action itself), or various forms of conventional mining
and excavating operations in lieu of the more common bench-blasting
principles. Of course, typical cratering calculations do not apply
to charges near a vertical free face, unless modified suitably. For
example, a long column of explosives near a free face, and parallel
to the face, could be said to represent a linear crater charge with
a depth of burial equal to the distance to the free face, with the
expectation of little rock breakage beyond the charge (that is,
"below" the crater).

Using cratering principles, one can expand the concept of a point
charge to two dimenions for line charges, and to three dimensions for
array charges, using simple dimensional analysis. In the case of a
line charge, an additional dimension is being added. The unit weight
of the charge now becomes the charge weight (W) divided by the charge
length (S), or (W/S). Similarly, the unit weight of a plane (array)
charge is the charge weight divided by the area of the plane. In a
square array, each side of the plane could be called (S) and the area
of the plane called $(S)^2$. Therefore, the unit weight of the charge is
(W/S^2).

Equivalent expressions for scaled depths of burial for point
charges, line (row) charges, and plane (array) charges are then:

Point charge: $(dob)_{pt} = \dfrac{DOB}{W^{1/3}}$

Row charge: $(dob)_{ln} = \dfrac{DOB}{(W/S)^{1/2}}$

$$\text{Array charge:} \quad (dob)_{pl} = \frac{DOB}{(W/S^2)^{1/1}}$$

Note that as a linear dimension is added to a point charge to produce a line charge, the equivalent scaling changes from the 1/3 to the 1/2 power; and as an additional dimension is added to produce a plane charge, the equivalent scaling changes from the 1/2 power to the 1/1 power. Therefore, in order to make a line charge equivalent to a point charge, we must take the 2/3 power of it. And, in order to make a plane charge equivalent to a point charge, we must take the 1/3 power of it. If the reader were to manipulate the numbers in accord with the above statements, he would discover a rule that is in agreement with field experience: for an increasing number of simultaneously detonating charges (point to row to array), one should increase the depth of burial for an equivalent mounding of the rock.

The important thing to remember is that there is no "theory" to determine the correct depth of burial. It is determined empirically, by trial and error in the field. It is highly site specific. A scaled depth of burial of 3.0 may produce ideal mounding at one site, yet be inappropriate at another site. The usefulnees of the above relationships comes in reducing the number of trials needed for the design of full-scale operations.

Similar rules of thumb and general basic principles apply to bench blasting. In both concepts, it is a question of acquiring enough field experience in different geologic settings to discover emerging guide lines. For example, it is common in bench blasting to use a spacing-to-burden ratio of the order of 2.0. It is common to use a depth of stemming of the order of 20 to 25 times the hole diameter, or in the range of 0.7 to 1.0 times the hole spacing, and it is also common to drill below the expected depth of excavation an amount which is of the order of 0.3 times the hole spacing. However, these "rules" are highly site specific. At any and all sites, it is essential to determine by observation whether or not these designs produce the desired results, and to make whatever changes are necessary to meet the project demands. Such "rules" are designed mainly to help the inexperienced user of explosives reduce the number of field trials to optimize his results.

Direct Damage From Blasting

The primary interest of this paper is that of blasting effects on excavation perimeters and nearby slopes. From the above comments, the reader can appreciate the difficulty in providing a concise rule to enable him to predict accurately the distance into a slope or final excavation perimeter that fractures might extend for any given size of explosives charge. Too much depends on other features of the blast design besides charge size, in addition to the rock characteristics. As an illustration, let us assume a relatively large charge

of explosives, say 450 kg in a single hole. Such a charge would have
the capability of mounding perhaps 750 cubic meters of "average"
rock. If the charge were a large-diameter crater charge (no bench
face), the rock might be broken for about 10 to 15 meters in all directions.   That action would be damaging if the charge were detonated
against a final wall. However, if the charge were detonated as part
of a bench blast, with a free bench face several meters away, the
breakage would be asymmetrical. There would be strong movement toward the free face and reduced breakage into the final wall. It
could be conceived as a linear crater charge turned 90 degrees, so
that the bottom of the linear crater becomes the final wall. We can
carry the concept farther by extending the depth of the holes, reducing the charge diameter and trimming a small burden of rock with
very little damage to the final wall. Carried even farther, this plan
evolves into cautious, controlled perimeter blasting, such as preshearing or smooth blasting. (See Figure 1.)

Direct rock damage of the type discussed above is not merely the
fragmentation of rock due to passing stress waves. One of the most
important physical effects occurring near the blasting source is that
of block motion or inelastic ground displacement, just beyond the
zone of fragmentation. Typically, the maximum range of such inelastic displacement will be the result of the venting of explosives
gases beyond the immediate crater zone, and not the result of vibration.   Identifying the true nature of such disturbance is important
for the reason that the methods for eliminating it depend very strongly on what is causing it. Too often there is an automatic conclusion
that ground displacement or block motion beyond the immediate crater
zone is the result of vibration when usually it is not.

Control of Rock Breakage

Control of rock breakage usually refers to the control of the perimeter of the excavation.   Greater control means a smoother, less
disturbed final surface. Of course, the word "control" can be used
with other meanings, such as fragmentation control, control of the
movement of displaced or ejected rock, or control of vibration. If
any misunderstanding is possible, a writer should specify his meaning
with additional comment for clarification.

To achieve control of the limits of a rock excavation, the explosives user must ensure that the spatial distribution of the explosives is proper for the soundness and smoothness of the final surface
that is desired. An example of relatively uncontrolled blasting
would be the use of large, concentrated charges, widely spaced. Such
blasting will produce an irregular perimeter. Perhaps the least controlled of all would be a single cratering charge. As a general principle, one could say that the least control is achieved with the
smallest number of largest concentrated charges; whereas the greatest
control is achieved with the largest number of smallest , spatially
distributed charges (Oriard, 1971). The accompanying graph portrays

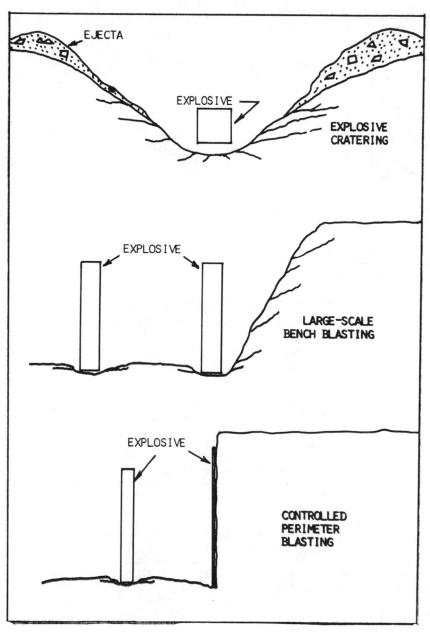

Figure 1 - Illustration of Effects of Spatial
Distribution of Explosives Charges

this concept in graphical form. (Figure 2.)

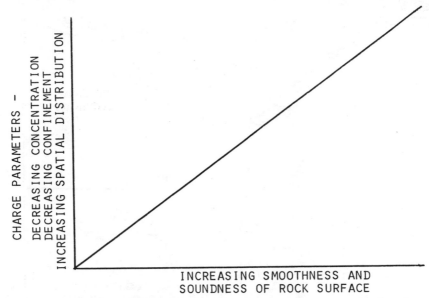

Figure 2 - General Relation of Charge Parameters to Soundness
of Rock Surface

Pre-splitting or pre-shearing is a very cautious technique for
the control of blast effects at the perimeter of an excavation. With
this technique, small-diameter cartridges are detonated in a larger-
diameter hole to decouple the charge from the rock surface and enhance the generation of a prominent fracture between the holes, while
reducing the development of cracks in other directions (Oriard, op.
sit.) In the pre-splitting method, the perimeter charges are detonated first, sometimes as a completely separate operation. If the
perimeter charges are detonated last in the blasting sequence, the
method is usually called cushion blasting, smooth blasting, or merely
trim blasting. Field experiences gained since the time of the last
conference in this series has shown that various modifications of
controlled perimeter blasting have proved to be worthwhile in preserving the integrity of open-pit slopes, and the practice is becoming widespread. It has been standard practice in structural excavations for a very long time. Of course, the appropriate extent of
such an effort is highly site-specific, depending not only on the
characteristics of the rock but also on the needs of the project. In
general, the writer prefers a blast design which provides a three-
stage approach to slope control. This design includes different details for (1) the pattern holes, (2) a buffer zone consisting of at
least one row of holes between the main pattern and the perimeter,
and (3) a line of perimeter holes, with increasing caution in the
blasting design as the perimeter is approached. Such a design will

help to preserve the bench faces in a large excavation.

Normally, the writer prefers that the perimeter row be detonated last in the detonation sequence, rather than as a pre-split blast. When the perimeter is detonated first, as a pre-splitting operation, the nearby buffer holes must be placed closer to the perimeter and detonated with heavier charges, in order to break and eject all rock back to the final perimeter surfaces. Thus, the buffer holes must be very precisely drilled and loaded, in order to avoid either under-break or overbreak. On the other hand, if the perimeter holes are detonated last, their function is not merely to generate a fracture plane, but to displace rock as well. Thus, they perform a work function which permits the buffer holes to be farther away and less crit-ically designed. Of course, the work function is a cost saving as well.

If financial considerations permit the use of highly controlled blasting techniques, it is possible to produce complex, "sculptured" structural excavations. For example, the photo (Figure 3) shows an excavation in which two adjacent openings were blasted in weak rock, leaving an undisturbed, narrow web between the two zones where the blasting took place. Although this particular excavation received attention for the unusually precise sculpture blasting that was done, it did not involve any new theory, - merely a highly site-specific application of existing technology. The underlining is added to emphasize for the reader the importance of fitting explosives tech-nology to the specific site conditions involved.

Perhaps the reader will be interested in some recent developments that have application to control of perimeter fracturing, as well as other special uses. The method is called fracture-control blasting. With this method, the sides of the blast holes are notched longitud-inally. The notches or grooves provide stress concentrations which promote fracturing at lower borehole pressures, and control the di-rection of fracturing, - being primarily within the plane of the notches. The experience of the writer in field tests suggests that there can be a reduction in explosives charge quantities of the order of 2 to 5 times (Oriard, 1981). The boreholes can be notched with either mechanical tools or high-pressure water jets. Of course, either of these notching methods adds time and expense to the project, which must be justified on the basis of improved results. However, it appears that at least two additional options are worth investigat-ing. One of these would be the development of a drilling tool which has the capability of drilling and notching in a single pass. Research is currently underway at the University of Maryland regarding the development of such a tool (Ravinder et al, 1980).

Another option is to make use of a linear shaped charge of explos-ives rather than the grooves along the borehole wall. The effective-ness of either of these methods would be highly dependent on the jointing characteristics between widely-spaced boreholes. In the case of large-diameter boreholes and large-scale blasting operations

These photos illustrate the precision that can be achieved when site-specific blasting designs are used for sculpture blasting in structural excavations, even in this poor-quality rock.
Note the undisturbed 40 cm web of rock between the two deeper excavations in the top photo.
The bottom photo shows several undisturbed complex rock monoliths on the far side of a trench which extends to a depth of nearly 15 meters.

Figure 3 - Site Specific
    Controlled Blasting.

in an open-pit mine, the latter option may be more attractive, at least within the framework of existing technology.

GROUND VIBRATIONS FROM BLASTING

If our interest were only that of predicting the intensity of ground vibrations from blasting, it would not be necessary to discuss wave types and propagation phenomena to any great extent. The subject may be approached empirically, with the inclusion of prediction formulae such as those of the writer, shown later. However, it is necessary for an understanding of the present topic to include sufficient detail of seismic wave phenomena to demonstrate how these relate to slope stability. (A portion of this discussion will appear in a forthcoming AIME volume on underground mining, - see Oriard, 1981A).

After the primary shock front or pressure pulse has passed beyond the zone in which shattering or fracturing of the rock occurs, it passes through the rock in the form of elastic waves or vibrations. As this energy passes through the rock, it takes on different forms which travel at different velocities and cause different types of deformation to occur in the rock. The fastest traveling wave was originally given the name "Primary" or P-wave. This is a compressional wave, sometimes called a radial or longitudinal wave, because the rock is deformed in the radial direction from the energy source. Following the P-wave is a slower traveling wave which was originally called a "Secondary" wave or S-wave. This is a shear wave, sometimes called a transverse wave. Although this wave travels in the same direction as the P-wave, the deformation of the rock is at right angles (transverse) to the direction of the wave travel. The P-wave and S-wave move through the main mass of the rock and have the general name "body waves".

When the body waves arrive at the ground surface, new waves are generated. Some continue through the body of the rock mass as new body waves. Another group travels along the surface and given the name "surface waves". Their motion is quite different from that of the body waves, being characterized by larger amplitudes, lower frequencies, and a lower propagation velocity. In most cases, these waves contain significantly more energy than the body waves, although they do not exist in most underground situations.

If one makes the usual assumption that there is an elastic half space that is homogeneous and iostropic, elastic wave theory describes the wave motions that can be anticipated. In practice, it is simpler and more reliable to determine particle motions by means of field measurements rather than through theoretical calculations. However, it is important to remember that the different wave forms are characterized by different particle motions and are propagated at different velocities. The compressional or dilational wave is propagated with the velocity

$$C_p = \left| E(1-\mu)/\rho(1-2\mu)(1+\mu) \right|^{1/2}$$
$$= \left| (\lambda+2G)/\rho \right|^{1/2}$$

where

$$\lambda = \frac{\mu E}{(1+\mu)(1-2\mu)}$$

and

$$G = \frac{E}{2(1+\mu)}$$

E is the modulus of elasticity, ρ is mass density, and μ is known as Poisson's ratio.   The constants λ and G are known as Lame's constants.   G is also known as the shear modulus.

Compressional wave transmission (propagation) velocities for most rock types fall in the range of about 1500 mps to about 6000 mps, - correspondingly less for weathered or decomposed rock.   Most soils fall in the range of about 150 mps to about 1200 mps.

The shear wave propagates at the velocity

$$C_s = \sqrt{G/\rho}$$
$$= \left| E/2\rho(1+\mu) \right|^{1/2}$$

The ratio of compressional and shear velocities is

$$C_p/C_s = \left| 2(1-\mu)/(1-2\mu) \right|^{1/2}$$

Poisson's ratio for most rock materials is very nearly 0.25.   Thus, the velocity ratio C_p/C_s is often very nearly $\sqrt{3} = 1.73$.

The Rayleigh wave is named after Lord Rayleigh who was the first to examine the case of this seismic wave traveling along the boundary of a free surface.   This wave is characterized by particle motion that is polarized in a vertical plane parallel to the direction of the wave propagation, and the particle motion is elliptical retrograde.   When Poisson's ratio is equal to 0.25, the velocity of the Rayleigh wave is 0.92 times the velocity of the shear wave.

Not only do these different wave forms travel at different velocities, but they have the additional characteristic of attenuating at different rates.   In the case of spherical symmetry in a nondispersive medium, such as the outward-advancing body wave, elastic theory shows that the amplitude is inversely proportional to the distance. In contrast, surface waves have an amplitude that is inversely proportional to the square root of the distance.   Thus, when the point of observation is close to the energy source, there will be a complex combination of several different wave forms.   However, as one moves farther from the source, the wave forms become separated, arriving at different times and producing different types of particle motion. The

more distant the point of observation is from the source, the more
prominent will the surface waves be compared to the body waves.
(There are other types of surface waves in additional to Rayleigh
waves, but Rayleigh waves are usually the most prominent).

Both theory and observation suggest that the particle motion transmitted to a free surface is more prominent than for the same wave
within the body of the solid. For a wave arriving at normal incidence
to a plane surface, the particle amplitude may be doubled. This is
of interest to seismic body waves passing through a hillside to an
opposite slope.

Theoretically, the stress generated by the passage of a seismic
wave is proportional to the product of the acoustical impedance and
the particle velocity. When a plane wave arrives at normal incidence
to a plane boundary, the partitioning of energy between transmitted
and reflected stresses takes place according to the relationship
between the acoustical impedances of the two materials, as

$$\frac{\sigma_t}{\sigma_r} = \frac{2\rho_2 c_2}{\rho_2 c_2 - \rho_1 c_1}$$

Kinematics of Particle Motion

The displacement or amplitude of the ground wave is the distance
from a particle at rest to its peak or trough as the wave passes.
Typical displacements for blasting vibrations of interest fall in the
range from about 0.025 to about 2.5 mm. The term amplitude is used
also to refer to the trace amplitude on the seismogram (recording of
the motion), and can, therefore be somewhat ambiguous.

The frequency of a vibration is the number of cycles that pass a
given point in unit time, usually expressed as cycles per second or
hertz. Frequencies of interest for blasting usually fall in the
range of 1 to 500 Hz, most often being 10 to 100 Hz. Period is the
inverse of frequency, and defines the length of time required for one
complete cycle of vibration.

Particle velocity is the time rate of change of particle displacement. It is the velocity of the motion of a particle during the
passage of the seismic wave beneath the particle. Particle velocity
is not the same as propagation velocity. Propagation velocity, or
transmission velocity, is the velocity with which a wave travels
through a given medium. The propagation velocity varies widely according to the elastic properties of the medium, whereas particle
velocity is a function of the vibration itensity. In the following
discussion relative to vibration intensity, we will be discussing
particle velocity.

Acceleration is the time rate of change of particle velocity. It
refers to the acceleration of a particle as the seismic wave passes

beneath this particle. For simple harmonic motion, the following relationships apply:

Defining: "x" is displacement at time "t"
"A" is maximum value of x which is equal to the zero-to-peak amplitude
"f" is the frequency
"v" is the particle velocity
"a" is acceleration
"w" is angular frequency

Then: $v = 2\pi f A$

$a = 0.1 \, f^2 A$ (approx.) in gravity units.

$w = 2\pi f$

$x = A \sin wt$

$v = w \cos wt = w \sin wt + \pi/2$

$a = -w^2 A \sin wt = w^2 A \sin (wt + \pi)$

$a = 4\pi^2 f^2 A$

Predicting the Vibration

   In order to predict the intensity of ground vibration from blasting, one must consider the influence of the blasting parameters and the influence of the geological setting. If we were always dealing with simple point charges of a single explosives type, the first question would be a relatively simple matter of data scaling, and the second would be a question of wave attenuation, though by no means simple. In reality, the two questions are often closely intertwined because of many departures from the ideal assumptions that are often made to simplify calculations.

   In order to compare blasts of different sizes at different distances, it is customary to scale the distance by some function of the explosives charge weight per delay (the amount detonating at any given instant of time), so that such diverse data can be plotted on a simple graph. If the charges were spherical, theory would dictate the use of cube root scaling, because the charge weight would vary as the cube of the radius of the sphere. Dimensional analysis has also been used to support the concept of cube root scaling, but dimensional analysis does not apply to multi-form wave propagation questions. If the charges were long cylinders, we would expect that the use of square root scaling would apply, because the charge weight would vary as the square of the radius of the cylinder. However, in the vast majority of cases involving blasting, neither of these ideal models is accurately duplicated in the field. Many times, the charge weight is increased merely by increasing the number of separate charges, although they are usually long cylinders of explosives. Thus, there are many geometrical complications involved in the question of data scaling. Statistical analysis will often show variations in the best fit for any of the scaling laws. However, the question is only of academic importance if the decision maker knows where a particular data point falls in the general range of experiences, and understands the consequences of scaling up or down the sizes of charges, or distances, using different scaling laws. There is a considerable amount of scatter in data points relating to blasting vibrations. If the investigator recognizes that a particular vibration were unusually low, he could anticipate that another test of the same design might give a considerably higher value the next time. Also, if he uses a small charge measured at close range to serve as a model for a much larger blast later, at the same scaled distance, he should be aware of the influence his selection of data scaling will have on his final prediction. There will be an important difference in his prediction according to his scaling methods. If he is not aware of the differences, he should try different methods, in accord with his experiences and the conservatism he wishes to incorporate into his predictions (Hendron and Oriard, 1972).

The writer has found it convenient to use square root scaling for prediction of the widest range of blasting conditions. His experience has shown this scaling method to more accurately portray a larger range of field conditions and blasting techniques than cube root scaling or other scaling. The writer has analyzed several hundred thousand vibrations from blasting, and has found the graph (Figure 4) of particle velocity versus scaled distance to fairly represent that experience. The relationships can be expressed mathematically in the form

$$V = H (D/W^{1/2})^{-1.6} k_1, k_2, k_3, \ldots$$

where V = peak particle velocity

            H = velocity intercept at unity scaled distance

            W = charge weight per delay

            k factors represent the variations in explosives, confinement, spatial distribution, geology and other parameters of interest.

For a typical upper bound prediction line

$$V = 242 (D/W^{1/2})^{-1.6} x (1.0)$$

The slope of (-1.6) represents the attenuation. It is not the same at all sites nor the same for all wave types, but in fact it is surprising how well this slope accurately represents most situations. The attenuation of blasting vibrations with distance is a complex function of the strain level, the various wave forms, and the geological setting. There are several theoretical reasons why one should anticipate a two-slope attenuation curve, rather than the straight line shown in the prediction curves of this writer. For waves which begin at very high strain levels, we should anticipate a more rapid

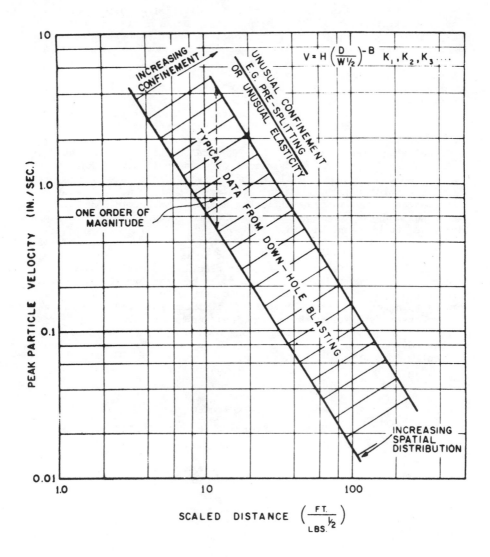

Figure 4 - Peak Particle Velocity vs Scaled Distance

attenuation initially, until low seismic strain levels are reached,
at which time the attenuation should remain relatively constant in
accord with the geological setting. Similarly, waves arriving at
locations near the source are a complex function of several wave
types, all combined. Because of different transmission velocities,
these waves then separate as they move away from the source, spreading
out between arrival times and generating different types of particle
motions. Thus, there should be a more rapid attenuation near the
source, and later an attenuation which is determined by the predominant wave, usually the Rayleigh wave. The above theories should hold
for a point source. Some data show agreement with such theories,
especially those taken from nuclear detonations, where it is clear
that both high strain levels and a point source are correctly modeled.
However, there is an interesting departure from this type of attenuation for most blasting operations. Usually, a straight extrapolation
serves the purpose. Sometimes it even reverses from theory and forms
a flatter slope near the source. This happens when a large number of
holes detonates simultaneously. There is no such thing as zero distance in that case, and the attenuation is distorted by the departure
from a point charge. Even in the case of very large blasts, such as
"coyote" blasts, (a tunnel filled with explosives), the condition de-
parts from theory by a failure to detonate instantaneously or from
the center outward, - two theoretical requirements. And in the case
of relatively low energy levels, there is some experience to demonstrate the validity of the same attenuation carried very close to the
source. A case in point is that of small charges detonated in old
concrete at Lock and Dam No. 1, Minneapolis. The writer's prediction
curves were found to represent accurately the attenuation extrapolated as close as 8 inches from the source. (Tart, Oriard and
Plump, 1980; Oriard, 1980). In order to measure these very unusual
vibrations, it was necessary to use accelerometers having a frequency
range up to 30,000 Hz and an acceleration range to 30,000 g. The
writer's data cover a distance range from 8 inches to 20 miles and
a charge weight range from 1 gram to nuclear devices. The curves
are intended to portray that range of experience.

The broad base of experimental data mentioned above provides a convenient means of making reasonable predictions of vibration intensity. However, it is helpful for the understanding of slope stability
questions to be aware of some of the factors that influence other
characteristics of the vibration besides intensity, such as frequency
(hence wave length) and displacement. Starting at the source, one
finds an inverse relationship between charge size and frequency, for
any given medium. The larger the charge, the lower the frequency and
the larger the displacement. Similarly, normal wave propagation phenomena bring about a decrease in frequency with distance. The higher
frequencies are more quickly attenuated. The more predominant factors
relating to the attenuation of seismic waves are (1) geometric spreading, (2) selective scattering, (3) absorption, and (4) dispersion.
As mentioned previously, geometric spreading is inversely proportional
to the distance for body waves, and inversely proportional to the
square root of the distance for surface waves. Scattering varies in-

versely as the fourth power of wave length (therefore directly with
frequency); absorption increases with the second power of frequency,
and dispersion varies with the first power of frequency.

The wave transmitting medium (geological setting) has a strong influence, also, on the frequency and displacement characteristics of
a seismic wave. For example, hard massive rock will be characterized
by smaller displacements and higher frequencies, whereas soil will be
characterized by larger displacements and lower frequencies. Both
the attenuation and the wave form characteristics are influenced by
such geological factors as layering, jointing and water content, as
well as the small-scale elastic properties of the medium. For example, in certain regions underlain by prominent horizontal layers of
sedimentary rock, it has been noted that surface waves appear to be
more prominent and persist to greater distances than is typical for
regions that are more heterogeneous and/or geometrically complex.

VIBRATION EFFECTS ON SLOPES

The experience of this writer suggests that there is a frequent
need for a fast, relatively simple evaluation of the stability of
slopes subjected to blasting stresses. Blasting operations themselves
frequenty produce slopes of sufficient height to be in need of evaluation, and they are often found in proximity to other slopes of concern, whether composed of soil or rock or some combination of materials. In the majority of cases when a question arises concerning
stability, a judgment must be made rather quickly because of the financial and scheduling needs of the project which is underway. Sometimes the evaluation is required in advance of project start-up, for
varying reasons, including that where there is a need to provide "documentation" that the future project will not generate public or private hazards. It happens quite frequently in such cases that there
is a specific demand for a calculation using some "standard" pseudo-
static method of analysis. With such methods, it is often assumed
that the estimated horizontal acceleration of the predicted vibration
will act as a static force in the horizontal plane in the direction of
the outer slope face. Those who have performed such exercises will
come to realize that they predict dire consequences in nearly every
case involving blasting vibrations, despite the long history of experiences to the contrary. In spite of this lack of correlation with
blasting experience, the use of such methods remains widespread.

Terzaghi's Method

The origin of pseudo-static methods of analysis of slopes and embankments subjected to vibration may have developed a very long time
ago. However, the "standardization" of an approach very likely began
with Terzaghi (1950),(Seed, 1979). Terzaghi described the method
as follows:

"An earthquake with an acceleration equivalent n_g produces a mass
force acting in a horizontal direction of intensity n_g per unit

of weight of the earth. The resultant of this mass force, $n_g W$, passes like the weight, W, through the centre of gravity O_1 of the slice abc.* It acts at a lever arm with length F and increases the moment which tends to produce a rotation of the slice abc about the axis O by $n_g FW$. Hence the earthquake reduces the factor of safety of the slope with respect to sliding from G_s, equation (1) to

$$G'_s = \frac{slR}{EW + n_g FW} \quad \cdots \cdots \cdots \cdots (2)$$

"The numerical value of n_g depends on the intensity of the earthquake. Independent estimates (Freeman, 1932) have led to the following approximate values

Severe earthquakes, Rossi-Forel scale IX $n_g = 0.1$
Violent, destructive, Rossi-Forel scale X, $n_g = 0.25$
Catastrophic $n_g = 0.5$

The earthquake of San Francisco in 1906 was violent and destructive (Rossi-Forel scale X), corresponding to $n_g = 0.25$.

"Equation (2) is based on the simplifying assumptions that the horizontal acceleration n_g acts permanently on the slope material and in one direction only. Therefore the concept it conveys of earthquake effects on slopes is very inaccurate, to say the least. Theoretically, a value of $G'_s = 1$ would mean a slide, but in reality a slope may remain stable in spite of G'_s being smaller than unity and it may fail at a value of $G'_s > 1$, depending on the character of the slope-forming material.

"The most stable materials are clays with a low degree of sensitivity, in a plastic state (Terzaghi and Peck, 1948, p. 31), dense sand either above or below the water table, and loose sand above the water table. The most sensitive materials are slightly cemented grain aggregates such as loess and submerged or partly submerged loose sand . . . . "

* Terzaghi's figure is not reproduced here because it is not needed for the purposes of this paper. Quotation was used to make sure that there were no misrepresentations of Terzaghi's statements.

In recognition of Terzaghi's eminent status in the field of soil mechanics, it is not surprising that this concept caught on and became widely applied. However, there are aspects of the question worth noting. One is that Terzaghi himself recognized the complexity of the problem, even as it related to the case of earthquakes, which is relatively simple when compared to blasting phenomena. Paradoxically, later advocates of this method often chose to use seismic coefficients which were much less conservative than those recommended by Terzaghi; yet the method normally is so dramatically over-conservative for blasting phenomena that it is usually quite misleading to use it. The reason for this apparent paradox is due to the differences between earthquake and blasting vibrations. Most blasting vibrations are characterized by relatively high frequencies compared to earthquakes. In turn, acceleration is proportional to the square of the frequency.

Thus, blasting vibrations generate relatively high accelerations for whatever particle velocity is involved (hence, strain). At the low frequencies associated with large earthquakes, an acceleration of 0.1 g is regarded as strong motion, and an acceleration of 1.0 g would be regarded as catastrophic. This is not at all true of blasting vibrations. For small charges at close distances in rock, the corresponding accelerations may be many tens of g's without necessarily being of concern. For example, the author has measured non-damaging accelerations approaching 1000 g in the walls of an operating powerhouse in Venezuela, and accelerations in the range of 20-30,000 g in the walls of an old concrete lock, as mentioned previously. It is quite clear that acceleration alone is not a diagnostic feature of the damaging potential of propagating seismic waves, or even of simple mechanical shaking. A small displacement at high frequency may have a higher acceleration than a larger displacement at low frequency, yet the latter may have more damage potential because of larger strains generated. The stress generated by a passing seismic wave is proportional to the product of the acoustic impedance of the material and the particle velocity. Therefore, there is no theoretical reason why particle acceleration should be used in evaluating strains in slopes due to passing seismic waves. Using the relationships for sinusoidal wave forms, the following relationships can be seen to illustrate the above comments:

Displacement	Frequency	Acceleration	Strain
0.001 in. (base case)	100 Hz	1 g	1 unit (base case)
0.1 in. (100 times increase)	10 Hz	1 g	10 units
10.0 in. (10,000 times increase)	1 Hz	1 g	100 units

Newmark's Method

In an effort to improve on earlier pseudo-static models of slopes subjected to seismic shaking, N. M. Newmark, in his Rankine Lecture of 1965, proposed a procedure for evaluating the potential deformations of an embankment subjected to earthquake shaking (Newmark, 1965). In this method, it is assumed that slope failure is initiated and movements begin when the inertia forces are large enough to overcome the yield resistance of the slide mass, and that movements stop when the inertia forces are reversed. The cycle may or may not be repeated. With this model, the investigator computes the acceleration at which the inertia forces become sufficiently high to cause yielding, then integrates the effective acceleration on the sliding mass in excess of the yield acceleration as a function of time to obtain velocities and displacements. The velocities are shown as functions of time for both the accelerating force and the resisting force. The maximum velocity for the accelerating force has the magnitude V given by the expression $V = Agt_o$. After the time t_o is reached, the velocity due to the accelerating force remains constant. The velocity due to the resisting force has the magnitude Ngt. At a time t_m, the two velocities are equal and the net velocity becomes zero, or the body comes to rest relative to the ground. Displacements are made in distinct, discrete steps if there is sufficient difference in the velocities of the base and the sliding mass.

For very long waves, such as those generated by typical earthquakes the slope motion can be said to be a very crude form of a mechanical shaking table, and indeed shaking tables have reproduced this type of step-by-step displacement in scale models of embankments tested in the laboratory (Seed, 1979, 1980). Such laboratory tests and actual earthquake experiences have shown that embankments can undergo substantial accumulations of discrete displacements without necessarily "failing". The Newmark method has been found to be quite useful where the yield resistance of the embankment can be reliably determined, where pore pressures do not change significantly, and where the materials do not lose more than about 15% of their original strength during the shaking (many clayey soils, some dense saturated sands and clayey sands), if the mass can initially tolerate an inertia force of the order of about 0.1 to 0.15 g without yielding, and crest accelerations are less than about 0.75 g (Seed, 1979, op sit.).

The phenomenon of accumulated displacements is normal for many rock slopes and soil slopes. For very small dynamic loads, no effect whatever may be noted. For somewhat heavier loads, small displacements may be initiated. As they accumulate, there is often ample opportunity for observation and the development of remedial measures. Most soil and rock slopes develop sufficient residual strength after the initial movements to have a controlling influence on later movements. Many slopes are not capable of undergoing sudden failure because of this residual strength. Exceptions are such cases as the first, sudden failures of rock wedges subjected for the first time to strong shaking. There is an increasing sensitivity as we proceed from the case of previously failed zones which have come to rest in new stable positions, to the case of still-moving masses, to those which have never failed but are potentially very unstable.

Unfortunately, the Newmark method does not model wave propagation phenomena. It assumes that the slope rests on a rigid base subjected to mechanical shaking. Of course, this is not at all true for blasting vibrations. There are different wave forms involved, and the wave lengths are often short compared to the slope length. Very often, there is a dramatic attenuation within the slope length of interest. There are additional complexities of a geometrical nature that are not significant in the case of earthquake shaking.

Some of the major differences which normally exist between typical earthquakes and typical blasting can be summarized as follows:

1. Boundary conditions for blasting are not usually those assumed for the slope model. Commonly, only a small portion of the

slide mass is subjected to a given motion at any given instant of time.

2. One cannot assume a single intensity of shaking, since the vibration will attenuate within the slope.

3. Not only will different parts of the slope undergo different intensities of vibration, but there are also different frequencies involved with attenuation, so that the different sections do not move in phase.

4. Different wave forms are involved, which separate with distance, so that even the duration of vibration changes, as well as the intensity and frequency.

5. The surface motion will be different from that at depth. Which motion should be considered to act on the slope? Body waves or surface waves?

6. The direction of travel and angle of incidence are important. Is the wave arriving at grazing angle of incidence to the slope surface (in which case there may be very little motion tending to stimulate sliding)? Or is it coming from an angle that may generate surface reflection of long waves? Is it realistic to consider that all possible angles of incidence need to be evaluated?

7. Considering not only the low stress levels usually involved, but the small particle displacements as well, small surface irregularities become more important in resisting slope movements.

8. Cycle duration is shorter, just as the wave lengths are shorter, thus providing less opportunity for displacements to occur.

9. The combination of small displacement and high frequency are quite significant. Mass dilation may occur without any slope displacement. The condition can be compared crudely to space tolerances in mechanical equipment subjected to vibration. If two parts are separated by a distance comparable to the particle displacement of the vibration, it is not likely that the vibration will affect them.

In consideration of the complexities of wave propagation phenomena, it is not desirable to use pseudo-static methods of analysis for the more common cases of blasting. Of course, it is possible that a history of experiences in a given geological setting, with repetitions of a given type of blasting, might permit the investigator to develop a special application of such methods by determining empirically the appropriate "artificial" seismic coefficients. However, such an approach would fall apart quickly if there were significant changes in blasting methods or site characteristics, including geometric considerations. It is not likely that such an approach would have any advantage over the simpler experience of comparing observed slope behavior to measured particle velocities.

Possible New Pseudo-Static Method of Analysis

Considering the attractive, convenient simplicity of pseudo-static methods like the Terzaghi method or Newmark method, it would seem worthwhile to pursue a similar approach that takes into account some of the physical parameters that are more characteristics of wave propagation phenomena than those considered in the former methods. Oriard and Yen (1977) presented a discussion of such an approach regarding blasting effects on unstable slopes of the Panama Canal. It was hoped that this project might offer an opportunity to gather some meaningful well-controlled field data during the proposed deepening program (Oriard, 1980A). However, the effort was cut back due to political changes. Consequently, the writer has not yet had the opportunity to gather any field data which could be used to evaluate the suitability of such a method. A considerable effort would be required to develop such a method and to check its validity against known performance of identified slopes.

As a beginning approach to such a method, the following concepts could be pursued initially, - perhaps modified later as needed.

F_{sta} is the static shearing force, primarily the downhill component of the gravitational force. It may include pore water pressure, if any exists. For a generic element in the slope, the static driving force could be expressed as

$$F_{sta} = (\rho \, g \, h \, \sin \beta) \, (\cos \beta \, dL)$$

where ρ = mass density

g = acceleration of gravity

h = vertical dimension of slope element

dL = element length along slope angle

β = slope angle

F_{dyn} is the dynamic load induced by blasting. The dynamic force is transient, cyclic and varies in direction and magnitude. There is no known closed-form solution for F_{dyn}. The dynamic force depends on the many variables previously mentioned. Although the degree of conservatism is not known for actual field conditions, perhaps a reasonable, though conservative, approximation could be represented by

$$F_{dyn} = (\rho \, c_s \, v) \, h$$

where c_s = shear wave velocity

v = peak particle velocity

F_{res} is the residual force that may exist in the rock slope, such as that which may be due to tectonic stresses, chemical stresses, and

the like. For a simplified analysis, F_{res} may generally be neglected.
For failure,

$$F_{sta} + F_{dyn} + F_{res} \geq \text{Resisting Force}$$

A further reasonable assumption for slope failure would be a requirement for the slope particles to move in unison, that is, an
in-phase velocity field, thus limiting the zone of interest to one
which has a dimension less than 1/2 wave length,

$$L \leq \lambda/2$$

For the purposes of an initial evaluation of this method, it could
be assumed that the rock strength under combined static and dynamic
stress should be less than its peak static strength. That is, we
could make a beginning assumption that

$$F.S._{dyn} = \frac{R}{F_{sta} + F_{dyn} + (F_{res} = 0)}$$

where R is the peak static shearing resistance

$$R = (\rho g h \cos^2\beta) \, dL \cdot \tan(\phi_r + i)$$

where ϕ_r = friction angle

i = equivalent friction angle increase to account for such factors as joint roughness.

Pursuing this concept, one can prepare families of curves such as
those in Figure 5.

Further research is needed to determine how well field experience
will agree with the predictions. As in all other aspects of explosives engineering, it is anticipated that there will be a need for
judgmental factors or "coefficients" to relate the calculations to
experience. Depending on the range of such needed coefficients, the
practicability of the method may then be assessed. Although there are
many obvious theoretical shortcomings to the method, there seems to
be a chance for somewhat better correlation than with previously used
pseudo-static methods. And it seems more likely that there would be a
possibility of better correlation with shallow slope failures than
with deeper ones. For deeper failures, the boundary conditions would
surely become increasingly more important. In all cases, it is very
important to study the slope responses to physical stimuli and to determine the time history of repeated responses. If the measured displacements are decelerating the failure is regressive and there may be
no need for further action at that time. If, however, the displacements are accelerating, the failure is progressive, and action may be
needed quickly, such as a change in blasting methods, or remedial
work on the slope, such as suggested in Figure 6 which represent
previous actual field experience. (Oriard, 1971)

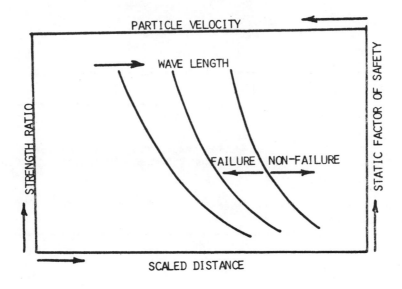

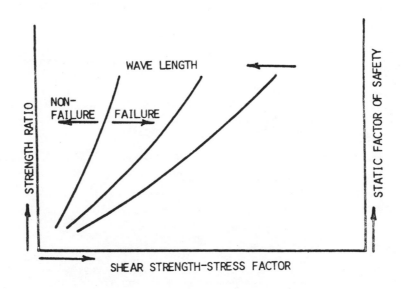

Figure 5 - Possible Slope Failure Criteria

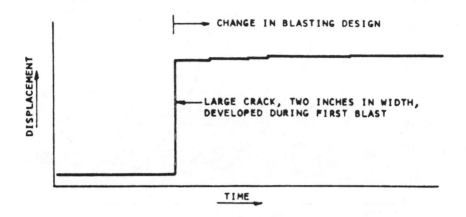

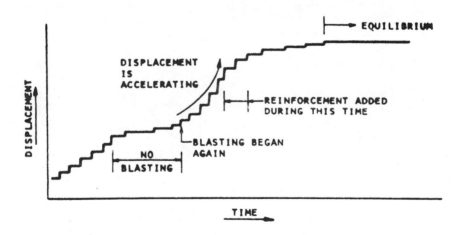

Figure 6 - Illustration of Progressive and Regressive
        Dynamic Responses to Blasting Vibrations.

Dynamic Methods of Analysis

Dynamic methods of analysis do exist, and have shown significant improvements in recent years. With the increasing sophistication of computerized analytical techniques now available, such as the finite element method and the finite difference method, it is now possible to analyze dynamically the simpler vibration models and most types of slope models. At least two types of vibration input can now be analyzed quite accurately. One is that of a rigid-base model subjected to any vibration history of interest. The other is that of a simple form of stress wave propagating through a continuum. Unfortunately, these methods are relatively time-consuming and expensive, and still face very formidable problems when dealing with blasting phenomena. Wave propagation models become extremely complicated, and the range of possible cases becomes discouragingly large. And, of course, the slope must be accurately modeled if the methods are to have reasonable validity.

On many projects, there is neither the time nor the financial resources available for the development of an acceptably accurate dynamic model. In such cases, judgements must be made without the benefit of such analysis. Two of the most common approaches to such cases are (1) program the blasting in order to limit stresses to conservative values, and/or (2) monitor slope behavior carefully to observe the first signs of any adverse reaction.

Blasting is Controllable

One very important distinction between blasting and earthquakes, in addition to those mentioned previously, is that blasting can be controlled by design. In most cases, it will be found possible to limit blasting stresses to acceptable levels without adding any significant financial burden to the project. If, however, a more precarious condition exists, it is normally a simple matter to begin blasting on a limited scale and build up to a larger scale on a programmed basis while monitoring slope behavior. Of course, this not true of earthquakes. The controllable aspects of blasting, whether the vibration intensities are controlled by distance or by design parameters, permit a close scrutiny and detailed instrumental monitoring of the slope at low vibration intensities, and at increasing vibration intensities, in accord with the wishes of the investigator. It is relatively rare that the question must be answered for the case of a single, large event, with no opportunity for preliminary observation, although the latter case sometimes arises (Oriard and Jordan, 1980).

Dynamic Stability

The dynamic stability of a slope is very closely related to its static stability. Those same physical properties (especially in-situ larger-scale mass characteristics, and properties of weak planes) that render a slope unstable under static loading conditions contribute to its lack of stability under dynamic loading. That is, the higher the static factor of safety, the higher the dynamic factor of safety, in general. However, some of these factors, such as surface irregularities on rock slopes, have varying relationships to stability in accord with the type of dynamic loading that occurs, not merely its numerical value of acceleration or velocity, as previously discussed.

It is anticipated that the state-of-the-art regarding static stability will be discussed at length in this conference. For the purposes of this paper on blasting effects, only a few general comments will be made regarding static stability, in order to complete the discussion of blasting.

Three of the most important factors relating to stability of rock slopes are (1) size, location and orientation of critical discontinuities, (2) the shear strength along these discontinuities, and (3) the pore pressures on these discontinuities. A person wishing to evaluate dynamic stability would proceed initially in the same manner as an investigator evaluating static stability. He would investigate such factors as

1. Geologic history, including weathering processes and profiles, geologic age, rates of steepening or flattening of slopes through natural processes and/or the activity of man.

2. Stress history and anticipated in-situ stresses.

3. Climatic and hydrologic history, past and present.

4. All factors relating to the present "mechanical" conditions, such as type of materials, bedding and jointing (frequency, orientation, fillings, openness, irregularity, etc. (see Goodman, 1981; Barton, 1981, and others).

5. Any previous dynamic history, such as earthquake activity, previous blasting activity, or steady-state vibration sources.

CASE HISTORY - PRECARIOUS SOIL SLOPES

The following case history will illustrate some of the problems that are often encountered when dealing with blasting effects and slope stability, and one of the approaches to dealing with such problems. The case involves certain unstable tailing dams. The slurry formed by the fines left over from ore milling, mixed with waste water, was pumped by pipeline to waste areas. As the slurry began drying at the perimeter, the dry, fine sand around the perimeter was reworked with bulldozers to form dams to contain additional slurry. The process was continued, simultaneously building up the dams and filling more tailings behind them. One of the dams had a crest height of approximately 43 m at the time of this investigation. Two or three local slope failures had occurred, and an investigation led to the conclusion that the dams were statically precarious and incapable of withstanding the shaking action of a moderate earthquake that might occur in the region at some time in the future. Consequently, a decision was made to place a rockfill buttress against the steeper, lower portion of each embankment (Figure 7). A suitable quarry site was found nearby, from which the rockfill material would be obtained. The rockfill was to be high-quality material, requiring blasting. Thus, a large-scale quarrying operation would be required in the vicinity of statically precarious tailings embankments. A total volume of about 1,300,000 tons of rockfill would be required. In addition, there was a need to blast drainage trenches immediately in front of the toes of the embankments. Thus, one of the interesting aspects of this case was the need to consider three different kinds of vibration, (1) low frequency vibration generated by an earthquake, (2) mid-frequency vibration generated by quarry blasting, and (3) high-frequency vibration generated by trench blasting.

One would consider the factor of safety of existing conditions to be about 1.0 $\pm$. Theoretically, the embankments were incapable of tolerating any vibration. Even after the construction of the buttresses, they would be capable, theoretically, of withstanding an acceleration of only 0.12 g (the design earthquake). If the usual pseudo-static models were considered valid, no remedial work could be done because the blasting would generate unacceptable vibrations. Fortunately, experience has demonstrated that certain vibrations can be tolerated under such circumstances, and that the higher the frequency of the vibration, the greater the acceleration that can be tolerated (for equivalent strain).

The following table illustrates the particle motion parameters of interest, showing the range in particle velocities and displacements for various frequencies, assuming a constant acceleration of 0.12 g.

Acceleration	Frequency	Velocity		Displacement	
0.12 g	0.1 Hz	75	ips	120	in
0.12 g	1.0 Hz	7.5	ips	1.2	in
0.12 g	10 Hz	0.75	ips	0.012	in
0.12 g	100 Hz	0.075	ips	0.00012	in
0.12 g	1000 Hz	0.0075 ips		0.0000012	in

The reader can see from the above figures that if we are given a constant acceleration as a limit, we then find that velocity is inversely proportional to the first power of the frequency, and that displacement is inversely proportional to the square of the frequency.

At limiting equilibrium, if the acceleration is limited to zero, due to instability, no vibration whatever can be tolerated. Hence, in theory, no blasting can be tolerated. Even if we assume that the blasting takes place after the completion of the work, at which time

Figure 7 - A rockfill buttress being placed on the lower
    slopes of a precarious tailings embankment. Blasting
    was required at the toe of the slope and in a nearby
    quarry.

the embankments can theoretically tolerate an acceleration of 0.12 g, there is still a serious limitation. There is no possible way that any trench blasting could be done at the toes of the embankments without greatly exceeding 0.12 g. If we were to believe that acceleration is a valid criterion, the exercise is self-defeating, because there is a rapid increase in frequency with reduction in charge size. Even though particle velocities and displacements might be reduced by reducing charge size at close distances, we might begin to pick up high frequencies that do not exist at greater distances. Charges of only several ounces of high explosives may generate accelerations of the order of 10,000 g to 30,000 g within the first several feet, as mentioned previously (Oriard, 1980). If we seriously believed in acceleration criteria, we would want to make every blast act like a small earthquake, and that would not be a wise approach.

The writer recommended using the observational approach in this case, believing that any of the "standard" methods of analysis would only be misleading. It was obvious that high accelerations would be generated by the trench blasting, and that moderate levels of acceleration would be generated by the quarry blasting. If neither moderate nor high levels of acceleration could be tolerated, nothing would be gained by making the calculations.

With the observational approach actually applied to this case, the latest techniques in blasting technology were used to control the vibrational particle velocities to levels considered to be conservative, and would still permit the work to proceed at a large scale. There was no measurable sacrifice due to blasting controls, beyond the first few days of initial trials while the embankment behavior was being very carefully monitored. The embankments were monitored for displacements, changing pore water pressures and phreatic water levels. If there were no significant increases in pore pressures, and the embankments did not undergo any displacement, no change in stability would take place. It was concluded that a series of smaller displacements would occur and accumulate before there would be any danger of a significant failure. In the case of the trench blasting, at higher stress levels, any single displacement could conceivably be greater, but would be limited to a small portion of the embankment immediately adjacent to the blasting area. Thus, even though the mechanisms could be different for the two types of blasting, there would still be an incremental development of any significant displacements.

Two types of piezometers were used. One type was the isolated-tip type, consisting of a porous tip installed at the specific point of interest. These are more sensitive and react more quickly than the other type which was used, - the open-well piezometer. With the first type, pore pressure changes could be monitored within a few minutes after a blast. The open-well piezometers provided information on changes in the over-all phreatic line in an embankment. Both assisted in the assessment of stability, or change in stability.

All blasting vibrations were monitored. Seismographs were placed in suitable locations to record bedrock vibrations and embankment responses. The embankment responses often showed an amplification of the order of 4 times greater than that of the bedrock base. Of course, for the trench blasting, there was a significant attenuation from the toe to the crest, although the same relative amplification could be detected between the tailings and rock.

Quarry blasting began cautiously and increased in scale while the embankments were monitored. Initially, there were very minor increases in pore pressures, but these were quickly dissipated (typically, in 15 minutes to several hours) with no observable longer-term effects of concern. Very close observations were continued until the quarry blasting had been increased progressively to a point beyond the level desired for long-term, continued operation, then reduced to that for the long-term program. The time intervals between embankment observations were then slowly increased as no disturbances were noted.

Figure 7 illustrates the field setting. Figures 8, 9 and 10 illustrate typical bench-mark readings, piezometer locations and piezometer readings.

Despite the obvious precariousness of the embankments, there were no adverse effects of any type observed during the six-month period of blasting, involving well over 200 blasts.

Vibrations were monitored with velocity gages rather than with accelerometers. Accelerations were neither measured nor calculated, because they were not regarded as being diagnostic of either shaking intensity or damage potential. However, quarry blasting designs were programmed in such a way as to keep the predicted range of accelerations at or below the 0.12 g design earthquake acceleration (not the theoretical zero acceleration assumed as a limit before the buttresses were placed). The trench blasting probably generated accelerations of the order of 50 g near the toe, of the order of 1.0 to 1.5 g 15 m within the embankments and about 0.5 g at 30 m (meters).

In this case, the observational approach was selected over analyses that were regarded as inappropriate. At the same time, however, the writer would like to repeat the opposite concern about being too liberal with such analyses when applied to earthquakes generating very low frequency ground shaking, recalling the recommendations of Terzaghi (1950) and Seed (1979) mentioned previously. The emphasis should be placed on recognizing the differences in these different types of vibrations, and in treating them accordingly.

The experience of observing very high accelerations (even high velocities and high strains) that are not damaging to slopes thought to be statically precarious raises some interesting questions. Some of these questions cannot be answered merely on the basis of short wave

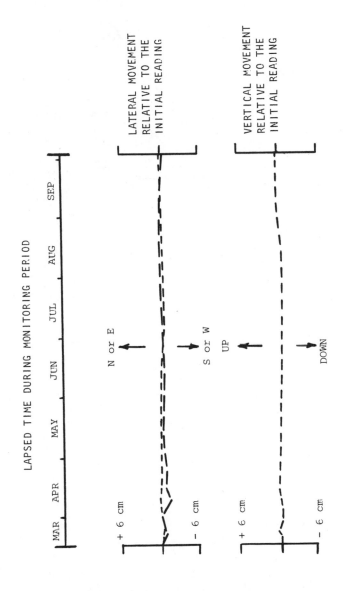

Figure 8 - Typical Slope Displacement Monitoring Data

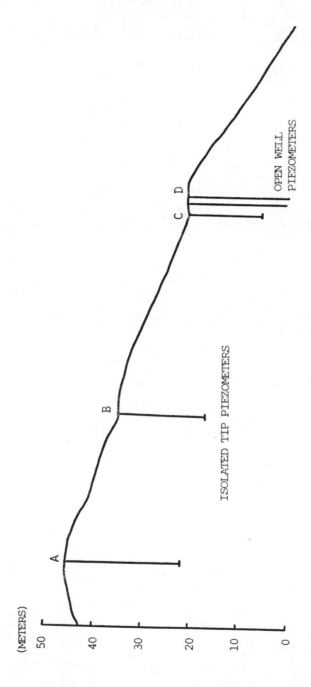

Figure 9 – Typical Tailings Embankment Section with Piezometers

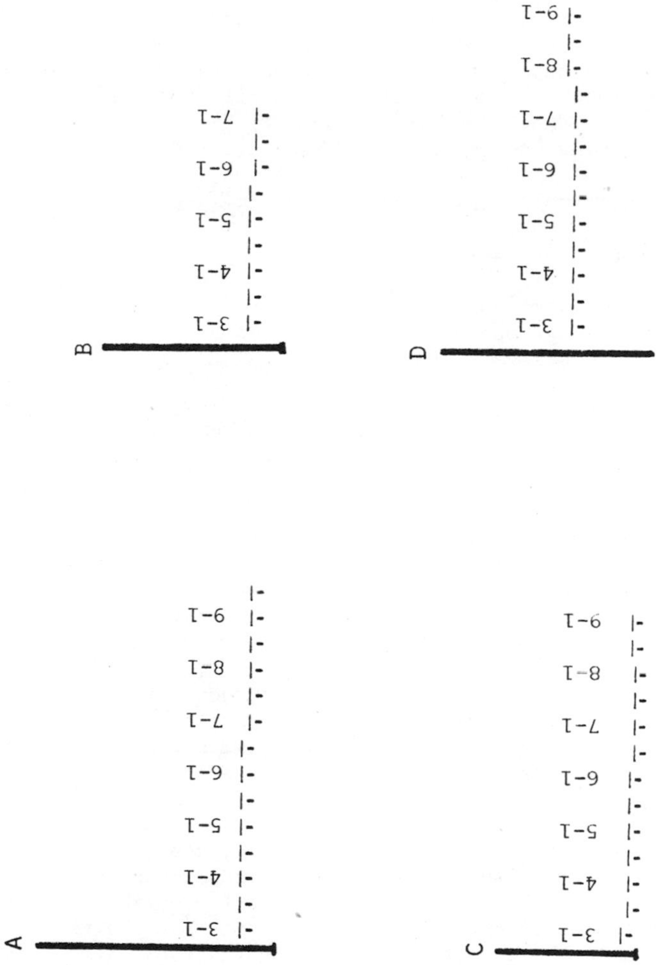

Figure 10 - Typical Long-Term Monitoring Data from Piezometers

lengths alone, although that is a factor of great importance. That
would lead only to the conclusion that the damaged zone should be of
limited dimension. Of course, one would expect the damage to occur
where the stresses were highest, - near the blasting. This type of
experience suggests that we need to look very carefully at boundary
conditions. For example, a failure might be possible if a small zone
in the embankment could be artificially bounded by failure planes,
but that it must be greatly strengthened by being bounded by a continuation of the same material well beyond the distance at which high
stress levels would be found. That is, the potential zone of failure
is supported or held by adjacent material not under the same level of
stress. Thus, the true boundary conditions are not necessarily the
bounds of some potential mass of sliding, but may be determined more
properly by understanding the character of the vibration.

CASE HISTORY - PRECARIOUS ROCK SLOPE

The following history was selected to illustrate a situation completely different from that of the previous case. In the following
case, steep rock slopes were subjected to the direct rupturing and
tearing actions of blasting operations, leaving damaged rock layers
in an unpredictably precarious situation.

The case involved a quarrying operation in steeply dipping layers
of limestone. Layers were typically from 20 to 30 meters in thickness, separated by prominent discontinuities with very little shear
strength. Blast holes were drilled vertically by drills which were
lowered down the slopes by ropes and cables. Drilling and blasting
began at the toe of a layer, breaking off sections which then tumbled
by gravity along the surface of the next underlying layer. The broken
rock was picked up at the toe of the slope and hauled away. Successive blasts continued up the slope to higher and higher elevations.
The operation is illustrated in Figure 11.

The experienced reader will recognize this procedure as being the
same as that which is used to bring down and dispose of precarious
rock wedges or other potential rock slide zones, with the difference
in the latter case that a larger portion of the precarious zone is
usually blasted to ensure the failure of the entire zone.

In the present case, a slope failure occurred at the time that a
drilling crew was working on the slope, killing and burying the men.
It is not known, in hindsight, all the factors that contributed to the
failure, and in what proportion. There is no doubt that the previous
primary blast caused damage in tearing away from the rest of the layer.
It is not known whether or not compressed air from the drilling operations might have been injected inadvertently into the parting between
the layers, nor how many drills were actually in operation at the moment of the failure. It is not known if there were any visual indications of slope loosening or displacement prior to the failure. Nevertheless, it is hoped that the reader will recognize the hazards in

Figure 11 - Precarious rock layers in a limestone quarry.
The massive limestone layers are unstable only when
undercut by the quarrying operations. Otherwise, they
remain stable (see text).

this type of operation, and take steps to avoid them.

The writer was asked to recommend a method for removing the remainder of the unstable, undercut layer, and to develop a new, safe quarrying plan. Briefly, the safe removal of the undercut area was accomplished by placing the drills to the side of the layer and drilling horizontal holes into the unstable rock, keeping men and equipment off it. With the proper design of hole length and sequence of detonation, it was possible to fragment the rock for product use, and avoid bringing the layer down merely as a slope failure (which would have required very expensive secondary blasting below). Future quarry development called for benching from the top down, although gravity could still be used in place of hauling units, merely by pushing the muck to the steeply dipping bedding planes and letting it slide to the bottom as before.

Because of the possibility of a sudden slope failure at any time, the writer did not consider it safe to use the approach described previously for the tailings embankments. Neither was there any attempt whatever to perform any type of analysis. In the judgment of the writer, it would not have been possible to determine the stability with sufficient accuracy to be meaningful, even though the slide surface was unusually accessible for examination and/or testing. The remaining rock might have failed at any time due to simple gravity loading, or it might have withstood the remainder of any blasting activities performed according to the previous quarrying methods, since there had been a very long history of use of the previous methods before this accident.

It was this writer's opinion that it would not have been wise to attempt to draw a very fine prediction line between failure and no failure in this case. On the other hand, had it been necessary to continue placing men and equipment on the precarious slope, it would have been possible to develop an observational approach different from that described for the tailings embankments. Water was not present in this slope, so pore pressures were not of interest. Displacements could not have been permitted to accumulate to the same degree as those in a soil embankment. Therefore, the monitoring would have to be tailored to a much greater degree of sensitivity to early warnings of displacement. For example, acoustic emissions and very sensitive displacement monitors (such as LVDT's) could have been used. The preferred approach, where personnel safety was so important, was to avoid the problem altogether.

Repeating for emphasis, it is this writer's opinion that methods of evaluation, methods of observation, and methods of blasting must be highly site specific.

SUMMARY

The existing state of the art regarding blasting effects on slopes relies heavily on the experience and judgment of individual special- ists. It might be called a technical art. It is not an exact science at all.

The physics of explosions is quite well understood, as is the rock breakage process. It is not expected that additional research will bring about any changes in fundamental concepts, but only in refine- ments in applications.

With the wide range of explosives products and methods now known, it is possible technically to exercise any degree of precision that may be desired in the rock blasting and excavation process. The lim- itation on this activity is not that of technology, but of cost. Ex- cessive costs, of course, may prohibit the use of certain methods on a particular project, or render a particular project impracticable. Unfortunately, the success of blasting techniques depends very greatly on the skill of the individual blaster, primarily on his ability to judge the many details of the site that influence the results, and how to adapt blasting technology to best suit those specific site condi- tions. In addition, virtually all sites are somewhat variable, and require adaptations as the work progresses. Thus, there are certain aspects of blasting which will remain a technical art for the fore- seeable future. The limiting factor is the inability to determine in advance all significant details of the site.

The physics of single seismic waves is quite well known theoretic- ally. For any given single wave type, in any given single material, computational procedures exist for developing synthetic seismograms which can be considered fairly representative. However, there are so many possible combinations of wave forms and particle motions for a blast detonated in even relatively simple field conditions that the problem becomes very complex in most cases.

Knowing the particle motion at a particular point is not the same as knowing how to model the complex motion within a much larger zone. There are different wave forms with different velocities and different particle motions (different wave lengths, frequencies and displace- ments), attenuating at different rates, with different laws relating to their transmission, refraction and reflection. This complexity is compounded by very important geometric relationships, such as angles of incidence to the zones of interest. Wave lengths are very import- ant because they limit any particular phase of motion to a zone which may be significantly smaller than a potential slide mass. In that case, the boundary conditions are not those determined by the slide planes of the slide mass, but by the dimensions of the traveling seismic waves. In many cases, it is not acceptable to assume that the particle motion is that of a rigid-base model.

For very long waves, such as those generated by earthquakes, it is possible to consider a slope to be a small model with a rigid base. Such a model lends itself to pseudo-static methods of analysis. Both Terzaghi's method and Newmark's method have been used successfully in such cases, when properly combined with experienced judgment. However, these methods are inappropriate for many blasting cases, and will often lead to very misleading conclusions, being increasingly conservative as the vibration frequency increases.

It is possible that a new pseudo-static method of analysis could be developed which would have a better chance of correlating with actual field experience relating to blasting phenomena. Such a method would have to consider some of the physical parameters which are more closely related to wave propagation, such as wave length and the strain induced by a passing wave. This paper suggests a beginning approach to such a method. Shortcomings of the method are recognized and it is expected that it will always be necessary to exercise a considerable amount of field judgment in the application of any such method.

Dynamic analytical procedures have been developed, and have been applied successfully to earthquake analysis. The most common method employed for such analysis is the finite element method, although finite difference methods are often employed for wave propagation phenomena. Future refinements of these methods may bring about suitable techniques for blasting analysis, although the complexities are formidable, and there is continuing doubt about our abilities to model a slope with sufficient accuracy to justify the time and expense for the use of such methods as applied to blasting.

As with many other aspects of the over-all question, the most serious limitation appears to be our inability to determine with sufficient precision all of the significant properties of a slope in advance of observing some aspect of its behavior. This limitation seriously impairs the development of any analytical procedure which would be suitable on a routine basis for blasting operations. At the present time, it is common to rely on the judgment of experienced specialists and to monitor slope behavior.

Fortunately, the vast majority of cases permits an observational approach which satisfies most project needs. One of the reasons is that blasting vibrations can be controlled technically to any level of interest. This permits the vibrations to begin at any level desired and to increase to any level desired, while one observes the slope behavior to the desired degree of precision. The monitoring procedures and degree of precision required are highly site specific. Items of common interest may include displacement, pore pressure, strain and acoustic emission. If the rate of response of the observed parameter is seen to accelerate, or to occur at an unacceptable level, remedial action of some sort is suggested.

REFERENCES

Barton, N., 1981, "Investigations to Evaluate Rock Slope Stability," Proceedings 3rd International Conference on Stability in Surface Mining, Vancouver.

Fourney, W.L., and Barker, D.B., 1979, "Effect of Time Delay on Fragmentation in a Jointed Model," NSF Grant No. DAR 77-05171, Dept. of Mech. Egnineering, Univ. of Maryland, College Park, MD.

Goodman, R.E., 1981, "Geology and Rock Stability," Proceedings 3rd International Conference on Stability in Surface Mining, Vancouver

Hendron, A.J., and Oriard, L.L., 1972, "Specifications for Controlled Blasting in Civil Engineering Projects," Proc. N. American Rapid Excavation and Tunneling Conference, SME of AIME, New York.

Newmark, N.M., 1965, "Effects of Earthquakes on Dams and Embankments," Geotechnique, Vol. 15, No. 2.

Oriard, L.L., 1971, "Blasting Effects and Their Control in Open Pit Mining," Proc. Second International Conference on Stability in Open Pit Mining, Vancouver.

Oriard, L.L., Ewoldsen, H.M., and Mahmood, A., 1976, "Rapid Dam Construction Using the Directed Blasting Method," Second Iranian Congress of Civil Engineering, Shiraz.

Oriard, L.L., and Yen, B., 1977, Unpublished report to The Panama Canal Co. regarding blasting effects on slopes, PC-2-1627.

Oriard, L.L., 1980, "Observations On The Performance of Concrete at High Stress Levels From Blasting," Proc. Sixth Conference on Explosives and Blasting Technique," Soc. Expl. Engrs., Montville, OH.

Oriard, L.L., 1980A, "Drilling, Blasting and Dredging Techniques for Deepening the Panama Canal," World Dredging and Marine Construction, Vol 16, No. 6.

Oriard, L.L., and Jordan, J.L., 1980, "Rockfill Quarry Experience, Ord River, Australia," Jour. Constr. Div., ASCE, New York, March.

Oriard, L.L., 1981, "Field Tests With Fracture-Control Blasting Techniques," Proc. 1981 Rapid Excavation and Tunneling Conference, San Francisco, SME of AIME, Littleton, CO.

Oriard, L.L., 1981A, "Blasting Effects and Their Control," Underground Mining Methods Handbook, SME of AIME, Littleton, CO.

Ravinder, C., et al, 1980, "Notched Boreholes for Fracture Control Applications in Soft Coal," ENME 404 Final Design Project Report, Dept. Mech. Engineering, Univ. of Maryland, College Park, MD.

Seed, H.B., 1979, "Considerations in the Earthquake-Resistant Design of Earth and Rockfill Dams, Geotechnique, Vol. 29, No. 1.

Seed, H.B., 1980, "Seismic Design Today," lecture presented at Los Angeles Section of ASCE, January.

Tart, R.G., Oriard, L.L., and Plump, J.H., 1980, "Blast Damage Criteria for a Massive Concrete Structure," <u>Minimizing Detrimental Construction Vibrations</u>, ASCE National Convention, Portland, ORE. February.

Terzaghi, K., 1950, "Mechanism of Landslides," The Geological Survey of America, Engineering Geology (Berkey) Volume.

Question

Can seismic refraction profiling provide parameters for static and/or dynamic slope stability modeling. How.

Answer

Yes, conventional seismic refraction profiling techniques can provide useful information relative to such factors as the weathering profile, such as the progression from soil through decomposed rock to fresh rock, for example, and thus assist in the identification of potential slide planes. Other forms of layering can be identified and quantified, also, as long as there is an increase in velocity with depth. A series of such profiles will reveal lateral variations in material properties as well. And, of course, other techniques can provide additional information. For Example, cross-hole techniques are commonly used to measure shear-wave velocities to calculate in-situ dynamic moduli, parameters that are commonly used in both static and dynamic analyses.

Some persons have used the ratio of P-wave velocities (laboratory versus field values) to evaluate the character and frequency of jointing, for example.

On the other hand, it can be very misleading to attempt a direct correlation between wave velocities and stability unless one were to include other important relationships. The two case histories in my paper are good examples of this fact. There would have been only a single P-wave velocity for each of the two cases, - a uniform, very high velocity for the hard rock site, and a uniform, very low velocity for the tailings embankment (except for a change at the water table). Velocities, per se, would not have been useful information in either of these cases.

Question

Although presplitting does not entirely retard transmission of vibration, it does allow preservation of the rock strength. Could you comment on this benefit in relation to open pit stability.

Answer

Some form of cautious perimeter blasting, including presplitting is indeed beneficial in preserving the integrity of final bench surfaces. For this reason, these methods have become widely used in open pit mining. In the majority of cases, I prefer cushion blasting or smooth blasting in preference to presplitting for the reason that it is more economical and offers more freedom in the drilling and blasting of the next row of holes.

There is a distinction between the effect that these methods have on preserving the integrity of bench faces and the question of benefit to the stability of the large-scale average slope behind the bench faces. Pre-splitting should not be relied upon to serve as any type of isolation device or barrier to the transmission of significant vibrations to the pit slopes.

Question

With large scale blasting in open pits is there a possibility of low frequency vibration, more characteristic of earthquakes, causing failure at pit walls distant from the shot.

Answer

Your concern is well founded theoretically. Fortunately, we are assisted by Nature in this question, however. The lower frequencies become more prominent at greater distances where the intensity of motion is lower. At most sites, the low-frequency surface waves are not well developed in the source vicinity, though they may become very pronounced at greater distances. This wave development is a function of both site geology and blasting design. Of course, we have no control over geology. We can expect more pronounced surface waves (lower frequency) in well defined soft-rock layers overlain by deep soil cover. Higher frequencies will be found at a hard-rock site. Two of the controls we should watch in blast design are (1) the maximum size of any single charge, or group of adjacent charges detonating simultaneously, and (2) the velocity with which the detonation sequence passes along a bench surface (so that we do not unwittingly provide constructive reinforcement of Rayleigh waves or flexural waves.

Question

When is a delay interval an effective delay interval for minimizing (a) overbreak, and (b) ground vibration.

Answer

Popular wisdom has it that the minimum delay interval should be 8 milliseconds or more. This delay interval is often specified as the minimum effective delay for vibration control in civil construction projects, and appears similarly in the OSM regulations. However, there is no sound technical basis for that specific number since, among other things, it does not consider other factors which are at least as important, such as the distance between consecutively firing charges, the size of the charges, or the elastic properties of the rock. The concern develops from the theory that two sine waves are partially additive if the second arrives during the first quarter cycle of motion generated by the first wave; - therefore, longer delay is better. However, the exact number is meaningless without other considerations, since reinforcement can also occur theoretically at any other whole-number multiple of period intervals and/or any distance interval that corresponds with the wave velocity through the rock. For example, 25 milliseconds for holes 25 ft. apart has the same relationship to constructive wave reinforcement as 5 milli-seconds for holes 5 ft. apart. To avoid this "apparent" dilemna, Langefors advises readers to wait for several oscillations of the significant energy to die out. Unfortunately, either recommendation (8 ms, or several oscillations) may become impractical or even pose serious difficulties or hazards on some projects. Fortunately, the concern is usually unwarranted. Constructive wave reinforcement is rare. It is even more rare for the reinforcement to be sufficient to be of concern. For further discussion of the theory and a review of field data on this subject, you may wish to read Oriard and Emmert, 1980.

Question

In view of the sensitive relationship between permeability and joint width and the opening of joints by the action of blasting, could you give any evidence for a local decrease in stability due to blasting (bench scale) but an over-all increase in stability (large scale) due to drawdown resulting from increased permeability.

Answer

No, I have no such evidence and would doubt that it exists for the large scale. The action of blasting in the opening of joints is restricted to a zone in very close proximity to the blasting. It does not extend into the slope a sufficient distance to affect large-scale drawdown, only that near the bench faces.

Question

It would seem feasible, using limit equilibrium methods with slices to attach a different vector acceleration to each slice and thus crudely model high-frequency, short-wave-length, high-acceleration, blast-induced vibrations. Has this been done. By whom. How.

Answer

I am not aware of this approach being used before, but that would not necessarily mean that it hasn't been done. Although such an approach seems attractive intuitively, it is my opinion that the judgements that would have to accompany such an analysis are as determining as the analysis itself, -just as Terzaghi's method will give the right answer if the analyst knows what seismic coefficients to use. Your suggestion is a step in the right direction because it would have the net effect of using a lower "effective" value of acceleration. But, how does one handle the boundary conditions, or determine them in the field? If the adjacent zone to the one under consideration is of equal mass and 180 degrees out of phase, the net effect is zero, assuming that the wave motion and gravity act in the same direction. This result may also be non-representative. Source location and direction of wave travel then become critical. We come back to the same dilemma: Which approach is more acceptable? To multiply the wrong answer by a judgment factor to get the right answer? Or merely to form an estimate of the right answer without the benefit of the wrong calculation? This dilemma forces us into the same operating mode that is common in static analysis, - that of observing the first sign of physical response to a known force. Of course, we must use a monitoring system that is appropriately sensitive for the site in question.

Question

Would you consider the effects of vibrational acceleration on the abrupt pore water pressure build-ups or increases which could de-crease the shear strength for stability analysis. If yes, how would you introduce this concept into stability analysis.

Answer

In my opinion, one should always be concerned about the presence of water in a slope or embankment. Interestingly, experience demon-strates that the duration of shaking may be at least as important as the intensity of the shaking. With soils, it has been demonstrated many times that the exact number of oscillations at a given strain level is a critical factor in determining whether or not a failure will occur. A few oscillations less and no damage occurs; a few more and there is a disaster. One of the most dramatic examples was the terrace failure near Anchorage, Alaska, during the Good Friday earthquake of 1964. An interesting contrast is that a densification rather than failure may occur when a saturated embankment is subjected to a transient, high-frequency vibration of short duration. The response in rock may be somewhat different, depending on individual block size, the prominence of jointing, etc. We know of the potentially damaging effects of abrupt pressure increases from explosive gases, hydraulic shock or compressed air, causing local block motion. We know also of the increase in hydraulic head due to the dilatancy generated by high strain levels from earthquakes. For an open pit mine slope, I would recommend the approach described in my paper for the tailings dams, but to a greater degree of sensitiv-ity, - monitoring the response to pre-programmed blasting loads.

Chapter 5

INFLUENCE OF EARTHQUAKES ON ROCK SLOPE STABILITY

Charles E. Glass
Associate Professor of Geological Engineering

University of Arizona
Tucson, Arizona 85721

ABSTRACT

A steadily growing body of evidence indicates that earthquake
ground motions can cause failure of rock slopes that are otherwise
stable under static loading conditions. As a result, the economic
optimum working slope angle is reduced when earthquake shaking is
considered in surface mine design, especially in regions of high
seismicity.

When designing surface mine slopes to withstand seismic motion it
is as important to consider the probability of experiencing various
levels of ground shaking as it is to consider the characteristics of
the ground motion. For most slope designs, simple probabilistic techniques such as extreme value or maximum likelihood techniques are
appropriate. These techniques, however, should be applied in such a
way that a probability value is assigned to characteristic ground
motion parameters at the mine rather than to nearby earthquake magnitudes.

Several techniques have been developed to analyze rock slopes subjected to dynamic loads. Most of these techniques do not adequately
account for the ground motion parameters of concern. The best technique appears to be one that utilizes actual earthquake time histories. A simple technique that calculates block displacement for any
input accelerogram, using a linear acceleration approximation between
time steps, has been developed that accurately accounts for rock block
response.

Results using this technique indicate that the following factors
are important for assessing rock slope stability in earthquake regions:

89

1. Earthquake hazard.
2. Amplitude, frequency, and duration of ground motion.
3. Slope geometry.
4. Joint properties, especially shear strength and roughness scaling coefficients.

INTRODUCTION

Landslides are a common secondary effect of earthquakes, and have caused enormous losses of life and property. The number of landslides triggered by an earthquake correlates with earthquake magnitude. The Richter magnitude 8.4 Alaska earthquake of 1964, for example, triggered probably more landslides than any other seismic event in recent history. In Alaska, landslides were reported from an area of more than 210,000 km^2 (National Academy of Sciences, 1968-1973; U.S. Geological Survey, 1965-1970). Prior to 1964, the 10 September 1899 Alaskan earthquake (M = 8.6) caused numerous landslides and triggered an avalanch 690 km from the epicenter (Morton and Streitz, 1967). The 31 May 1970 earthquake off the coast of Peru (M = 7.7) and 4 February 1976 Guatemala earthquake (M = 7.5) triggered numerous landslides at distances exceeding 150 km from the epicenters (Algermissen, 1970; Espinosa, 1976). The smaller 9 February 1971 San Fernando, California earthquake (M = 6.4) caused several destructive landslides; one landslide on Kagel Mountain, 1 km east of Pacoima, extended over 1/2 km along a mountain side composed of dacite and gneiss (Morton, 1975).

Even though examples of seismically induced slope failures are numerous, seismic loads are seldom considered in studies of the economics of mining operations. This may be due to a common misconception that seismically induced landslides occur only in the immediate epicentral area of large earthquakes. This misconception has been reinforced by the Modified Mercalli Intensity Scale of 1931, in which the first description of induced landslides is associated with an Intensity of X (Wood and Newmann, 1931). The association of seismically induced landslides only with violent shaking underestimates the importance of dynamic loads. As the above examples demonstrate, low levels of ground motion have triggered landslides at considerable distances from earthquake epicenters and in areas having little or no record of slope instability under non-seismic conditions. The Fortuna-Rio Dell, California earthquake in 1975 and the San Francisco, California earthquake of 1957 are examples of moderate earthquakes (M = 5.7 and 5.25, respectively) that have caused landslides, rockfalls and earth slumps (Nasson et al., 1975; Pestrong, 1976). The 9 January 1966 earthquakes near Hope, British Columbia triggered a landslide on Johnson Peak, even though the Richter magnitudes were only 3.2 and 3.1 (Mathews and McTaggart, 1969). Table 1, modified from Keefer et al. (1978), shows earthquakes that have caused numerous landslides. Of the landslides listed in Table 1, rockfalls and shallow disintegrating rock slides, slumps and block slides in rock, and cut-slope failures are more common and important to mining.

Rock falls and shallow, disintegrating rock slides are the most abundant slope failures listed in Table 1, and are common in rock that is poorly cemented, closely jointed, chemically altered, or highly weathered. Slumps or block slides occur less often during earthquakes than rock falls and shallow, disintegrating landslides and usually occur on steep slopes, although some have continued to move on very gently dipping slide planes. Two block slides triggered by the 1968 Inangahua earthquake, for example, moved on slide planes having dips of less than four degrees (Keefer et al., 1978; New Zealand Department of Scientific and Industrial Research, 1968; U.S. Geological Survey, 1963; New Zealand Society for Earthquake Engineering Bulletin, 1969). Cuts and excavation of open-pit mines increase landslide susceptibility by locally increasing slope steepness and by disturbing rock material adjacent to the excavated slopes. As a result, cut-slope failures are widespread during earthquakes (Keefer et al., 1978; Seismological Society of America Bulletin, 1963; New Zealand Department of Scientific and Industrial Research, 1933; Seismological Society of America Bulletin, 1971; Harp et al., 1978; New Zealand Department for Earthquake Engineering Bulletin, 1969; Morton, 1971; Morton, 1975; Castle and Youd, 1972; Earthquake Engineering Research Institute Conference Proceedings, 1973).

EARTHQUAKE FACTORS CONTROLLING PIT DESIGN

Probability

Problem solving is often attempted in terms of absolutes; an alternative is either *"safe"* or *"unsafe"*, a decision is *"right"* or *"wrong"*. The way in which one arrives at such an absolute description of an alternative or decision, however, is obscure at best. Because one can never know all of the parameters and facts that influence a project absolutely, it is more realistic to deal in terms of *"risk"* rather than *"safety"* in approaching complex engineering problems (Wiggins, 1973). An analysis based on the concept of risk permits a more efficient decision-making process and an opportunity to realistically balance conflicting objectives that might otherwise create economically disadvantageous overdesign or underdesign.

In assessing the economic feasibility of an open pit mine, for example, the conflicting objectives of steep slope angles, which create economic benefits due to reduction in waste stripping, should be balanced against the decreased risk of slope failure inherent in lower slope angles. This balance can be accomplished in practice by selecting a "working slope" and modeling the mine for a specified period of time. This modeling allows the extraction of information relating to annual ore and waste tonnages, average yearly grades, operating costs, and detailed pit geometry (Kim et al., 1976).

Using a base of geotechnical data, probabilities of slope instability can be calculated for each sector of the open pit during each operating period by determining potential failure modes and analyzing their stability (Call et al., 1976). The probability of instability can then be incorporated into a benefit-cost analysis to determine an optimum slope angle.

Numerous techniques have been used in earthquake engineering practice to generate design seismic parameters. The most common practice is to design structures to withstand shaking from a postulated *"maximum"* earthquake. The concept of a *"maximum expectable earthquake"*, *"safe shutdown earthquake"*, *"design basis earthquake"*, *"operating basis earthquake"*, or *"maximum probable earthquakes"*, can be justified in civil engineering construction where the potential for life loss may be high. The *"maximum earthquake"*, however, may actually have a low probability of occurring during the period of time for which a structure is designed. For this reason an analysis based on ground motions from a postulated maximum earthquake can lead to excessively conservative design for mining operations. An analysis to determine an economic optimum slope design should consider the probability of occurrence for different magnitude earthquakes, the maximum ground motion at the site from these earthquakes, the probability of experiencing a given acceleration at the site within a given period of years, and the predominant period and duration of strong motion.

Once the probability distributions of earthquake ground motions are calculated, the distribution can then be sampled using a Monte Carlo technique in a benefit-cost model to determine if an earthquake occurs during a specified mining period. If the sampling indicates an earthquake motion at the mine, a dynamic slope stability analysis can be used to calculate a probability of slope instability schedule for that time period.

Extreme Value Technique

The theory of extremes developed by Gumbel (1958) provides a convenient method for obtaining estimates of earthquake hazard. This technique treats earthquakes as a stochastic process $F(x,t)$ where x is the variable of interest for design. For example, x may correspond to earthquake magnitudes recorded within a specific region, or to earthquake accelerations or intensity values at a particular site. Often the engineering design depends less on an accurate knowledge of $F(x,t)$ than on the largest value that x can assume within a given design period. If the entire earthquake catalogue $[F(x,t)]$ is accurately known, then the maximum values of x are likewise known. The complete data needed for precise definition of $F(x,t)$, however, are generally unavailable for most regions. Because the larger events are usually recorded, even in regions having poor instrumentation, the extreme value technique, which uses these maximum values, provides a useful tool for such stochastic processes.

First, a time scale is divided into equally spaced intervals. Only the extreme value y, which the variable x reaches within each interval, is considered in the analysis. The extreme value y forms a regular point process within the original process F(x,t). Gumbel found only four mathematically distinct distributions of y. His "Type 1" distribution takes the form:

$$G(y) = \exp(-\alpha e^{-\beta y}) \tag{1}$$

where α and β are found from a least squares fit (Lomnitz, 1974).

When the parameters α and β have been determined, the probabilities of occurrence of earthquakes having magnitudes greater than the extreme value y are calculated from:

$$P(y) = 1 - \exp(\alpha D e^{-\beta y}) \tag{2}$$

where D is the number of years over which the probability is to be assessed.

The intensity distribution of ground motion at the site can be estimated by applying representative attenuation relationships (Seed and Idriss, 1969; Cornell, 1971) either to (1) the magnitude distribution derived from the Gumbel analysis, or to (2) each individual earthquake in the earthquake catalogue and performing a Gumbel analysis on the resulting intensity distribution.

Probabilistic Analysis Utilizing Entire Earthquake Catalogue

Most probabilistic earthquake analyses that utilize the entire earthquake catalogue partition the region surrounding a site into seismogenic zones (Cornell, 1971; Algermissen and Perkins, 1972); however, instead of selecting only the maximum yearly magnitude, all earthquakes within a zone are considered. Average occurrence rates are calculated, then assigned to each zone. These occurrence rates are assumed to be statistically independent for calculation of the probability distribution N_y (the number of earthquakes causing site motion with a given intensity y). Finally, the total probability at the mine site is calculated by summing the probabilities from each individual zone.

The earthquake hazard analyses above determine the probability of exceeding a given level of ground motion in a specified period of years at a mine site. The maximum ground velocity, or acceleration, is the most frequently used measure for slope design. Recent experience, however, suggests that frequency and duration of motion are also very important to rock slope stability and emphasizes the importance of using the entire time-history of motion to assess the response of rock slopes to earthquakes.

ROCK SLOPE RESPONSE

Regardless of the techniques that are used to determine the earthquake hazard, the major problem faced by those involved in rock slope
design has been to determine the way in which the ground motion parameters are to be used in a dynamic slope analysis. Finite element and
finite difference numerical techniques have been used to calculate
slope stability under dynamic loads, but these techniques tend to be
cumbersome and overly expensive when many slopes are to be considered.

Newmark's Technique

One technique that has been used for rock slope design was developed by Newmark (1965). Newmark's technique is a pseudo-static technique in which the downslope displacement of a rock block is considered to be proportional to the ground particle velocity squared.
Newmark's equation for block displacement is

$$D = \frac{V^2}{2g\,\underline{N}} \text{ Max } (6, \sqrt{\tau}) \qquad (3)$$

where V is the maximum earthquake particle velocity, g is the acceleration of gravity, N is a constant function of joint shear strength (ϕ)
and slope angle (α) given by

$$N = (\frac{\tan \phi}{\tan \alpha} - 1) \sin \alpha, \qquad (4)$$

and τ is a constant equal to the duration of strong ground motion.
Newmark's equation is compatible with other suggestions that earthquake and blast-induced damage is proportional to the maximum ground
particle velocity and not to ground particle acceleration.

Laboratory shaking tests conducted by Wilson (1979) using granite
blocks subjected to harmonic motions showed that the Newmark equation
approximates block displacements for frequencies greater than approximately 20 hz. Unlike the impulse loading used by Newmark, however,
the block displacement under cyclic loading exceeds that predicted by
the Newmark equation by approximately 60 percent, as shown in Fig. 1.
At frequencies below 6 hz, block displacements calculated using Eq.
(3) overestimate actual block displacements as shown in Fig. 2. The
results of laboratory shaking tests suggest that block displacements
are sensitive to the frequency of dynamic loads as well as to peak
particle parameters and joint roughness.

Linear Acceleration Dynamic Response of Slopes (LADRS)

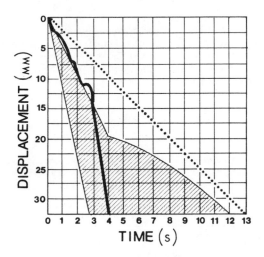

Figure 1.   Shaking table response of a rock block to harmonic motion.
The range of displacements of the 8 kg granite block
is shaded.   Newmark's equation predicts the dotted displacement, LADRS predicts the displacement represented by
the solid line.   Frequency = 5.23 hz, Amax = 0.34 $\underline{g}$, and
slope = 23°.

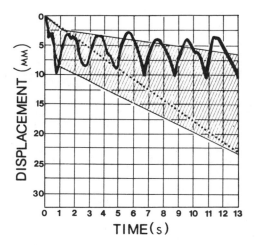

Figure 2.   Shaking table response of a rock block to harmonic motion.
Comparisons are the same as Figure 1.   Frequency = 3.6 hz,
Amax = 0.12 $\underline{g}$, and slope = 26°.

The linear acceleration dynamic response of slopes (LADRS) is a simple technique that enables a realistic description of both the rock slope and ground motion parameter to be included in the analysis. In this technique a rock block is modeled as shown in Fig. 3. The equation of motion is given by

$$\ddot{x} + 2\beta\omega\dot{x} + \omega^2 x = a(t) \pm f(t) \tag{5}$$

where x is the displacement parallel to the slope, β is the fraction of critical damping, ω is the angular frequency ($\sqrt{K/M}$), K is the spring stiffness, M is the block mass, and a(t) and f(t) are the

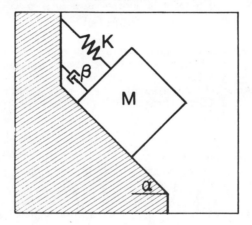

Figure 3.   Geometry for LADRS technique

ground and frictional acceleration, respectively. If we assume that the acceleration varies linearly wihin each timestep,

$$\ddot{x}_1 = \ddot{x}_0 + A\Delta t, \tag{6}$$

$$\dot{x}_1 = \dot{x}_0 + \ddot{x}_0\Delta t + 1/2\, A\Delta t^2, \tag{7}$$

and

$$x_1 = x_0 + x_0\Delta t + 1/2\, x_0\Delta t^2 + 1/6\, A\Delta t^3, \tag{8}$$

where x_0 is the displacement at time t_0, x_1 is the displacement at time $t_1 = t_0 + \Delta t$, from Eq. (6) $A\Delta t = \ddot{x}_1 - \ddot{x}_0$, and the dot superscript denotes a derivative with respect to time. Substituting $A\Delta t = \ddot{x}_1 - \ddot{x}_0$ from Eq. (6) into Eqs. (7) and (8) yields:

$$\dot{x}_1 = \dot{x}_0 + 1/2\,\ddot{x}_0\Delta t + 1/2\,\ddot{x}_1\Delta t \qquad (9)$$

and

$$x_1 = x_0 + \dot{x}_0\Delta t + 1/3\,\ddot{x}_0\Delta t^2 + 1/6\,\ddot{x}_1\Delta t . \qquad (10)$$

Using Eqs. (9) and (10) in Eq. (5), one obtains

$$x_0 = a_0 - f_0 - \omega^2 x_0 - 2\beta\,\dot{x}_0 \qquad (11)$$

and

$$\left(\omega^2 + \frac{6\,\beta\omega}{\Delta t} + \frac{6}{\Delta t^2}\right) x_1 = a_1 - f_1 + 2\beta\omega b_0 + d_0 , \qquad (12)$$

where

$$b_0 = \frac{3}{\Delta t}\, x_0 + 2\dot{x}_0 + \frac{\Delta t}{2}\,\ddot{x}_0,$$

$$d_0 = \frac{6}{\Delta t^2}\, x_0 + \frac{6}{\Delta t}\,\dot{x}_0 + 2\ddot{x}_0,$$

a_0 and f_0 are the earthquake acceleration and frictional acceleration parallel to the slope at time t_0, and a_1 and f_1 are equivalent accelerations at time $t_1 = t_0 + \Delta t$.

The linear acceleration method proceeds as follows:

1. Compute x_0 from Eq. (11).

2. Compute b_0 and d_0.

3. Compute x_1 from Eq. (12).

4. Compute $\dot{x}_1$ from $\dot{x}_1 = \frac{3}{\Delta t}\, x_1 - b_0$.

5. Compute $\ddot{x}_1$ from $\ddot{x}_1 = a_1 - f_1 - \omega^2 x_1 - 2\beta\omega\dot{x}_1$.

The input to the calculations is the digitized accelerogram (a(t)) representing the earthquake, which enables accurate inclusion of all ground motion parameters, including frequency and earthquake duration. Other input includes spring stiffness, damping factor, and a friction model. Once block movement begins, the stiffness and damping have little effect compared to the friction model and consequently can be set to zero.

The character of overall block displacement is controlled by the dynamic forcing function a(t) (earthquake) and friction model. In LADRS, the forcing function is assumed to be a horizontal accelerogram with a displacement vector in the plane of the rock slope as shown in Fig. 4. This input motion is then decomposed into two orthogonal vectors parallel and normal to the slope. If three-component accelerograms are available, the resultant vector can be decomposed into component parallel and normal accelerations. Static accelerations on the block due to gravity are familiar and are given by

$$D_s = \underline{g} \sin \alpha$$

and

$$R_s = \underline{g} \cos \alpha \cdot f, \qquad (13)$$

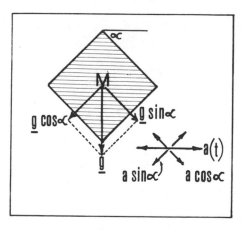

Figure 4.   Freebody diagram for LADRS technique

where D is the driving acceleration (driving force/mass), R is the resisting acceleration, and f is the frictional coefficient. The addition of an earthquake produces two effects. When the earthquake vector moves the slope toward the block the normal force increases, which increases the resisting acceleration and decreases the driving acceleration yielding

$$D_d = D_s - a(t) \cos \alpha$$

and

$$R_d = R_s + a(t) \sin \alpha \cdot f = (\underline{g} \cos \alpha + a(t) \sin \alpha) f \qquad (14)$$

$$= J \cdot f$$

Conversely, when the earthquake vector moves the slope away from the block, the normal force decreases and the driving force increases. Eq. (14) also describes this situation for the acceleration amplitude $a = -a$. The resultant acceleration of the block is then

$$\ddot{x}_1 = D_d \pm R_d \qquad (15)$$

where the sign of R_d is opposite to the sign of the block velocity. Eqs. (13), (14), and (15) indicate that block response depends on the dynamic forcing function $a(t)$, the slope angle α, and the frictional coefficient f. In order to compute realistic block response, a realistic friction model is required. Figs. 1 and 2 show that block response is strongly dependent upon surface roughness. In Fig. 1, the slopes of the displacement curves are consistent for the first 20 mm of displacement.   At approximately 20 mm, the rock block slowed due to a rough spot on the slope. After several seconds, the block again proceeded down-slope at a velocity consistent with the first 20 mm of sliding.

The friction model used in LADRS was initially proposed by Barton and Bandis (1980) and is given by

$$f = \tan \left(T \cdot JRC \; Log_{10} \frac{JCS}{J} + \phi_r \right), \qquad (16)$$

where JRC is the joint roughness coefficient, JCS is the joint compressive strength (approximately equal to the unconfined compressive strength), ϕ_r is the residual friction angle, and T is a scaling factor that depends upon the ratio of block displacement δ to the distance to peak joint strength mobilization δp. The values of T (suggested by Barton (1980) are given in Table 2. The distance to peak shear mobilization δp is a function of joint roughness and joint

length, and is approximately equal to one percent of the joint length
up to a limiting length of five meters for undulating joint surfaces
and three meters for planar surfaces. Shear-deformation curves that
represent unique situations at individual mines can be obtained using
results from laboratory direct shear tests on rock joints.

Table 2. Suggested Values for Scaling Factor T

Ratio$_\delta$*	T
1.0	1.0
$1.0 < \text{Ratio}_\delta \leq 2.5$	$\frac{1}{6}$ (7.0 - Ratio)
$2.5 < \text{Ratio}_\delta \leq 10.0$	$\frac{1}{30}$ (25.0 - Ratio)
$10.0 < \text{Ratio}_\delta \leq 100.0$	$\frac{1}{80}$ (100.0 - Ratio)
> 100.0	0.0

*Ratio$_\delta$ = $\delta/\delta p$

The LADRS technique was applied using an input motion that duplicated
the shaking table studies represented by Figs. 1 and 2.   The computed response duplicates the shaking table response remarkably well
even though the joint roughness parameters were not measured for the
shaking tests and had to be estimated.

The response of a rock block to harmonic motions over a range of
frequencies from 1 hz to 60 hz was computed using LADRS for a constant
maximum acceleration of 0.39 g for the first suite of frequenices and
a constant maximum velocity of 15 cm $\cdot$ s^{-1} for the second suite. The
results of these calculations are shown in Fig. 5. Fig. 5 shows
that either maximum acceleration or velocity should provide consistent
scaling over a wide range of frequenices from approximately 20 hz to
60 hz. Below approximately 20 hz, however, maximum particle motion
parameters become poor estimators of rock slope response and below
10 hz are unreliable.

It is informative to compare pseudostatic and dynamic techniques.
This can be done coveniently using Fig. 5. Choosing an earthquake
frequency of 5 hz (common for earthquakes and the frequency at which
the peak acceleration of 0.39 g corresponds to peak velocity of 15
cm $\cdot$ s^{-1}), Fig. 5 indicates that the block displaced approximately
2 cm. A limit equilibrium analysis, applying the peak acceleration as
a static acceleration, predicts that the driving forces exceed the
resisting forces and the slope is in a condition of failure. It is
clear that the block has displaced; however, describing a 2 cm block

displacement as a failure is clearly questionable when large blocks
are of interest. In this case, the earthquake would probably have
caused small rock blocks (∿0.5 m) to tumble down pit slopes, but
would not have resulted in a large failure. The amount of displacement is extremely important to know when assessing the probability
of failure of open pit mine slopes.

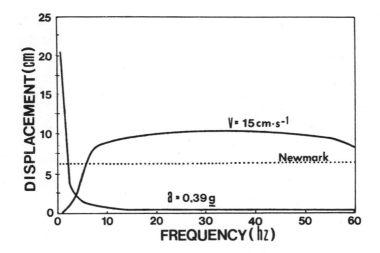

Figure 5.   Response of a rock block to harmonic motion having constant
            maximum acceleration and maximum velocity (computed using
            LADRS). Newmark's equation is represented by the dotted
            line.

     Fig. 6 shows common spectral ranges for vibrations affecting
slopes. Earthquake ground motion has most of its energy in frequencies between 0.5 and 10 hz, a range in which psuedostatic techniques
perform very poorly. The higher frequencies associated with construction and mining blasting probably explain the success of velocity and
acceleration scaling laws used to predict blast damage and also the
reason why open-pit blasting may not affect slopes to the extent that
do earthquakes.   Figs. 5 and 6 imply that pseudostatic techniques
should not be used to predict slope response to vibrations rich in
frequencies below about 20 hz, a region characteristic of earthquakes
and large-scale blasts, and a region in which most landslides are
triggered. For earthquake loading of slopes, the earthquake time-
history should be used to calculate slope stability.

Table 1. Landslides Triggered by Earthquakes (Altered from Keefer et al., 1978)

EARTHQUAKE NAME, DATE, AND MAGNITUDE	ROCK FALL - AVALANCHES	SLUMPS AND BLOCK SLIDES	FALLS AND SHALLOW, DISINTEGRATING SLIDES	CUT-SLOPE FAILURES	LIQUEFACTION-INDUCED LANDSLIDES IN ARTIFICIAL FILLS	LANDSLIDES IN ARTIFICIAL FILLS NOT CAUSED BY LIQUEFACTION
	A. LANDSLIDES IN ROCK			B. LANDSLIDES INVOLVING ARTIFICIAL CUTS OR FILLS		
Friuli, Italy 1976 M 6.5			M			
Olympia, WA 1949 M 7.1			M			
Long Beach, CA 1933 M 6.3					S	
Fortuna-Rio Dell, CA 1975 M 5.7					S	
Managua, Nicaragua 1972 M 6.2				S	S	S
Borrego Mountain, CA 1968 M 6.4		S		S	S	S
San Fernando, CA 1971 M 6.4		M	M	S	S	S
Inangahua, New Zealand 1968 M 7.0		S	M			S
Hebgen Lake, MT 1959 M 7.1	E		E	E	S	S
San Juan, Argentina 1977 M 7.4		M	E	M	S	S
Guatemala 1976 M 7.5	E	S	E	M	S	S
New Madrid, MO 1811 MMI XII				M		
Peru 1970 M 7.7		S	E	M		S
Hawke's Bay, New Zealand 1931 M 7.9	M	M	M	S	S	S
San Francisco, CA 1906 M 8.3	E	E	E	M	S	S
Chile 1960 M 8.4	M	E	M	S	S	S
Alaska 1964 M 8.4	E	M	E	S	S	S
Kansu, China 1920 M 8.6	E					

LANDSLIDE TYPE[1]

[1] Classifications after Varnes (1978)

Explanation

E – Large number of landslides
> 5,000 small landslides, or
> 100 large landslides, or landslides
common over > 1000 km² area

M – Moderate number of landslides
500 to 5000 small landslides, or 10 to 100
large landslides, or landslides common
over 100 to 1000 km² area

Explanation (continued)

S – Small number of landslides
< 500 small landslides, or < 10 large
landslides, or landslides common over
< 100 km² area

No symbol – No landslides of this type
reported

References - Table 1

Kansu, China: (Close and McCromick, 1922)
(Meyers and Von Hake, 1976)

Alaska: (Nat. Academy of Sc., 1968-1973)
(U.S. Geol. Survey, 1965-1970)

Chile: (Billings Geol. Society, 1960)
(Seis. Soc. of Amer. Bulletin, 1963)

San Francisco: (Youd and Hoose, 1978)

Hawke's Bay: (N.Z. Dept. of Scientific and
Industrial Reasearch, 1933)

Peru: (Browning, 1973)
(Earthquake Eng. Research Inst.
Earthquake Report Comm., 1970)
(Seis. Soc. of Amer. Bulletin, 1971)

New Madrid: (Fuller, 1912)

Guatemala: (Harp, Wieczorek and Wilson, 1973)
(Hoose, Wilson and Rosenfeld, 1978)

San Juan, Argentina:
(Youd, Keefer and others, unpublished
data)

Hebgen Lake: (Billings Geol. Society, 1960)
(Christiansen and Blank, 1972)
(Fischer, 1960)
(Richter, 1958)
(Seis. Soc. of Amer. Bulletin, 1962)
(U.S. Coast and Geod. Survey, 1961)

Inangahua: (N.Z. Dept. of Scientific and
Industrial Research, 1968)
(N.Z. Soc. for Earthquake Eng.
Bulletin, 1969)
(Seis. Soc. of Amer. Bulletin, 1970)
(U.S. Geol. Survey, 1963)

San Fernando:
(Marachi, 1973)
(Morton, 1971)
(Morton, 1975)
(Youd, 1971)

Borrego Mountain:
(Castle and Youd, 1972)

Managua: (Earthquake Eng. Research Institute
Conference Proceedings, 1973)

Fortuna-Rio Dell:
(Nason,Harp,La Gresse and Malley, 1975)

Long Beach: (Morton and Streitz, 1967)

Olympia: (Newmark, 1967)

Friuli, Italy:
(Govi, 1977)

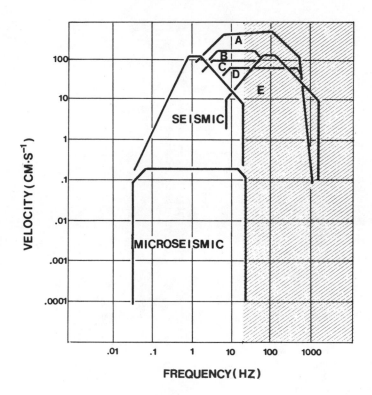

Figure 6.   Spectral ranges for common vibratory ground motion. A =
nuclear and large chemical explosions; B, C, and D correspond to, coal, quarry, and construction blasts; and E
corresponds to other industrial vibrations.

REFERENCES

Algermissen, S.T., 1970, "The Peruvian Earthquake of May 31, 1970",
    Mineral Information Service, Volume 23, Number 10, pp. 200-207.

Algermissen, S.T., and Perkins, D.M., 1972, "A Technique for
    Seismic Zoning - General Considerations and Parameters," Pro-
    ceedings of the International Conference on Microzonation for
    Safety Construction Research and Application, Seattle.

Barton, N.K., and Bandis, S., 1980, Technical Note, "Some Effects
    of Scale on the Shear Strength of Joints," International
    Journal of Rock Mechanics, Mining Science and Geomechanics
    Abstract, Volume 17, pp. 69-73.

Billings Geological Society, 1960, Earthquake Papers, Billings,
  Montana, Eleventh Annual Field Conference Guidebook, pp. 24-85.

Browning, J.M., 1973, "Catastrophic Rockslide, Mount Huascaran,
  North-Central Peru, May 31, 1970," American Association of
  Petroleum Geologists Bulletin, Volume 57, Number 7, pp. 1335-
  1341.

Call, R.D., Savely, J.P., and Nicholas, D.E., 1976, "Estimation of
  Joint Set Characteristics from Surface Mapping Data," Rock
  Mechanics Applications in Mining, 17th U.S. Symposium on Rock
  Mechanics, Ed. by W.S. Brown, S.S. Green, and W.A. Hustralid,
  AIME, New York, pp. 65-73.

Castle, R.O., and Youd, T.L., 1972, "Engineering Geology, in the
  Borrego Mountain Earthquake of April 9, 1968," U.S. Geological
  Survey Professional Paper 787, pp. 158-174.

Christiansen, R.L., and Blank, H.R., 1972, "Volcanic Stratigraphy
  of the Quarternary Rhyolite Plateau in Yellowstone National
  Park," U.S. Geological Survey Professional Paper 729-B, p. 18.

Close, U., and McCromick, E., 1922, "Where the Mountains Walked,"
  National Geographic, Volume 41, Number 5, pp. 445-472.

Cornell, C.A., 1971, "Probabilistic Analysis of Damage to Struc-
  tures under Seismic Loads," Dynamic Waves in Civil Engineering,
  Proceedings of a Conference, Swansea, Wales, John Wiley and
  Sons, Ltd., London, England.

Duke, C.M., and Leeds, D.J., 1963, "Response of Soils, Foundations, and Earth Structures," Seismological Society of America
  Bulletin, Volume 53, Number 2, pp. 309-357.

Earthquake Engineering Research Institute Conference Proceedings,
  1973, Managua, Nicaragua Earthquake of December 23, 1972,"
  Volume 1, pp. 8-264.

Earthquake Engineering Research Institute Earthquake Report Committee, 1970, Peru Earthquake of May 31, 1970, preliminary
  report, Earthquake Engineering Research Institute, p. 55.

Espinosa, A.F. (ed.), 1976, "The Guatemalan Earthquake of February
  4, 1976, a preliminary report," U.S. Geological Survey Professional Paper 1002, p. 89.

Fischer, W.A., 1960, "Highlights of Yellowstone Geology with an
  Interpretation of the 1959 Earthquakes and Their Effects in
  Yellowstone National Park, p. 62.

Fuller, M., 1912, "The New Madrid Earthquake," U.S. Geological
Survey Bulletin 394, p. 118.

Govi, M., 1977, "Photo-Interpretation and Mapping of Landslides
Triggered by the Friuli Earthquake (1976)," Inter. Assn. of
Engrg. Geol. Bulletin, Number 15.

Gumbel, E.J., 1958, Statistics of Extremes, Columbia University
Press, New York, 1958.

Harp, E.L., Wieczorek, G.F., and Wilson, R.C., 1978, "Earthquake-
Induced Landslides from the February 4, 1976, Guatemala Earth-
quake and Their Implications for Landslide Hazard Reduction,"
International Symposium on the February 4, 1976, Guatemalan
Earthquake and the Reconstruction Process, Volume 1.

Hoose, S.N., Wilson, R.C., and Rosenfeld, J.H., 1978, "Liquefac-
tion-Caused Ground Failure During the February 4, 1976, Guatemala Earthquake," International Symposium on the February 4,
1976, Guatemalan Earthquake and the Reconstruction Process,
Volume 2.

Keefer, D.K., Wieczorek, G.F., Harp, E.L., and Tuel, D.H., 1978,
"Preliminary Assessment of Seismically Induced Landslide Susceptibility," Proc. Second International Conference on Micro-
zonation, Volume 1, p. 281.

Kim, Y.C., Cassun, W.C., and Hall, T.E., 1976, "Economic Analysis
of Pit Slope Design," Paper presented at the Rock Mechanics
Symposium, Vancouver, B.C., Canada.

Lomnitz, C., 1974, "Global Tectonics and Earthquake Risk," in
Developments in Geotectonics 5, Elsevier Scientific Publishing
Company, New York.

Marachi, N.D., 1973, "Dynamic Soil Problems at the Joseph Jensen
Filtration Plant in Murphy, L.M., Scientific Coordinator,
San Fernando, California Earthquake of February 9, 1971,"
National Oceanic and Atmospheric Administration, Volume 1,
Part B., pp. 815-820.

Mathews, W.H., and McTaggart, K.C., 1969, "The Hope Landslide,
British Columbia," The Geological Association of Canada
Proceedings, Volume 20, pp. 65-75.

Meyers, H., and von Hake, C.A., 1976, "Earthquake Data Summary,"
Boulder, Colorado, National Oceanic and Atmospheric Administra-
tion Environmental Data Service, Key to Geophysical Records
Documentation, Number 5, p. 54.

Morton, D.M., 1971, "Seismically Triggered Landslides in the Area above the San Fernando Valley, in the San Fernando, California Earthquake of February 9, 1971," U.S. Geological Survey Professional Paper 733, pp. 99-104.

Morton, D.M., 1975, "Seismically Triggered Landslides in the Area Above the San Fernando Valley, in the San Fernando, California Earthquake of February 9, 1971," California Division of Mines and Geology Bulletin 196, pp. 145-154.

Morton, M., and Streitz, R., 1967, "Landslides," Mineral Information Service, Volume 20, Number 10, pp. 123-129.

Nason, R., Harp, E.L., La Gresse, H., and Malley, R.P., 1975, "Investigations of the 7 June 1975 Earthquake at Humbolt County, California," U.S. Geological Survey Open-File Report 75-404, p. 37.

National Academy of Sciences, 1968-1973, "The Great Alaska Earthquake of 1964," Washington, D.C., National Academy of Sciences Printing and Publishing Office, Volume 1, Geology, and Volume 3, Hydrology.

Neumann, F., 1935, "United States Earthquakes 1928-1935," U.S. Coast and Geodetic Survey. pp. 10-28.

Newmark, N.M., 1965, "Effects of Earthquakes on Dams and Embankments," Geotechnique, Volume 15, pp. 139-160.

New Zealand Department of Scientific and Industrial Research, 1933, "The Hawke's Bay Earthquake of 3rd February, 1931," New Zealand Journal of Science and Technology, Volume 15, Number 1, pp. 1-116.

New Zealand Department of Scientific and Industrial Research, 1968, Preliminary report on the Inangahua earthquake, New Zealand, May, 1968, Research Bulletin 193, p. 39.

New Zealand Society for Earthquake Engineering Bulletin, 1969, Special Inangahua reports, Volume 2, Number 1, p. 148.

Pestrong, Raymond, 1976, "Landslides - The Decent of Man," California Geology, July, pp. 147-151.

Richter, C.D., 1958, Elementary Seismology, W.H. Freeman and Co., San Francisco, p. 768.

Seed, H.B., and Idriss, I.M., 1969, "Rock Motion Accelerograms for High Magnitude Earthquakes," Earthquake Engineering Research Center Report No. EERC 69-7, University of California, Berkeley.

Seismological Society of America Bulletin, 1962, "The Earthquake at Hebgen Lake, Montana on August 18, 1959 (GCT)," Volume 52, Number 2, pp. 153-273.

Seismological Society of America Bulletin, 1963, Special issue on Chilean earthquakes of May, 1960, Volume 53, Number 6, pp. 1123-1441.

Seismological Society of America Bulletin, 1970, The 1968 Inangahua Earthquake, Report of the University of Canterbury Survey team, Volume 60, Number 5, pp. 1561-1606.

Seismological Society of America Bulletin, 1971, Special issue on Peru earthquake of May 31, 1970, Volume 61, Number 3, pp. 511-640.

U.S. Coast and Geodetic Survey, 1961, Abstracts of earthquake reports for the Pacific coast and western mountain region, MSA-103, July, August, September, 1959, pp. 51-89.

U.S. Geological Survey, 1963, The Hebgen Lake, Montana earthquake of August 17, 1959, U.S. Geological Survey Professional Paper 435, p. 242.

U.S. Geological Survey, 1965-1970, "The Alaska Earthquake, March 27, 1964," U.S. Geological Survey Professional Papers 542-A, 542-B, 542-C, 542-D, 542-E, 542-F, 542-G, 543-B, 543-E, 543-F, 544-A, 544-B, 544-D, 545-A, 545-C, 545-D.

Varnes, D.J., 1978, "Slope Movement Types and Processes," Chapter 2 in Schuster, R.L., and Kryzek, R.S., eds., Landslides - Analysis and Control, Transportation Research Board Special Report 176, National Academy of Science.

Wiggins, J.H., 1973, "The Risk Imbalance in Current Public Policies," Risk Acceptance and Public Policy Proceedings of Session IV International System Safety Society Symposium, Denver, Colorado, July 17-20.

Wilson, J.A., 1979, "Physical Modeling to Assess the Dynamic Behavior of Rock Slopes," Masters Thesis, University of Arizona.

Wood, H.O., and Newmann, F., 1931, "Modified Mercalli Scale for 1931," Seis. Soc. Amer. Bull., Volume 21, pp. 271-283.

Youd, T.L., 1971, "Landsliding in the Vicinity of the Van Norman Lakes in the San Fernando, California Earthquake of February 9, 1971," U.S. Geological Survey Professional Paper 733, pp. 105-109.

Youd, T.L., and Hoose, S.N., 1978, "Historic Ground Failures in
   Northern California Triggered by Earthquakes," U.S. Geological
   Survey Professional Paper 993, p. 177.

Question

How would you suggest that one gather and use earthquake data in
pit slope design. What level of earthquake activity is dangerous
for liquefaction in earth embankment dams.

Answer

There are several types of data needed for assessing the dynamic
response of rock slopes. First, it is important to remember that
earthquakes occur only occasionally. Because earthquake forces
are not constant, such as gravity forces, we need to represent
this periodicity in some meaningful way. A parameter that does
this is the earthquake hazard, the probability of exceeding a
given earthquake or ground motion in a specified period of years.
For economic pit slope designs the earthquake data available
from National Oceanic and Atmospheric Administration is sufficient.
These data can be acquired in either card or magnetic tape format
from Dr. Carl Von Hake, at code D622, Environmental Hazards Branch,
NOAA/EDIS, Boulder, Colorado 80303.

Second, the characteristics of the earthquake motion at the site
are needed. To approximate the amplitude, frequency content, and
duration of the design seismic ground motions, attenuation
formulae should be used that correspond to the tectonic setting
of the region comprising the mine site.

Third, data that characterize the geometry and strength properties
of pit slopes and rock discontinuities must be acquired. I
recommend using field techniques as suggested by Call et al. (1976)
for collecting spatial data. Shear behavior of rock joints can be
acquired using the techniques of Barton and Bandis (1980), and
laboratory direct shear tests should be used to determine
suitable shear-displacement curves for the critical rock
discontinuities.

The general procedure is to assume a working slope for a particular
section of the mine for a typical operating period. Strength and
spatial probability distributions are sampled to calculate the
stability under static conditions. If the slope is stable, a
dynamic analysis is conducted using LADRS. If the slope is stable
the iteration continues; if not, the smallest earthquake motion
causing slope failure together with its probability of occurrence
is used in the probability of failure calculation.

I must preface my answer to your second question by stating that
I do not consider myself an expert on embankment dams. I know of
no absolute guidelines regarding liquefaction susceptibility; each

dam should be considered as a unique structure with its own design peculiarities.

It is worth noting, however, that, because of the contrast in material properties between embankment dams and their foundations, the dams respond as structures with their own fundamental modes of vibration. Consequently, the peak crest acceleration can be much higher than the peak earthquake acceleration, and the frequency content of the earthquake is very important.

Question

In wet slopes, earthquake duration is very critical due to strain-induced pore pressure rise, but is duration just as critical in dry or free-draining rock masses.

Answer

Yes. Once we overcome our preoccupation with static analyses and their reliance on the concept of factor of safety, we find that a slope does not "fail" under dynamic loads merely because the driving forces temporarily exceed resisting forces. A five centimeter displacement, for example, may have little effect on a large block or wedge (on the order of 50 m) but may cause a smaller block (on the order of 0.5 m) to slide into the pit. People have observed that earthquakes often cause rocks to slide and roll down open pit and natural slopes, even when no major slides occur. This observation does not indicate that no displacement occurred on larger slides, only that the displacement was not significant. An earthquake having a longer duration will cause larger displacements, increasing the size of rock blocks that fail.

Question

Apart from the questionable case of Chuquicamata, do you know of any major open pit slide (in contrast to natural slopes) that was caused by an earthquake. How do you explain the experience at Bougainville where both high pit walls and high rock waste dumps were subjected to the design earthquake with only minor surface damage.

Answer

I know of no major open pit mine slide caused by an earthquake.

Answering your second question is difficult because I have no specific knowledge of the conditions at Bougainville or of the "design earthquake" to which you refer. There could be a number of reasons for the "Bougainville experience":
1.   Slope angle was too shallow.
2.   The seismic design could have been a static analysis that predicted failure based on limit equilibrium, whereas actual displacements during the earthquake were significant only to smaller blocks (minor surface damage) and not to larger masses.
3.   No significant structures were daylighted at the time.

There are probably many more possibilities; it is easier to explain why a failure occurred than to explain why no failure occurred.

Question

Do you know of any data which is available for large slope deformational response for different distances from the free face as a result of high magnitude earthquakes.

Answer

I'm not sure what you mean by deformational response. If you mean particle displacement of the propagating seismic waves, the doubling of displacement caused by the wave impinging on a free surface occurs only within ½ to ¼ of one wavelength of the surface. If the wave velocity is 3,000 m·s^{-1} and the frequency is 5 hz, the wave amplitude decreases to approximately ½ its surface value at a depth of approximately 150 to 300 m into the slope.

Other than that I know of no empirical data available for large slopes during earthquakes.

Question

What is the effect of earthquake vibrations on pore pressures.

Answer

Water pressure will increase as the slope moves toward the rock wedge decreasing the total normal stress across the failure surface. When the slope moves away from the wedge, or dilatency occurs during shearing, water pressure will decrease (perhaps becoming negative) increasing the total normal stress. The actual degree to which the effective stresses and pore pressures change is unknown. Because water pressure increases tend to affect a larger volume than water pressure decreases, the overall effect would be to decrease stability. Unfortunately, the large quantity of excellent dynamic tests performed on soils helps little and essentially no data exist on dynamic behavior of rock joints.

Question

Joint roughness resists down-slope shearing of a block. When an earthquake acceleration record is "discretized", and some up-slope shearing occurs, this is resisted by lower JRC (initially) than in the down slope direction. This may explain the very positive effect of roughness on stability - do you agree.

Answer

The difference in upslope and downslope JRC values certainly has an effect; hoever, the downslope movement of a rock block depends on three things, which are: 1. Slope angle.
                                                                      2. Forcing function.
                                                                      3. Friction Model.

Even if the upslope and downslope JRC values were equal the friction model still has a profound effect on the block response.

Question

Water filled rock joints resist seismic displacements by generating negative pressures when shearing/dilation tries to occur. What is the time-scale of these displacement reversals under a sizeable rock-slope.

Answer

I can't answer this question as I have not run LADRS on a large slope and no field data is available. My guess is that reversals will be extremely small, both temporally and spatially, even on relatively flat slopes.

Question

Would you consider the effects of vibrational acceleration on the abrupt pore water pressure build-ups or increases which could decrease the shear strength for stability analysis. If yes, how would you introduce this concept into stability analysis.

Answer

Dynamic triaxial and direct shear tests have had a profound effect on our understanding of the behavior of soils and the effects of pore pressures under dynamic loads. Unfortunately, no tests are available to improve our understanding of the effect of water pressure in rock fractures during earthquakes or blasting. Intuitively, they must be important.

This is a neglected area of research. Perhaps the recently published report of the National Research Council entitled "Rock-Mechanics Research Requirements for Resource Recovery, Construction, and Earthquake-Hazard Reduction" (National Academy Press, 1981), which recommends dynamic rock testing, will provide some encouragement to individuals and institutions to help fill this gap in our knowledge.

Chapter 6

MECHANICS OF ROCK SLOPE FAILURE

by Douglas R. Piteau and Dennis C. Martin

D. R. Piteau & Associates Limited
West Vancouver, B.C.

ABSTRACT

Instability of rock slopes may occur by failure along pre-existing structural discontinuities, by failure through intact material or by failure along a surface formed partly along discontinuities and partly through intact material. Although certain fundamental failure modes are recognized, the mechanisms of slope failure are varied and complex. Such mechanisms are governed by the engineering geology conditions of the rock mass, which are almost always unique to a particular site. An understanding of failure mechanisms requires a knowledge of the physical, mechanical and strength properties of the intact material and discontinuities which make up the rock mass, as well as the structural geology and hydrogeology. These engineering geology parameters also must be evaluated with respect to the slope geometry to determine failures which are kinematically possible. Only after obtaining a reasonable appreciation of the possible failure modes can a rational mechanical stability analysis be carried out.

Both two-dimensional and three-dimensional failure mechanisms should be considered in assessing the design of rock slopes. Simple analyses methods are used initially to identify those possible failures which could control slope stability. More complex and detailed analyses usually are required for a few critical failure modes which could control stability, or where a more complex type of failure mechanism is envisaged. Complex failure mechanisms are usually identified when assessing previous slope failures. More detailed information concerning slope geometry and engineering geology parameters must be acquired where analyses must be more rigorous.

113

In assessing the potential failure mechanisms of a particular slope, consideration must be given to the size of failure and its effect on operations. The potential for large deep-seated failures involving the whole slope as well as small local failures involving only one or two benches must be critically assessed separately. Generally, it will be found that either one or the other controls the design of the intermediate slopes and, consequently, the overall slope.

INTRODUCTION

Instability of rock slopes may occur by failure along pre-existing structural discontinuities, by failure through intact slope forming material or by failure along a surface formed partly along discontinuities and partly through intact material. The mechanism of failure of a particular slope is governed by the engineering geology conditions of the rock mass and is almost always unique to a particular site. A clear understanding of the failure mechanisms of a slope requires a knowledge of the mechanical and strength properties of intact rock and discontinuities which make up the rock mass. Most important of the factors which contribute to these properties are the characteristics of discontinuities and hydrogeological conditions of the slope. The interrelationship of all the relevant engineering geological parameters must be correctly evaluated in order to determine the failure mechanism(s) which could be operating in the slope.

Failure mechanisms are easiest to assess if they can be represented two dimensionally. In these cases the failure surface is assumed to be subparallel to the strike of the slope so that analysis can be carried out for a unit width of slope. Failures which are assessed in this manner generally require the presence of upper boundary release surfaces or lateral surfaces of separation which do not provide any resistance to failure. Two-dimensional failures which involve a single discontinuity or a single set of discontinuities consist of plane, toppling and buckling failures. In some cases a second set of discontinuities in the slope could result in tension cracks or, stepped failure surfaces which accordingly modify both the geometry of the failure surface and failure mechanism.

Wedge failures develop when two intersecting discontinuities form a tetrahedral failure block which could slide out of the slope along either one or the other of the discontinuities or along both discontinuities. Assessment of these types of failures generally requires a three-dimensional analysis technique or some method of correctly resolving forces from the three-dimensional model to a two-dimensional model for analysis purposes. Presence of additional discontinuities can considerably alter the wedge geometry, resulting in a complex shape for a possible failure surface.

Rock mass failure involving rotational shear failures and/or block flow could develop when the slope is sufficiently high and/or steep that the shear strength of the rock (or soil or weathered rock) mass is exceeded due to high stresses in the slope. The state of the art for analysis of these types of failures relies heavily on soil mechanics principles, for uniform materials. The reason for this is the application of more suitable rock mechanics analysis methods for weaker rock masses which are usually involved in this type of failure as yet are not fully understood. A two-dimensional or three-dimensional approach may be used for analysis of these types of failures.

In addition to the discrete failure mechanisms discussed above, assessment of rockfalls and ravelling are also important in terms of safe operation and maintenance of rock slopes.

Analyses of slope failure problems should generally be carried out using simple limit equilibrium techniques to evaluate sensitivity of possible failure conditions to slope geometry and rock mass parameters. More detailed limit equilibrium techniques, finite element analyses methods, statistical techniques and/or related probability assessments are carried out for those cases where failure mechanisms, as well as operating parameters, are sensitive to slope stability.

PLANE FAILURE

Plane failure involves sliding of a failure mass on a single surface as shown for typical cases in Figs. 1(a) and 1(b). Because only one surface is involved, a two-dimensional analysis may be carried out using limit equilibrium techniques for a "sliding block" having a unit width. Because of the two-dimensional nature of the problem, plane failure analysis is easiest to understand and, provided that certain geometrical conditions are met, it is easiest to apply to simple slope design problems.

Geometric Conditions For Plane Failure

In order for failure to occur by sliding on a single plane, the following geometric conditions should be met:

1. A continuous plane on which sliding occurs must strike parallel or nearly parallel (within approximately $\pm 20^{\circ}$) to the slope face.
2. The failure plane must "daylight" in the slope face. This means that its dip (ψ_p) must be shallower than the dip of the the slope face (ψ_f).
3. The dip of the failure plane (ψ_p) must be steeper than the angle of friction (ϕ) on this plane unless, groundwater pressure exists, in which case failure can develop for $\psi_p < \phi$.

Fig. 1(a). View plane failure parallel to foliation at a copper
           mine in central British Columbia.

Fig. 1(b). Plane failure along major joint at a mine in
           northwestern Ontario.

4. Surfaces of separation which provide negligible resistance to sliding must be present in the rock mass to define the lateral boundaries of the failure block.

The factor of safety against two-dimensional failure along a possible failure surface is controlled by:

1. the height and inclination of the slope, inclination of the failure surface which governs the weight of the failure mass and the direction of failure;
2. the shear strength of the failure surface; and
3. external influences such as groundwater, vibrations, seismic activity, rock anchors, climate, rock stress etc.

Method of Analysis

Plane failure analysis is carried out using a vertical cross section of the slope which illustrates the slope geometry, failure plane geometry and forces acting on the failure plane for a unit width as shown in Fig. 2. Thus the area of the sliding surface can be represented by the length of the failure surface and the volume of the sliding block by the cross-sectional area.

Simple limit equilibrium analysis for a "sliding block" is used to resolve the various forces acting about the failure plane. The presence of a vertical tension crack in the slope face generally tends to reduce the factor of safety. Therefore, it is good practice to consider several tension crack locations in the analysis to determine a minimum factor of safety.

Summary of Forces. Forces which may be considered to be acting on the failure block are described as follows:

1. *Weight of the Block*: weight of the block is calculated from the slope geometry and failure geometry.
2. *Normal Water Pressure*: normal water pressure or uplift force acting on the failure plane is generally obtained by direct scaling from the groundwater profile in the slope as shown in Fig. 2. Water pressure may be somewhat irregular, but for analysis purposes, a triangular or uniform curve of pressure distribution is assumed to act normal to the failure plane.
3. *Crack Water Pressure*: hydrostatic force in a tension crack is calculated from the assumed height of water in the crack. This water pressure acts horizontally (see Fig. 2).
4. *Dynamic Forces*: dynamic forces due to earthquakes or blasting may be of extreme importance for permanent slopes. Earthquake triggered slides or rockfalls are known to have occurred. The possible dynamic forces, as determined from seismic risk analysis, are generally expressed as a percentage of the gravitational force (g), and are presumed as a horizontal force due to ground acceleration and an oscillating upwards and downwards vertical force due to the ground motion. For the plane

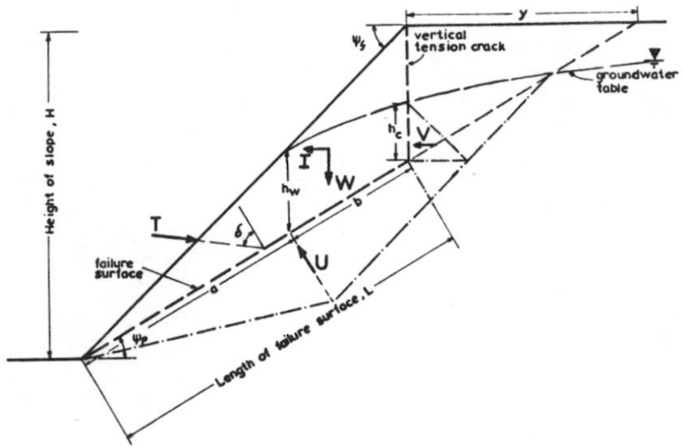

SUMMARY OF FORCES

1. Weight of block, W

$$W = \frac{\gamma H^2}{2}\left[\tan\psi_p - \frac{1}{\tan\psi_{fd}}\right] - \frac{\gamma Y^2}{2}\tan\psi$$

where γ = unit weight of rock

2. Normal water force, U

$$U = \frac{1}{2}\gamma_w \frac{a}{2} + \gamma_w\left[\frac{h+h_c}{2}\right]b$$

where γ_w = unit weight of water

3. Crack water force, V

$$V = \frac{1}{2}\gamma_w \frac{h_c^2}{2}$$

4. Restraining force due to anchors, T.

5. Force due to seismic vibrations, $I = W\frac{\ddot{x}}{g}$

where $\ddot{x}$ is the horizontal seismic acceleration expressed as a fraction of the acceleration due to gravity, g.

RESOLUTION OF FORCES

1. Forces along the failure surface.

Force due to W = $W\sin\psi_p$
Force due to U = $-U$
Force due to V = $V\cos\psi_p$
Force due to T = $a-T\sin\delta$
Force due to I = $I\cos\psi_p$

2. Forces normal to the failure surface

Force due to W = $W\cos\psi_p$
Force due to U = $-U$
Force due to V = $-V\sin\psi_p$
Force due to T = $T\cos\delta$
Force due to I = $-I\sin\psi_p$

CALCULATION OF FACTOR OF SAFETY

1. Disturbing forces = Σ forces along the failure surface.

$$= W\sin\psi_p + V\cos\psi_p - T\sin\delta + I\cos\psi_p$$

2. Resisting forces $= \Sigma\left[\text{Normal Forces}\right]\tan\phi' + (L)(c')$

$$= \left[W\cos\psi_p - U - V\sin\psi_p + T\cos\delta - I\sin\psi_p\right]\tan\phi + (a+b)(c')$$

3. $$F = \frac{\left[W\cos\psi_p - U - V\sin\psi_p + T\cos\delta - I\sin\psi_p\right]\tan\phi' + (a+b)(c')}{W\sin\psi_p + V\cos\psi_p - T\sin\delta + I\cos\psi_p}$$

ASSUMPTIONS

1. Unit thickness perpendicular to the failure surface.

2. The tension crack and failure surface strike subparallel to the slope crest.

3. The assumption of limiting equilibrium applies.

4. No internal deformation of the failure block occurs.

5. Groundwater conditions in the slope are steady.

6. Overturning moments due to water forces are not considered.

7. The shear strength of the failure surface is described by the effective cohesion, c', and effective angle of friction, ϕ'.

Fig. 2. Slope geometry, distribution of forces and calculation of factor of safety for stability analysis of plane failure.

failure analysis purposes the horizontal force I, is resolved
about the failure plane as shown in Fig. 2.

5. *Forces from Artificial Support*: if artificial support in the
form of rock bolts or anchors is to be used, the total support
force (T) required on the failure plane can be determined
from the proposed inclination (δ) of the anchors with respect
to the failure plane. The optimum inclination (δ) has been
defined as: tan ($\delta + \psi_p$) = tan ϕ. This total force should
be evenly distributed on the failure plane to determine the
number of anchors, spacings and loads required to maintain
adequate stability for the slope in question.

Resolution of Forces and Calculation of Factor of Safety. The disturbing and resisting forces above are resolved into component forces
normal to the failure plane and component forces parallel to the
failure plane. The factor of safety, F is expressed as:

$$F = \frac{\text{Sum of largest forces which may be}}{\text{Sum of disturbing forces}}$$

The resisting forces which constitute the shear strength along the
failure plane of length, L, are calculated from the friction (ϕ)
and cohesion (c) along the failure plan as:

Resisting Forces = (Σ normal forces) tan ϕ + (L)(c)

The disturbing forces are simply the sum of the forces acting
downwards along the failure plane.

The factor of safety is calculated as summarized in Fig. 2.
Stability analysis results may be evaluated by plotting graphs of
factor of safety vs slope angle for varying slope conditions. These
curves plus other information are used as a basis for evaluating
the stability condition of the slope and for considering the optimum
slope angle, bench design and other slope design aspects.

Sensitivity of Parameters in Plane Failure Analysis

The sensitivity of factor of safety with respect to variations in
individual slope parameters evaluated by varying each parameter for
a hypothetical slope at the same time that all other parameters are
kept constant. Piteau et al(1979) has carried out a sensitivity analysis for a hypothetical slope with a single discontinuity as shown in
Fig. 3(a). The effects of varying individual parameters are assessed
for each case by keeping all parameters constant except the one which
is being varied. Typical sensitivities for the various parameters
are shown in Fig. 3(b) to 3(k). The sensitivity of each parameter
will depend to a certain extent on the other parameters which are
kept constant. However, this example provides an appreciation of the
relative effects of the various parameters for the case examined.

INITIAL SLOPE PARAMETERS

Slope Height, H = 30.48 m (100 feet)
Slope Angle, ψ_f = 50°
Dip of failure plane ψ_p = 35°
Cohesion on failure plane, c = 68.95 k Pa (10 psi)
Friction angle, ϕ = 30°
Unit weight of rock γ = 26.71 kN/m³ = 170 (pcf)
Location of Tension Crack, y = 0
Height of water above toe of slope, h_w = 0
Height of water in Tension Crack, h_c = 0
Seismic acceleration due to earthquakes, a_e = 0.0g
Anchor Force, T = 0 kN

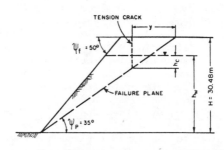

a)   Initial slope condition and parameters investigated for sensitivity.

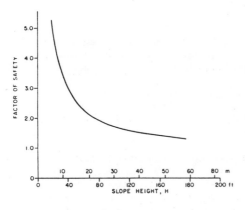

b)   Slope height, H

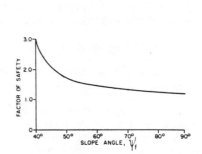

c)   Slope angle, ψ_f

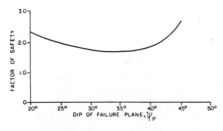

d)   Dip of failure plane, ψ_p

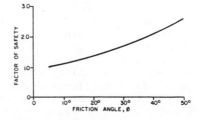

e)   Friction angle, ϕ

Fig. 3 Sensitivity of slope parameters in plane failure analysis.

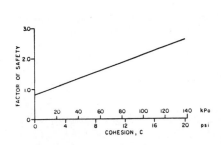

f) Cohesion, C.

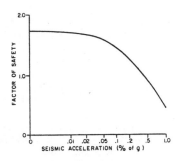

g) Seismic acceleration due to earthquake.

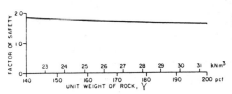

h) Unit weight of rock, γ

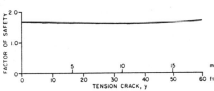

i) Location of Tension Crack, y

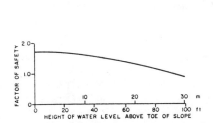

j) Height of water above toe of slope, h_w.

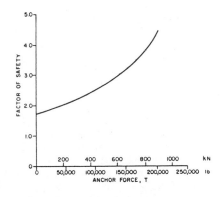

k) Horizontal Anchor Force, T

Fig. 3 (continued) Sensitivity of slope parameters in plane failure analysis.

Effects of Intact Rock and Surface Asperities on Plane Failure
Mechanisms

The analytical method described for plane failure is only realistic if failure occurs along well defined continuous, planar, non-
undulating discontinuities, such as undisturbed bedding joints or
planar faults. If engineering geology investigations indicate that
discontinuities are not throughgoing or planar, the above analyses
may not be realistic. In these cases, in fact a planar failure surface would occur partly along discontinuities and partly through
intact rock.

Discontinuous Joints. If no continuous failure plane exists in the
slope, the prospect still exists that failure may result from sliding
on discontinuous joints of relatively low shear strength and from
failure through intact rock bridges between joints (see Fig. 4(a)).
In this case the resisting forces are increased in the analysis from
additional shear resistance provided by the intact rock. The contribution due to the intact rock is calculated based on the total
length of the intact rock bridges which occurs on the failure plane
and the shear strength of the rock (Jennings, 1970). Simple
analyses by Martin (1978) indicate that even for high slopes greater
than 300m high in moderately hard rock occurrence of less than 10%
intact rock along a prospective failure surface would provide suitable resisting forces to achieve limit equilibrium against shear
failure.

Development of the Step Failure Mechanism. In general, it is expected
that the occurrence of large amounts of intact rock along a possible
failure surface would result in the failure geometry being altered
so that failure occurs along a path of least resistance. Because
a rock mass generally contains one or more sets of relatively uniformly distributed discontinuities with a relatively consistent
length and spacing, the failure surface can generally be
expected to form along a path of minimum resistance due to a combination of sliding and separation along discontinuities and failure
through small amounts (for reasons explained above) of intact rock
(see Fig. 4(b) and (c).

Where a single set of discontinuities exists in a slope, it is
likely that the plane failure mechanisms will be replaced by a step
failure mechanism, involving shear sliding along the undercut discontinuities and tensile failure of the intact rock between discontinuities as shown in Fig. 4(b). A small portion of the failure surface may occur by shear failure along rock bridges between adjacent
discontinuities. If an additional set of discontinuities occurs at
a steep angle to the undercut discontinuities, the step failure
mechanism is developed more easily by sliding on the undercut discontinuities and separation across the steeper dipping "cross" joints
as shown in Fig. 4(c). In all step failure mechanisms, the minimum
resistance path would be expected to involve a minimum amount of
intact rock. It is noteworthy also, that the stepped failure path

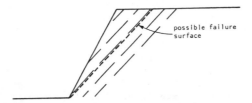

a) Development of plane failure by shear sliding along discontinuities and shear failure through intact rock bridges.

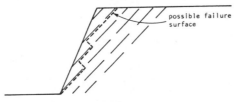

b) Development of step failure involving shear sliding along discontinuities and tensile failure or shear failure of intact rock.

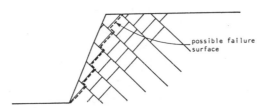

c) Development of step failure involving separation across steep dipping cross joints and shear sliding along undercut discontinuities.

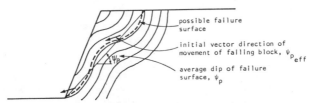

d) Decrease in the effective angle of dip of the failure surface due to waviness.

Fig. 4. Effects of intact rock and surface asperities on plane failure mechanisms.

is always steeper than the dominant set of discontinuities along
which shear failure takes place.

Effects of Waviness on Slope Stability. The occurrence of uniformly
distributed waves or undulations on the failure surface will result
in an initial vector direction of movement of the failure mass which
is at a shallower angle than the average dip of the failure surface
as shown in Fig. 4(d). The shallower "effective angle of dip of the
failure surface due to waviness" is used to resolve the disturbing
and resisting forces along the failure surface. A case record
documented by Martin and Piteau (1976) indicates that the presence
of waviness on possible failure surfaces could possibly justify an
increase in overall slope angle of 2.5° to 6° for several 150 to
300m high slopes examined. In this case the rock mass was relatively
weak with the result that it was felt that the waves could not be
significant unless careful control blasting was carried out to maintain the integrity of the rock.

Failure Mechanisms for Thin Slabs on a Slope

   Application of plane failure analysis techniques to most slopes
would indicate that the slope would be stable if excavated subparallel to the major discontinuities so that those discontinuities
were not undercut or daylighted by the excavation (see Fig. 5(a)).
In certain cases, particularly in coal bearing rocks, the bedded
nature of the rocks can lead to formation of long thin slabs of rock
on the footwall slope which are controlled by weaker horizons. A
good example is the presence of narrow uneconomic coal seams below
a footwall slope in coal. In addition to sliding along undercut
bedding surfaces (Fig. 5(a)) the main failure mechanisms of a slope
of this type are shown in Figs. 5(b) to (f) and are described as
follows.

Planar Sliding Along Coal-Shale Seams in Combination with Cross
Joints. Instability possibly could develop involving planar sliding
along bedding surfaces or coal-shale seams in combination with cross
joints or thrust faults as shown in Fig. 5(b) and 5(c). In order
for this failure mechanism to develop, continuous cross joints,
thrust faults or cross bedding surfaces with a similar strike would
have to occur in the slope. Shallow dipping cross joints would
result in a compound sliding block mechanism as shown in Fig. 5(b).
Moderate or steeply dipping cross joints could result in a *ploughing*
or *block lifting* mechanism as shown in Fig. 5(c). Analyses of these
mechanisms could be carried out by assuming that the slope was made
up of two sliding blocks. Forces would be resolved for each block
and a factor of safety calculated accordingly.

Failure of Intact Rock in the Toe of the Slope. In extremely high
slopes, instability could develop if the compressive stress at the
toe of the slope exceeds the compressive strength. This failure
mode is illustrated in Fig. 5(d). In weak rock, shear and/or tensile

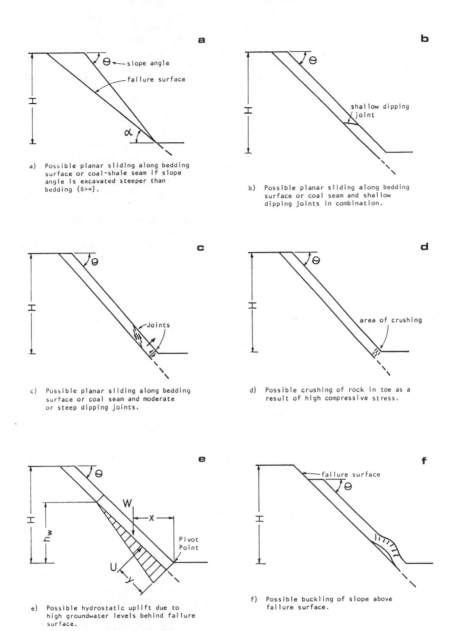

a) Possible planar sliding along bedding
surface or coal-shale seam if slope
angle is excavated steeper than
bedding ($\theta > \alpha$).

b) Possible planar sliding along bedding
surface or coal seam and shallow
dipping joints in combination.

c) Possible planar sliding along bedding
surface or coal seam and moderate
or steep dipping joints.

d) Possible crushing of rock in toe as a
result of high compressive stress.

e) Possible hydrostatic uplift due to
high groundwater levels behind failure
surface.

f) Possible buckling of slope above
failure surface.

Fig. 5.   Possible mechanism of failure on a footwall with a narrow
coal-shale seam behind the slope.

failure of the intact rock could also occur along a failure surface
similar to those shown for cross joints in Fig. 5(b) and (c).
Analysis of this type of failure would be the same as for the compound sliding blocks described above, except that the shear strength
of intact rock would be substituted for discontinuity shear strength
where appropriate.

Hydrostatic Uplift. If high groundwater pressures exist behind the
coal-shale seams in a high footwall slope, it is possible that hydrostatic uplift or heave could develop as shown in Fig. 5(e). The
factor of safety for this mechanism is calculated by resolving the
various restoring moments, the weight of the block and overturning
moments due to uplift forces about a pivot point at the toe of the
slope.

Buckling. Buckling of thin slabs of rock such as shown in Fig. 5(f)
has been documented by Walton and Taylor (1977). Analysis of this
failure mechanism is carried out by determining the critical buckling
load, P_{cr}, using the "Euler Buckling Solution" for a column with
free ends. This solution is described in most introductory text-
books on mechanics as follows:

$$P_{cr} = \frac{\pi^2 EI}{4 \ell^2}$$

where E = modulus of elasticity,

$\quad$ I = moment of inertia of the cross-section of the beam = $\dfrac{Zb^3}{12}$

$\quad$ ℓ = length of the column over which buckling occurs.

The factor of safety, F, against buckling is calculated as:

$$F = \frac{P_{cr}}{P_{dr}}$$

where, P_{cr} is the critical buckling load, and P_{dr} is the sum of
forces acting on a column of length ℓ due to the rock above the
column and shear resistance acting along the base and sides of the
column of rock considered. In this analysis the effects of gravity
are ignored. The slope may be assumed to be fully drained or water
pressure may be incorporated into the analyses if required.

In evaluating the potential for buckling using the Euler solution,
the factor of safety considered to be practical for purposes of
slope design is considerably higher than for the more conventional
analyses of the other failure modes. The reason for this is to
account for the effects of eccentricity, discontinuities, defects
in the rock, etc. The analyses prepared by Walton and Taylor (1977)
indicate that a F $\approx$ 5.0 appears to be appropriate for design for the
buckling analyses.

An analysis of the various mechanisms described above by Piteau and Martin (1979) indicated that hydrostatic uplift mechanism was much more important than simple buckling in the evaluation of a proposed high footwall in a coal mine in the Canadian Rockies. In this case stringent control of groundwater in coal-shale seams behind the footwall slope was critical for stability of the slope.

TOPPLING

Toppling is possible wherever a set of steeply dipping reasonably well developed discontinuities with moderate to close spacing occurs subparallel to the slope. The basic conditions for sliding and toppling of a single block on an inclined plane are shown in Fig. 6 (see Hoek and Bray, 1977). The conditions for sliding are related to the friction angle and dip of the base plane. Whenever the dip of the base plane is greater than the friction angle i.e. $\psi > \phi$, sliding could occur. Toppling is related to the ratio of width to height, b/h for the block which regulates the location of the resultant force due to the weight of the block with respect to a pivot point at the lowest corner of the block. Whenever b/h < tan ψ the resultant force occurs outside of the toe of the block and an over-turning force develops about the pivot point which could lead to toppling. Combination of the two conditions, i.e. $\psi = \phi$ and b/h=tan ψ, results in four areas on the diagram in Fig. 6 within which the block may be stable or may fail by sliding, toppling or a combination of the two.

These basic conditions and resultant forces with respect to the pivot point could be significantly altered by additional external forces due to water acting along the block, surcharge loads or artificial support. The principles illustrated by this simple analysis are used as the basis for more detailed assessment of the various toppling mechanisms.

Flexural Toppling and Related Mechanisms

Basic Mechanisms of Flexural Toppling. Flexural toppling develops when a set of well developed or throughgoing discontinuities dips steeply into the slope. In this type of failure, long thin columns or slabs of rock formed by discontinuities bend or break in flexure due to loss of support or oversteepening. This results in the centre of gravity of individual blocks acting outside the toe of the blocks, the result being that the overturning moments exceed the resisting moments. Typical flexural toppling mechanisms are shown in Fig. 7(a) and 7(b). Goodman and Bray (1976) have documented numerous types of flexural toppling, block toppling and associated mechanisms which are related to this basic type of toppling mechanism.

In addition to flexure between columns of rock, the toppling mechanism also involves tensile and compressive failure along the

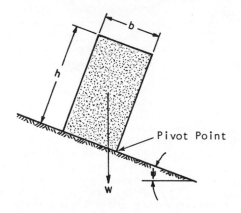

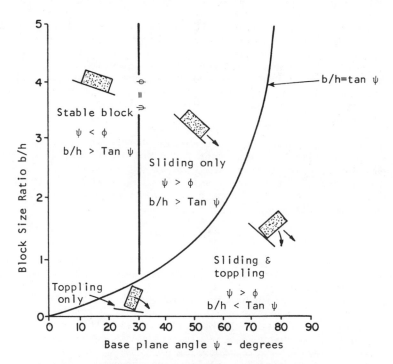

Fig. 6. Conditions for sliding and toppling
of a block on an inclined plane(after
Hoek and Bray, 1977).

Basal Failure Surface

Fig. 7(a). Flexural toppling failure at a mine in northwestern Ontario. Note well defined basal failure surface.

Fig. 7(b). Complex flexural toppling at a mine in northwestern Ontario. Note highly broken nature of rock after failure.

columns as well as along the base of the columns. In most documented
cases this tensile failure occurs above a clearly defined basal
failure surface (see Fig. 7(a)). Basal failure surfaces may form
by tensile failure of intact columns or more commonly along cross
joints. Field observations indicate that the basal failure surface
is generally consistent within a particular rock type and may be in
part controlled by orientation of cross joints.

Analysis Method. Analysis of flexural toppling is carried out using
a limit equilibrium analysis method developed by Goodman and Bray
(1976). This method assumes that a given slope contains continuous
joints which have a uniform dip (90-α) and spacing (Δx) and form
the boundaries of discrete columnar blocks as shown in Fig. 8(a).
As toppling occurs, a stepped basal failure surface at an angle, β,
is developed below which toppling of the individual blocks does not
occur. In order to calculate the water forces on each block, a
simplified configuration of the water table is constructed as shown
in Fig. 8(a).

    Analysis is performed on the highest block in the slope first and
all resultant forces are transferred to subsequent blocks. The
forces due to gravity, water pressure, and frictional properties
along the basal failure surface and between the blocks are calculated
and resolved. Each block is analyzed for possible failure modes involving both planar sliding on the basal failure surface and/or toppling about the lowest corner of the block. The resultant force,
P_n, (if any), for each block is determined as shown in Fig. 8(b).
The failure mode with the largest resultant force, P_n, is assumed
to define the failure mode of the block, and that force is assumed
to be transferred to the lower adjacent block. Thus, the net force
acting on each block is computed to determine whether each particular
block is stable, or if it could fail by toppling or sliding. The
failure mode and resultant force determined for each block is recorded
as shown in Fig. 8(c).

    Instability of the slope occurs when there is a positive resultant
force, P_n, for the lowest block at the toe of the slope. Where slopes
are unstable, the possible support required to insure stability is
estimated from the resultant force P_n. The block by block analysis
also may be used to assess the effect of removing one or more blocks
from the toe of the slope. Similarily the effects of binding blocks
together, such as with bolts or dowels, may also be assessed. A
resultant force of $P_n = 0$ on the bottom block would be equivalent to
a factor of safety of 1.0. A factor of safety could be calculated
by assessing the required friction angle (ϕ_{req}) for equilibrium, i.e.

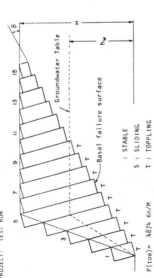
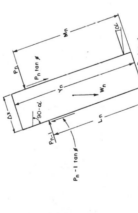

GEOTOP-LIMIT EQUILIBRIUM ANALYSIS OF TOPPLING FAILURE IN JOINTED ROCK
PROJECT: TEST RUN

P(toe)= 4874 Kn/M

: STABLE
S : SLIDING
T : TOPPLING

a) Model for limit equilibrium analysis of toppling.

$$\text{Sliding} \quad P_{n-1_S} = P_n - \frac{W_n(\cos\alpha \cdot \tan\phi - \sin\alpha)}{1 - \tan^2\phi}$$

$$\text{Toppling} \quad P_{n-1_T} = \frac{P_n(M_n - \Delta x \cdot \tan\phi) + (W_n/2)(Y_n \sin\alpha - \Delta x \cos\alpha)}{L_n}$$

b) Conditions for toppling and sliding of the nth block (after Goodman and Bray 1976).

GEOTOP - LIMIT EQUILIBRIUM ANALYSIS FOR TOPPLING FAILURE IN JOINTED ROCK

PROJECT: TEST RUN

MATERIAL DENSITY= 2700 Kg/M3
FAILURE SURFACE ANGLE= 30 DEG
SLOPE HEIGHT= 50 M
BEDDING PLANE SPACING= 7 M
WATER TABLE DEPTH= 25 M
ADJUSTED BEDDING PLANE SPACING= 6.1

INTERNAL FRICTION ANGLE= 30 DEG
FRICTION ANGLE= 35 DEG
SLOPE ANGLE= 70 DEG
DIP ANGLE= 75 DEG

N	PN,T	PN,S	PN	RN	SN	SN-PN MODE
15	-386	-491	0	676	181	.268	T
14	-317	-861	0	1187	318	.268	T
13	-249	-1232	0	1698	455	.268	T
12	-181	-1602	0	2208	592	.268	T
11	-112	-1973	0	2719	729	.268	T
10	-44	-2343	0	3230	865	.268	T
9	25	-2714	25	3726	977	.262	T
8	124	-2969	124	4068	1006	.247	T
7	287	-3103	287	4353	1029	.236	T
6	500	-3173	500	4646	1065	.229	T
5	986	-3194	986	4809	877	.182	T
4	1698	-703	1698	3563	1065	.299	T
3	2585	433	2585	2535	493	.194	T
2	4436	1640	4436	1664	-920	-.865	T
1	4874	3810	4874	965	42	.044	T

c) Typical output for toppling analysis.

Fig. 8. Limit equilibrium analysis method for flexural toppling (after Piteau et al, 1981).

$$F = \frac{\tan \phi \text{ available}}{\tan \phi \text{ req}}$$

The analysis method has been developed so that a large number of iterations may be carried out. The results are recorded in terms of the various parameters and the resultant force, P_n, on the toe block.

Analyses may be carried out for different discontinuity dips, spacing, slope heights slope angles and groundwater conditions. Comparative plots which show the variation of toppling conditions for the various parameters can accordingly be prepared. Further discussions of this analysis method are given by Piteau et al (1981).

Presentation of Analysis Results for Benches at a Mine in Meta-sediments. Toppling analyses were carried out for different bench face angles, dips and spacings of foliation joints for a 10.67m (35 foot) high single bench. For each case, the dip and spacing of foliation joints corresponding to zero resultant force (i.e. $P_n = 0$) at the toe was determined from the stability analysis. The results are shown on a graph of foliation dip versus spacing of foliation joints as shown in Fig. 9(a) for a 10.67m (35 foot) high bench in a dry slope with a 70° bench face angle. The resulting line represents $P_n = 0$ for that bench face angle. Below this line the foliation dip and spacing of foliation joints increases and the slope is stable with respect to toppling. Above this line the foliation dip and spacing of foliation joints decrease so that the slope is unstable.

Bench face angles were varied to develop a set of lines corresponding to analyses of various bench face angles as shown in Fig. 9(b) for a single 10.67m (35-foot) high bench in a dry slope. These curves indicate the variation in bench face angle required to prevent toppling for various combinations of foliation dip and spacing of foliation joints.

Comparison of Analysis Results with Documented Slopes at a Mine in Metasediments. In order to obtain a better appreciation of the rock mass conditions when toppling failure develops, a careful study was carried out for stable (not toppled) and unstable (toppled) slopes on benches at a mine in steeply dipping metasediments. Empirical evaluation of slopes in this manner proved extremely useful in assessing the basic mechanisms of toppling failure that were operating and for providing a basis of comparison of the results of the toppling analysis described above. It was assumed that all slopes had been excavated vertically using similar blasting practices. Relevant characteristics, such as lithology, structural geology parameters and slope geometry parameters, were recorded for over thirty locations on the west side of the open pit.

The distribution of documented slopes in terms of plotting dip versus spacing of foliation joints is shown in Fig. 10. Shaded

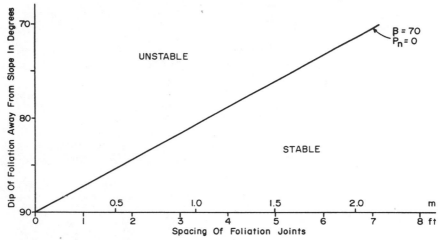

a) Toppling analysis results for a 10.67 m high drained slope
    at an angle of 70°.

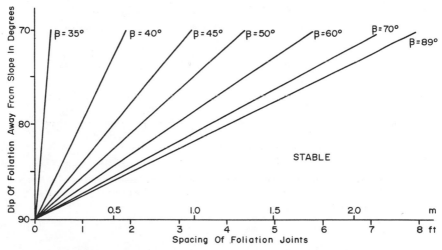

b) Design chart showing bench face angles required to prevent
    toppling for 10.67 m high single benches for a dry or
    drained slope.

Fig. 9.   Presentation of toppling analysis results and development
          of design charts for 10.67m high benches for a dry or
          drained slopes at a mine in metasediments.

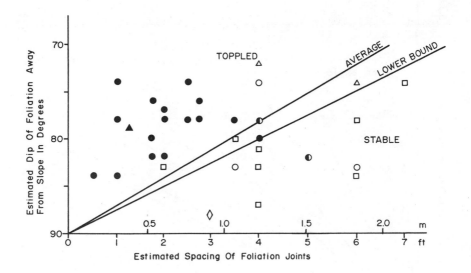

Fig. 10.   Distribution of stable and toppled slopes for 10.67m high
benches examined in an open pit in steeply dipping metasediments.

symbols represent toppled slopes, whereas unshaded symbols represent stable slopes as indicated. The importance of the interrelationship of dip of foliation joints and spacing of foliation joints in terms of stability of the bench face is clearly shown in Fig. 10. It can be seen in Fig. 10 that the boundary conditions for stable and un-stable (or toppled benches) with regard to dip and spacing of folia-tion joints appear to be demarcated by a well defined line. Both an average and a lower bound line has been estimated for stable slopes compared to toppled slopes. The estimated lower bound of toppled slopes shown in Fig. 10 is in close agreement with the theoretical analysis results for an 89° bench face angle shown in Fig. 9(b). This remarkably close correlation between the theoretical model and empirical data from documented stable and unstable slopes indicates the validity of these theoretical analyses results as a reliable method for design.

Effects of Slope Height and Groundwater on Flexural Toppling. The analysis method described may be used to develop comparative plots for different bench heights or different groundwater conditions. Increase in slope height increases the potential for toppling by re-ducing the relative spacing to length ratio of individual toppling blocks. As a result the design lines for $P_n = 0$ for the various bench face angles are shifted downwards. Thus, slopes with a partic-ular combination of foliation dip and foliation joint spacing which are stable for a single bench could possibly topple for a double bench unless excavated at a shallower bench face angle.

Toppling is also sensitive to effects of groundwater along the foliation surfaces. As mentioned earlier, different groundwater conditions may be included in the analysis to prepare the appropriate comparative plots. Groundwater increases the over-turning moments which reduces stability. Hence, the respective design curves for $P_n = 0$ move downwards compared to the case of dry slopes. Further details of the effects of the various parameters in the toppling analysis are given in Piteau et al (1981)(a).

Toppling Involving Discontinuities Which Dip Steeply Toward the Slope

Condition for Overturning. When discontinuities are vertical or dip steeply toward the toe of the slope, overturning moments due to the weight of the block alone do not develop. However, presence of external loads due to groundwater pressure, ice jacking or blasting vibrations could produce overturning forces which could result in toppling. Typical failures of this nature are shown in Fig. 11(a) and 11(b).

Analysis Method. Analysis for this type of toppling is carried out assuming that the potential failure blocks are triangular in shape and completely isolated from the surrounding rock, as shown in Fig. 12(a). The forces of gravity and water pressure are considered to be acting on the boundaries of the failure blocks and forces are

Fig. 11(a). View of slope
with foliation joints which
dip steeply towards the slope.
Area in foreground toppled
due to probable adverse
effects of groundwater and
blasting.

Fig. 11(b). Toppling failure involving discontinuities which dip
         steeply towards the slope. Effects of blast damage are
         considered to have contributed significantly to
         instability in this case.

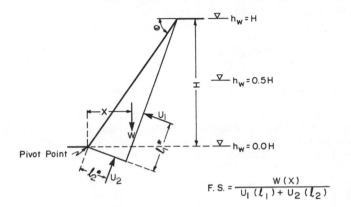

$$F.S. = \frac{W(X)}{U_1(l_1) + U_2(l_2)}$$

a) Analysis Method

Note: *Assumes water table is at the surface.

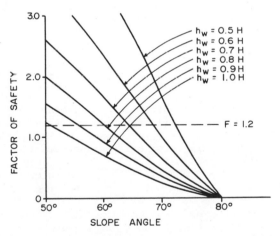

b) Typical analysis results for discontinuity
dipping towards the slope at $80°$.

Fig. 12. Analysis method and typical analysis results for toppling
when discontinuities dip steeply towards the slope.

resolved in terms of moments about a pivot point at the toe of the
slope, as shown in Fig. 12(a). Friction and cohesion along the plane
of separation are not considered, thereby making the analysis somewhat conservative. The factor of safety is calculated as:

$$F = \frac{\Sigma \text{ restoring moments due to gravity}}{\Sigma \text{ overturning moments due to water pressure}}$$

Analyses are carried out for different discontinuity dips, spacing
and groundwater conditions. Groundwater pressures are calculated
based on a horizontal phreatic surface which is expressed as a ratio
of the slope height. The factor of safety is independent of slope
height. Results may be plotted in terms of factor of safety vs slope
angle for various discontinuity dips and phreatic levels as shown
for a typical case in Fig. 12(b).

The toppling analysis described is considered to be somewhat
conservative because the effects of shear resistance along the edges
of the toppling blocks are not considered. The analysis also assumes
that the plane of separation along the bottom of the triangular
block is continuous and through-going and has zero shear strength.
In addition, the effects of lateral constraint due to the rock
adjacent to the possible toppling block are not considered. In
failures of this nature, which rely on hydrostatic uplift, it is
likely that after small movements the slopes would drain. In the
case of overturning failure the slope could restabilize relatively
quickly after a small amount of movement. Subsequent buildups of
water could lead to recurring movement.

Documentation of Case Records. Analysis for this type of toppling
was carried out for benched slopes in steeply dipping metamorphic
rocks at a large open pit mine in Tazmania. Fig. 11(b) shows typical
toppling at the mine. Analyses were carried out as summarized in
Fig. 12(a). Field investigations indicated that blasting vibrations
and damage in the toe area, as well as water pressures, were contributing to the overturning mechanism. Typical analyses results for
various height of water for a schistosity dip of 80° are shown in
Fig. 12(b), which clearly indicate the effects of high water on the
behaviour of the slope. It is of interest to note that overturning
moments are not likely to develop unless the water levels in the
slope are greater than about one-half the height of the slope above
the toe, i.e., $h_w \geq 0.5H$. In this particular case, groundwater
control, modification of blasting practises and adoption of inclined
bench faces would be useful to control the toppling problem.

Remedial Measures to Control Toppling

One of the most effective methods of stabilizing a toppling slope
is to provide adequate drainage to prevent excessive groundwater
pressures in the slope or prevent water from entering the slope.
Unloading crest areas, thus reducing the overturning moments, has

Fig. 13.   View of Hell's Gate Bluffs, British Columbia , showing
            area of crest of large toppling failure which has been
            unloaded to increase stability.

also been used successfully in controlling toppling failures.
Remedial measures consisting of unloading the crest, placement of
a sealed coating, and installation of drainholes as shown in Figs.
13 and 14 were successfully used to stabilize slopes which were
undergoing toppling failure at Hells Gate Bluffs in the Fraser
Canyon, British Columbia (Piteau et al, 1976).

    Artificial support in the form of dowels, rock bolts and/or rock
anchors can be used to effectively bind the individual rock columns
together, thus providing a wider base and increasing the ratio of
base to height (b/h) of the rock columns. Use of buttresses or
other passive support at the base of the slope may also be used to
advantage to counteract the overturning action. In this regard
development of inclined bench faces using preshearing or other suitable controlled blasting techniques provides a buttress in the toe
area to reduce overturning forces. Preshearing effectively reduces
the possibility of development of unfavourable overturning forces
as blasted rock is removed in front of the bench face.

Fig. 14. View of polyethylene sheeting used in conjunction with
light mesh and steel dowels to provide a temporary cover
at crest of toppling failure at Hell's Gate Bluffs;
British Columbia.

WEDGE FAILURE

Description and General Conditions for Wedge Failure. In the
simplest cases, wedge failures occur by sliding on a combination
of two planes such as shown in Fig. 15. In these cases the failure
block is considered as a tetrahedron, the sides of which are defined
by the two planes, the slope face and slope crest. The cross
sectional area of the tetrahedron varies laterally along the slope.
Hence, the two dimensional model which is acceptable for plane
failure must be replaced by a three-dimensional model for wedge
failure analysis. Stereographic projection techniques can be used
to great advantage in analyzing the three-dimensional model.

   In actual mining situations the simple wedge failure tetrahedron
may occur at any orientation in the slope and as a result failure
may occur in any of the following ways:

   1. failure by sliding along both planes in a direction along
      the line of intersection as shown in Figs. 15 and 16.
   2. failure by sliding along one plane with separation across the
      other plane as shown in Figs. 17 and 18.
   3. failure by rotational sliding on one plane and separation
      across the other plane as shown in Fig. 19.
   4. failure by progressive ravelling of rock along planes formed
      by the wedges in highly jointed rock as shown in Fig. 20.

Fig. 15. Typical wedge failure involving sliding on two
          discontinuities denoted A and B.

Occurrence of tension cracks and/or additional discontinuities
in the slope could also alter the wedge geometry resulting in failure
on more than two surfaces. Analysis methods for these types of
failures become increasingly more complex.

In a particular rock mass there are often several sets of discontinuities. Intersection of discontinuities of one set with
discontinuities of other sets could result in numerous possible
wedge failure combinations. The basic approach to assessment of
possible wedge failures is to identify all wedge combinations which
are significant to stability for a particular slope orientation
and dip. Having identified the wedge combinations which daylight
on the slope, the next step is to carry out simplified analyses to
identify those wedges which are unstable or are in a marginal state
of stability. Slope design will normally be controlled by a few
"critical wedges" which are in a marginal state of stability. Hence,
detailed stability analyses are carried out for these critical wedges
to prepare a rational slope design.

Assessment of Wedge Failure Mechanisms

Identification of Wedges Which are Undercut on Slope. Stereographic
projections are extremely useful for preliminary assessment of
possible wedge combinations. Great circles corresponding to the
various discontinuities in a rock mass are shown in Fig. 16(b).

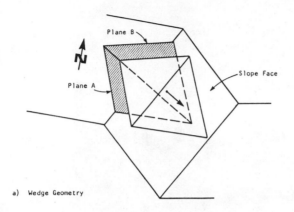

a) Wedge Geometry

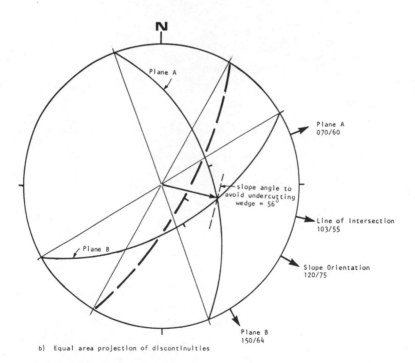

b) Equal area projection of discontinuities

Fig. 16. Parameters for typical wedge failure which involves
          sliding along two discontinuities.

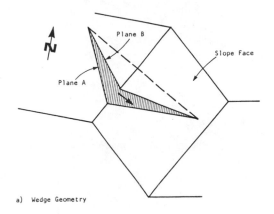

a) Wedge Geometry

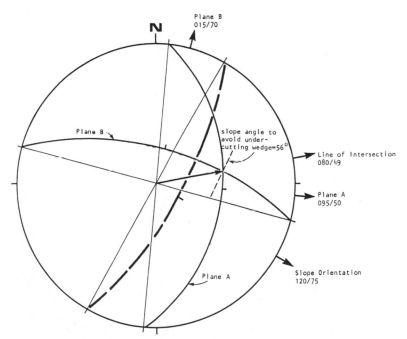

b) Equal area projection of discontinuities

Fig. 17. Parameters for typical wedge failure involving sliding
along one discontinuity and separation across a second
discontinuity.

Fig. 18.   Typical wedge failure which involves sliding primarily
           along a single discontinuity plane. Note that the line
           of intersection is oblique to the slope.

Fig. 19.   Large wedge failure which involves rotational sliding
           primarily along one plane. Progressively increased
           amount of movement along tension crack from left to
           right indicates rotational mechanism for failure.

Fig. 20.   View of highly fractured rock at a copper mine in
           northern British Columbia. This type of rock mass
           is susceptable to continuous ravelling of small
           wedges and rockfalls as well as possible deep seated
           failure through the rock mass.

The orientation (trend and dip) of the slope is indicated by the
heavy line in Fig. 16(b). Any discontinuity intersections which
occur outside this slope are daylighted or undercut. By usual
convention it is these wedges which are undercut which are assumed
to be wedge failures which are kinematically possible. The orientation and plunge of the line of intersection of each wedge can be
determined from the stereonet. For particularly critical wedges
the slope angle which would be required to avoid undercutting any
particular wedge may also be determined directly from the stereonet
by determining the dip of a plane with the same trend of the slope
which would be required to prevent daylighting as shown in Fig. 16(b).
It is important to note that if wedges are oblique to the slope, the
slope angle required to avoid undercutting the wedges is steeper than
the plunge of the wedge.

Determination of Wedge Failure Mechanism. A simple test to determine
whether a wedge failure could occur by sliding on a single plane or

by sliding along the line of intersection of two planes is described
by Hocking (1976). As shown in Figs. 17(a)and(b), if the dip
direction of either plane occurs between the dip direction of the
slope and the line of intersection of the wedge, sliding will occur
on that plane. In this case the other plane will act as a release
surface or surface of separation. If the dip direction of the slope
and the line of intersection occur between the dip direction of the
two discontinuities then sliding will occur on both planes in a
direction along the line of intersection as shown in Figs. 16(a)
and (b).

It is noteworthy that sliding on a single plane could occur by
simple plane shear and/or by rotational sliding due to hinging of
one side of the wedge. An example of rotational sliding of large
wedges at a mine in Tasmania is shown in Fig. 19.

Complex Wedges and Multi Plane Failures. Numerous failures are documented which involve more than two planes. A simple case is the
occurrence of a subvertical joint which could form a tension crack
or back scarp of a wedge. These features would reduce the size of
the failure block and, accordingly, alter the wedge geometry. Additional planes with different orientation could lead to an even more
complex failure geometry, the effects being that possible sliding
could occur on several planes. These types of failures are usually
recognized in the field only, in existing slope failures. In general,
the complex wedge mechanism would not be used for design of slopes
unless previous empirical information concerning previous slope
failures indicated that these mechanisms could develop.

In some cases if numerous planes could be involved in a wedge
failure a more generalized failure mechanism involving a failure
surface along joints as well as intact rock may be applied. Analysis
for this failure mechanism is presented in the section dealing with
rotational shear failure.

As described for planar sliding, the occurrence of discontinuous
joints, offset of joints or joint waviness could significantly affect
the failure geometry of wedges. In certain specific cases these
modifications could be incorporated into the failure mechanism. An
extremely detailed and complex analysis, based on reasonably extensive engineering geology information, would be required for analysis
of these types of failures.

Analysis Methods

Numerous methods have been presented for analysis of wedge failure. Most of these methods employ the use of stereographic projections and analytical techniques of static equilibrium. With a clear
understanding of the spatial relationships of the planes forming the
wedge, the various angular relationships which are required for the

analytical solutions for stability assessment can be readily
determined from the stereonet. The analytical solution for assessing the stability of the wedge, on the other hand, is much more
complex. However, by making certain basic assumptions about the
shear strength of the discontinuities and the groundwater characteristics of the slope, the problem can be simplified. With the simplified version of the analysis a rapid assessment of the stability
conditions may be obtained. Methods of wedge failure analysis are
discussed by John (1968), Calder (1970), Hendron et al (1971), Hoek
et al (1973) to name only a few. The basic approach and methodology
for assessing wedge failure problems are presented in the following.

Simplified Analysis Methods. Several authors have developed wedge
failure analyses which consider that resisting forces are developed
due to friction along planes only. If the effects of cohesion and
water pressure along the planes are not included the analysis is
considerably simplified, because the effects of size or weight of
the failure block are ignored.

1. Markland's Test

   A simple analysis is described by Markland (1972). In this
   approach the failure slope direction and surface friction angle
   may be represented by a great circle and friction circle as
   shown in Fig. 21. Any line of intersection which occurs within
   the shaded area shown in Fig. 21 defines a possible wedge which
   daylights in the slope and which has planes that are steeper
   than the friction angle. Simply speaking, any wedge intersection which lies in the shaded area defines possible unstable
   wedges.

2. Simplified Analytical Solution

   Detailed analytical methods for wedge failure analysis involve
   inclusion of slope geometry, failure geometry, strength properties and water forces. When simplified by ignoring cohesion
   and water pressure, these analyses methods may be expressed
   in terms of one or two equations to determine the factor of
   safety. Analyses of this nature have been presented by Hendron
   et al (1971) Calder (1970) and Hoek and Bray (1977) to name
   only three. Stability charts have been developed for some of
   these simplified analyses.

   As described previously, the simplified analyses methods are
generally used for a preliminary assessment of possible wedge failures. Rigorous analysis methods for specific slope failures or
detailed slope stability analysis and design considerations are given
in the following.

Rigorous Analysis Methods. For back analyses of failures and analyses
of specific wedge type problems it is often justified to carry out a

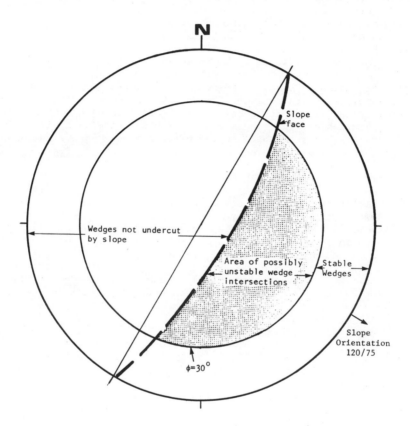

Fig. 21.   Determination of possible unstable wedges using the
Markland test (after Markland, 1972).

detailed wedge analysis which incorporates the effects of cohesion
and water forces along failure planes. These analyses require an
accurate knowledge of wedge geometry, including the volume and
weight of the wedge, orientation and surface area of the planes involved, cohesion and friction properties of the planes and groundwater conditions. Accurate determination of geometric parameters
of a particular wedge comprises a large part of the calculations for
the analyses. A comprehensive description of the basic methods for
rigorous analysis of wedges is given by Hoek et al (1973).

1. Algebraic Solution

   The algebric approach requires resolution of the forces acting
   on the various failure planes, using simple algebraic and
   trigonometric methods. In this case wedge forces are resolved
   into components normal and parallel to the line of intersection.
   These component forces are expressed in terms of the component
   normal forces across each plane. If forces are resolved about
   the line of intersection the factor of safety may be expressed
   as:

   $$F = \frac{C_1 A_1 + C_2 A_2 + N_1 \tan\phi_1 + N_2 \tan\phi_2}{W \sin \psi i}$$

   where N_1 and N_2 are the component forces normal to the failure
   planes, C_1, C_2, ϕ_1 and ϕ_2 are the cohesion and angle of friction
   along the planes, A_1 and A_2 are the area of the planes, W is the
   weight of the wedge and ψi is the plunge of the line of intersection. Addition of water forces and forces on a tension crack
   or other planes further complicates the solution of this problem.

2. Engineering Graphics Solution

   The analytical solution described above may be solved by use
   of scaled drawings and engineering graphics solutions. This
   approach essentially requires completely developed views of all
   planes and lines on the wedge. The magnitude of forces acting
   on each surface is calculated and scaled graphically on the
   diagram. All forces are plotted on a force diagram and the
   resultant forces and factor of safety are calculated graphically.

3. Use of Stereographic Projection

   Analytical solutions may be calculated considerably more easily
   if the basic angular relationship and resolution of forces is
   carried out using stereographic projection techniques. Using
   this method the shear strength along the planes and forces
   acting on the planes involved in the wedge are resolved on a
   stereographic projection, and a factor of safety is calculated

from the magnitude and angular relationship of the resolved
forces.

4. Vector Solution

The analytical solution may also be considerably simplified
if the analysis is carried out using vector machanics. All
forces acting on a wedge are described in terms of vectors.
Vector equations, such as those developed by Bray and Brown
(1976), are used to assess the failure mechanism of a possible
wedge and calculate the factor of safety. The vector solution
is particularly useful for computer applications where a large
number of possible wedges require assessment.

Statistical Analysis of Wedge Failures

It is noteworthy that based on the writers' experience, it can be
said that of the order of 90 to 95 percent of all possible failure
modes in a slope are wedge failures. By virtue of the fact that peak
concentrations of both joint sets and joint subsets are considered
in the analysis, the great circles representing these joint set
concentrations define a whole array of potential wedges for various
shapes and orientations. It is also noteworthy, however, to
recognize that it is the apparent plunge of these wedges and not
the true plunge that ultimately must be considered in rational slope
design.

For design of high walls in many open pits a statistical evaluation of possible wedge failures is required. A unique solution,
combining limit equilibrium and statistical or probability techniques,
has been developed by Piteau et al (1981)(b) for the assessment of
innumerable possible wedge failures in this manner. Using this
method, the orientations of possible wedge failure surfaces are given
a specific number related to the intensity of the joint set which
describes a particular failure plane. Each failure plane is
evaluated with respect to all other failure planes to determine if
a possible wedge failure could develop. Each possible wedge combination is given a number value relating to the specific numbers
related to the percentage concentration of each plane forming the
wedge. Statistical evaluations are made by plotting the relative
(not actual) cumulative percentage or frequency of possible wedges
against apparent dip or plunge of the specific wedges with respect
to the slope orientation being considered. A typical cumulative
frequency plot is shown in Fig. 22. The "S" shaped nature of these
curves can be used to advantage to select an optimum plunge angle for
slope design. Experience by the authors on several projects have
indicated that plunge angles corresponding to a relative frequency
of possible unstable wedges of between 10% and 25% are practical for
design. For example, on one project back analysis of over 300
possible and existing wedge failures were considered to determine
the relative frequency which was most appropriate. Details of the

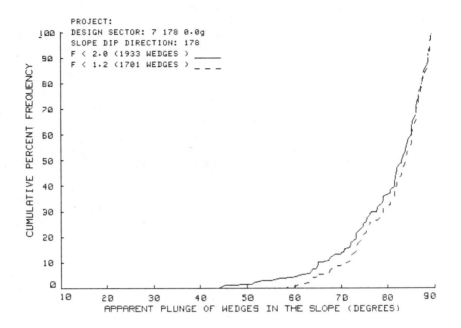

Fig. 22.   Typical plot of cumulative percent frequency of kin-
           ematically possible unstable wedges vs apparent plunge
           of wedges in the slope for wedges with a factor of
           safety less than 2.0 and 1.2.

application of this technique for slope design in several large open
pit mines are presented by Piteau et al (1981)(b).

ROCK MASS FAILURE

The rock mass is defined as the anisotropic discontinuous and
heterogeneous medium made up of intact material and discontinuities
somewhat analogous to partitioned solid bodies. Analysis of failure
mechanisms which involve one or more discontinuities are based on the
concept that the strength of the rock mass is governed primarily by
the strength of the discontinuities and not the strength of intact
rock itself. However, if the intact rock is sufficiently weak, the
rock mass is sufficiently highly fractured and/or the slope is sufficiently high that failure could occur through intact material or
the rock mass itself, assessment of stability must be approached
differently. In this case, failure surfaces could develop through
intact material, the rock mass in general or along a surface which
occurs partly through intact rock and partly along discontinuities
somewhat more easily. In this case the strength of the rock mass

could be considered to be made up of a component due to the strength
of the intact rock and a component due to the shear strength of
discontinuities. A reliable estimate of rock mass shear strength may
often be obtained from back analysis of slope failures.

Rotational Shear Failure

   If slopes occur in a relatively weak material which has uniform
strength properties throughout, failure could possibly occur along
a circular or irregular failure surface. This type of failure has
been applied to the analysis of slopes excavated in soils and
solution of these types of problems is based primarily on soil
mechanics principles. It is important to recognize that this type
of failure mechanism is only valid for homogeneous materials with
uniform strength properties such as uniform soils, unjointed rock
masses or very highly jointed and/or very weak altered rock masses
(see Fig. 20).

   Rotational sliding may occur generally by a process of yielding
and then rotation. When soil and rock masses are sufficiently plastic
and ductile to yield without excessive loss of strength at locations
of high stress, stresses become more evenly distributed along the
potential sliding surface. Because of this yielding, the shear
strength of the material is maintained even though some zones have
been overstressed. Failure will occur by shearing along a circular
surface, as shown in Fig. 23(a). In material where a substantial
part of the strength is derived from internal friction, the curved
shear surface will pass through the toe.

Conditions for Rotational Failure. The conditions under which
rotational sliding occurs arise when the individual particles of
soil or rock mass are very small compared with the size of the
slope and when these particles are not interlocked, so that a
reasonably continuous failure surface may form. The total shear
force causing sliding is assumed to be equal to or greater than the
total shear strength along the surface of sliding. It is further
assumed that the strength parameters in the material are the same in
all directions.

   The strength of the material is generally based on the Mohr-
Coulomb Failure Criterion:

$$\tau = c' + (\sigma - \mu) \tan \phi'$$

where τ is the shear strength, c' is the apparent effective cohesion
of the material, σ is the normal stress on the sliding surface, μ
is the pore water pressure on the sliding surface and ϕ' is the
apparent effective residual angle of friction of the material.

   Information to establish that sliding would be by rotational shear
is provided either by the behaviour of previous slopes or by

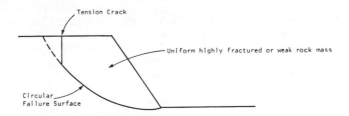

a) Rotational Shear Failure.

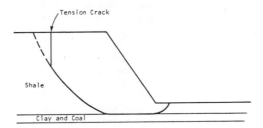

b) Failure along an irregular failure surface which may be
controlled in part by lithology and/or a major structural
feature.

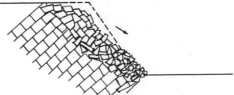

c) Block Flow which involves general breakdown of rock mass
due to progressive transfer of stresses and failure within
the slope.

Fig. 23. Typical Mechanisms of rock mass failure.

structural geology investigations and rock mechanics tests. Rotational shear failure could be anticipated in rock masses in which the ductility ratio (i.e. the ratio of uniaxial compressive strength at strains greater than at maximum strength to the maximum strength) is greater than about 0.6. There is clearly a relationship between slope height, slope angle, shear strength of the rock mass and potential for rotational failure.

Even strong rock masses could possibly fail by rotational failure if the slopes are high enough that shear strength of the rock mass is exceeded.

Methods of Analysis. The most common method of analysis of rotational sliding is to calculate the factor of safety against failure based on the static equilibrium of the failing block along a circular failure surface. A circular failure surface is used to simplify the moment equilibrium calculations, thus providing a system for comparing a suite of possible failure surfaces. The factor of safety, F, is calculated as:

$$F = \frac{\text{total shear resistance along the slip surface}}{\text{total shear force along the slip surface}}$$

The factor of safety is generally computed by using moment equilibrium of the forces on a slip circle about the centre of the circle.

Simplified analyses may be carried out using total stresses in which the internal friction angle is assumed to be 0 and the strength of the mass is attributable to the undrained strength, Cu only. By assuming the slope fails as an intact block on a circular failure surface, a conservative lower bound solution for undrained strength conditions is determined. Stability charts based on this approach were originally developed by Taylor (1948). These charts are suitable for checking the sensitivity of the factor of safety of a slope over a wide range of conditions. Because they are based on undrained strength only, they are of limited use for design of rock slopes.

For specific cases where pore pressures are not evenly distributed throughout the slope or the strength characteristics along the failure surface are not uniform, an effective stress analysis using the method of slices is generally used for rotational failure analysis. In this method the slope is divided into 5 to 11 slices and the various forces acting on each slice are calculated and summed using moment equilibrium equations about the centre of the slip circle. These calculations are carried out for several slip circles to determine the slip circle with the lowest factor of safety which controls the stability of the slope. The process is iterative and time consuming, although good computer programs are available for the analyses. The two most common method of slices analyses are the Conventional or Fellinius Method and the Simplified Bishop

Fig. 24.   Large rock mass failure at Anvil Mine in the Yukon.
            Failure surface is irregular and partly controlled
            by bedding faults which dip at about 30°.

Method. The Simplified Bishop Method generally gives factors of
safety which are slightly higher than the Conventional Method and
is generally accepted as being more accurate. Details of the various
procedures for analysis by the method of slices are given in several
advanced soil mechanics text books and need not be discussed further.

Shear Failure Along an Irregular Failure Surface

    For many soil or rock masses it is often more realistic to
consider an irregular or non-circular failure surface, such as is
shown in Fig. 23(b) or Fig. 24. Solution of this type of problem
is based on the Janbu analysis method, which applies force equilibrium equations to a series of vertical slices in the slope. The
Janbu analysis is more flexible than the slip circle analysis in that
it allows for irregular failure surfaces which may be constructed
according to the engineering geological conditions. Iterations of
the irregular failure surface may be carried out to assess the
sensitivity of variation in factor of safety.

Block Flow

    In some steep slopes instability may develop by breakdown of the
rock substance. In such cases geologic structural conditions are not
amenable for failure to develop totally along discontinuities, and
also the rock mass is not sufficiently ductile to permit rotational
sliding. This type of behaviour is particularly prevalent in brittle
rock masses with low ductility ratios (i.e. less than or equal to

0.6). Consequently, crushing of the rock at points of highest stress
will occur. After local crushing develops, the load is transferred
to adjacent zones, which in turn are subjected to excessive stress
leading to further crushing. This progressive action, which can be
observed when slopes are deforming before major movement develops,
continues until a general breakdown of the rock mass occurs with
a flow of broken rock as shown diagrammatically in Fig. 23(c).

Zones of high stress in the slope can arise from various causes
and can be represented by stress trajectories. In homogeneous rock
masses, deflection of stress around the toe of the slope results
in a concentration in this area, somewhat similar to the notch effect
in structural members. This stress concentration may result in
crushing, leading to instability. Analysis of block flow mechanisms
is generally approached in terms of the stress-strain relationships
of individual units of the slope. This technique involves finite
element analysis methods, using computer techniques such as those
described by Coates et al (1977).

SMALL FAILURES, ROCKFALLS AND RAVELLING

In many open pit mining and rock excavation situations the
concerted effort in slope design is given primarily to preventing
major failures. Although the overall slope may be stable, numerous
small failures or rockfalls may develop which could have a significant effect on cleanup and safety. Small failures or rockfalls
are often particularly pronounced on benched slopes. This is
particularily the case where numerous joints daylight on the bench,
or where blasting has damaged the rock with a concomitant loss of
cohesion. In addition, ravelling of loose or weak material may often
occur in fault zones, weak or altered rock, slaking shales etc. which
requires special design considerations.

In most open pit mines, intermediate slopes between haulroads
are seldom steeper than 55° to 60° due to geological conditions,
operating constraints, local mining regulations or other factors.
In many cases, particularly those in hard rock with favourable
geological characteristics where steep slopes are feasible, the
possible occurrence of small failures, rockfalls or ravelling may
govern the stability and operational safety of the slope to a large
extent. In most cases the nature and extent of these small failures
are such, however, that control, rather than prevention, is the most
feasible approach. Hence, the bench geometry of the slope must be
examined in considerable detail and the benches should be designed
to accommodate the various types of failures for a typical bench
geometry as shown in Fig. 25.

Control of Small Failures With Bench Design

In hard rock, failures involving discontinuous joints are generally

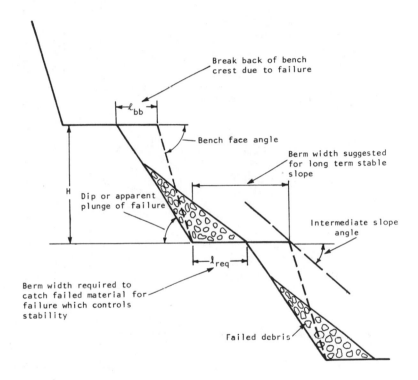

Fig. 25. Typical bench geometry parameters showing interrelationship of bench geometry and geometry of failures on benches.

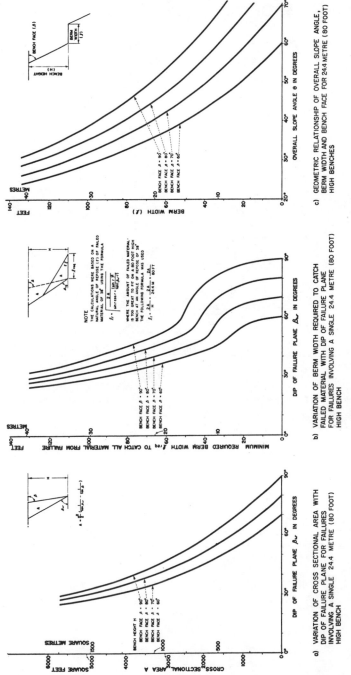

Fig. 26. Typical bench design charts showing interrelationship of slope geometry and failure geometry for 24.4 metre (80 foot) high benches. (after Martin and Piteau, 1977).

small in comparison to the size of the slope. If the slope is
benched many of these small failures may become unstable and fail on
individual benches. In most cases these failures would be caught
on berms and, therefore, controlled on the slope. A simple graphical
technique for design of benches to control failures has been presented by Martin and Piteau (1977). This technique relates the dip
or plunge of a possible bench failure to the possible volume of
failed debris for the particular bench height. The berm width
required to catch the failed debris is calculated assuming the debris
sits at a natural angle of repose on the benches as shown in Fig.
25. Hence, the bench height, bench face angle and berm width required to catch a particular volume of failed debris is used to
determine the slope geometry and acceptable slope angle to accommodate bench failures on the slope. A series of charts may be
developed as shown in Fig. 26 which enable a rational slope design
to be prepared for the particular slope or mining situation being
considered.

Application of Bench Design Techniques to Slope Design. Application
of the technique at Cassiar Mine in northern British Columbia has
been described by Piteau and Martin (1977) for the upper 700-ft-high
section in argillite of an 1,100-ft. high hanging wall slope.
Numerous potential wedge failures, formed by the combination of two
joint sets, controlled slope geometry and the stability of the
benches (see Fig. 27).

Fig. 27. View of hanging wall slope at Cassiar Mine showing
          numerous small wedge failures which control slope
          geometry.

The graphical technique was used in combination with probability theory to determine the optimum bench geometry required to contain a reasonable number of wedge failures on berms. The plunge of the wedges varied mathematically as a normal distribution about the mean plunge value. Probability techniques were applied to assess the likelihood of unstable wedges occurring whose plunge would be greater than that determined by probability criteria. The probability of an unstable wedge spilling over a berm, for a particular overall slope angle, was determined as shown in Fig. 28. Mine management was then in a position to evaluate alternative slope designs.

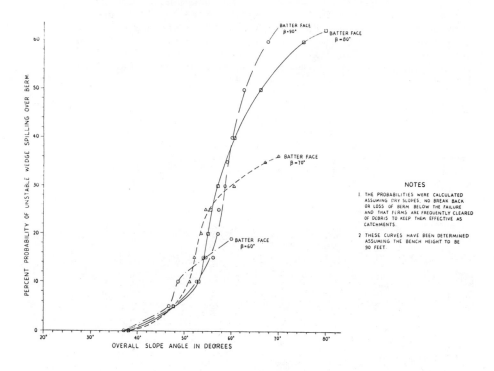

Fig. 28.   Probability of possible wedge failures formed by Joint Sets A and C of spilling over berms for various overall slope angles (after Piteau and Martin, 1977).

Rockfalls

In many situations, as mentioned above, rockfalls may develop due to the highly fractured nature of the rock, blasting damage, frost action, etc. Prediction of the nature of rockfall hazards is difficult, although the effects of rockfall could have serious consequences, particularly on access roads or near critical installations. Most work concerning rockfalls has been carried out in relation to design of highway slopes.

Ritchie (1963) used emperical techniques to determine if rockfalls would drop vertically, bounce or roll down a slope. The slope height, slope angle and mode of failure have been used to prepare a set of design criteria for width of fall out area and possible depth of ditches as summarized in Fig. 29. This work is based mainly on rock- fall trials on slopes of varying heights.

A rockfall model by Piteau and Clayton (1977) based on computer techniques was developed to simulate several hundred rockfalls for a slope of specified geometry. A coefficient of restitution is used for the rockfall in the bouncing mode, and increased friction is used for the rolling mode to determine the paths of the rockfalls for different input velocities and heights. The slope profile is divided into straight segments or slope cells (Fig. 30), which are numbered consecutively from the top to the bottom of the slope. Rockfalls are introduced at different locations above the slope as desired. Probability factors can be assigned to these locations to recognize areas that are either more likely or less likely to be the source of the rockfall. Catch walls, at various positions and heights can be assumed in the model to evaluate their effectiveness. In the same manner, the effect of varying the slope geometry by including ditches, benches, and berms can be evaluated.

Although obviously not applicable as a general analysis technique in open pit mines, this approach might be considered for specific rockfall design problems. On long high footwalls, for example, where the possible rockfall distribution on the slope may be required to assess the location of catch fences, catch berms or interception ditches, such an approach may be worthy of consideration. Details relating to the rockfall model theory are described by Piteau et al (1979).

Ravelling

Ravelling is best described as the degradation of a slope due to gradual breakdown of the rock material or the rock mass. Good examples of ravelling most often occur in fault zones and associated highly fractured zones, weak rock and altered rock. Ravelling is often aided by exposure of the slope material to weathering or frost action. Because ravelling debris is small, failures of this nature are often of nuisance value, even though they do not significantly affect slope stability. In some cases, such as for ravelling of slaking shale which overlies a hard sandstone, such ravelling may severely undercut and may result in failure of large blocks (see Fig. 31). Ravelling problems are generally treated by slope flatten- ing or remedial measures. Use of shotcrete and/or wire mesh and bolts on the slope are often used to control ravelling problems. Catch ditches and walls at the toe of the slope help to contain ravelling material and protect haulroads or working areas. Slope protection and stabilization techniques to remedy rockfall and

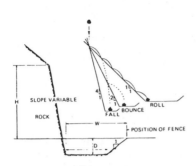

Fig. 29 Path of rock trajectory for various slope angles and design criteria for shaped ditches (after Ritchie, 1963)

Rock Slope		Fallout Area Width (m)	Ditch Depth (m)
Angle	Height (m)		
Near vertical	5 to 10	3.7	1.0
	10 to 20	4.6	1.2
	>20	6.1	1.2
0.25 or 0.3:1	5 to 10	3.7	1.0
	10 to 20	4.6	1.2
	20 to 30	6.1	1.8[a]
	>30	7.6	1.8[a]
0.5:1	5 to 10	3.7	1.2
	10 to 20	4.6	1.8[a]
	20 to 30	6.1	1.8[a]
	>30	7.6	2.7[a]
0.75:1	0 to 10	3.7	1.0
	10 to 20	4.6	1.2
	>20	4.6	1.8[a]
1:1	0 to 10	3.7	1.0
	10 to 20	3.7	1.5[a]
	>20	4.6	1.8[a]

[a]May be 1.2 m if catch fence is used.

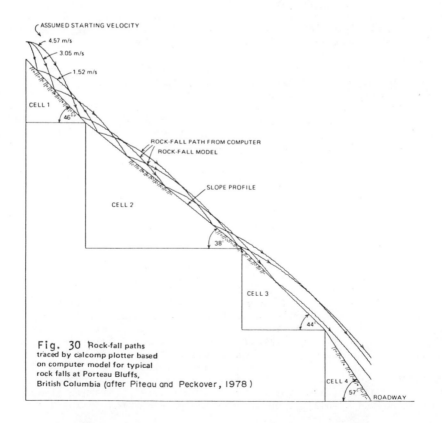

Fig. 30 Rock-fall paths traced by calcomp plotter based on computer model for typical rock falls at Porteau Bluffs, British Columbia (after Piteau and Peckover, 1978)

Fig. 31.   Small rockfalls and ravelling caused by weathering of
weak shale underlying hard sandstone in a highway slope
in Tennessee.

ravelling problems are discussed by Piteau and Peckover (1978).

CONCLUSIONS

Mechanisms of slope failure are varied and complex. In terms
of design of slopes, simple two dimensional and three dimensional
failure mechanisms are generally examined to assess the potential
for instability. Because a particular slope may have numerous
possible failure modes, simple analysis methods are used to identify
the main failure modes and most importantly those possible failure
modes which control stability. Detailed and complex analyses are
normally only carried out for the few critical failure modes which
control slope stability.

In terms of assessing pre-existing slope failures, more detailed
information concerning the geometry and engineering geology parameters is generally available. Consequently, detailed or rigorous
analysis methods may be applied to back analyze the specific failure
cases. In this way a better appreciation of the engineering
behaviour and strength parameters of the rock mass may be obtained
for design of future slopes in similar conditions.

In assessing the possible mechanisms of failure of a slope,
consideration must be given to deep seated failure involving the
whole slope or large sections of the slope, as well as failure of
local individual benches. Optimization of slope geometry to control
both these types of failure is essential to prepare a rational slope

design.

Assessment of possible failure mechanisms requires an accurate knowledge of the engineering geology aspects and parameters relevant to rock mechanics assessment of the rock mass. Information concerning rock type, structural discontinuities, shear strength of both intact rock and discontinuities and hydrogeological conditions is essential. Also effects of anomalous factors such as blasting, seismic vibrations, surcharge loading, flooding etc. must be assessed as these factors often contribute significantly to instability. All geotechnical information must be assessed with respect to the slope orientation and the slope geometry before stability analysis and design can be carried out on a rational basis.

ACKNOWLEDGEMENTS

The authors wish to dedicate this paper to Dr. Donald F. Coates, the former Director General of CANMET, Department of Energy, Mines and Resources in Canada and the author of numerous outstanding publications, including the book "Rock Mechanics Principles". The authors acknowledge with appreciation and thanks the outstanding contributions Dr. Coates has made towards the understanding and application of mechanics in the analysis of both surface and underground openings in rock. The authors also acknowledge with thanks the staffs of the various mines and workers in the field who are too numerous to mention for their contributions and assistance in one way or another in the development of the contents of this paper.

REFERENCES

Bray, J.W. and Brown, E.T., 1976, "A Short Solution for the
    Stability of a Rock Slope Containing a Tetrahedral Wedge,"
    Int. J. Rock Mech Min Sc & Geomech. Abst., Vol. 13 pp 227-229.

Calder, P.N., 1970, "Slope Stability in Jointed Rock," CIM Bulletin,
    May.

Coates, D.F. ed, 1977, "Pit Slope Manual Chapter 5-Design," Canmet
    Report 77-5, 126p, March.

Goodman, R.E. and Bray, J.W., 1976, "Toppling of Rock Slopes,"
    Rock Engineering for Foundations and Slopes, ASCE Specialty
    Conference, Boulder Colorado, August 1976, Vol II, pp 201-233.

Hendron, A.J. Cording, E.J. and Aiyer, A.K., 1971, "Analytical
    and Graphical Methods for Analysis of Slopes in Rock Masses,"
    U.S. Army Engineer Explosive Excavation Research Office, Technical
    Report No. 36., 126 p.

Hocking, G. 1976, "A Method for Distinguishing Between Single and Double Plane Sliding of Tetrahedral Wedges," Int. J. Rock Mech. Min. Sci. & Geomech. Abstracts. Vol. 13, pp 225-226, Pergamon.

Hoek, E., Bray, J.W. and Boyd, J.M., 1973, "The Stability of a Rock Slope Containing a Wedge Resting on Two Intersecting Discontinuities," Quart Jour of Eng. Geol., Vol. 6, No. 1, January, 1973.

Hoek, E. and Bray, J.W., 1977, "Rock Slope Engineering," 2nd edition Institution of Mining and Metallurgy, London.

Jennings, J.E., 1970, "A Mathematical Theory for the Calculation of the Stability of Slopes in Open Cast Mines," Planning Open Pit Mines Van Rensburg ed. SAIMM Open Pit Mining Symposium, Johannesburg, 1970, Balkema. pp 87-102.

John, K.W., 1968, "Graphical Stability Analysis of Slopes in Jointed Rock," Jour. Soil Mech. and Found. Div. ASCE, Vol. 94, SM2 pp 497-526.

Markland, J.T., 1972, "A Useful Technique for Estimating the Stability of Rock Slopes when the Rigid Wedge Sliding Type of Failure ie Expected," Imperial College Rock Mechanics Research, Report No. 19, 1972, 10 pages.

Martin, D.C. and Piteau, D.R., 1976, "Application of Waviness of Structural Discontinuities to Slope Design," Proc. 29th Canadian Geotechnical Conference, Vancouver, B.C., October, pp VIII-I to VIII-17.

Martin, D.C. and Piteau, D.R., 1977, "Select Berm Width to Control Local Failures," Engineering and Mining Journal, McGraw Hill, June.

Martin, D.C., 1978, "The Influence of Fabric Geometry and Fabric History on the Stability of Rock Slopes," M.Sc. Dissertation Imperial College, London, Unpublished, 99p.

Martin, D.C. and Piteau, D.R., 1981, "Design Charts for Design of Benches and Interramp Slopes in Open Pit Mines" (in preparation)

Piteau, D.R., McLeod, B.C., Parkes, D.R. and Lou, J.R., 1976, "Overturning Rock Slope Failure at Hell's Gate Bluffs" in Geology and Mechanics of Rock Slides and Avalanches, B. Voight, Ed., Elsevier, Amsterdam, Chapter 10.

Piteau, D.R. and Clayton, R. 1977, "Discussion of Paper Computerized Design of Rock Slopes Using Interactive Graphics for Input and Output of Geometric Data," Proc. of 16th Symposium on Rock Mechanics, Minneapolis, 1975, pp. 62-63.

Piteau, D.R. and Martin, D.C., 1977, "Slope Stability Analysis and Design Based on Probability Techniques at Cassiar Mine," CIM Bulletin, March.

Piteau, D.R. and Peckover, F.L., 1978, "Engineering of Rock Slopes," Chapter 9, Landslides: Analysis and Control. Special Report 176, Transportation Research Board, Washington, D.C.

Piteau, D.R., Martin, D.C. and Stewart, A.F., 1979, "Rock Slope Engineering Reference Manual," Part A to Part H incl. Report No. FHWA-TS-79-208 NTIS Accession Numbers PB80 103-294, -310, -328, -336, -349, -351, -369, -377. Course reference manuals.

Piteau, D.R. and Martin, D.C., 1979, "Footwall Stability Study for a Large Canadian Coal Mine," Unpublished report.

Piteau, D.R., Stewart, A.F. and Martin, D.C., 1981 (a) "Design Examples of Open Pit Slopes Susceptible to Toppling," Third International Conference on Stability in Surface Mining, Vancouver, B.C. June 1-3, 1981.

Piteau, D.R., Stewart, A.F., Martin, D.C. and Trenholme, B.S., 1981 (b), "A Combined Limit Equilibrium and Statistical Analysis Technique for Design of Rock Slopes," AEG Bulletin (in preparation).

Ritchie, A.M., 1963, "Evaluation of Rockfall and Its Control," Highway Research Board Highway Research Record, 17, pp 13-28.

Taylor, D.W., 1948, "Fundamentals of Soil Mechanics," John Wiley, New York, 700 p.

Walton, G. and Taylor, K.K., 1977, "Likely Constraints on the Stability of Excavated Slopes Due to Underground Coal Workings," Conference on Rock Engineering Newcastle-Upon-Tyne, England, April, 1977, pp 329 to 349.

Question

Would you analyze deep open pit slope stability mechanisms the same way you would for shallow open pit wall or bench wall.

Answer

The basic approach to design of high slopes is to assess the possibilty of deep seated failure of the entire slope or large sections of the slope, as well as the effects of numerous possible bench failures. Assessment of deep seated failure requires an assessment of failure mechanisms, i.e. whether failure occurs on throughgoing major discontinuities or by failure of the rock mass. In addition to deep seated failure, design of slopes must be adequate to control failures of individual benches to provide a safe slope with reliable access. It is clear that, in many cases, design of benches may control the overall slope design. Optimization of slope design indicated from deep seated failure analyses as well as bench design is essential therefore, to prepare a rational slope design. In some cases deep seated failure criteria will be found to control the slope, whereas in other cases it will be found that it is the optimum bench geometry which dictates what the inter-ramp slope angles, and hence what the overall slope angle, will be.

Question

What kind of extra load/stress would you set at toe of deep/high wall - how will it influence the failure mechanisms - could you elaborate on this in contrast to shallow wall.

Answer

Load and stress relationships within the slope effect the rock mass failure or deep seated failure mechanism as well as the bench failure consideration. These parameters depend on the slope geometry and stress strain relationships in the slope and may require modelling using physical models or stress analysis models to fully evaluate slope conditions.

Question

You touched briefly on rock mass strength assessment which is probably one of the most difficult geotechnical parameters to assess or predict. You mentioned back analysis as the best method, but of course such failures are not always available. Would you discuss the empirical methods available for assessing rock mass strength.

Answer

If it is anticipated that the rock slope will behave as a uniform "soil-like" material, an assessment of rock mass strength is required. Probably the most reliable method of estimating rock mass strength is by documentation and back analysis of existing slopes; this should include both stable as well as failed slopes. This approach is much more applicable in an area of the open pit which is excavated in a series of phases, whereby the results and experience gained on each phase may be applied to design of succeeding phases. Such back analysis is particular appropriate where careful documentation of rock mass conditions from surface mapping and drill core information can be applied to the behavior of actual mined slopes. Extrapolation of known conditions to future behaviour of mining slopes is the ultimate objective of any detailed geotechnical investigation program. If the rock mass model and related parameters are fully considered at the outset of a project, and continually refined as mining proceeds, the prospects of achieving a reasonably rational slope design, whether for intermediate or final walls, will increase relative to the efforts expended in this aspect of the work.

Question

How do you consider the scatter in orientation of planes of weakness potentially forming wedges or other unstable rock masses.

Answer

The key to assessing scatter of joint orientations is to determine the scatter due to measurement and scatter due to natural variation of joints. If a particular joint population is defined in terms of a mean orientation and range of orientation of the joint sets making up that population, the probability of obtaining a structure with a different orientation relative to the value of the mean which has been determined is a relatively simple statistical excerise. Based on these probabilities various combinations of orientations may be evaluated and appropriate analyses carried out to determine their significance in terms of rational slope design. Hence it is possible to evaluate the factor of safety corresponding to the probability of occurrence for a failure with a certain geologic structural geometry. Either a limit equilibrium analysis or probability analysis may be developed based on the statistical information relating to geologic structural geometry and geometry of kinematically possible failures.

Question

In bench design, after calculating the volume of failure mass in the bench, what correction is used for the failed volume anticipated on the bench below and how is the correction arrived at.

Answer

A correction factor for bulking may be applied to the volume of failed material. In general, an angle of repose of failed material may be selected based on field experience. The bench design charts may be developed for any bulking factor or any angle of repose of failed material. It must be appreciated that debris will be removed during excavation. Hence all of the material assumed to be available to fill the bench may in fact, only partly exist. In this regard the analysis using these design charts can prove to be somewhat conservative. For this reason we have generally not used a bulking factor in the analysis, thus tending to counter the conservative aspects of the analysis and making the results more realistic in a practical sense.

Question

In your rock fall analysis you mention using a coefficient of restitution, have you any comments on how this coefficient should be determined for use in analyses.

Answer

Coefficient of restitution varies for hard rock as opposed to talus
or scree. This was determined experimentally with field rockfall
trials and a movie camera. In this way the rockfall projectile paths
could be determined for a specific slope geometry (Porteau Bluffs,
B.C.). The same slope was then assessed using the mathematical
rockfall model and by fitting the rockfall model to actual performance in the field. The coefficient of restitution values which apply
to rockfall was determined by simple comparative analysis methods.
These values are probably site specific depending on rock type,
degree of weathering and to some extent on size of rockfall.
Assuming the coefficient of restitution for the fully elastic system
is 1.0, these experiments have indicated that a coefficient of
restitution of hard rock bouncing on hard rock should be about
0.7 to 0.8. Where rockfall must bounce or roll through slide debris,
talus, gravel terraces, etc. this value probably should reduce to
about 0.5. An important point to remember is that large rockfall
blocks essentially not only bounce, but they also can have a ploughing action in loose overlying material which introduces another
aspect altogether, namely frictional resistance. In a recent analysis where potential rockfall blocks bouncing down a slope about one
mile in length at the Hope Slide, B.C. were evaluated to determine
whether such blocks would reach the highway, both coefficient of
restitution and frictional resistance was used in the analysis.

Chapter 7

SHEAR STRENGTH INVESTIGATIONS FOR SURFACE MINING

Nick Barton

Terra Tek, Inc.
Salt Lake City, Utah

ABSTRACT

Simple methods for estimating the shear strength of rock joints
and waste rock are reviewed. For the case of rock joints, the methods are based on a quantitative characterization of the joint roughness and the joint wall strength. Size-effects are found to reduce
the peak strength of large joint samples to values below the ultimate
or so-called "residual" values measured in the laboratory. Tilt
tests and surface profiling on natural size blocks within the jointed
rock mass are recommended for obtaining scale-free properties. The
joint parameters obtained can be used to model complete strength-dis-
placement-dilation behavior if this level of input is required. Large
scale tilt tests can be performed with advantage on both rock joints
and waste rock. The behavior of these two materials is surprisingly
similar. Both are influenced by the size-effects on the compression
strength of the rock, and both have similar log-linear relationships
between effective normal stress and the peak drained friction angles.
The resulting high values of friction near the toe or close to a
slope face in either material can be misleading.

INTRODUCTION

It is now known with reasonable certainty that tests on small
samples of rock produce artificially high values of strength. In the
past, arguments have been put forward to explain size-effects by
changed stress distributions, changed machine stiffness, etc. Such
arguments cannot be invoked to explain scale effects observed on
joints. A simple but convincing demonstration is the tilt test.
Tilt angles measured during self-weight gravity sliding tests of a
large slab of jointed rock are found to be many degrees less than

the steep tilt angles measured when the same jointed slab is sawn into small samples and these are tilted individually or as an assembled "rock-mass". Tests of this kind performed by Barton and Choubey (1977) and Bandis et al. (1981) indicated tilt angles increased from 59° to 69°, and from 47° to 62° respectively.

Thought of objectively, it is remarkable that rock engineers have entertained the possibility that "laboratory-size" samples could ever represent the characteristics of a failure surface perhaps one thousand times larger. Similar optimism has been displayed in the design of large rockfill dams. The pioneering work of Pratt et al. (1974) in investigating scale effects on rock joints, and of Marachi et al. (1969) in the field of rockfill, have emphasized the importance of large scale tests, and the importance of extra conservatism and adequate safety factors in the absence of such tests.

In their general report at the Denver Rock Mechanics Congress, Hoek and Londe (1974) suggested that when a very large structure such as an arch dam or major pit slope is being designed for long term stability, the design should be based on zero cohesion and residual friction, "which can be determined in small scale laboratory tests".

This philosophy probably results in more than adequate conservatism for these major slopes. However, recent work has shown that the peak strength of a large joint sample may be lower than the "residual" or ultimate strength measured after large displacement of a small sample. It is very difficult to reach the true residual strength of a non-planar joint surface since dilation persists during surprisingly large displacements.

QUANTITATIVE JOINT DESCRIPTION

The requirement for careful characterization of individual joint sets may be elevated to a high priority task, if a preliminary structural analysis indicates the potential for failure of a planned or existing pit slope. Since the mechanics of slope failure are treated elsewhere, this article will be directed towards the estimation of appropriate input data concerning the shear strength, for use in some form of stability analysis. The methods proposed are of a simple, practical nature, but they can be used to produce sophisticated strength-deformation formulations of joint behavior, if this level of input is required.

A simple though quite complete method of characterizing the shear behavior of rock joints was developed some years ago (Barton, 1973). It consists of three components: ϕ_b, JRC and JCS. A basic or residual friction angle (ϕ_b or ϕ_r) for flat non-dilatant surfaces in fresh or weathered rock, respectively, forms the limiting value of shear strength. To this is added a roughness component (i). This is normal stress dependent and varies with the magnitude of the joint wall compressive strength (JCS), and with the joint roughness coefficient

(JRC). The latter varies from about 0 to 20 for smooth to very rough surfaces respectively. The peak drained angle of friction (ϕ') at any given effective normal stress (σ'_n) is expressed as follows:

$$\phi' = \phi_r + i = JRC \log(JCS/\sigma'_n) + \phi_r \tag{1}$$

Example

$\phi_r = 25°$, JRC = 10, JCS = 100 MPa, $\sigma_n' = 1$ MPa

Equation 1 gives $\phi' = 45°$.

Examples of the strength envelopes generated with JRC values of 5, 10 and 20 are illustrated in Figure 1. The compression strength of the joint walls (JCS) has increased influence on the shear strength as the joint roughness increases. Values of JCS and its variation with weathering are measured with the Schmidt (L) hammer. Experimental details are given by Barton and Choubey (1977).

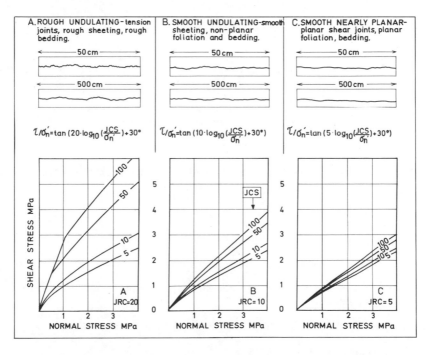

Figure 1. Method of estimating the peak shear strength of rock joints, based on the joint roughness coefficient JRC (20, 10 or 5), and on the joint wall compression strength JCS (100, 50, 10 or 5 MPa), after Barton (1973)(1 MPa = 145 psi).

The residual friction angle (ϕ_r) of weathered joints is very difficult to determine experimentally due to the large displacements required, particularly if only small joint samples are available. A simple empirical approach has been developed as shown below.

$$\phi_r = (\phi_b - 20°) + 20 \ r_1/r_2 \tag{2}$$

where ϕ_b = basic (minimum) friction angle of flat unweathered rock surfaces (obtained from tilt tests on sawn blocks, or from triple core tilt tests - see Figure 2)

r_1 = Schmidt rebound on saturated, weathered joint walls

r_2 = Schmidt rebound on dry unweathered rock surfaces (i.e., saw cuts, fresh fracture surfaces, etc.)

Example:

$\phi_b = 30°$, $r_1 = 30$, $r_2 = 40$

Equation 2 gives $\phi_r = 25°$

The joint roughness coefficient (JRC) can be estimated in several different ways. For example, Barton and Choubey (1977) show a set of 10 increasingly rough joint profiles measured on 10 cm long specimens, which can be physically compared with profiles measured on other joints. However, a more reliable method of determining JRC is by conducting tilt tests on jointed core, or on jointed blocks extracted from existing slopes, as illustrated in Figures 2 and 3.

The value of JRC is back-calculated directly from the tilt test by rearrangement of the peak strength equation:

$$JRC = \frac{\alpha° - \phi_r}{\log(JCS/\sigma_{no}')} \tag{3}$$

where $\alpha°$ = tilt angle when sliding occurs ($\alpha° = \arctan \tau/\sigma_{no}' = \phi'$)

σ_{no}' = effective normal stress acting across joint when sliding occurs

Example: $\alpha = 75°$, $\phi_r = 25°$, JCS = 100 MPa, $\sigma_{no}' = 0.001$ MPa

$JRC = (75° - 25°)/5 = 10$

The values of JRC, JCS and ϕ_r are used to generate peak shear strength envelopes over the required range of stress. The table of

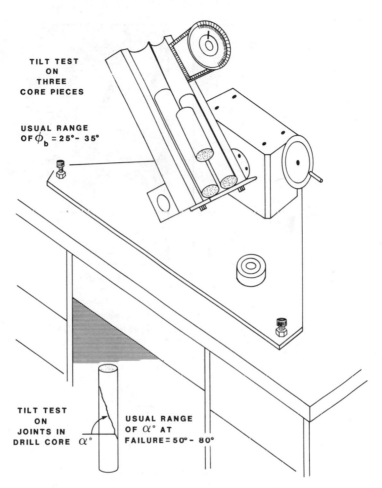

TILT TEST
ON
THREE
CORE PIECES

USUAL RANGE
OF ϕ_b = 25° - 35°

TILT TEST
ON
JOINTS IN
DRILL CORE $\alpha°$

USUAL RANGE
OF $\alpha°$ AT
FAILURE = 50° - 80°

Figure 2. Tilt tests can be used to measure ϕ_b (of flat surfaces) and to measure the friction angle of joints intersecting drill core. These low stress tests are readily extrapolated to design stress levels.

values below indicates how the value of ϕ' varies inversely with the log of effective normal stress. This is a fundamental result for rock joints, rockfill, gravel, etc. (Barton and Kjaernsli, 1981).

Example: JRC = 10, JCS = 100 MPa, ϕ_r = 25°:

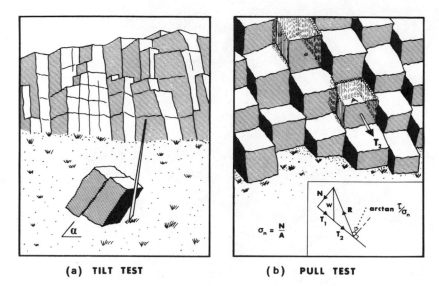

(a) TILT TEST   **(b)** PULL TEST

Figure 3. Tilt tests (or pull tests) can be performed relatively
inexpensively on large jointed blocks of natural size. Only
gravity loading is required, thereby removing the need for
heavy jacking equipment. Test results are readily extrapolated to design stress levels.

	$\sigma_n{}'$	ϕ'
Approx. lab tilt test	0.001	75°
Approx. field tilt test {	0.01	65°
	0.1	55°
Approx. design {	1.0	45°
loading	10.0	35°

SIZE-DEPENDENT JOINT PROPERTIES

Large scale shear tests of joints in quartz diorite (Pratt, et al.,
1974) and a comprehensive series of tests performed by Bandis (1980)
have indicated that larger shear displacements are required to mobilize peak strength as the length of joint sample is increased. This
means that larger but less steeply inclined asperities tend to con-

trol peak strength as the length of sample is increased. Due to the change of significant asperity size and inclination angle, the change of sample size reduces both the dilation component (d_n) and the asperity failure component (S_a), shown in Figure 4. However, the sample size does not apparently affect the magnitude of ϕ_r or ϕ_b, but it does affect the shear displacement needed to reach these values.

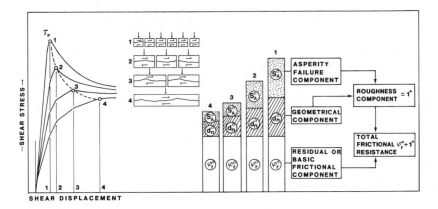

Figure 4. Peak strength, "residual strength", and the shear displace-
         ments needed to mobilize them are all dependent on the
         length of joint tested, after Bandis et al. (1981).

In practice both JRC and JCS are lower when the size of joint sample is increased. A useful method of allowing for this double scale effect has recently been developed as a result of numerous shear tests on different sizes of model joint replicas, as shown in Fig. 5.

Example: <u>Laboratory test</u>, L_0 = 15 cm, JRC = 10, JCS = 100 MPa,

<u>In situ test</u>, L_n = 90 cm, JRC $\cong$ 7, JCS $\cong$ 50 MPa,
(of same joint)

Taking our earlier example with σ_n' = 1 MPa, the laboratory value of ϕ' of 45° would reduce to approximately 37° if measured on a 90 cm long joint sample in situ. The potential effect of sample size on slope stability is too large to be ignored, unless joints are un-usually planar (low JRC). Planar joints have many of the character-istics of a residual surface, and sample size appears to have only a minor influence on strength.

Surface Profiling

Shear box tests or tilt tests performed on small samples of a joint, such as those obtained in carefully preserved drillcore (Figure 2) may not produce reliable strength data, even after approx-

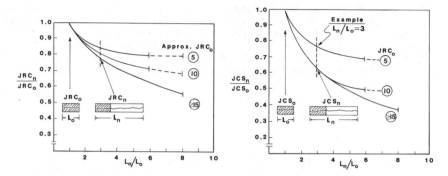

Figure 5. Approximate method for extrapolating the results of small scale laboratory shear tests to in situ scale, after Bandis et al. (1981).

imate correction for the scale effects on JRC and JCS. Joints can have considerable roughness on a small scale, yet be rather planar over a length of meters. Conversely, a joint may be rather smooth on the scale of a drillcore sample, but quite undulating over a length of meters.

Where possible, attempts should therefore be made to sample large scale exposures of the relevant joint set, using a simple straight edge (taut wire) and offset method. Values of maximum amplitude (a) measured over a sample length (L) of 1 meter and up to several meters can be used to obtain a rough estimate of JRC at the appropriate scale, as shown in Figure 6.

The method of estimating (JRC) shown in the figure was developed from the following approximate relationships:

$$JRC \cong 400. \; a/L \quad \text{for} \quad L = 0.1 \text{ m}$$
$$JRC \cong 450. \; a/L \quad \text{for} \quad L = 1.0 \text{ m}$$
$$JRC \cong 500. \; a/L \quad \text{for} \quad L = 10 \text{ m}$$

These were obtained from an analysis of some 200 roughness profiles measured on 0.1 m long joint samples (Barton and Choubey, 1977; and Bandis, 1980), and from tests on model replicas of joints of different roughness reported by Bandis (1980).

In practice, it is recommended that as many as possible of the above methods of estimating shear strength are utilized concurrently to improve reliability:

1. Tilt tests on joints sampled in drillcore (Figure 2) (estimate large scale JRC and JCS from Figure 5).
2. Tilt (or pull) tests on blocks of natural size (Figure 3).
3. Roughness profiling at different scales (Figure 6).

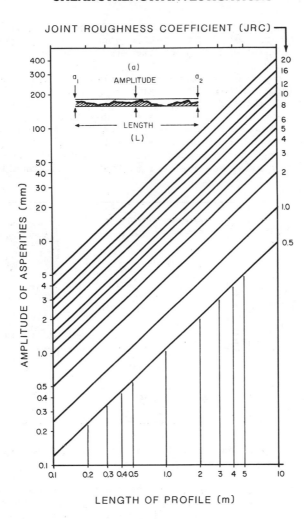

Figure 6. Measurement of joint roughness amplitude on various lengths of joint provides estimates of JRC, and its variation with sample size.

APPROPRIATE TEST SIZE

Up to this point the shear strength scale effect has been discussed without suggesting a specific sample as the "correct" size for tilt testing, profiling or extrapolating towards. Unfortunately, an inherent limitation of direct shear testing of individual jointed blocks is that the response of the surrounding rock mass is absent. A key question is whether the stiffness of the rock mass overlying and underlying a plane or zone of potential shear failure is "soft" enough to allow the blocks to follow the individual shear paths required to maintain contact with the smaller scales of roughness.

Test results reported by Bandis et al. (1981), shown schematically in Figure 7, indicate that the shear strength of a densely jointed mass (small block size) may be higher than that with a more massive block size. Small blocks have greater capacity to rotate slightly and maintain contact with small-scale features of roughness. In effect, any joints intersecting a potential failure plane are potential hinges, giving the rock mass just the degree of freedom necessary to suffer a similar scale effect to that of individually jointed blocks.

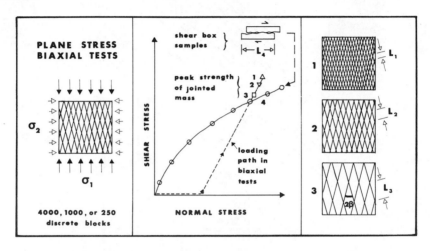

Figure 7. Size effects on individual joint samples also extend to a jointed mass of interlocked blocks.

In the light of these arguments, the appropriate test size would appear to be the natural block size. Simple tests like those depicted in Figure 3 are designed to provide precisely that scale of joint characterization. It is believed that the extrapolation and profiling procedures outlined earlier should also focus on the natural block size. Profiling of longer joint exposures (if available) may help to determine whether a larger scale undulation angle (i) needs to be added to the block-size ϕ' value to account for local changes in joint dip.

NUMERICAL MODELING OF STRENGTH-DISPLACEMENT BEHAVIOR

A limiting equilibrium analysis of a potentially unstable pit slope, using rigid block assumptions, and a convenient stereographic analysis to account for a three dimensional joint structure, gives a useful estimate of the factor of safety of the slope. However, it does not indicate to mine management whether they should withdraw personnel and valuable plant when the slope monitoring equipment indicates a large displacement in the central part of the slope, after development on a new level at the toe. The alternative, numerical modeling, has now reached the stage where it can assist in evaluating stability of a jointed rock slope. The familiar complaint concerning inadequate input data may no longer be valid.

Mobilization of Shear Strength

A review of some 650 test results on joint surfaces ranging from 40 mm to 12 metres lengths indicates a quite consistent trend for increased displacement to mobilize peak strength as sample size is increased. For convenience results have been grouped into the four surface categories shown in Figure 8, and the following three size ranges: 0.03-0.3 m, 0.3-3.0 m and 3.0-12.0 m. The fifteen data points on earthquake fault slip magnitudes were average values derived from a review paper by Nur (1974).

An analysis of the data (Barton, 1981) indicates that the following equation gives a reasonable approximation to the observed values:

$$\delta = \frac{L}{500} \cdot \left(\frac{JRC}{L}\right)^{0.33} \tag{4}$$

where δ = slip magnitude required to mobilize peak strength, or to remobilize residual strength

   L = length of joint or faulted block (meters)

Example 1. <u>Laboratory Specimen</u>. Assume: JRC = 15 (rough), L = 0.1 m

Equation 4 gives δ = 1.0 mm

Example 2. <u>Natural jointed block</u>. Assume: JRC = 7.5, L = 1.0 m
Equation 4 gives δ = 3.9 mm.

Recent developments allow the modelling of not only peak shear strength using equations 1 and 4, but the complete shear stress-displacement history of a shearing event including dilation. This is made possible by the observation that the joint roughness mobilized at any given shear displacement (δ') follows a consistent trend for a great variety of joint surfaces. The dimensionless ratios JRC(mobilized)/JRC(peak) and δ'/δ(peak) are used to generate the desired shear stress-displacement data. Figure 9 illustrates how the method produces a realistic simulation of shear test results, spanning a wide range of surface roughness. Changes in normal stress, and changes in scale are readily simulated.

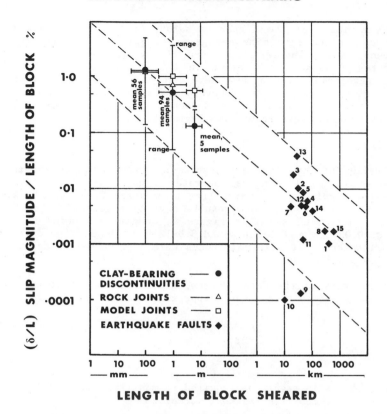

LENGTH OF BLOCK SHEARED

Figure 8. Slip magnitudes depend on the length of surface sheared. Number of samples are relevant to the clay bearing discontinuities. A total of 650 test results are incorporated in the four classes of surface. (After Barton, 1981).

The potential consequences of the scale effect on shear stress-displacement behavior is illustrated in Figure 10. A rough joint in competent rock may show quite artificial behavior at laboratory scale. The key feature to be noted is that the ultimate or so called "residual" shear strength of a laboratory-scale joint sample may be higher than the peak strength of a large sample of the same joint, such as a naturally jointed block. This important result, which was also shown diagramatically in Figure 4, has been observed in numerous shear tests spanning a variety of surface roughness morphologies (see Bandis et al., 1981).

Design Values of Shear Strength

A numerical analysis of pit slope stability that incorporates stress-displacement modelling of the joints as illustrated in Figures

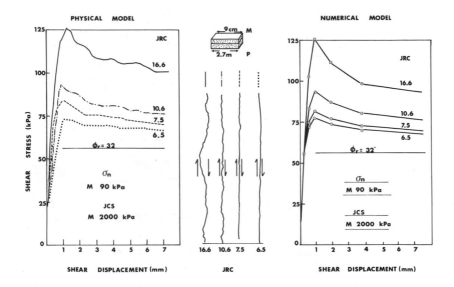

Figure 9. Numerical models of shear tests on a variety of surfaces illustrate the potential for generating shear stress-displacement data for use in slope design studies. The tests on physical models of rock joints are reported by Bandis et al. (1981).

9 and 10, will automatically mobilize pre-peak, peak, and possibly post-peak shear strengths in different parts of the slope, as appropriate.

However, in a rigid block equilibrium analysis, a whole failure surface is artificially allocated an appropriate value (or values) of strength. The design engineer who adopts true residual strength (ϕ_r) for his whole failure surface is simultaneously implying that the slope would be equally stable were it excavated to heights of 10, 100 or 1000 meters. This level of conservatism is probably correct when applied to a persistent clay-filled discontinuity.

However, for the case of rock joints that are substantially free of clay fillings, it appears appropriate to consider using the <u>full-scale</u> peak strength for at least that part of the failure surface where stresses do not exceed the available strength (Barton, 1974). It will be noted that full-scale peak strength is mobilized and exceeded in a "stable" manner, very different to the "unstable" post-peak behavior observed in laboratory-scale tests. Those parts of the failure surface where peak strength is exceeded (generally the central slope region where overburden is maximum) would be allocated the minimum shear resistance (ϕ_r).

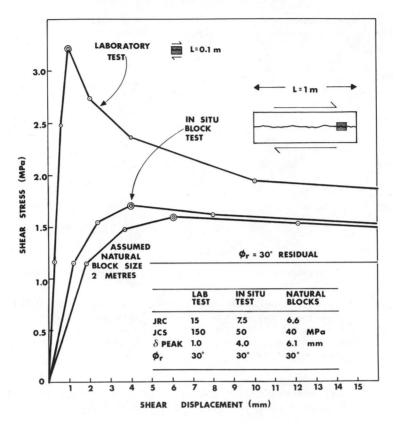

Fig. 10. An illustration of the importance of large scale test data
for rock joints. Scaling of JRC and JCS was based on Figure 5, and scaling of δ_{peak} on equation 4. A constant
normal stress of 2 MPa was assumed.

STABILITY OF WASTE ROCK DUMPS

The linear reduction of ϕ' with the logarithm of effective normal
stress observed in rock joints (equation 1) is also observed in triaxial tests of angular rockfill and gravels (Leps, 1970). Consequently, the adoption of a single value of c and ϕ to design a major
waste rock or tailings dump gives an erroneous factor of safety, just
as it does for a jointed rock slope.

In fact, for both rock joints and waste rock (rockfill), values of
ϕ' are dependent on sample size, stress level, surface roughness, and
on the uniaxial compression strength of the rock. Friction angles
are therefore higher for smaller samples, and very high where stresses are low, as at the toe or near the face of a slope. Barton and

Kjaernsli (1981) show that the value of ϕ' for rockfill can be quantified by an equivalent roughness (R), and an equivalent particle strength (S), as illustrated in Figure 11. The value of (R) depends on the porosity of the fill and on the particle angularity and surface roughness.

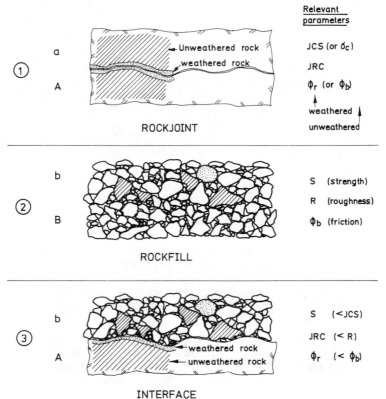

Fig. 11. Empirical approach to shear strength estimation for rock joints, rockfill (or waste rock) and any interface between the two.

The equivalent roughness (R) which is analogus to the JRC value of a rock joint, can be estimated from Figure 12. For example, dumped rock with an inplace porosity of about 35% will probably have an equivalent roughness (R) of about 5 to 6 (sharp, angular particles).

An empirical method of estimating the equivalent strength (S) of rock particles is shown in Figure 13. This parameter is analogous to the joint wall compression (JCS) of rock joints and is also scale dependent. The peak drained friction angle (ϕ') of rockfill or waste rock can be estimated from equation 5, which is exactly analogous to equation 1 for rock joints:

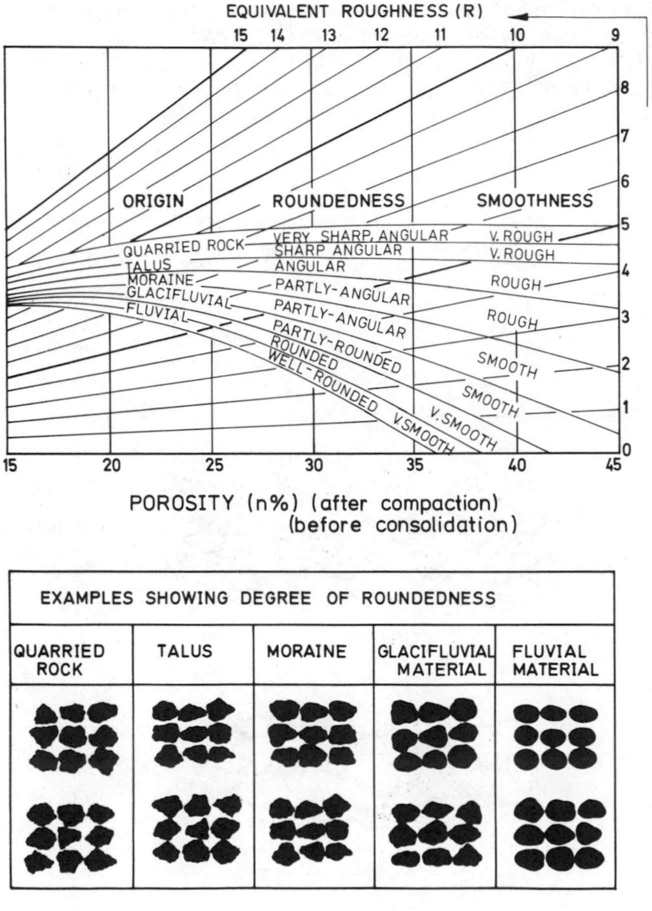

Fig. 12. Method of estimating the equivalent roughness (R) of rockfill or waste rock, based on the porosity of the dump, and on the angularity of the particles, after Barton and Kjaernsli, 1981.

$$\phi' = R.\log\,(S/\sigma_n{}') + \phi_b \qquad (5)$$

Example: porphyry waste dump:

σ_c = 150 MPa, d_{50} = 250 mm, S ≅ 30 MPa (Figure 13)

n = 35%, sharp angular particles, R ≅ 6 (Figure 12)

ϕ_b = 30° (obtained from tilt tests on sawn blocks)

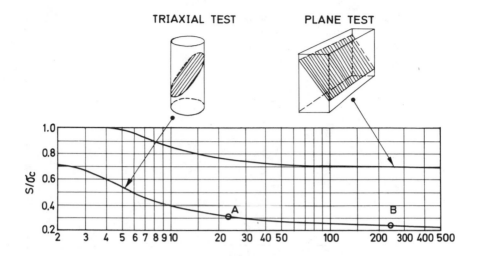

Fig. 13. Method of estimating the equivalent strength (S) of rockfill or waste rock, based on the uniaxial compression
strength (σ_c), and on the d_{50} particle size, after Barton
and Kjaernsli, 1981.

The following values of ϕ' would be obtained from equation 5.

σ_n' (MPa)	ϕ
0.1	45°
1.0	39°
10	33°

It will be noticed from this tabulation and an earlier one for
rock joints that the value of ϕ' varies by R or JRC degrees, for each
ten-fold change in stress level. Comparisons with published data
(Barton and Kjaernsli, 1981) indicate that this degree of stress-
dependency extends over at least five orders of magnitude. Figure 14
illustrates how this stress dependency influences the values of ϕ'
available in different parts of a waste dump, when assuming a simple
triangular (self-weight) distribution of vertical stress.

The lesson to be learned from this type of example is that the
excellent surface stability of a dump, and the steep angles toler-

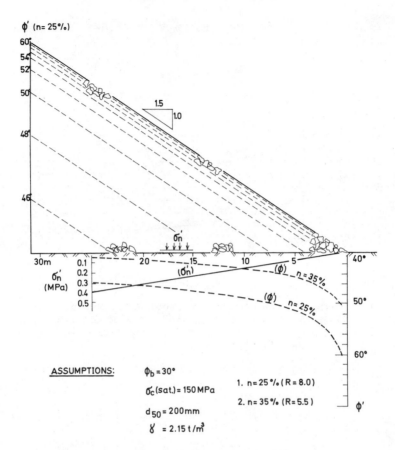

Fig. 14.   Estimated variation of ϕ' under the toe and beneath the
slope of a waste dump or dam.

ated when slopes are of moderate height, may each be misleading.
Some tens of meters beneath a waste dump the available shear strength
may be 10-15° lower than at the surface or at the toe. A deep seated
failure through the waste rock (or partly along an underlying founda-
tion-waste interface (Figure 11) is therefore more likely than surface manifestations of instability.

Tilt-Test for Waste Rock

A serious limitation of laboratory shear strength investigations
is the inability to test full-scale rockfill or waste rock samples
with as-built gradings and porosities. Development of this equivalent
roughness method, with its potential for accurate extrapolation of
strength over many orders of magnitude of stress, provides a means of

interpreting large scale tests on wasterock performed under extremely low stress, in a similar manner to tilt tests on jointed blocks of rock. The principle of a method for tilt-testing waste rock is illustrated in Figure 15. A tilt frame of several meters length is suggested. The one end can be elevated by hydraulic rams, in the manner of a dumper truck.

The maximum angle of tilt ($\alpha°$) tolerated before failure of this size of sample will probably be of the order of 55° to 65° under the extremely low effective normal stress operating across the failure surface. The angle $\alpha° = \phi'$, can be extrapolated to design stresses by estimating values of S and ϕ_b from index tests, and back-calculating the value of R from equation 5. The inevitable errors in estimating S and ϕ_b are automatically compensated by the values of R back-calculated. Under-estimates of S or ϕ_b are compensated by over-estimates of R, and vice versa. Final estimates of ϕ' are therefore unusually accurate. This error compensation is also common to the tilt tests on rock joints shown in Figures 2 and 3. Tilt tests have the added advantage that the non-uniform strain and progressive failure common to conventional shear box tests on rockfill are much reduced, due to the inherently more uniform nature of gravitationally induced shear and normal stress.

CONCLUSIONS

1. Shearing between the two interlocked walls of a rock joint, and between the interlocked particles of dumped waste rock or rockfill, results in the mobilization of quite similar values of peak shear strength. However, rock joints generally reach their peak strength at much smaller strains than rockfill or dumped waste rock.

2. The peak drained friction angles of rock joints and rockfill (or dumped waste rock) can be quantified by an equivalent roughness (JRC or R), an equivalent asperity or particle compression strength (JCS or S), and by the residual or basic friction angle (ϕ_r or ϕ_b) respectively. The latter are generated by non-dilatant surfaces of the given rock types, following large displacements. Each of these six parameters can be estimated by simple index tests, using the Schmidt hammer and various forms of tilt test.

3. The analogous behavior of jointed rock and fragmented waste rock extends to size-effects on strength, and to log-linear relationships between the effective normal stress and the peak drained friction angles (ϕ') of each material. Values of ϕ' tend to be high at the toe and close to the face of rock slopes and waste-dumps. This apparent stability should not be misinterpreted.

4. Size-effects can probably be almost eliminated by conducting self-weight tilt tests (or regular direct shear tests) on jointed rock blocks of natural size. In the case of waste rock or rockfill, tilt tests should be performed on full-size gradings whenever possible.

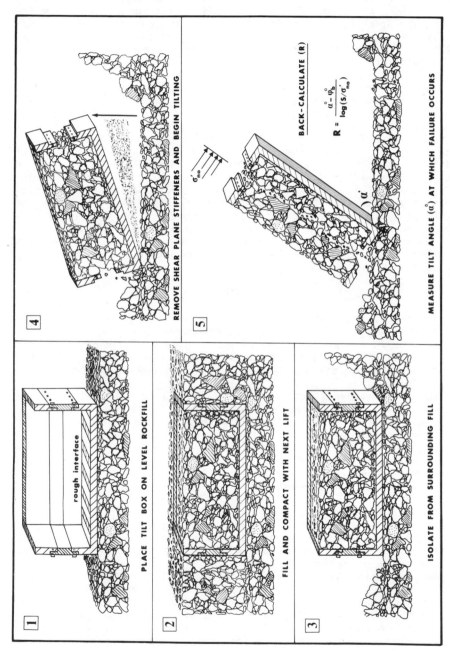

Fig. 15. A method for measuring the shear strength of dumped waste rock after Barton and Kjaernsli, 1981.

5. In cases where large samples are unavailable, suggested methods are described for scaling down the values of JRC and JCS for rock joints (Figures 5 and 6) and for scaling down values of S for waste rock or rockfill (Figure 13).

6. It is suggested that the full-scale value of peak shear strength for rock joints can be used in those parts of a potential failure surface where the shear stress does not exceed peak strength. Recent tests indicate that the full-scale value of peak strength is often lower than the ultimate or assumed "residual" strength measured on laboratory-scale samples.

7. Complete shear strength-displacement and stress-strain modeling of rock joints and rockfill can be achieved using the concept of roughness mobilization. The term JRC (mobilized) is used for rock joints, and the term R (mobilized) is used for rockfill. Good agreement with experimental strength-displacement and stress-strain curves are achieved with these methods.

8. A major uncertainty remains in the analysis of rock slope stability. The continuity of individual joints is difficult to estimate and the possibility of limited interaction between joint sets hard to quantify. The assumption of zero cohesion (no intact rock "bridges") is probably a wise precaution when joints have adverse orientations.

REFERENCES

Bandis, S., 1980, "Experimental Studies of Scale Effects on Shear Strength, and Deformation of Rock Joints," Ph.D. Thesis, University of Leeds, England, 385 pp.

Bandis, S., Lumsden, A. and Barton, N., 1981, "Experimental Studies of Scale Effects on the Shear Behavior of Rock Joints," International Journal of Rock Mechanics and Mining Sciences and Geomechanics Abstracts, Pergamon, Vol. 18, pp. 1-21.

Barton, N., 1973, "Review of a New Shear Strength Criterion for Rock Joints," Engineering Geology, Elsevier, Vol. 7, pp. 287-332.

Barton, N., 1974, "Rock Slope Performance as Revealed by a Physical Joint Model," Proceedings, 3rd Congress of the International Society for Rock Mechanics, Denver, Vol. IIB, pp. 765-773.

Barton, N. and Choubey, V., 1977, "The Shear Strength of Rock Joints in Theory and Practice," Rock Mechanics, Springer, No. 1/2, pp. 1-54.

Barton, N. and Kjaernsli, B., 1981, "Shear Strength of Rockfill," Journal of the Geotechnical Engineering Division, Proceedings, American Society of Civil Engineers, (in press).

Barton, N., 1981, "Modelling Rock Joint Behavior from In Situ Block Tests: Implications for Nuclear Waste Repository Design," <u>Office of Nuclear Waste Isolation</u>, Battelle, (in press).

Hoek, E. and Londe, P., 1974, "Surface Workings in Rock,"Proceedings, 3rd Congress of the International Society for Rock Mechanics, Denver, Vol. I.A, pp. 613-654.

Leps, T.M., 1970, "Review of Shearing Strength of Rockfill," <u>Proceedings</u>, American Society of Civil Engineers, Vol. 96, No. SM4, pp. 1159-1170.

Marachi, N.D., Chan, C.K., Seed, H.B. and Duncan, J.M., 1969, "Strength and Deformation Characteristics of Rockfill Materials," Report, University of California, Berkeley, 139 pp.

Nur, A., 1974, "Tectonophysics: The Study of Relations Between Deformation and Forces in the Earth," <u>Proceedings</u>, 3rd Congress of the International Society for Rock Mechanics, Denver, Vol. I.A, pp. 243-317.

Pratt, H.R. Black A.D. and Brace, W.F., 1974, "Friction and Deformation of Jointed Quartz Diorite," <u>Proceedings</u>, 3rd Congress of the International Society for Rock Mechanics, Denver, Vol. II.A, pp. 306-310.

Question

The joint roughness as measured by you appears to assume dry surfaces. Have you tried any tests using a wet surface. Would it make a significant difference to the roughness coefficient by acting as a lubricant.

Answer

Our tilt tests were performed dry to avoid negative pore pressures (suction) delaying the tilt failure. In the field, moisture is probably present anyway, so tilting should be performed slowly, i.e. drained. However, as a point of definition, the value of JRC is back-calculated using the saturated value of the joint wall compression strength (JCS), which is usually 10-15% lower than the dry value. I believe water might act as a lubricant for some layer-lattice minerals, i.e. rocks rich in mica, etc. It would then presumable tend to reduce JRC, but I suspect most of the water effect would be seen in strongly reduced JCS values. Rocks rich in quartz with massive crystal lattices, apparently suffer an anti-lubrication effect. I believe they also show the least change in JCS, with saturation. The 140 rock joints that I developed the tilt test hypothesis on, were all tested saturated in the shear box, following the index tests. The correlations were therefore developed between dry tilt or pull tests, and wet shear box tests. I don't recall any correlation problems specifically related to the gneiss or slate.

Question

If ϕ_r reduces with increase of $\sigma_n{}'$, and ϕ_r reduces with increase of size of shear surface, can you recommend a $\sigma_n{}'$ for testing laboratory samples.

Answer

Firstly, a point of definition. Maybe we should talk of ϕ_u (ultimate) as measured at the end of a given test, because the true minimum ϕ_r (residual) should not change with σ_n or size of shear surface. So answering the question with regard to ϕ_u; yes, I think a suitable testing level for $\sigma_n{}'$ could be recommended which would force the laboratory size joint to give the same lower ϕ' (peak) value as a large sample. The necessary value of $\sigma_n{}'$ could be calculated if JRC and JCS were known - it would depend on both. However, I don't think this would be the right way to proceed. Wouldn't it be more logical to test at the design level of $\sigma_n{}'$, and extrapolate JRC and JCS to the correct size values? Best of all, do tilt or pull (self-weight) tests on the correct size of surface in the first place, where this is practical.

Question

Does your tilt box test for aggregate require a particular ratio between the size of aggregate and the size of the box. Can a similar approach be used for evaluating a heavily jointed cliff face with several sets of Joints.

Answer

The objective of the large scale tilt test for aggregate (wasterock) or rockfill) is to overcome the normal test limitations. We specifically want to be able to test full-scale material without the usual requirement of using model gradings. A box length of at least $5 \times d_{max}$ or at least $30 \times d_{50}$ is perhaps reasonable. It would be worth experimenting with a small box and model gradings before designing the large tilt box frame. The upper parts of a waste dump will probably have an entirely different grading curve from the lower parts, so the size of tilt frame will not be ideal for either case.

The heavily jointed cliff face with several sets of joints - if sampled "undisturbed" - with joints still interlocked - will possibly display too high a ϕ'(peak) value; the tilt angle at which failure occurs might be too high (i.e. 75°) such that tensile rather than shear failure occurs. The empirical relationship between ϕ'(peak) JRC and JCS might otherwise work reliably since a modified version involving equivalent roughness (r) and equivalent strength (S) works well for rockfill. Judging by the steep and strongly curved envelopes obtained by Jaeger for Bougainville(Panguna) andesite, I would expect quite high values of JRC (or R), unless failure is joint controlled.

Question

How do your JCS and JRC shear strength relationships vary with
different rock types.

Answer

This is a difficult question to give clear-cut answers to. Joints of
different roughness (JRC) obviously occur in the same rock type, i.e.
bedding joints and cross joints in sandstone, foliation joints and
tension joints in gneiss. Similarly, in the same rock type one joint
may be tightly interlocked, "non-conducting" to water, and
essentially unweathered (high JCS), compared to the major weathered,
water-conducting set with lower JCS. The study we conducted in
Norway some years ago involved fifteen different types of joints
sampled from seven rock types. A wide range of behaviour was exhibited - from ϕ'peak of about 80° to about 30°, for exactly the above
reasons. The chief cause of this scatter was roughness, i.e. JRC ≈ 18
for bedding joints in hornfels, and JRC ≈ 1 for cleavage joints in
slate. Joint compression strength (JCS) ranged from about 140 MPa in
granite to 20 MPa in soapstone, according to the Schmidt hammer
testing.

Recent work on size effects by Bandis et al. (see references in
paper) show that these wide differences in JRC and JCS are effectively reduced at large scale, but will still give ϕ' (peak) values as
different as 10 - 30°, depending on joint type.

Question

Did all small blocks fail at the same (or nearly the same) tilt angle.
Is there a fundamental basis for your shear strength equation.

Answer

The eighteen small blocks cut from the large jointed slab give a tilt
angle of 69° (3 x 18 tests in all). Some were too strong to tilt test,
so were pulled horizontally under self-weight. From memory, they
showed quite a wide range of tilt (or pull) angles; perhaps 55° - 80°.
Beside experimental scatter, some small samples were essentially
shearing down the dip of major asperities, and others up-dip. So
there would need to be a big difference. It was noticeable that the
large sample (before division) gave a tilt angle of 59°each of the
three times. Lower and more consistent JRC values seem to be a
fundamental feature of larger size samples. That is one of the reasons
we recommend tilt (or pull) tests on blocks of natural (large) size.

The "fundamental basis" question would take time to answer fully.
I'll try a shortened version. First of all, roughness (JRC) is related to asperity height (a) and sample length (L) (Figure 6). Values of
$\Delta a/\Delta L$ obviously represent crude approximations to Patton's (i)
values - which can be added to ϕ_r in a fundamentally justifiable
manner to give ϕ' (peak). Secondly, we have found from our and
other's results (Iwai, Bandis, et. al.) that the ratio of true to
assumed contact areas (A_1/A_0) at or prior to peak strength, are

approximately equal to the particular value of σ_n'/JCS. In other words, asperities are just about at failure.

A review of a large number of high pressure triaxial tests that I undertook some years ago, indicated that the Mohr failure envelopes for a wide range of rocks were horizontal when the "critical" stress state $\sigma_1' = 3\sigma_3'$ was reached. This condition is the same as $(\sigma_1 - \sigma_3)/\sigma_n' = 1$ on the failure plane. The "confined strength" $(\sigma_1 - \sigma_3)$ has proved to be useful estimate of JCS at stress levels higher than $\sigma_n' = \sigma_c$ (unconfined compression strength). By implication, our high stress "critical" state with JCS/$\sigma_n' = 1$ also implies total crushing of asperities, with $A_1 = A_0$, and, intuitively, total suppression of dilation. These more or less logical patterns in the JRC, JCS behaviour constitute something like a "fundamental basis", at least for me. Most important is the fact that the equation works, over many orders of stress magnitude, for rock joints, rockfill, and for interfaces between the two. So far, I haven't been able to understand why ϕ' (peak) for these materials changes by approximately JRC or R degrees ($^\circ$) for each ten-fold change in effective normal stress. This is a simple, fundamental result that needs to be explained.

Question

Do I understand you to mean that a highly fractured model (2D) has higher strength than a less fractured model. What happens when you tesselate a volume as opposed to a plane. Does the relationship still hold?

Answer

The highly fractured (4000 block) models do have higher strength than the less fractured (1000, and 250 block) models. But they also have lower deformation modulus, which seems to be the reason why the small blocks have the freedom necessary to slightly rotate and register the smaller scale (higher JRC) roughness. I think the result is chiefly a useful experimental justification for saying that block size effects shear strength. In the real world, highly fractured rock masses will often not have as rough joints as the more massively jointed rock masses, so the same result is less likely to be seen. But when you get down to the case of a heavily fractured, multiple joint set material, where the shear strength is no longer anisotropic (perhaps), I suppose interaction of blocks comes into the picture to cause rather high strength, low modulus behaviour. This was not the case in my tension crack model tests, with just two intersecting sets.

I believe the block-size controlled result is likely to persist to the three-dimensional case, but with the same provisos as I've mentioned.

Question

As one of the main authors of the rockmass Q-system approach in tunneling design, could you elaborate on your hope, or the potential in the long term, to develop the same approach to rock slope stability.

Answer

As you probably know, Bieniawski has suggested that his RMR system can be applied to both tunnels and slopes, by modifying the original orientation weighting factor. If one has a good eye for slope instability, I suppose this could work. As far as the Q-system is concerned, I would prefer to see it used in estimating the relative strengths of individual discontinuities (J_r/J_a gives a realistic estimate of tan ϕ), or for helping to estimate the shear strength of a randomly jointed mass (as done by Hoek and Brown). The orientation question is so important for slopes that are structurally controlled, that I feel it may be dangerous to use classification systems on their own, to classify good, poor, or very poor (presumably failed) slopes. Where does one draw the line between stability and instability? However, I would encourage use of the Q-system parameters for a general classification of the degree of jointing, clay fillings, etc. The first four parameters RQD/J_n, J_r/J_a give a very detailed description of rockmass conditions, useful for communication in reports, and understood by increasing numbers of engineers. As in the case of tunnels, judgment would need to be exercised in classifying J_r and J_a for the discontinuities most likely to allow failure to initiate.

Chapter 8

SLOPE STABILITY ANALYSIS TECHNIQUES
INCORPORATING UNCERTAINTY IN CRITICAL PARAMETERS

D.L. Pentz

Group Vice President, Golder Associates
Seattle, Washington

ABSTRACT

The use of "probability" methods in slope stability analysis has been practised by geotechnical engineers for some time. This paper questions whether the term probability has been properly used since in many cases the term is used in a classical statistical manner only to describe frequency distributions of data. This paper proposes a definition of the term probability which includes the engineers' subjective state of confidence for all critical parameters. Such parameters may include the slope stability model, undetected structural features, strength, groundwater pressure distribution, and geological framework. This process does not make a "better" analysis but it does force the geotechnical engineer to state his opinion in such a manner that others can understand and if necessary, criticize the assumptions. Further, this technique forces the question, "What impact would failure of a slope have on the production schedule or costs of a mine?" The level of risk taken in adopting a particular slope design strategy in a mine is not a question of what Factor of Safety should be selected but, rather, what trade-offs result from steeper slopes and the associated costs of failure.

This paper also presents several simple probabilistic techniques which, though widely used in other fields have been all but ignored in geotechnical slope stability analyses. These include the use of stratified sampling to optimize Monte Carlo uncertainty approaches, and the use of simple functional approximations rather than complex stability models in probabilistic analysis.

The author presents a case example of this technique and discusses the results in comparison with more traditional methods.

197

INTRODUCTION

The stability of slopes forming open pit mines is not entirely a function of geotechnical considerations but is also influenced by the constraints imposed by production requirements and the economics of the overall operation at any instant of time.   This paper evaluates the tools which geotechnical and mining engineers have available to them, which affect their ability to predict the behavior of slopes which are at or near a critical state of stability.

All available methods of slope analysis presuppose at least a partial knowledge of the following components:

1. The geomechanical framework, i.e., the structural geology and its relationship to mechanical strength.

2. The groundwater pressure distribution, i.e., the distribution of water pressures within existing rock masses and how they will naturally or artifically change with time and excavation of the open pit.

3. The influence of vibrations, i.e., the effects of blasts and earthquakes on the mechanical and possibly groundwater pressure properties of a slope.

4. The prediction models, i.e., the methods by which the components summarized above can be coupled to arrive at mine planning decisions.

It should be stated that when "probabilistics" were first introduced to the geotechnical profession several years ago, they were regarded by some as an opportunity to achieve a more rational basis for design than conventional "Factor of Safety" methods.

The prime purposes of this paper are to examine why this potential has been only partially achieved and to present practical solutions to several of the problems encountered thus far.The paper is divided into the following broad parts:

1. Discussion of the differences between probabilistic and deterministic methods of analysis.

2. Analysis of the implications of probabilistic slope analysis in mine evaluation and planning.

3. Discussion of the differences between statistical and probabilistic approaches. Most statistical approaches assume that analysis and sampling methods are correct and

incorporate statistical or measurement uncertainty. In
reality, however, statistical uncertainty as derived from
laboratory or field tests has very little to do with
probability of failure or in some cases, even the actual
uncertainty of the parameter in question. A probabilistic
rather than statistical approach requires the explicit
incorporation of engineering judgement in the evaluation of
uncertainty throughout the analysis process.

4.   Discussion of alternative approaches to the performance of
     probability analysis. The Monte Carlo methods generally used
     to evaluate probability of failure can be expensive and
     inefficient. Improved Monte Carlo methods used in other
     disciplines are discussed.

5.   Finally, a case example of probabilistic design for a portion
     of a major slope is presented which uses improved Monte Carlo
     methods.

DETERMINISTIC AND PROBABILISTIC METHODS OF ANALYSIS

It is not the intent of this paper to discuss the variety of well
documented procedures for analyzing the stability of open pit slopes
such as those developed by Bishop (1955), Sarma's (1973) circular
methods or three-dimensional planar wedge analysis. These analysis
methods are based on the principles of limiting equilibrium and the
resulting degree of stability is expressed as a factor of safety.

Normal practice is for the engineer or geologist to select
appropriate values for all of the governing parameters or variables
and then design the slope with a selected "model" or analysis procedure to achieve a specific factor of safety. The selection of a
factor of safety greater than unity is an arbitrary process reflecting
both the overall state of uncertainty of the parameters and methods
used and also the importance of a particular slope to the mine. Often
a band of slope height/slope angle relationships with varying factors
of safety is passed to the mine owner or operator for his decision.
This method of analysis, which uses point estimates, is deterministic.

The alternative approach to slope stability analysis procedures is
known as a probabilistic method in that the basic uncertainties and
variabilities in the governing parameters and the analysis model are
recognized. Factors including strength, water pressure distributions,
etc., are defined in the analysis in a probabilistic manner. The
results are therefore expressed so that the safety factor is defined
in a probabilistic sense. Thus, the predicted performance of a slope
can be expressed as the probability of a factor of safety less than

unity (see Figure 1). Since the probabilistic definition of factor of safety is different from the conventional deterministic description we suggest the use of the term "stability coefficient". It should be clear that the absolute value of the stability coefficient is of little interest in probabilistic analyses: only the values less than unity are of concern. However, if we wish to further modify the results by incorporating uncertainty in the analysis procedure, for example, the absolute values of the stability coefficient distribution should be retained.

MINE PLANNING AND SLOPE DESIGN

The part that mine slope design plays in open pit design must be understood by the geotechnical engineer if he is to be a useful part of the overall planning and evaluation process. It is true that slope stability recommendations are only part of the problem of slope design. Other considerations relate to production constraints, variables in location, grade of the ore and market price and waste stripping ratios.

The geotechnical engineer conveys the results of slope stability analyses to mine planners or owners in the form of an index, either as a Factor of Safety or as the probability of failure. In either case, the engineer must be able to incorporate this index into the planning process and to do this he must understand the consequences of failure on the feasibility or operations of the mine. The geotechnical engineer will therefore be required to describe the nature of failure, e.g., high velocity failures or progressive "stick slip" failures, volume and geometry of material, lateral extent, etc.

The use of Factor of Safety as an index requires that a rational means of evaluating the significance of this be established so that mine planning decisions can logically proceed. The advantage of continuing to use this index is that there is a considerable body of experience built up in the methods of judgement used to calculate Factors of Safety.

Probabilistic analysis, on the other hand, forces rational cost/benefit analyses to be made of the consequences of failure. These, as stated above, are not trivial and will heavily tax the judgment and the systems employed by the geotechnical and mining engineers. Such problems as when a failure could occur within the life of a slope, and thus in a production schedule, must be addressed. While significant strides have been made in the evaluation of the probabilistic cost of failures, these are still relatively simplistic. It may not be reasonable to evaluate all the mine planning variables in a probabilistic format.

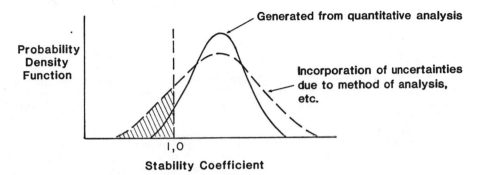

Fig. 1   Stability Coefficient Density Function, including
Uncertainties Related to Analysis Method, etc.

The main advantage of expressing the results of slope stability analyses in a probabilistic manner is that it truly conveys to the mine planning process the engineer's judgment of the slopes in question and, provided the cost/benefit analysis can be adequately executed, expresses the factor of real interest to the mine engineer, as well.

STATISTICAL VERSUS PROBABILISTIC METHODS OF ANALYSIS

The goal of probabilistic methods of analysis is to quantify the geotechnical engineer's uncertainty about any given slope and then to allow this uncertainty to be rationally incorporated into the mine planning process.

It is perhaps important at this point to attempt to clarify any confusion which may exist between the meaning of the terms probability and statistics. The use of statistics is related solely to the examination of the frequency distribution of samples of data. We may apply statistical treatment and analysis to, for example, laboratory shear strength data or geological field mapping. We may, on the basis of such treatments, make statements about the spread or variability of such data. The terms probability should be applied only when we desire to express the nature of our uncertainty related to the real world.

The term "probability", therefore, should be restricted in use to description of the state of confidence in a parameter, event or model. The statistical term "confidence level" as applied to a population distribution of sample data is misleading since it implies that the data represents the real world, when in most cases, and certainly for most geotechnical problems, such a statement represents only a part of our state of knowledge. Engineers, in particular, are trained to make estimates (more accurately point estimates) based on experience and data. An example of a point estimate is, "the cohesion of a clayfilled fault is 4 psi and the associated friction angle is 20 degrees". Such a statement represents only part of the engineers state of confidence or judgement since he knows there is variation in the field and as such the point estimate is only part of the truth. If, however, he adds a value judgement expressed numerically to the quantitative data and then expresses the whole of his knowledge or judgment in the form of a probability distribution, he can express accurately what he really believes about that parameter, event, or model.

In the succeeding sections of this paper an attempt to clarify a probabilistic method of slope stability analysis will be made. Before proceeding to this, two important statements should be made.

1. Sensitivity analysis using upper, lower bound and best estimates has been routinely carried out by many slope stability engineers. While this procedure is an integral part of probabilistic analysis methods, rather than characterizing the range of parameters as a series of point estimates, specific probabilities are assigned to each of the input parameters.

2. Engineers, it appears, have considerable difficulty in expressing their confidence or uncertainty in parameters quantitatively. The comment is often made, "That is only guessing!" The precise advantage of the probabilstic method is that it forces the engineer to fully evaluate the implications of field and laboratory data. If, in fact, such an evaluation genuinely suggests that there is an equal chance that a particular parameter will have a value which could fall anywhere within a range of values, then the probability distribution should express that view (see Figure 2). The result of such statement of judgement will then, if it is an important parameter, be reflected as part of the final result. If no further field or laboratory testing will refine the probability distribution of the parameter(s), then this distribution is regarded as a residual uncertainty. The influence of such a statement may have a profound effect on the strategy of an open pit mine plan.

Variability and Uncertainty

It is important to differentiate between variability of parameters and uncertainty in the performance of the slope due to the same parameters. In this section we attempt to explain this difference by discussing one of the four important parameters categorized in the Introduction, i.e., the geomechanical framework which leads to an estimation of rock mass strength.

The strength of a rock mass may be determined from two sources:

1. Laboratory testing

2. Performance of actual slopes where other governing parameters are also known.

If the geological interpretation indicates that a particular portion of a pit slope can be treated as a single zone, a series of relatively small-scale samples are tested--in the laboratory or the field. The point-by-point strength results will typically show a

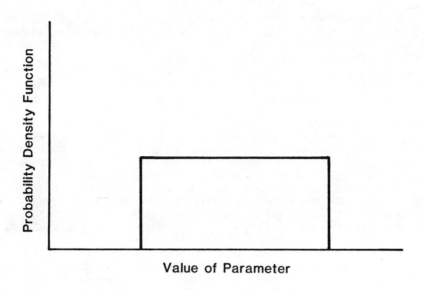

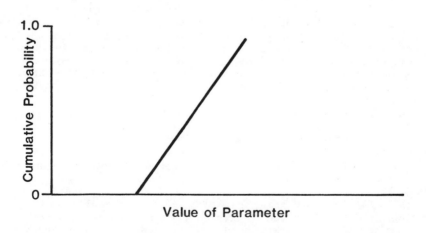

Fig. 2   Probability Representation of Equal Chance

spread of behavior which reflects real variability in the point-by-point results within the so-called "homogeneous" zone (see Figure 3).

The effects of size of sample and orientation, spacing, and nature of discontinuities within the sample are such that for most geological environments the distribution of sample test results rarely fully describes the mass strength. This is even true in cases where we may be able to test larger samples which may contain several sets of discontinuities (Jaeger 1970; Lockhart, Lumsden 1980). Thus, application of statistical tests such as the "t" test to a normally distributed population of test results is only part of the process of attempting to establish rock mass strength. As we increase the size of sample, the "probability-density function" will tighten for such results, i.e., reduce the uncertainty associated with estimated mean (see Figure 4). However, there is still considerable uncertainty of both the mean and variance of such populations related to mass strength. This includes:

1. Uncertainty in validity of laboratory testing. For particularly simple failure modes such as a wedge failure, this level of uncertainty may not be great provided appropriate testing has been carried out. However, if the failure mode involves discontinuity sliding and failure of intact material, then extrapolation from laboratory to in situ rock mass is far less clear and therefore more uncertain.

2. If we recognize the ability of a full-scale failure mechanism to find a path of least resistance through the rock mass, it may be concluded that the overall mass strength might be less than the mean sample strength, not withstanding the uncertainties summarized above.

3. Finally, there is uncertainty associated with disturbance of test samples, and the effect on strength of deformations induced during mine development by blasting or earthquakes.

In summary, it should be clear that concepts of variability and associated statistical treatment of uncertainty have quite separate meanings. The probabilistic description of any input parameter cannot be mathematically determined and a pragmatic approach requires that engineering and geological judgement with appropriate experience must be incorporated explicitly. Subjective factors can be stated explicitly. The "pitfalls" of this are discussed in the final section of this paper.

If we examine the geological framework, we know that the use of orientated core, or better still, line mapping of exposures can establish real variability of joint attitudes (e.g., McMahon 1971 and 1975). Equally, the persistence or planar extent of such features may

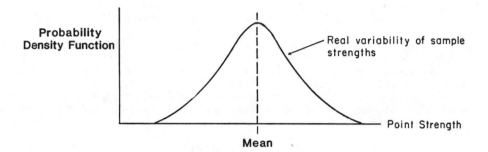

Fig. 3   Probability Density Function for the
Population of Individual Sample Strengths

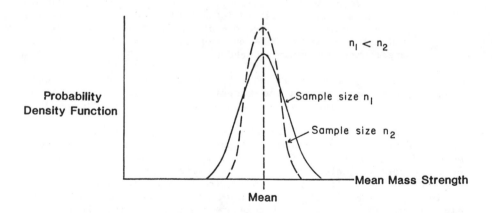

Fig. 4   Interval Estimates of the Mean Strength on
the Basis of Samples of Different Sizes

also be partially sampled in exposures and statistical methods used on the results to examine the frequency distribution of joints of particular sets (e.g., Call, et al., 1976, Canadian Pit Slope Manual, 1977, discusses procedures for generating potential failure mechanisms based on this data). The application of such techniques for small-scale failures such as bench failures is clearly useful provided the statistical frequency data is modified for the effect of undetected features. However, for large-scale failures of interramp or overall slope dimensions, real variability and uncertainty arises from the nature and extent of cohesive resisting forces generated by intact rock bridges or mega asperities. We are therefore faced with two alternatives:

1. Explicitly describe the variability and subsequently the uncertainty of the structure and then provide appropriate uncertainty description of entire strength by coupling the discrete probabilistic strength components and all other input factors, or

2. Based on a thorough evaluation of the structural geology, utilize engineering judgments to identify potential failure mechanisms and then incorporate the influence of intact rock failure along the potential failure surface into the probabilistic statement of the rock mass strength. Thus, the probability of each mechanism is assumed to be 100 percent and the results can then be modified explicitly to reflect the uncertainty in the selection of the mechanisms.

PROBABILISTIC SAMPLING TECHNIQUES

The first requirement for probabilistic analysis of slope stability is characterization of all uncertainties involved in the prediction process, as explained in the preceding section. The second requirement is the incorporation of those uncertainties into the analysis to determine a probability of failure. For simple functions, it is possible to evaluate the probability distribution of a function of a random variable analytically. For a one-variable function, for example:

$$f[g(x)] = \frac{f_x(y)}{g_x^1(y)}$$

Where $f_x(x)$ = distribution of random variable x

$g_x(x)$ = slope stability function of random variable x

$g_x^1(x)$ = derivative of $g_x(x)$ with respect to random variable x

$f_x[g(x)]$ = distribution of function g(x) of random variable x

and $f_x(y) = f_x(x)$ expressed in terms of y

$g_x^1(y)$ = magnitude of $g_x^1(x)$ expressed in terms of y

For a rigorous treatment of closed form uncertainty analysis see Papoulis (1979). In most real situations, however, analytic solutions are intractible, and iterative Monte Carlo sampling techniques must be used. These involve the sampling of all input parameters according to their probability distribution and the sorting of model output to develop output distributions.

Stratified Sampling

The only theoretical requirement for the use of Monte Carlo methods is that samples must be taken uniformly throughout distributions of the parameter (Zaremba, 1968). Thus, uniform, random, or systematic samples could equally well fulfill the theoretical requirements of Monte Carlo sampling. At this time, a random, uniformly distributed sampling pattern is used almost exclusively in geotechnical engineering. This approach requires very little thought on the part of the engineer, and is sufficiently simple to be intuitively satisfying. However, other sampling approaches such as stratified sampling, Latin Hypercube, and reduction of variance provide considerably better convergence with far fewer iterations.

Systematic sampling techniques are well covered in several papers--notably Murthy (1967); McKay, Beckman, and Conover (1979); Yakowitz, 1979. Stratified sampling involves division of parameter distributions into several groups, or strata, and then random sampling from within those strata in such a manner as to optimize sampling coverage of region of interest. In the example shown in Figure 5b, stratified sampling is used to assure sampling from every part of the distribution with a small number of samples. In the example in Figure 5C, increased detail is obtained in one stratum by increasing the sampling intensity by a factor of 10, and weighing the resulting samples accordingly to maintain a uniform sampling distribution. This method, extended to many parameters, is referred to as Latin Hypercube Sampling (Iman, 1980). Convergence in regions of interest is considerably more rapid using stratified sampling techniques, and the costs of Monte Carlo simulation can thus be reduced substantially.

Functional Approximation of Geomechanical Models

Geomechanical slope stability models are frequently quite complex. As a result, models which are quite reasonable for deterministic analyses become unrealistic to use in Monte Carlo analyses requiring thousands of iterations. The solution to this problem is the replacement of the actual geomechanical model by a simple functional approximation for use in the simulation. This approximation can be obtained by interpolation of the results of sensitivity analyses using the geomechanical model (Figure 6). In some cases, a curve can be fitted to the model results and that curve can then be sampled. In other cases, a linear interpolation can be used between sensitivity

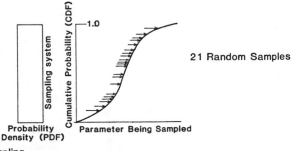

21 Random Samples

(a) Random Sampling

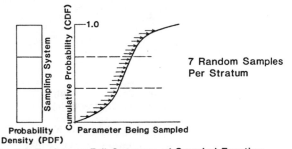

7 Random Samples
Per Stratum

(b) Stratified Sampling To Assure Full Coverage of Sampled Function

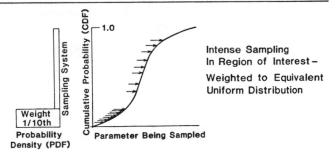

Intense Sampling
In Region of Interest –

Weighted to Equivalent
Uniform Distribution

(c) Stratified Sampling To Provide Increased Accuracy in a Specific Stratum

Fig. 5   Stratified Sampling Techniques

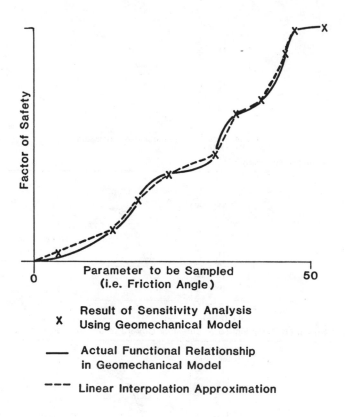

Factor of Safety

0   Parameter to be Sampled   50
          (i.e. Friction Angle)

X   Result of Sensitivity Analysis
     Using Geomechanical Model

——   Actual Functional Relationship
     in Geomechanical Model

- - - Linear Interpolation Approximation

Schematic Only

Fig. 6   Functional Approximation
              of Geomechanical Model

analyses results. These Functional Approximations or Transfer Functions can then be used in Monte Carlo simulation without having to re-run the actual model. This approach can be implemented with any number of variables, and allows calculation of failure probabilities as a function of the cost of analyses using the actual models. For additional discussion of the use of functional approximations, see Box and Jenkins (1970).

PROBABILISTIC ANALYSIS PROCEDURES FOR EAST PIT SLOPES AT TWIN BUTTES

The pit slopes at the Anamax Twin Buttes Mine in Arizona which form the east wall were found to be highly fractured and a major program of slope depressurization has been undertaken (Pentz 1979). A comprehensive slope stability analysis was carried out for this portion of the pit to provide a reliability schedule for the overall slopes and the interramp components.

The geology of the pit slopes is complex but can be summarized as follows. The bedrock is overlain by a capping of alluvium which reaches some 130 m. in thickness. This is underlain by approximately 30 m. of weak oxidized porphyries and arkoses. Beneath this oxide layer the bedrock is divided into two halves by an approximately vertically-dipping fault zone striking northwest-southeast. Northeast of the fault are strong, tough Paleozoic metamorphosed sedimentary sequences of limestones and sandy siltstones. This series does contain faults of limited persistence dipping as flat as 55 degrees towards the pit bottom. Pit slopes in this material will ultimately reach approximately 190 m. in vertical height. The bedrock to the southwest of the fault zone is composed of highly fractured Mesozoic sediments of arkose and intrusive porphyries. The ultimate height of this portion of the slope will be approximately 160 m. A typical design section is shown in Figure 7.

A probabilistic analysis of the Paleozoic sections of the pit was carried out and with the exception of the possibility of an effect from low angle undetected throughgoing gouge clay-filled faults, interramp slopes of 45° were shown to have reliability of between 6 and 7 percent. In addition, slope stability analyses were carried out on the interramp slopes in the Mesozoics and the overall pit slope.

A general outline of steps invoved for the overall analysis is summarized below.

Step 1   Establishment of Uncertainty Distributions of Input Parameters

For this particular part of the slope stability analysis, two parameters were required to be identified in a probabilistic manner.

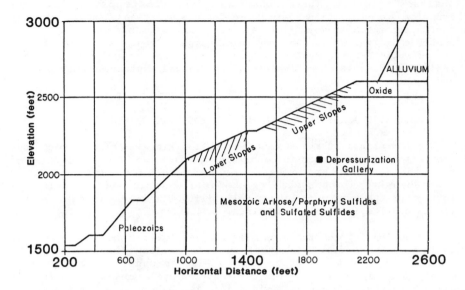

Fig. 7   Section

1. Rock mass strength
2. Groundwater pressures

Rock Mass Strength

A substantial number of triaxial tests were completed on six-inch and NX core on both porphyry and arkose samples. In addition, direct shear tests on core discontinuities were carried out and, finally, a number of artifically prepared aggregate triaxial tests were also carried out.

The absolute minimum value of the mass strength is the residual strength of heavily sheared discontinuity surfaces. This strength would be defined approximately by a friction angle of 30 degrees for porphyry or arkose. Experience would suggest that the residual strength values do not represent realistic lower limits to the rock mass strength and the following values have been specified:

TABLE 1

Normal Stress Range (psi)	Cohesion (psi)	Friction Angle (Degrees)
0-120	3	35
120-350	19	30

A three-dimensional rock quality model was constructed for the design slopes which was related to fracture frequency and the degree of alteration. The worst zones were limited to the immediate vicinity of the Fault Zone. While the strength parameters of these materials were determined as having mean values of a friction angle of 24 degrees and a cohesion of 5 psi, they were not considered to be significant from an overall strength standpoint.

A realistic (5 percent) lower bound of the rock mass strength is defined by the mean of results of direct shear tests on fresh clean surfaces and recompacted samples. The following results were used:

TABLE 2

Normal Stress Range (psi)	Cohesion (psi)	Friction Angle (Degrees)
0-120	5	38
120-350	30	30

The fresh sulphide arkose and porphyry was expected to exhibit somewhat higher rock mass strength parameters than given in Table 2 above. The median values used in design, (i.e., those with a 50 percent chance of being exceeded) were also influenced by empirical results of Barton (1973) and back analysis of similar slopes in other mines. The latter lay above the 50 percent levels. The 50 percent values were selected as follows:

TABLE 3

Normal Stress Range (psi)	Cohesion (psi)	Friction Angle (Degrees)
0-120	8	42
120-350	48	30

Finally, the complete specification of the probabilistic description of the mass strength of the sulfide porphyry/arkose requires that maximum strength values also be considered. The strength levels above which there is virtually no chance of occurrence have been specified as follows:

TABLE 4

Normal Stress Range (psi)	Cohesion (psi)	Friction Angle (Degrees)
0-120	15	45
120-350	65	30

These results are shown conventionally on Figure 8 and presented as cumulative probability plots in Figures 9A and 9B.

It is important to note, however, that they should generally be expressed initially as population density functions (i.e., "bell" curves) for accuracy and then converted to cumulative plots. Also, the cohesion and friction angle cumulative distribution plots are coupled (i.e., correlation coefficient = 1) as shown in Figure 9. A third point of interest is that these distributions are a result of a very substantial testing program of large diameter triaxial tests, direct shear tests, back analysis of failures at Twin Buttes and performances of slopes in similar material elsewhere.

Groundwater Pressure

Similarly, the potential groundwater pressures were described probabilistically. This was done using two parameters. The first uncertainty refers to the probability of the existence of a low horizontal permeability zone or "dam" created by the fault zone separating the Mesozoics and Paleozoics. On evaluation of the geology and extensive piezometric data it was considered that there is a 50 percent probability that the hydrologic state was represented by the "dam" scenario and a 50 percent probability that the hydrologic state lies within the range of states represented by the "dam" and "no dam" cases. This is shown quantatively in Figure 10A.

As part of a depressurization program, a tunnel was driven beneath the Mesozoic slopes and was monitored by some 75 piezometers. An attempt was then made to model the water pressure distribution as the pit was lowered. Continuum model predictions, unfortunately, were unable to predict in situ measurements as a result of the complexity of the hydrogeological structure.

In recognition of this fact, an additional uncertainty was assigned to all of the hydrologic states represented in Figure 10A.

Figures 10B and 10C show diagramatically how the model results of groundwater pressure values throughout the slope were modified explicitly to incorporate in situ variability and uncertainty. The figures shown on 10C are typical for a particular location within the slope.

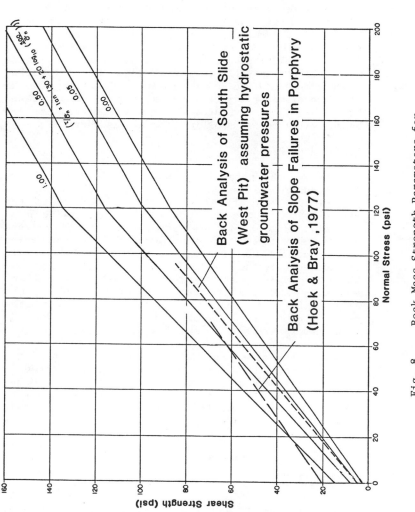

Fig. 8   Rock Mass Strength Parameters for
Porphyry/Arkose Section

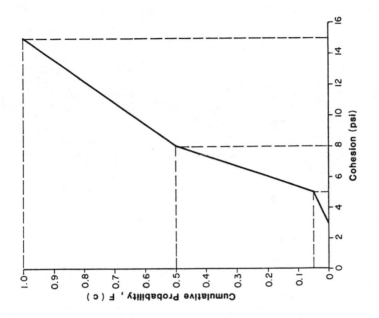

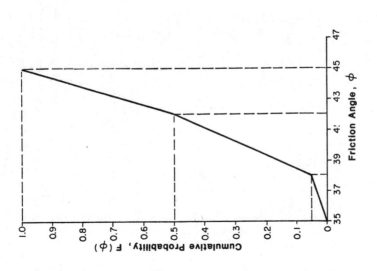

Fig. 9A   Uncertainty Description of Rock Mass Strength
(Normal Stress Rnage 0-120 PSI)

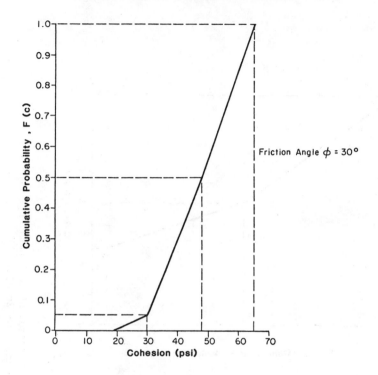

Fig. 9B   Uncertainty Description
of Rock Mass Strength
(Normal Stress Rnage 110–350 PSI)

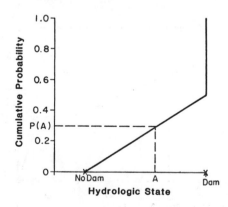

Fig. 10A   Uncertainty Description of
Hydrologic State

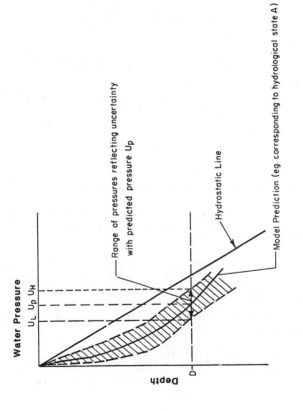

Fig. 10B & C   Uncertainty Description of Water Pressure
at a Fixed Point for a Particular Hydrologic State

Step 2   Selection of a Failure Mechanism(s)

The most obvious failure mechanism for the Mesozoic
sediments at Twin Buttes which is intensely fractured,
is a pseudo circular slip surface (Bishop 1955, Sarma
1973). The degree of fracturing and jointing, either
pre-existing or excavation-induced, was presumed
sufficiently intense for the rock mass to be regarded,
for the scale of slope being considered, as a continuum.
All known clay-filled faults are steeply dipping.

In defense of this choice, the following comments can be
made. A jointed rock mass can fail in modes which are
not simply related to structure as a result of progressive intact material fracture and bulking and
dilation of the original closely interlocked system.
Thus, the prime question should not be applicability of
the failure mechanism but suitability of the rock mass
strength (see Step 1 above). Indeed, it can be
demonstrated that the results of the analyses are
relatively insensitive to exact failure mechanism if the
appropriate mass strength is selected.

Step 3   Establish the Relationship (Transfer Function) of
Stability Coefficient with each Input Parameter

A limited number of deterministic stability analyses
were undertaken in order to establish the relationship
between the stability coefficient and a simple, interpolated Transfer Function. This is carried out by the
following variables:

1. The rock mass strength characteristics
2. The dam/no dam hydrologic state
3. The range of possible water pressures.

This transfer function could then be used in probabilistic analyses to develop probability distributions for
the Transfer Functions.

In all cases the other possible variables including the
geotechnical properties of the oxide and alluvium were
considered deterministic since considerable practical
experience of their performance was available, i.e.,
their uncertainty was considered for practical purposes
to be minimal.

It should be noted that the deterministic analyses must
be carried out sufficiently often to clearly define the
Transfer Functions. Also, there is no limit to the

number of parameters which can be included in a Transfer
Function. It is important, however, to be able to
extrapolate between one computed value of the stability
coefficient and the next. In summary, this step for
every set of slope geometry is precisely the same as
carrying out sensitivity analysis with particular care
being given to definition of the transfer functions with
values at unity or below.

Step 4   Establishment of Stability Coefficient Uncertainty
          Distributions

In this step, the cumulative probability distributions
of input parameters are repeatedly sampled according to
a uniform distribution, (i.e., between 0 and 1.0).
Then, for each sampled probability the associated value
of the input parameter can be selected from the cumulative probability relationship. Uniform sampling of the
probability values of the input variables provides
values of all the input variables. Associated with each
value of the parameter the value of the stability
coefficient can be interpolated from the transfer
functions with the appropriate assigned probability for
a variety of slope geometries. A typical result of this
step is shown in Figure 11.

The sampling process of the probability distributions of
the input variables must be done carefully. Coupled or
dependent variables such as the values of cohesion and
friction angle for two normal stress ranges must be
sampled with the same selected probability.

The sampling of each probability distribution can be
done using a completely random Monte Carlo method or
more systematic sampling methods. The advantage of
systematic sampling methods is that the accuracy of the
tails of the distribution can only be established if
sampling is sufficiently efficient. This is important
since it is the lower tail of the cumulative distribution which defines the risk of failure. Whatever
sampling method is used, it must be uniform for each of
the input-variable distributions.

Step 5   Incorporating Qualitative Uncertainty in the Analysis
          Procedure

As a result of our doubts about the analysis procedure
itself and about some factors such as the specific

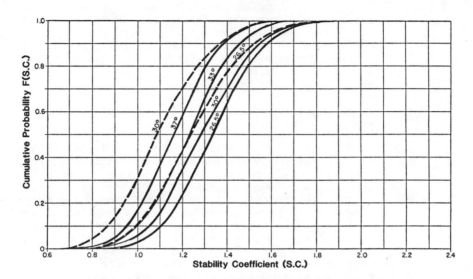

Upper Inter-ramp Slope Angle (Depressurized Slopes)
Upper Inter-ramp Slope Angle (Undrained Slopes)

Fig. 11   Stability Coefficient Distribution Functions
for Overall Slopes

weight of the rock mass which were treated deterministically, (i.e., point estimates) it was assessed that
modelled values of the computed stability coefficient
could vary by as much as ten percent. Our subjective
uncertainty in the analysis procedure was thus expressed
quantitatively in exactly the same manner as the input
variables such as rock mass strength. Thus, a correction factor to the stability coefficients was expressed
probabilistically by a simple log linear relationship
(see Figure 12). Thus, modified cumulative distributions
were calculated by simply multiplying the original
stability coefficient distributions (i.e., results of
Step 4) by this multiplier, thus increasing the probability of failure, i.e., the number of cases with a
stability coefficient less than 1.0.

Step 6   Produce Reliability Schedules

In this particular case example, an overall slope
reliability schedule was constructed for the Mesozoics
from the results of Step 5 (see Figure 13). The results
were calculated with and without the effects of the
depressurization system. The overall slope angle in the
figure is toe to crest inclusive of a single ramp or
haul road.

DISCUSSION OF PROBABILITY SLOPE STABILITY METHODS

Briefly, the advantages of full probabilistic methods of slope
stability analysis may be summarized as follows:

1. All parameters must be rigorously described in terms of the
    engineer's confidence. This allows other engineers to follow
    the path of reasoning and debate specific technical
    estimates.

2. The dilemmas concerning the choice of appropriate factors of
    safety required by traditional methods are eliminated. The
    reliability schedules produced, however, can only be
    rationally used if some sort of cost/benefit mine planning
    risk analysis is carried out. The factor of real interest
    is, however, calculated, i.e., the risk of failure.

3. This method can be used to estimate the cost benefit of
    further field or laboratory geotechnical work in terms of
    reducing the uncertainty of key input parameters to a
    residual level.

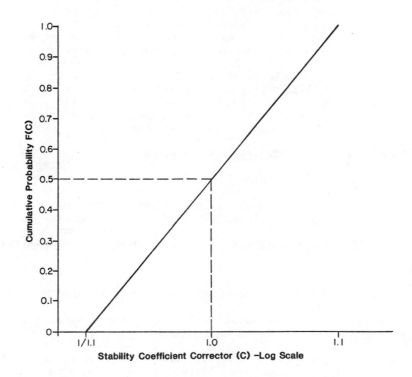

Fig. 12   Distribution Function for Parameter Reflecting
            Uncertainty in Stability Coefficient
            Arising from Analysis Procedure

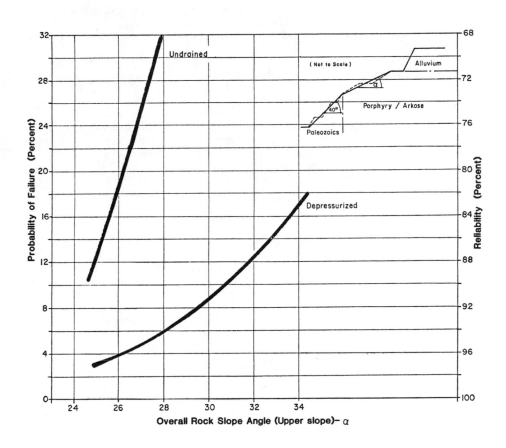

Fig. 13   Summary of Reliability Schedules for Overall Slope

4. As the pit is mined, so the uncertainty in key parameters should be gradually reduced to residual levels. Thus, from a geotechnical standpoint, the reliability schedules can be progressively optimized.

5. The methodology applied in this case example avoided expensive, inefficient traditional Monte Carlo sampling of the probability distributions of input parameters by use of the Transfer Function interpolation. It should be noted that if there are many variables, random sampling or stratified sampling methods may be better, particularly if the deterministic computer model is very efficient.

6. Probabilistic events such as earthquakes can be readily included in this type of analysis.

The disadvantages of the method are seen as follows:

1. Cost benefit analysis of failures is not an easy problem to tackle. Cost and consequence of failures are not simply a matter of deciding how much material will be required to be removed immediately to continue operations or how much stripping will be required to flatten the slope. Complex mine planning questions related to timing and alternatives at or immediately preceeding failure must be addressed. Some of these beg interesting geotechnical questions concerning rates of displacement, the extent to which ore will be irrevocably lost, etc. A conventional financial index, (e.g., Return on Investment) for evaluating feasibility studies would also have to be expressed in risk or probability terms. Even if only the cost of production, recovery, and marketing, is calculated as an index of the viability of the mine, then a complete probabilistic cost benefit analysis will be required.

2. In the case example, we attempted to include the pro-babilistic evaluation of the effect of undetected throughgoing adversely oriented clay-filled faults on the interramp slopes in the Paleozoic section. Although this approach was actually used, the resulting reliability schedules were viewed with some suspicion since the mechanism of failure and appropriate mass shear strength parameters were radically different from the base case assumption.

ACKNOWLEDGEMENTS

Several portions of this paper includes work carried out by several Golder Associates personnel. These include Dr. J. Bryne and Mr. W. Dershowitz whose advice and criticism are appreciated. The author also acknowledges our indebtedness to the Anamax Mining Corporation for permission to publish a portion of the work carried out in 1979.

REFERENCES

Barton, N.R., 1973, "Review of a New Shear Strength Criterion for Rock Joints," Engineering Geology, v. 7.

Bishop, A.W., 1955, "The Use of the Slip Circle in the Stability Analysis of Earth Slopes," Geotechnique, v. 5, pp. 7-17.

Box, G., and Jenkins, G., 1970, "Time Science Analysis Forecasting and Control," San Francisco: Holden Day.

Call, R.D., Savely, J.P., and Nicholas, D.E., 1976, "Estimation of Joint Set Characteristics from Surface Mapping Data," Proc. 17th U.S. Symposium on Rock Mechanics. Salt Lake City: University of Utah Engineering Experiment Station.

(Canada) Centre for Mineral & Energy Technology (CANMET), 1977, Pit Slope Manual, Chap. 5, Design. Ottawa, Canada: Energy, Mines and Resources, Canada.

Iman, R.L., et al., 1980, "Latin Hypercube Sampling," Sandia Report No. SAND79-173.

Jaeger, J.C., 1970, "The Behaviour of Closely Jointed Rock," Proc. 11th Symposium on Rock Mechanics, pp. 57-68. New York: AIME.

Lockhart, C.W., and Lumsden, A.M., 1980, "A Successful Method of Sampling and Testing Highly Fractured Material Using Large Diameter Core," Presented at 1980 AIME Annual Meeting, Las Vegas, Nevada; Reprint No. 80-30.

McKay, M.D., Beckman, R.J., and Conover, W.J., 1979, "A Comparison of Three Methods for Selecting Values of Input Variables in the Analysis of Output for a Computer Code," Technometrics, v. 21, no. 2.

McMahon, B.K., 1971, "A Statistical Method for the Design of Rock Slopes," Proc. of the 1st Australia-New Zealand Conference on Geomechanics, Melbourne, pp. 314-321.

McMahon, B.K., 1975, "Probability of Failure and Expected Volume of Failure in High Rock Slopes," Proc., 2nd Australian-New Zealand Conference on Geomechanics, Brisbane, pp. 308-313. The Institution of Engineers, Australia.

Murthy, M.N., 1967, "Sampling Theory and Methods," Calcutta: Statistical Publishing Society.

Papoulis, A., 1965, "Probability, Random Variables, and Stochastic Processes," Ne York: McGraw-Hill Book Co.

Pentz, D.L., 1979, "Case Examples of Open Pit Mine Drainage," Mine Drainage: Proc. of 1st International Mine Drainage Symposium, Denver, Colorado. Edited by G.O. Argall, Jr., and C.O. Brawner. San Francisco: Miller Freeman.

Sarma, S.K., 1973, "Stability analysis of Embankments and Slopes," Geotechnique, v. 23, no. 3, pp. 423-433.

Yakowitz, S.J., 1979, "Computational Probability and Simulation," Reading Mass.: Addison-Wesley Publ., Inc.

Zaremba, S.K., 1968, "The Mathematical Basis of Monte Carlo and Quasi-Monte Carlo Methods," SIAM Review, vol. 10, no. 3.

Chapter 9

MONITORING PIT SLOPE BEHAVIOR

Richard D. Call

President, Call & Nicholas, Inc.
Tucson, Arizona

ABSTRACT

In any open pit, some slope instability can be expected, varying
from bench sloughing to large-scale slope movement. Major slope displacements are preceded by small, but measurable, displacements and by
other indicators of instability, such as tension cracks, rock noise,
and changes in groundwater levels. A comprehensive monitoring program, capable of measuring and assimilating displacement related data,
is essential for sound pit operation.

The objectives of a pit slope monitoring program are

1) to maintain safe operational procedures for the protection of
personnel and equipment;
2) to provide advance notice of instability so that mine plans can
be modified to minimize the impact of slope displacement; and
3) to provide geotechnical information for analyzing the slope
failure mechanism, for designing appropriate remedial measures, and
for conducting future re-design of the slope.

Surface displacement measurement using conventional survey equipment and extensometers has been the most widely used method, and it is
still the most cost-effective. Tiltmeters and borehole inclinometers
are also useful tools, and there are promising developments in microseismic monitoring. A monitoring system should have redundancy in
both type and number of measurements, and be capable of rapid and effective dissemination of displacement information to those affected.

INTRODUCTION

In any open pit, some slope instability can be expected, varying
from bench sloughing to large-scale slope movement. Because of the

229

inherent variability of rock strength and geologic structure, the uncertainties associated with sampling and measuring rock characteristics, and the mathematical and geometric approximations of the stability analysis, even a "safe" slope, designed to some customary safety
factor, has a finite probability of instability.

Rather than attempting to design a permanently stable slope, the
current trend in slope design is to estimate the probability of failure by quantifying the variability of the stability analysis input
parameters and to utilize this probability of failure in a cost-benefit analysis in order to determine economic optimum slope angles.
Analyses of this type, which compare the cost of stripping to the cost
of slope instability, indicate that the economic optimum slope angle
may, in some cases, have probabilities of instability as high as 30
percent.

Acknowledging that slope instability can occur leads to commitment
to a monitoring program to ensure safe working conditions. The objectives of any slope monitoring program are

1) to maintain safe operational practices for the protection of
personnel and equipment;
2) to provide advance notice of instability, thus allowing for the
modification of mine plans to minimize the impact of slope displacement; and
3) to provide geotechnical information useful for analyzing the
slope failure mechanisms, for designing appropriate remedial measures,
and for conducting re-design of the slope.

SLOPE FAILURE

Defining slope failure is not as simple as it would first appear.
From a theoretical standpoint, if the rock is considered to be an
elastic material, any displacement beyond recoverable strain constitutes failure. This, however, is not a satisfactory definition for a
mine operator who often successfully mines a pit slope that has
"failed" from an elastic standpoint. Displacement of several feet,
which would be failure in a mechanical sense, may or may not cause
difficulties for a mine operation, depending on the rate of movement,
the type of mining operations, and the relationship of the moving material to the mining operation.

In a truck and shovel operation, which has considerable operational
flexibility, a displacement rate of 1 to 2 cm/day may present no major
problems because material is removed from the mining area at a faster
rate and any offsets in the haulroads can be smoothed over by routine
maintenance. The real hazard for this type of displacement is not the
existing rate of displacement but the potential of a greatly accelerated rate of movement.

On the other hand, a few cm of displacement of track in a rail pit
or in the foundation of a building adjacent to the pit would require

extensive realignment and repair. Thus, it is useful to distinguish between theoretical and operational "failure." When the rate of displacement is greater than the rate at which the slide material can be economically mined, or the movement produces unacceptable damage to a permanent facility, it is an operational failure.

Varnes (1958) used a similar economic concept to distinguish between creep and landslides. He restricted the lower limit of the rate of movement of landslide material "...to that actual or potential rate of movement which provokes correction or maintenance."

Most techniques used to calculate slope stability are static, rigid block, limiting equilibrium analysis. If the driving forces exceed the resisting forces, the slope is considered unstable.

These analyses cannot be used to predict post-failure deformation because the dynamic energy relationships of a moving block are not considered in this type of analysis. Therefore, our knowledge of the behavior of unstable slopes is largely empirical. Broadbent and Ko (1971), postulated a rheologic model which shows a good fit to observed displacement, particularly the cyclic displacement shown in Figure 1:

$$\mu = \frac{f}{K} \left(1 - e^{\frac{-Kt}{N}}\right) + \mu_o;$$

where

μ = displacement;
μ_o = initial displacement;
f = force difference;
K = elastic coefficient;
N = viscosity coefficient; and
t = time.

Because of the difficulty in determining the values of K and N from material properties, the model is limited to prediction of movement after sufficient displacement has occurred to obtain values of K and N by empirical curve fitting (Zavodni and Broadbent, 1978).

Predicting slope movement is further complicated by changes in the excess driving force, which can occur with displacement. The failure surface is rarely a smooth plane, and shearing of asperities can reduce the resisting force. Where the failure surface approximates a rotational shear, the driving force decreases as the toe heaves and the top of the slide drops.

The cyclic nature of slope displacement can also be explained by changes in pore pressure. For example, in the case of a water-filled tension crack, displacement will cause the crack to widen, and the water level will drop, reducing the driving force. As the crack fills again, the driving force will increase.

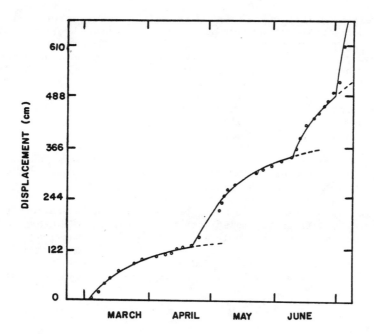

Fig. 1.   Cyclic Slope Displacement (from Broadbent & Ko, 1971).

   Given the complexity of slope displacement, no single mathematical
relationship is sufficient for predicting slope behavior. This does
not mean, however, that safe working conditions cannot be maintained
or that rapid slope movement will occur without warning. Major displacements are preceded by small but measurable displacement and other
indications of instability, such as tension cracks, rock noise, and
changes in pore pressure. As stated by Terzaghi (1950)".if a landslide comes as a surprise to the eyewitnesses, it would be more accurate to say that the observers failed to detect the phenomena which
preceded the slide."

   The Lavender Pit in Bisbee, Arizona, offers a good example of the
need for and benefits of monitoring (Metz,1974).Through the use of
simple displacement monitoring, it was possible to anticipate impending failures, with sufficient accuracy to move men and equipment from
the areas three days to sixteen hours prior to major displacement.

SURFACE DISPLACEMENT MEASUREMENT

   Surface displacement measurement using conventional survey equipment and extensometers has been the most widely used method of monitoring. It is still the most cost-effective.

Survey Network

A survey network consists of targets on the pit slope and instrument stations from which angles and distances to the targets are measured (Figure 2). Either triangulation with a theodolite or trilateration with an EDM can be used.

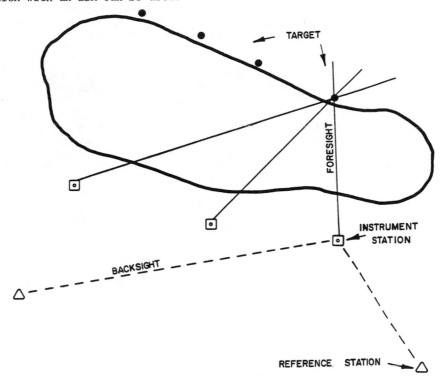

Fig. 2. Illustration of Survey Net for Monitoring.

In planning the network, care should be taken to ensure that a sufficient number of targets are established for collecting the necessary data, as long as periodic readings can still be taken with a minimum number of instrument set-ups. If a total station EDM instrument is used, approximately three minutes will be required for each reading; thus, about 30 to 40 prism targets can usually be surveyed in a half day. If a distance-meter EDM and a theodolite are used in combination, each reading would take about five minutes.

The survey network has several primary functions:

1) it establishes a surveillance system to detect initial stages of slope instability;

2) it provides a detailed movement history in terms of displacement directions and rates in unstable areas; and

3) it defines the extent of the failure areas.

The observation (instrument) stations should have stable bases because deviations in survey data can result from the inability to repeatedly set up exactly in the same position at the stations. Stable bases are best established with concrete or metal monuments. An instrument base plate is affixed to the top of the monument to serve as an instrument platform.

Primary survey points, used to tie the observation stations to the mine grid baseline, should be located on stable ground, beyond the influence of pit excavation. These relatively permanent stations (solid monuments) are needed to determine whether movement of the observation stations has occurred as a result of slope instability.

Prism targets should be attached to bench faces, if possible. A location 6 to 8 ft above the bench toe is usually preferred. Minor raveling may dislodge the prism if it is located near the crest. If the target is near the toe, it could be covered relatively quickly by raveled rock debris. In some areas, the prism reflectors will have to be mounted on sturdy tripods at selected monitoring points. It is best to allow newly installed targets to stabilize for one week before readings begin. Initially, the readings may be somewhat erratic, but an overall trend should soon become apparent.

An adjustment capability is usually needed for each target on a moving slope because it may be necessary to adjust the prism for proper instrument alignment if significant slope displacement occurs or if the instrument station is relocated. The prism can tolerate a misalignment up to 14° and still return the signal to the EDM.

The accuracy of survey measurements is a function of the precision of the instruments and the distance measured, as shown in Figure 3.

Tension Crack Mapping

One early, obvious indication of slope instability is the development of tension cracks. By systematic mapping of these cracks, the extent of the unstable area can be established (Figure 4). The ends of the cracks should be flagged so that on subsequent visits new cracks or extensions of existing cracks can be identified.

Wire Extensometers

Portable wire extensometers can be used to provide monitoring in areas of active instability and to provide backup for the survey system. These monitors can be quickly positioned and easily moved. A simple extensometer, which can be fabricated in the mine shop, is shown in Figure 5.

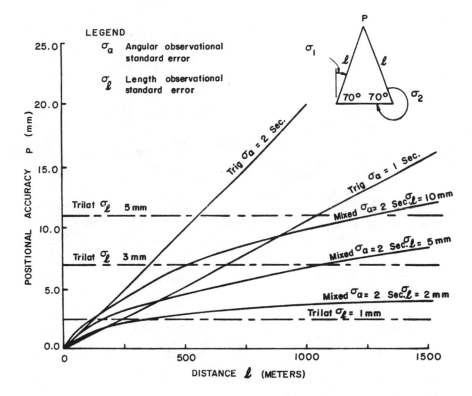

Fig. 3. Positional Accuracy of P by Triangulation, Trilateration, and Triangulateration (after V AShkenazi, 1973)

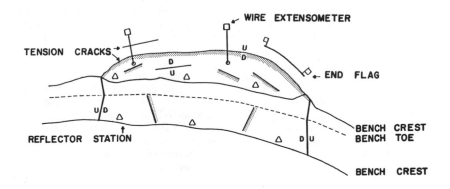

Fig. 4. Tension Crack Map.

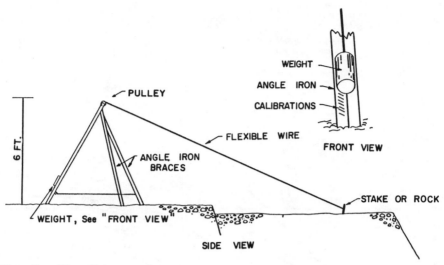

Fig. 5. Wire Extensometer.

For backup to the survey system, an extensometer should be positioned on stable ground behind the last visible tension crack, and the
wire should extend out to the unstable area (Figure 4). For warning
devices, or for information on deformation within the sliding mass,
wire extensometers can also be placed at any strategic location. Anyone working in the area can make an immediate check on slope movement
by inspecting the instruments.

A wire extensometer can be set up as a warning device by affixing
a switch several cm above the displacement weight; significant displacement will trip the switch. Lights or sirens powered with a 6-
volt dry cell battery wired to the switch will warn of slope activity.
A continuous drum-type recorder adapted to an extensometer can provide
a continuous record of slope movement (Figure 6). It will provide
excellent data regarding the sensitivity of the slope to blasting,
production, and rainfall.

The length of the extensometer wire should be limited to approximately 60 m because sag can produce inaccurate readings. Usually
15 to 20 kg of counterweight are needed for such a length, but this
depends on the tensile strength of the wire. Aircraft control cable,
or similar wire, which is manufactured to have very little tensile
stretch, is recommended for this type of monitoring device. The flexibility and durability of steel cable, compared to the rigidity and
brittleness of INVAR wire, outweigh the benefits of the thermal properties of INVAR wire. Temperature fluctuations can be measured, and
corrections can be made, if necessary.

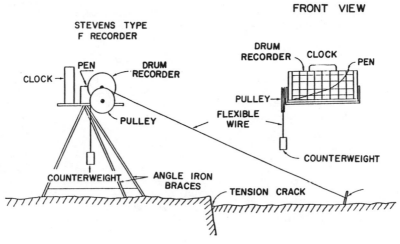

Fig. 6. Wire Extensometer with Continuous Recorder.

Other Surface Displacement Devices

When tension crack displacement is predominantly vertical, a tilt-
meter, consisting of a bar across the crack with a protractor and pen-
dulum, can be used to measure displacement. A mercury switch on the
bar can be attached to a warning light. Conversely, tiltmeters can
also be used to monitor angular displacement.

Manometers can also be used to monitor vertical displacement across
tension cracks.

SUBSURFACE DISPLACEMENT

Surface displacement measurements do not determine the subsurface
extent of instability, although it is possible to make inferences from
displacement vectors. There are situations, though, where subsurface
data is needed.

Shear Strips

Shear strips in a borehole will help to locate the position where
the hole is cut off. Either commercial segmented strips or a coaxial
cable with a fault finder can be used. These systems have the limitation of being go/no go devices.

Borehole Inclinometers

A borehole inclinometer that measures the angular deflection of the hole will give the deformation normal to the hole.

Borehole Extensometers

Borehole extensometers will give the deformation parallel to the borehole. But, because they are costly and difficult to use in locating the hole to effectively measure displacement, borehole extensometers are usually special application devices.

Piezometers

The correlation between pore pressure and slope stability is well established, both in theory and in practice. Measuring groundwater levels is an important part of monitoring, and simple standpipe piezometers are usually sufficient. There are situations, however, where low permeability or confined aquifers require pneumatic or electric devices.

Rock Noise

Experiments with microseismic recordings have established that there is a correlation between rock noise and slope movement. The cost and complexity of rock noise monitoring, though, has made it non-competitive with direct displacement measurements. However, the lessening cost of electronic equipment and its increased reliability make its potential effectiveness greater.

PRECISION, RELIABILITY AND COST

The number of different devices that can be used for monitoring, as well as the precision and sophistication of the devices, are a function of the ingenuity, time, and budget of the engineer in charge of monitoring. Since none of these factors is infinite, hard choices must be made. Some general guidelines for decision-making follow:

1) Measure the obvious things first. Surface displacement is the most direct and most critical aspect of slope instability.

2) Simpler is better. The reliability of a series system is the product of the reliability of the individual components. A complex electronic or mechanical device with a telemetered output to a computer has significantly less chance of being in operation when needed than do two stakes and a tape measure.

3) Precision costs money. The cost of a measuring device is often a power function of the level of precision. Measuring to ±1 cm is inexpensive compared to measuring to ±.0001 cm. A micrometer is unnecessary for monitoring slope movement that has a velocity of 5 cm per

day.

4)   Redundancy is required. No single device or technique tells
the complete story. A single extensometer or survey point cannot indicate the area involved in the instability, and, if it is destroyed,
the continuity of the record is lost.

5)   Timely reporting is essential. The data collection and analysis must be rapid enough to provide information in time to make decisions. Reducing last week's survey data and telling the mine superintendent that the slope was moving Thursday when a shovel was buried
Sunday does not lead to pay raises.

MONITORING SCHEDULE

A definite monitoring schedule should be established. If shooting
in the monitoring points is left up to the mine surveyor to do when he
gets the time, chances are nothing will be done.

The frequency of monitoring is a function of the precision of the
system, the rate of movement, and how critical the area is. Table 1
shows a suggested schedule. If there is a heavy rain or a large blast
in the area, additional measurements should be made.

TABLE 1.   Suggested Monitoring Schedule

| | Velocity | | Visual | | | | |
Mining	Ft/Day	Cm/Day	Inspection	Extension	Crack Map	Survey[3]	Piezometers
Active	0	0	Daily[1]	---	Monthly	Monthly	Monthly
	<0.05	<1.5	Daily[1]	Daily[2]	Weekly	Monthly	Weekly
	0.05 - 0.17	1.5 - 5.0	Each Shift[1]	Each Shift[2]	Daily	Weekly	Daily
	0.17 - 0.30	5.0 - 10.0	2 x Shift	2 x Shift	Daily	Daily	Daily
Inactive	0	0	Monthly	---	Monthly	Quarterly	Monthly
	<0.05	<1.5	Monthly	Monthly	Monthly	Monthly	Monthly
	0.05 - 0.17	1.5 - 5.0	Daily	Daily[2]	Weekly	Weekly	Weekly
	0.17 - 0.30	5.0 - 10.0	Daily	Daily[2]	Daily	2 x Week	Daily
	<0.30	<10	2 x Day	2 x Day[2]	Daily	2 x Day	Daily

1. Some mining codes require inspection of working face at beginning of each shift.
2. Extensometers should have warning lights.
3. If extensometers are not installed, survey observations should be on extensometer schedule.

DATA REDUCTION AND REPORTING

The following measurements or calculations should be made for each
survey reading:

1) date of reading, incremental days between readings, and total
number of days the survey point has been established;
2) coordinates and elevation;
3) magnitude and direction of horizontal displacement;
4) magnitude and plunge of vertical displacement;
5) magnitude, bearing, and plunge of resultant displacement vector;

and
    6) rates of horizontal, vertical, and resultant (total) displace-
ments.

    Both incremental and cumulative displacement values should be de-
termined. Calculating the cumulative displacement from initial values
rather than from summing incremental displacements minimizes the effects of occasional survey aberrations.

    Slope displacements are best understood and analyzed when the moni-
toring data are graphically displayed. For engineering purposes, the
most useful plots are

    1) horizontal position (Figure 7);
    2) vertical position (elevation vs. change in horizontal position,
plotted on a section oriented in the mean direction of horizontal displacement, Figure 8);
    3) displacement vectors (Figure 9);
    4) cumulative total displacement vs. time;
    5) incremental total displacement rate (velocity, usually in cm/
day) vs. time; and
    6) Schmidt plots of total displacement vectors.

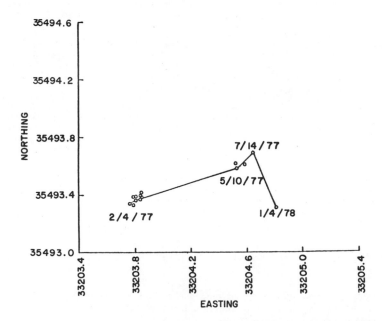

Fig. 7.   Example of Horizontal Position Plot for One Monitoring Point.

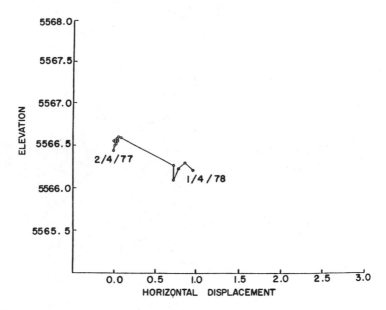

Fig. 8. Example of Vertical Position Plot for One Monitoring Point.

All graphs should be kept up-to-date and should be easily reproducible (for ease of distribution). By studying several graphs simultaneously, the movement of history of a particular slope can be determined.

The velocity-versus-time plot is usually constructed on log paper rather than on a linear scale (Figure 10). This allows a greater range of displacements to be plotted without losing the precision required for small measurements. Also, this type of graph is compatible with current monitoring techniques and analyses of slope failure kinematics (Zavodni and Broadbent, 1978).

Precipitation data should also be recorded in order to evaluate possible correlations with slope displacement. A gage located at the mine site can be used to measure occurrences and amounts of precipitation. In addition, measurement of the average daily temperatures will provide some indication of freeze and thaw periods.

The location of mining areas and the number of tons mined should also be recorded on a regular basis, because slope displacements are often associated with mining activity. One method of cataloging this information is to plot the mining area and then note the number of tons mined and the date on a plan map of the pit. A histogram can be made of tons mined versus time, and this plot can then be compared to the total displacement graphs.

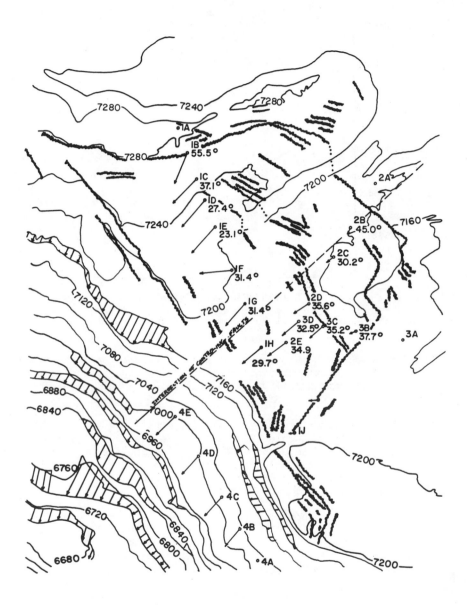

Fig. 9.   Displacement Vectors (from Miller, 1978)

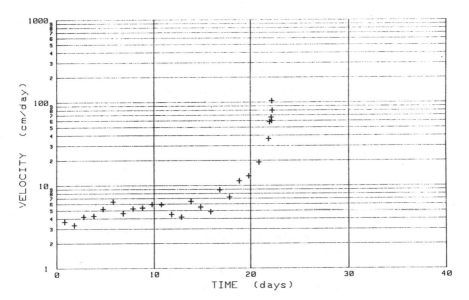

Fig. 10. Extensometer Displacement.

Monthly Slope Stability Report

    A formal monthly slope stability report should be prepared, con-
taining the data listed in Table 2. This ensures that mine management
receives the appropriate information and provides the discipline to
document slope behavior. Direct, informal communication should also
be maintained with pit operations, on a daily basis in the case of
mining in an active slide area.

Interpreting Displacement Data

    Often there are several possible failure geometries for a pit
slope, and it may not be clear, particularly at the onset of movement,
which failure geometry is active. The displacement vectors are useful
in determining the failure geometry. Figure 12 is a hypothetical example showing a possible plane shear along a fault, F_1, and a possible
wedge of faults, F_2 and F_3. The difference between the two would be
significant since the F_1 plane shear would affect the building while
the wedge would not. By plotting the displacement vectors on a
Schmidt plot, that the displacement is in the direction of the wedge,
not the plane shear, can be seen.

    On the basis of an examination of slope failure and displacement re-
cords, Zavodni and Broadbent (1978), have postulated two failure stages:
a regressive stage, during which the slope will restabilize if some

TABLE 2.   Monitoring Data Presentation

Graphs

Cumulative Displacement vs. Time

Velocity vs. Time (cm/day, Log Plot)

Precipitation vs. Time

Water Levels vs. Time

Mining vs. Time

Maps & Sections

Pit Map with Location of Unstable Areas (Figure 11)

Location of Monitoring Points with Displacement
   Vectors

Tension Crack Map

Horizontal Plot of Location with Time

Vertical Plot of Location with Time

Inclinometer Displacement

Map of Piezometric Surface

Cross-Section of Unstable Area

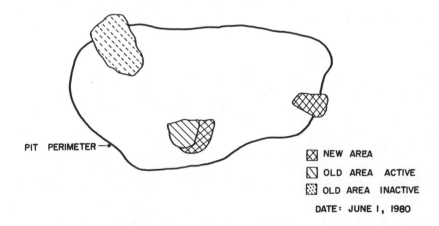

PIT PERIMETER

NEW AREA
OLD AREA ACTIVE
OLD AREA INACTIVE
DATE: JUNE 1, 1980

Fig. 11.   Slope Instability Location Map.

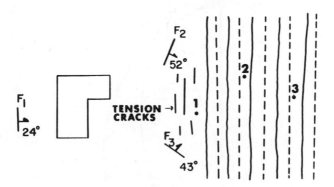

PLAN VIEW

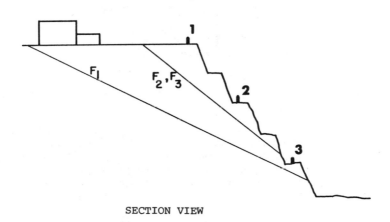

SECTION VIEW

Fig. 12.   Stereographic Plot of Displacement Directions.

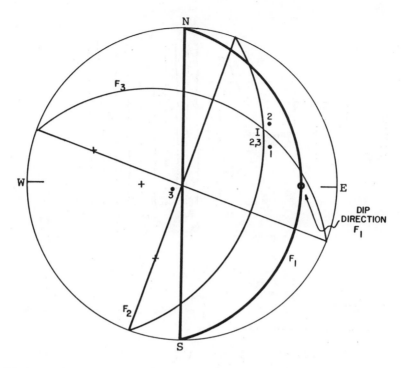

Fig. 12 (cont.).   Stereographic Plot of Displacement Directions.

external disturbance is removed; and a progressive stage, where the
failure will progress to the point of total collapse unless active
control measures are taken .   The displacement record appears to be
of an exponential form such that the velocity plots as straight line
segments on seismology graph paper with a change in slope at the onset
of the progressive stage (Figure 13).

Assuming T = 0 at the onset of the progressive stage, the equation
for the progressive stage would be

$$V = V_o e^{St};$$

where

        V = velocity;
        S = slope of line;
        t = time; and
        Vo = velocity at T (onset of progressive stage).

They postulated that the velocity of the collapse point could be

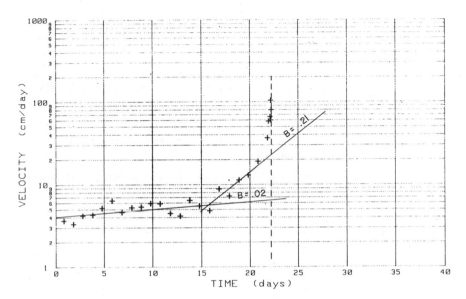

Fig. 13.   Extensometer Displacement - Regressive and Progressive
           Stages

estimated by

$$V_{col} = K^2 V_o ;$$

where K is an empirical constant. From the slope failures they obtained K values ranging from 4.6 to 10.4, with a mean of 7.21.

As can be seen in Figure 13, there may not be an abrupt collapse
point but a contained acceleration. The projections of velocity are
still useful estimators of future displacement rates, particularly if
new projections are made as new data points are obtained.

Consideration must be given to the geometry of the failure. For an
unstable area to continue to accelerate, there must be freedom to displace (a failing slope will stop if it hits the appropriate side of
the pit). Thus, the predictors of slope behavior must be made on the
basis of geometry and the potential changes in the forces acting on
the unstable mass, as well as the velocity record.

REFERENCES

1) Varnes, D.J., 1958, "Landslide Types and Processes," Landslides
   and Engineering Practice, E. B. Eckel, ed., U.S. Highway Research
   Board, Special Report 29, pp. 20-47.

2) Broadbent, C. D., and Ko, K. B., 1971, "Rheologic Aspects of Rock Slope Failures," Proceedings, 13th U.S. Symposium on Rock Mechanics, University of Illinois, Urbana, Aug. 30 - Sept. 1.

3) Zavodni, Z. M., and Broadbent, C. D., 1978, "Slope Failure Kinematics," Proceedings, 19th U.S. Symposium on Rock Mechanics, Mackay School of Mines, Reno, Nevada, May 1 - 3.

4) Terzaghi, Karl, 1950, "Mechanism of Landslides," Application of Geology to Engineering Practice, Berkey Volume, Sidney Paige, ed., Geol. Soc. America, pp. 83-123.

5) Metz, H. E., 1974, "Ground Movement and Slope Stability Problems in the Lavender Pit," Ann. Mtg., Arizona Section AIME, Tucson, Dec.

6) Miller, S. M., 1978, "Spatial Dependence of Fracture Characteristics Determined by Geostatistical Analysis," unpublished M.S. Thesis, University of Arizona.

7) Ashkenazi, V., 1973, "The Measurement of Spatial Deformation by Geodetic Methods," Symposium of British Geo. Tech. Eng. Soc. on Field Inst. in Geo. Tech. Eng., month of May.

Chapter 10

ARTIFICIAL SUPPORT OF ROCK SLOPES

Ben L. Seegmiller
Principal Consultant

Seegmiller International
Salt Lake City, Utah

ABSTRACT

Artificial support of rock slopes in mining applications may consist of rock anchors with and without auxiliary support systems, buttresses or shotcrete. The use of artificial support systems is not applicable in all cases of slope instability. Some pit slope problems simply cannot be cured with a support system. In other cases, a support system which would solve the problem would be uneconomical. Rock anchor design may be undertaken using an active or tensioned system or a passive or untensioned system. Failure problems to which artificial support systems are most commonly applied include planar, wedge, rotational and toppling failures. Design of auxiliary support systems and rock anchor peripheral items may be initially estimated using analytical concepts, but field trials are commonly used to determine final requirements. Field procedures usually require a close coordination between mining personnel and the support installation team. Great care must be exercised during installation to prevent numerous potential problems including poor anchorage, excess bearing and corrosion. Displacement monitoring of the completed system is a must if the effectiveness is to be closely judged by means other than total slope and/or support system failure. The cost of a support system is relatively high when compared to other remedial measures that may be taken to solve a slope instability problem. Therefore, the system costs play a vital role in determining whether or not artificial support should be considered beyond the initial assessment stage. Case histories are few, but do include examples of each type of support system.

INTRODUCTION

Artificial support of rock slopes in mining applications only received passing attention until 1969(Seegmiller, 1975). In that year the idea of using mechanical support techniques was put into practice

in an iron mine in eastern Canada. While the use of buttresses and
shotcrete is considered artificial support, the main technique that is
referred to as a mechanical support method is the rock anchor device.
Rock anchors may be used by themselves or they may include such auxiliary support techniques as concrete horizontal stringers between anchors, wire mesh/chain link fencing, rock bolts and/or shotcrete.

   The basic method by which a rock anchor provides slope support is
twofold. First, the anchor can apply a positive clamping force across
a discontinuity and thereby increase the natural shear strength, and
second, direct slope restraint can result. One of the major advantages of rock anchors is that a mine slope, which is almost certain to
fail, can be supported by a system requiring very little space. However, in order to be economically feasible, such a support method requires detailed analysis and planning. Potential planar shear mode
failures in hard, strong rock are the most amenable to rock anchor
support in mining applications. Other failure modes are supportable,
but add further complexity to an already involved design and implementation procedure.

   Artificial support techniques should be applied to potential
problems, not existing slope failures. Many times the use of artificial support is considered only after the problem has developed. Once
the pit slope begins to displace, it is very difficult, and probably a
violation of regulations, to put the required men and equipment on the
failure zone. A good, ongoing slope stability program should recognize the potential slope problems that may be amenable to support long
in advance of adverse displacements.

   A discussion of applicable support techniques is first presented
in this paper, followed by state-of-the-art design concepts and procedures. Next, the installation and monitoring of rock anchor systems
is described. System costs and some case histories are then reviewed
and a summary of the status of the art of slope support completes the
paper.

APPLICABLE TECHNIQUES

OVERVIEW

   Various artificial support techniques exist including rock bolts,
rock anchors, horizontal stringers, shotcrete, wire mesh, chain link
fencing, buttresses and retaining walls. Rock bolts, rock anchors and
shotcrete may be used individually to stabilize a slope, but horizontal stringers, wire mesh or chain link fencing are always used with
anchors or bolts. Buttresses placed at the pit slope toe have been
used to increase stability in some mines, but retaining walls have
found very specialized or limited applications.

Rock Bolts

These units commonly exist in diameters of 19mm and 22mm and are typically 1.5 to 3m in length. They may make use of an expansion shell to hold a force, they may use resin to secure a bond over their entire length or they may be the split tube variety which gives frictional support over their entire length. Rock bolts by themselves are not usually used to support large pit slopes, but they may be used on a small scale, such as adjacent to permanent in-pit structures such as a crusher. Use on a very large scale in civil works is common, particularly adjacent to large dams. Rock bolts are most applicable where the prevention of rock movement on a small scale is the key to stability. Therefore, by themselves, they cannot generally be used to prevent overall slope instability.

Rock Anchors

Basically, a rock anchor may be thought of as a very large rock bolt. The clamping action of the device provides an increase in normal stress across a discontinuity and thereby, in part, increases the frictional resistance to sliding in a slope. In addition, some anchors can provide a direct restraining force on the slope. Two varieties of anchors are commonly found, including the solid rod variety and the cable variety. Length of the units may be as short as 7m or as long as 80m depending on the specific application. Cable units may have higher tension loads than solid rod units due to the properties of the steel strand which is used. The large cable varieties may be tensioned to as much as 4.5MN and find common application in dam tiedown projects. For smaller slope stabilization projects, the cost of the rod variety is generally less and the ease of handling and fabrication is much better. Rock anchors as a stabilization tool may be used on a small or large scale. The size versatility lends itself to overall slope failure prevention as well as small scale bench support. Untensioned anchors may also be used. These devices would consist of cables or rods grouted into a borehole. Upon slope movement or rock dilation, the anchor would resist displacement and thus increase the normal stress across a discontinuity. With increased normal stress, the discontinuity would have a higher frictional resistance and the rock slope would be strengthened. Movement with the slope does, however, create cracks in the borehole grout and corrosion of the rod or cable could occur. Untensioned or passive anchors are therefore considered inferior to tensioned or active anchors. A schematic of an active cable anchor is shown in Figure 1, while the head and blockout of an actual field unit is shown in Figure 2.

Horizontal Stringers

These devices are either reinforced concrete beams or simply reinforcing bars. When they are used, they are always used in conjunction with rock anchors. Their purpose is to provide support for the material between adjacent rock anchors and they are fabricated or placed when the rock anchors are placed. A typical concrete horizon-

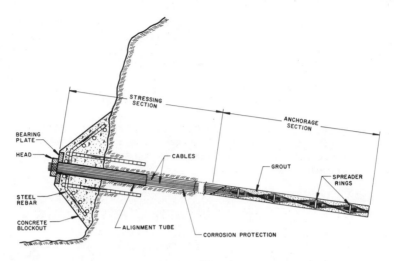

FIGURE 1
ROCK ANCHOR SCHEMATIC

FIGURE 2

IN-PLACE ROCK ANCHOR

tal stringer may be 0.5m wide, 0.5m high and 7m long and be reinforced with No. 10 bars. When reinforcing bars are used as horizontal stringers, No. 11 bars are typical. A concrete horizontal stringer with a rock anchor at the end is shown in Figure 3.

FIGURE 3

HORIZONTAL STRINGER

SHOTCRETE

Use of shotcrete as a slope stabilization tool has had limited application by itself. Where it has been used it is quite effective for the prevention of rock falls and even small bench failures to some degree. However, large scale overall slope failure would be practically unaffected by the use of shotcrete. In all cases, adequate drainage must be provided or adverse groundwater pressures will build up behind the shotcrete and cause partial or complete slope failure.

WIRE MESH

Mesh or welded wire fabric such as that used in concrete reinforcements may be used to prevent rock falls for pit stability projects. A typical application would be to use mesh with a 100mm X 100mm or 150mm X 150mm opening and a wire size from 9 to 4 gauge. Mesh use is particularly applicable to hard rock mines where the rock would tend to dislodge in blocks larger than 0.3m. Of necessity, rock bolts or rock anchors are required to keep the wire mesh in place. Use of wire mesh to contain and support an entire pit bench may be undertaken as shown in Figure 4, but large rock anchors are then required to restrain the wire mesh.

FIGURE 4

WIRE MESH

CHAIN LINK FENCING

Chain link fencing or steel drapes have been commonly used in civil projects to prevent rock falls in the same manner as wire mesh. Owing to the fact that fencing may be obtained with openings as small as 25mm, it may be used in applications where the rock is thinly bedded or where it tends to fragment into relatively small pieces. As in the case of wire mesh, rock bolts or rock anchors are required to maintain the fencing in place. Shotcrete may be used in conjunction with fencing, particularly where the rock may tend to ravel or air-slake.

BUTTRESSES

A buttress is a massive weight placed at the toe of a pit slope. The function of the buttress is to provide horizontal restraint and to increase the normal stress across a potential toe failure plane. The most common buttress used in open pit mines is stripped waste rock, as demonstrated schematically in Figure 5. In a few cases, wood cribs with rock fill have been used.

RETAINING WALLS

Retaining walls are commonly used in civil applications and they may consist of a reinforced concrete wall, steel pilings or timber supports. These devices are relatively expensive and their use would be restricted to special applications such as an in-pit crusher site.

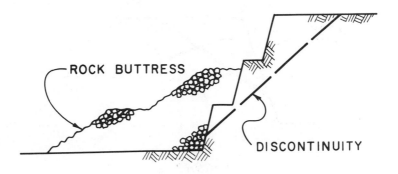

FIGURE 5
ROCK BUTTRESS SCHEMATIC

DESIGN CONCEPTS AND PROCEDURES

GENERAL CONSIDERATIONS

Supportable Failure Modes. Certain modes of slope failure are more amenable to successful support than other modes and some modes are not amenable to any type of support. Failure modes to which support can generally be successfully accomplished include planar shear, multi-block shear, wedge and toppling. Certain cases of rotational shear and massive block flow may be supported although, in general, they should not be considered supportable by mechanical means.

PLANAR SHEAR: This mode of slope failure is perhaps the best mode for which anchor support may be used. An example of planar shear mode with support anchors is shown in Figure 6. The most ideal planar shear case would be one where a single plane undercuts the slope in the upper portion of the open pit. Slope heights which must be supported over vertical heights of much more than 150m may require excessively long anchors which would be very expensive, if not impossible, to install. Each individual case will, in practice, be quite different and each must be evaluated in terms of its own characteristics.

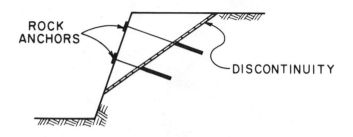

FIGURE 6

PLANAR SHEAR MODE

MULTI-BLOCK SHEAR: Multi-block shear instability is more complex than the planar shear mode, but it may be stabilized if only a
few separate blocks exist. The most simple case would be the two-
block situation as demonstrated in Figure 7. As in the case of planar
shear, the vertical height of the slope instability may be very important in the amenability of mechanical support. The best cases for
support by mechanical means would be pit slopes with multi-block modes
existing over only one or two bench heights or a total height of about
30m. Larger size multi-block situations may become very complex, both
from a design standpoint and an implementation standpoint.

WEDGE: The classical wedge failure mode which involves one or
two bench slope instabilities is quite amenable to mechanical support.
A typical wedge instability mode is shown in Figure 8. Pit slopes
having single or multiple wedges superimposed on each other may become
too large or too complex for mechanical support mechanisms. In contrast to other failure modes, the wedge mode requires a more extensive
feasibility study because (1) more than one geologic discontinuity is
involved and (2) such a failure mode is often more complex than first
realized, particularly where multiple wedges exist in a single pit
slope.

TOPPLING: The toppling failure mode may be readily supported
by mechanical means provided the size of the toppling blocks are not
too small. In open pit mining, toppling failures are not a common
happening except in some cases in limited areas on single benches. An
example of a simple case which would be amenable to anchor support is
shown in Figure 9. Other, more complex cases, may exist in actual
mining practice and they would have to be considered on an individual
basis.

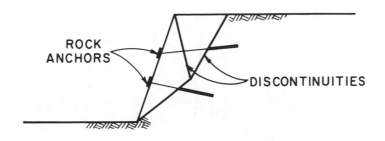

FIGURE 7

MULTI-BLOCK SHEAR MODE

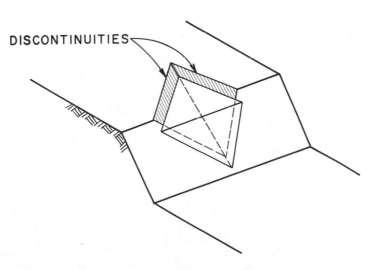

FIGURE 8

WEDGE MODE

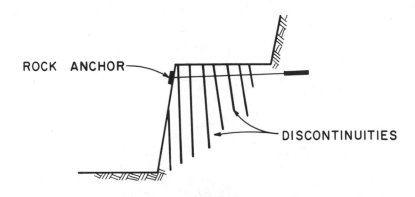

FIGURE 9

TOPPLING MODE

OTHER FAILURE MODES: Rotational shear and complex massive block flow are, in general, not supportable by mechanical means. A general rule of thumb is that if the failure mode will occur on a single plane, a series of parallel planes or two or three easily definable surfaces, the slope may be amenable to mechanical support. Complex failure modes with multiple non-parallel shear planes or shear planes which are undefinable prior to failure, cannot be effectively supported by rock anchors, even if horizontal stringers, mesh and/or shotcrete are included. Many times the complex modes or even the modes amenable to mechanical support can be supported by rock buttresses. The major limitation to using massive rock buttresses is one of space. Open pit mines are ideally sized no larger than is required for removal of the ore. A rock buttress may have to be 25 to 100m or more in width directly in front of the potential or existing instability and such an amount of space is very rarely available.

Stability Data Requirements. A prerequisite to the design of any artificial support system is a complete understanding of the stability data (Seegmiller, 1976 and 1979). The data pertinent to the slope stability include the desired mine plan, the geologic discontinuities, the groundwater, the shear strength of the potential failure plane, the density of each material in the slope and the seismic character of the site. The complete design of a mechanical support system will further require the rock-bearing capacity and rock-grout bonding strength to be measured or estimated.

Design Equation Development. The design of an artificial support system for a pit slope depends most basically on the mode of potential instability. Rock bolts, mesh/chain-link fencing and shotcrete may be effective only for small, simple, very-near-surface planar, wedge or

toppling failures. For larger potential slope failures, either buttressing or rock anchors must be used. Buttresses have their space
limitations as previously mentioned and therefore, are used in actual
practice only in isolated cases. Their design is fairly straightforward using standard design fundamentals (Richards and Stimpson, 1977)
common to each failure mode. The actual mathematical design of a mechanical stabilization system for a pit slope primarily centers on the
design of the rock anchors themselves. Procedures (Barron, K., et al.,
1970) have been formulated for design of horizontal stringers and wire
mesh systems when used in conjunction with rock anchors. Design of
chain link fencing may be done in a manner similar to that for wire
mesh. Shotcrete is not designed in a mathematical sense, per se, although the increase in shear resistance may be computed for a specific
shotcrete thickness across a specific surface where differential movement may occur. Rock bolts are generally used to provide support to
the wire mesh or chain link fencing and consequently their design is
usually based on experience. In cases where rock bolts are used by
themselves, design procedures (Burman,1969;Coates and Cochrane,1970;
Lang,1966;Reed,1974) do exist. Design of rock anchor systems have received much attention (Barron et al.,1970;Littlejohn,Bruce,1977;Major
et al.,1977) in the past 10-or-so years. Actual pit slope support problems are almost always complex and each problem must be considered individually. Few cases exist where derived (Barron et al.,1970; Hoek and
Bray,1977; Major et al.,1977; Sage,1977) design equations may be directly applied. However, to demonstrate the standard approach that is
used,some design equations will be developed for very simple cases of
planar shear and toppling. Derivation of design equations for multi-
block shear, wedge, rotational shear and other modes amenable to rock
anchor support involves the same basic principles, but they become
much more complex. (Hoek and Bray, 1977; Sage, 1977)

SIMPLE PLANAR SHEAR

Basic Assumptions. Several simplifying assumptions are necessary
to derive an equation for the safety factor against failure in the
case of simple planar shear. These assumptions in reference to Figure
10 are as follows:

1. There is a single potential failure plane which passes
    through or above the toe of the slope. This plane is inter-
    sected at some distance behind the slope face by a vertical
    tension crack forming a block.

2. Rock situated between the slope face and the potential fail-
    ure plane is potentially unstable, but may be regarded as a
    rigid body. Rock occurring in the zone beyond the potential
    failure plane may be regarded as a stable rock mass.

3. Forces of water uplift, water thrust, earthquake (assumed to
    be pseudo-static and equivalent to some proportion, α, of the
    weight of the rigid block) and the anchors pass through the
    centroid of the block and thus do not tend to cause rotation.

Further, the water uplift and water thrust may be approximated by triangular force distributions.

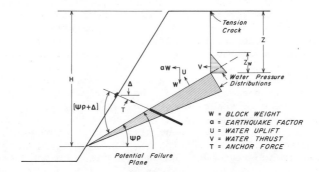

FIGURE 10

PLANAR SHEAR — WATER, EARTHQUAKE
AND ANCHOR FORCES

Stability Analysis. The derivation of the safety factor equation is based on the concept of balanced forces. The forces tending to resist slope movement are determined and summed. The forces tending to cause failure are also determined and summed. The ratio of resistive to disruptive forces is defined as the safety factor against failure. A safety factor of 1.00 indicates the state of limiting equilibrium, less than 1.00 suggests failure, and greater than 1.00 suggests stability.

The safety factor equation for planar failure is then:

$$\text{Safety Factor} = \frac{\text{Shearing Strength}}{\text{Shearing Force}} = \frac{Ss}{Sf} \qquad \{1\}$$

The shearing strength is assumed to be the strength that is defined by the equation:

$$\tau = \sigma_n \, \text{Tan} \, \phi + C \qquad \{2\}$$

where τ = shear stress

$$\sigma_n = \text{normal stress}$$

$$\phi = \text{sliding friction angle}$$

$$C = \text{cohesion}$$

Equation {2} may be rewritten in terms of forces and substituted into Equation {1} to produce the following equation for safety factor:

$$\text{S.F.} = \frac{F_n \; \text{Tan} \; \phi + CA}{Sf} \qquad \{3\}$$

where A = failure plane area

$$F_n = \sigma_n \; A$$

The resolution of forces which may affect the potential failing block in a slope is shown in Figure 11. Summing the forces normal to the failure plane (the resistive forces) gives:

$$F_n = W \; (\text{Cos} \; \psi\rho - \alpha \; \text{Sin} \; \psi\rho) - U - V \; \text{Sin} \; \psi\rho + T \; \text{Sin} \; (\psi\rho+\Delta) \qquad \{4\}$$

where

W = weight of failing block

U = water uplift force

V = horizontal water thrust force

T = total equivalent anchor force

α = earthquake loading factor

$\psi\rho$ = failure plane angle

Δ = angle between rock anchor and horizontal (negative if above horizontal, positive if below horizontal)

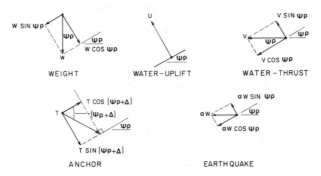

FIGURE II
RESOLUTION—FORCES AFFECTING PLANAR
SHEAR BLOCK

Summing the forces acting parallel to the failure plane (the disruptive forces) gives:

$$Sf = W (Sin \psi\rho + \alpha Cos \psi\rho) + V Cos \psi\rho - T Cos (\psi\rho + \Delta) \qquad \{5\}$$

Substituting Equation {4} and Equation {5} into Equation {3}, yields the basic safety factor equation for simple planar shear:

$$S.F. = \frac{\{W(Cos\psi\rho - \alpha Sin\psi\rho) - U - V Sin\psi\rho + T Sin(\psi\rho+\Delta)\}Tan \phi + CA}{W(Sin\psi\rho + \alpha Cos\psi\rho) + V Cos\psi\rho - T Cos(\psi\rho+\Delta)} \qquad \{6\}$$

Safety Factor Considerations. The basic safety factor equation that has been derived is for the case where the anchors are tensioned or active. That is, the force T Cos(ψ_ρ + Δ) actively lessens the disruptive forces. When a non-tensioned or passive anchor system is used the force T Cos(ψ_ρ + Δ) does not give a supporting force to the slope until displacement or dilation takes place. Therefore, in a passive system, the anchor force T Cos(ψ_ρ + Δ) behaves in a manner similar to the cohesive force CA and should be added to the resistive forces. In its simplest terms (i.e., without groundwater or seismic forces), the safety factor equation for the active planar shear case would be as shown in Equation {7}.

$$S.F._1 = \frac{\{W \cos \psi\rho + T \sin (\psi\rho+\Delta)\} \tan \phi + CA}{W \sin\psi\rho - T \cos(\psi\rho+\Delta)} \qquad \{7\}$$

Similarly, the safety factor equation for the passive case is

$$S.F._2 = \frac{\{W \cos \psi\rho + T \sin(\psi\rho+\Delta)\} \tan \phi + CA + T \cos(\psi\rho+\Delta)}{W \sin\psi\rho} \qquad \{8\}$$

A method of defining a safety factor for an anchored slope has been developed by CANMET (Barron, 1970). The model assumes for a safety factor of unity, the applied anchor force should exactly eliminate the maximum excess shear forces on the failure plane which is inclined at $\psi\rho$ to the horizontal. The shear force is defined as

$$S_F = W \sin \psi\rho \qquad \{9\}$$

The maximum excess shear force is equal to the shear force minus the natural resistive forces or

$$S_{MEF} = W \sin \psi\rho - \{W \cos \psi\rho \tan \phi + CA\} \qquad \{10\}$$

The applied anchor force mobilizes a total shear resistance of

$$S_R = T \{\cos (\psi\rho+\Delta) + \sin (\psi\rho+\Delta) \tan \phi\} \qquad \{11\}$$

The safety factor equation may then be stated as

$$S.F._3 = \frac{T \{\cos (\psi\rho+\Delta) + \sin (\psi\rho+\Delta) \tan \phi\}}{W \sin \psi\rho - \{W \cos \psi\rho \tan \phi + CA\}} \qquad \{12\}$$

An interesting comparison of these three methods may be made by considering a typical example pit slope. The pit slope is assumed to have a single plane upon which sliding could take place. The properties of the section of slope, which will be examined, are assumed as follows:

$$\phi = 20°$$
$$C = 25kPa$$

$$A = 930m^2$$
$$W = 735MN$$
$$\psi_\rho = 30^\circ$$

The anchors will be assumed to have the following characteristics:

$$\Delta = 10^\circ$$
$$T = nP = n1.8MN$$

where n = number of anchors

A plot of Equations {7}, {8}, {12} in terms of the value of the safety factor and the number of anchors is presented in Figure 12.

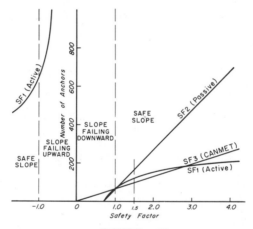

FIGURE 12
SAFETY FACTOR RELATIONSHIPS

Where no anchors are used, the active and passive cases have the identical safety factor of approximately 0.7 while the CANMET safety factor has a value of 0.0. All three have the same value of 1.00, as we would obviously expect, when approximately 64 anchors are used. However, if we were to use a 1.5 safety factor as our design minimum, the active equation (S.F.$_1$) would require 120 anchors, the passive equation (S.F.$_2$) would require 170 anchors and the CANMET equation (S.F.$_3$) would require only 95 anchors. If only 95 anchors were used, as would be acceptable with the CANMET equation, the active equation would indicate a safety factor of approximately 1.22. Therefore, it is most important to understand the differences in numbers of anchors and corresponding safety factors that could result as a consequence of the choice of a design equation. It is interesting to note that higher safety factors will result for the same number of anchors if they are active or tensioned as compared to passive or untensioned anchors. Of academic interest is the fact that, as demonstrated by the plot of $S.F._1$, for more than 655 anchors, the pit slope could actually fail upward because too much anchor force is applied to the unstable block. The active case is the only one which could theoretically give negative safety factors, which in the cited example occurs when more than approximately 265 anchors are used. One final note that should be mentioned in choosing a safety factor equation is that perhaps it would be best to use a different safety factor for each parameter. This suggestion was advanced by Londe (1965,1969,1970) indicating the magnitude of the safety factors should reflect the uncertainty attached to accurately defining them. That is, the highest values should be assigned to cohesive strength and water pressure while lower values could be used for frictional strength, weights and forces. Such an approach has merit and in cases where the end results are sensitive to parameters for which moderate uncertainty exists, it may be the most logical approach.

Optimum Anchor Inclination. The angle that the rock anchor makes with the horizontal is defined as the angle Δ. For the general situation the optimum Δ_o angle may be found by differentiating the safety factor equation, setting it equal to zero and evaluating Δ_o. For the passive case ($S.F._2$) and the CANMET case ($S.F._3$), the optimum angle is

$$\Delta_o = \phi - \psi_\rho \qquad \{13\}$$

In other words, if the anchor inclination is at an angle $\phi - \psi_\rho$, the safety factor equation is maximized for a given number of anchors. Such an inclination would minimize the support required to achieve a given safety factor. In the active case ($S.F._1$), the mathematics are much more complex and a trial and error solution is easier. The optimum angle is not necessarily the same as in the other cases. For example, in the slope example previously described, the optimum angle for the passive and CANMET cases would be

$$\Delta_o = 20° - 30° = -10°$$

This indicates an upward angle from the horizontal of 10°. A trial and error solution for the active case when 100 anchors are used, indicates that the optimum angle is approximately -15° or 15° above horizontal. The optimum angle, however, is not necessarily the most practical angle to use in the field. Experience indicates that positive Δ angles or downward directed anchors are both easier to emplace and easier to grout. For average field conditions it is recommended that the inclination angle be between 5° and 10° downward.

Anchor Length - Stressing Section. The length of the stressing section is dependent upon the location of the anchor in the slope, the slope geometry and the anchor angle (Δ). Referring to Figure 13, the stressing length for any given distance, h, above the potential failure plane's intercept with the bench face is

$$l_h = h(Cot\ \psi\rho - Cot\ \psi_f)\{Sin(-\Delta)\ Cot\ (\psi\rho + \Delta) + Cos(-\Delta)\} \qquad \{14\}$$

where Δ = anchor angle(negative if above horizontal and positive if below horizontal)

From experience, it has been found that the stressing section should have a fixed minimum length of about 7m. Stressing lengths much shorter than this obviously do not lengthen as much during tensioning. Consequently, the loss of force during lockoff will be proportionately much greater and may result in tension forces less than the design value. Therefore, the minimum stressing length may be stated as

$$l_h\ (minimum) = 7m \qquad \{15\}$$

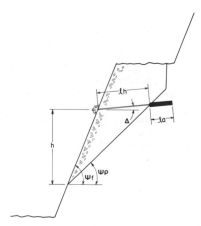

FIGURE 13
STRESSING AND ANCHORAGE LENGTHS

Anchor Length - Anchorage Section. The length of the anchorage section depends on the force of the anchor, the hole diameter and the bonding strength along the grout-rock interface. Experience has shown that the bonding strength along the grout-cable interface is significantly greater than the strength of the cable itself. For a given safety factor against anchorage failure, the minimum anchorage length is

$$l_a = \frac{U_m \, F_a}{\pi d \tau_b} \qquad \{16\}$$

where F_a = safety factor against anchorage failure

U_m = maximum anchor force

d = hole diameter

τ_b = grout-rock bonding strength

Bonding strengths depend on the rock type and may typically range from 200kPa for soft shale (Albritton,1974) to 2750kPa for hard limestone (Seegmiller, 1974). Most moderately competent rocks will have a bonding strength of at least 1500kPa but if there is any question of the bond strength, testing should be undertaken.

Anchor Spacing. The vertical anchor spacing, v, as shown in Figure 14, is usually determined by the bench height. That is, cost and ease of installation considerations will usually dictate that the anchors be placed on the bench floor against the bench face. If the installation is made during the mining operation, a half bench spacing can easily be used. This is accomplished by mining each bench in two lifts and installing the anchors after each lift is mined. The horizontal spacing, ℓ, is determined by considering the total length of the stabilized section, L, the number of rows of anchors, n, and the total number of anchors, N, necessary for slope stabilization. The number of rows is

$$n = \frac{Z_T}{v} \qquad \{17\}$$

where Z_T = total slope height

The number of anchors per row, η is

$$\eta = \frac{N}{n} \qquad \{18\}$$

Assuming that an anchor is placed near each end of the slope whose length is L, the horizontal spacing will be

$$\ell = \frac{L}{\eta - 1} \qquad \{19\}$$

or

$$\ell = \frac{L}{\dfrac{Nv}{z_T} - 1} \qquad \{20\}$$

To obtain the best distribution of forces across the potential failure plane, the vertical and horizontal spacings should be approximately the same. However, this may not always be possible, particularly when the bench heights are large. In these cases, consideration should definitely be given to mining the benches in a multi-lift fashion and installing anchors during each lift. Another method which would help distribute the forces more ideally is to offset or stagger the anchors on adjacent benches.

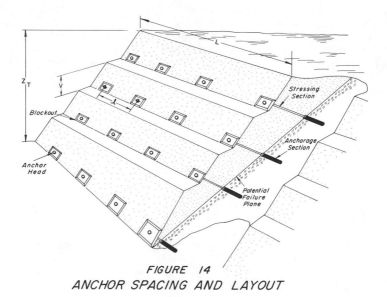

FIGURE 14
ANCHOR SPACING AND LAYOUT

Blockout Size. The blockout is a concrete block placed at the surface between the head of the anchor and the ground surface. Its purpose is to reduce the anchor bearing load on the rock to an acceptable level. The bearing capacity of brittle rock has been shown (Coates and Yu, 1971) to be at least three, and more likely, 8 to 21, times its unconfined uniaxial compressive strength. Therefore, if a minimum safety factor of three against bearing failure is acceptable, the allowable bearing pressure would be equal to the uniaxial compressive strength of the rock substance. The size of the blockout in terms of bearing area may then be obtained as follows:

$$A_b = \frac{U_m}{U_c} \qquad \{21\}$$

where U_m = maximum anchor force

U_c = uniaxial compressive strength of surface rock substance

Particular care must be taken in selecting specific sites for the anchors to insure that open joints or soft materials are not located immediately under the blockout.

Procedural Summary. A general procedure that could be used to produce a stabilization system design could be outlined as follows:

1 - Develop a section through the subject area showing the mining profile, the water pressure distributions and various mechanical properties of each geologic unit.

2 - Analyze the section and determine the most likely failure mode and the most likely failure plane. Standard analysis techniques (Hoek and Bray, 1972) may be employed as well as specialized techniques (Barron, 1970; Coates and Yu, 1971) which have been developed to determine the most likely failure plane.

3 - Determine the optimum anchor angle, A_o, which would be used if it could be achieved. Modify A_o to reflect the prevailing conditions and to avoid uphole emplacement and grouting, if possible. In some cases the use of several anchor angles may be the best procedure.

4 - Determine the total equivalent anchor force T for a particular safety factor.

5 - Select an anchor size and total number of anchors such that:

    (a) The anchor size times the total number of anchors will produce the desired total equivalent anchor force.

(b) The anchor size will not cause the bearing capacity of
the rock to be exceeded nor cause the anchorage section
to fail.

(c) The anchor spacing is such that the anchor forces are
spread out as evenly as possible over the entire slope.

6 - Adjust the interrelated variables of anchor size, spacing,
number, required bearing capacity and required anchorage
length such that the lowest overall costs may be achieved in
terms of hole drilling, anchor hardware and blockouts.

7 - Design any related support techniques such as horizontal
stringers, wire mesh/fencing, rock bolts and/or shotcrete.

Example Calculation. To demonstrate the design of an artificial
support system for a rock slope with a potential planar shear failure
mode, a simple example is presented. The pit slope to be supported
may be envisioned as the schematic example shown in Figure 15 with the
following characteristics:

Slope Height: 60m; Slope Length: 150m
Individual Benches: 4 benches each 15m high
Bench Width: 12m; Bench Face Angle: 78.7°
Overall Slope Angle (Bench to Bench): 45°
Overall Slope Angle (Pit Floor to Slope Crest): 51.3°
Potential Failure Plane: Inclined 32° above horizontal
Average Shear Strength Along Failure Plane: $\phi = 25^\circ$, C = 50kPa
Water Conditions: Slope is dry
Tension Cracks: None
Seismic Conditions: $\alpha = 0$
Average Unit Weight of Rock: 1875 Kg/m^3
Uniaxial Compressive Strength of Rock: 30MPa
Bonding Strength (Rock-Grout): 2500kPa
Important Geologic Discontinuities: Only potential failure plane

It will further be assumed that a permanent truck shop installation is
located adjacent to the pit crest and a minimum safety factor of 1.3
is desired against planar shear slope failure. The weight of the
truck shop will be assumed negligible when compared to the weight of
the potentially unstable block. The question to be answered is, "What
are the characteristics of the rock anchor system which would give the
required safety factor?"

An active rock anchor support will be chosen. Owing to the fact
that the potential sliding block is reasonably rigid because no
major discontinuities have weakened it, the active rock anchor
system should provide adequate support. Therefore, no auxiliary support in the form of horizontal stringers or wire mesh is
needed. The optimum anchor angle, Δ_c, would be at least 7° upward. For practical considerations (grouting, emplacement,
etc.) an anchor angle of +5° will be selected. Using Equation {7}

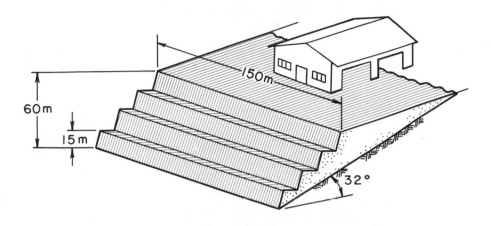

FIGURE 15

SCHEMATIC DRAWING—EXAMPLE PROBLEM

the safety factor against failure is computed to be approximately 1.15 if no anchors are used. Therefore, the safety factor
can probably be increased to 1.30 with a reasonable number of
moderately sized rock anchors. Equation {7} may then be set
equal to 1.3 and the total anchor force determined as follows:

$$1.3 = \frac{\{3975 \; Cos \; 32 + T \; Sin \; 37\}Tan \; 25 + 0.05(17000)}{3975 \; Sin \; 32 - T \; Cos \; 37}$$

$$T \simeq 241MN$$

To achieve this total force, 12-strand, 1.31MN units will be
selected. The total number of units required is

$$N = \frac{241}{1.31} \simeq 184 \; units$$

These units would be evenly placed with 46 units on each of the four benches. The horizontal spacing will be

$$\ell = \frac{150}{46-1} = 3.33m$$

If the units are placed 7.5m above each bench in a single row, the stressing length would be approximately:

Top Row   - 33.3m
Second Row - 25.3m
Third Row - 17.3m
Bottom Row - 9.3m

It will be assumed that a 127mm diameter hole is required for emplacement of the 12-strand 1.31MN units. The anchorage length (distance beyond the potential failure plane) must be determined using the maximum anchor load. The lockoff load is 1.53MN (0.7 ultimate load) and the maximum anchor load during stressing is 1.75MN (0.8 ultimate load). Using a safety factor of 3 against anchorage failure, the anchorage length is therefore

$$l_a = \frac{1.75 \times 10^6 \ (3)}{\pi(0.127) \ 2500 \times 10^3}$$

$$l_a = 5.3m$$

Using a four row layout, the total drilling requirements would be

$$46(38.6 + 30.6 + 22.6 + 14.6) \simeq 4900m$$

A preferred layout would be to place alternate anchors at the base of the bench and at 7.5m above the bench. Such a layout would yield eight rows of anchors with a spacing of 6.67m between anchors on the same row. The vertical and horizontal spacing would then be approximately the same and a reasonably even force distribution would exist. The bottom two rows should be placed slightly higher because the potential failure plane exists at the base of the bottom bench. Recommended row locations for the bottom two rows would be 5m and 10m above the slope toe, respectively. The stressing length would then be approximately

Top row   - 33.3m
Second row - 24.2m

> Third row   - 25.3m
> Fourth row - 16.2m
> Fifth row   - 17.3m
> Sixth row   - 8.2m
> Seventh row - 12.3m
> Bottom row - 7.0m

The total drilling requirements would become

$$23(38.6+29.5+30.6+21.5+22.6+13.5+17.6+12.3) \simeq 4285m$$

or approximately 615m less than the four-row layout. The block-out would require a minimum rock bearing surface area of

$$A_b = \frac{1.75 \times 10^6}{30 \times 10^6} = 0.06m^2$$

This area could be achieved with a square blockout approximately 0.25m on edge. This is a relatively small size requirement for the blockout and the stresses in the concrete will be the governing factor. The concrete design can be accomplished by considering the blockout to be a high capacity pile and then follow standard concrete design principles.

Summary. The basic principles that have been outlined may be used to formulate a slope support plan as shown in the example calculation. Unfortunately, most real field problems are much more complex. Groundwater is generally present, the potential failure plane may not be planar nor strike parallel to the slope and other geologic discontinuities may complicate the slope model.

SIMPLE TOPPLING BLOCK

Basic Assumptions. The toppling mode of slope failure is a very complex mode to accurately model. Many actual field problems have numerous factors that markedly deviate from the conditions that must be assumed in most models. Recent studies (De Freitas, et al., 1973; Goodman and Bray, 1976; Heslop, 1974) have made great progress in describing and mathematically evaluating the toppling mode. For purposes of demonstrating how such a failure mode may be artificially supported, a very simple and restricted model will be presented. The model is shown in Figure 16 and has the following qualifying assumptions:

1 - The block is fixed at its lower downhill corner such that it cannot slide down the plane. Rotation or toppling around the point "O" is the only movement permitted.

2 - The reaction force, R_N, is a point force and is acting at the point "O" at the beginning of rotation or toppling.

3 - A column of water exerts a thrust force, V, on the uphill
    side of the block and it may be approximated by a triangular
    force distribution.

4 - The column of water causes an uplift force, U, on the base of
    the block which is maximum on the uphill edge and decreases
    to zero on the downhill edge. It may be approximated by a
    triangular force distribution.

5 - The anchor force, T, is applied normal to the block and some
    fixed object exists upslope to which the anchor is attached.

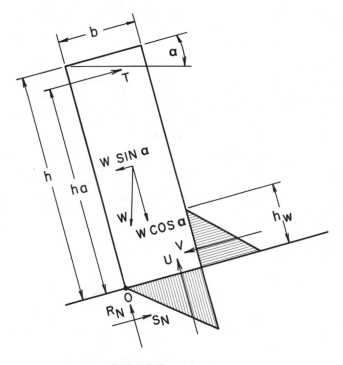

FIGURE 16

FORCES — SINGLE TOPPLING BLOCK

Stability Analysis. Summing moments about the point "O" in Figure 16 and equating them to zero would give the required anchor force at toppling limiting equilibrium of the block. The moments are

$$\Sigma M_O = 0 = W(\tfrac{h}{2}\text{Sin } \alpha - \tfrac{b}{2}\text{Cos } \alpha) + \tfrac{1}{2}h_w^2\gamma_w(1/3\ h_w) + \tfrac{1}{2}h_w\gamma_w b(2/3\ b) - Th_a$$

Simplifying and solving for T, the equation becomes

$$T = \frac{3W(h\ \text{Sin } \alpha - b\ \text{Cos } \alpha) + h_w\ \gamma_w(h_w^2 + 2b^2)}{6\ h_a} \qquad \{22\}$$

A further analysis may be made to determine what width the block should be in order to prevent toppling. Such a block width may be created by using rock anchors to tie together a number of smaller blocks into a single block. In other words, the length of the required anchors may be determined such that toppling will not occur. Toppling will be prevented when the overturning moment is zero. The overturning moment may be equated to zero with a zero anchor force and then the width may be determined. The moments are

$$\Sigma M_O = 0 = W(\tfrac{h}{2}\text{Sin } \alpha - \tfrac{b}{2}\text{Cos } \alpha) + \tfrac{1}{2}h_w^2\gamma_w(1/3\ h_w) + \tfrac{1}{2}h_w\gamma_w b(2/3\ b) - Th_a$$

$$\text{where} \quad W = hb\gamma_R \text{ for a unit width}$$

$$Th_a = 0$$

The moment equation will then take the form

$$b^2[\tfrac{1}{3}h_w\gamma_w - \tfrac{1}{2}h\gamma_R\ \text{Cos } \alpha] + b[\tfrac{1}{2}h^2\gamma_R\ \text{Sin } \alpha] + \tfrac{1}{6}h_w^3\gamma_w = 0$$

or

$$b^2[A] + b[B] + C = 0$$

The solution to this quadratic equation, which is the minimum required anchor length to prevent toppling, is

$$b = \frac{-B \pm \sqrt{B^2 - 4AC}}{2A} \qquad \{23\}$$

$$\text{where} \quad A = \frac{1}{3}h_w\gamma_w - \frac{1}{2}h\gamma_R \cos\alpha$$

$$B = \frac{1}{2}h^2\gamma_R \sin\alpha$$

$$C = \frac{1}{6}h_w^3\gamma_w$$

This solution assumes that blocks farther up the slope will not add any downslope force to the block. Obviously, such will normally occur and therefore, depending on the number and characteristics of the upslope blocks, a safety factor greater than 1.00 will be required.

Example Calculations. The design of the support system for a simple toppling problem may be demonstrated using the following data:

$$
\begin{array}{lll}
\text{Block Height:} & h & = 10\text{m} \\
\text{Block Width:} & b & = 1.5\text{m} \\
\text{Block Unit Weight:} & \gamma_R & = 2500\text{Kg/m}^3 \\
\text{Block Inclination:} & \alpha & = 10° \\
\text{Water Height:} & h_w & = 4\text{m} \\
\text{Anchor Height:} & h_a & = 8\text{m}
\end{array}
$$

For a unit width slope

$$T = \frac{3(10\text{x}1.5\text{x}1\text{x}2500\text{x}9.81)(10\sin10-1.5\cos10)+4(1000\text{x}9.81)(4^2+2(1.5)^2)}{6(8)}$$

$$T = 22.7 \text{ kN/m required anchor force}$$

If 110kN anchors were used, a safety factor of 1.00 would exist with an anchor spacing of

$$\text{Spacing} = \frac{110}{22.7} = 4.84\text{m}$$

Using a spacing of 4.5m, would yield a safety factor of

$$S.F. = \frac{110/4.5}{22.7} = 1.08$$

The required width of the block or length of the anchors to yield a non-toppling block would be calculated as follows

$$A = \frac{1}{3}(4)(1000)(9.81) - \frac{10(2500)(9.81)}{2} \cos 10 = -108 \times 10^3$$

$$B = \frac{(10)^2(2500)(9.81)}{2} \sin 10 = 213 \times 10^3$$

$$C = \frac{1}{6}(4)^3(1000)(9.81) = 105 \times 10^3$$

$$b = \frac{-213 \pm \sqrt{(213)^2 - (4)(-108)(105)}}{2(-108)} = 2.38m$$

Therefore, if more than 2.38m of the slope were anchored to-gether as a single unit, the block would not topple under the assumed forces. If other blocks exist upslope which will add downslope forces, they must be taken into account by using a larger safety factor or a more complex analysis model.

Summary. The use of the simple model that has been presented and the associated examples demonstrate the limitations in analyzing real field problems. While the simple model may not provide a complete solution, it can be seen that a rock anchor support system can stabi-lize the toppling failure mode.

FIELD PROCEDURES - ROCK ANCHORS

INSTALLATION

Hole Drilling. An air-track or other similar drilling machine which is capable of drilling relatively flat holes is used. Hole sizes depend on the size of the anchor and the ease of insertion. For example, an 8-strand cable unit may normally be placed in a 90mm dia-meter or larger hole, while a 16-strand unit usually requires a 115mm diameter, or larger, hole. Following the drilling, each hole should be cleaned of all loose material and mud coatings to obtain maximum rock-to-grout bonding and ease of anchor emplacement. This may be done with water and a blowpipe. Holes which experience circulation problems during drilling may have to be grouted and redrilled to ensure a reliable anchorage section. Water testing with packers and flow meters is a reliable way to test the hole tightness and determine the necessity of grouting and redrilling.

Fabrication. Threaded rod units are manufactured in specific lengths and are attached end-to-end by a special coupling sleeve. The stressing section is sheathed and taped to prevent bonding to grout. Cable units are fabricated by pulling 12.7mm diameter wire rope out of a cable pack and cutting it to the desired length. The anchorage section is in part formed by taping the strands to three or four spreader ring assemblies at approximately 2m intervals. These devices allow the end of the anchorage to be expanded and contracted much like sausage links. A grout tube runs the entire length of the tendon and passes through the center of each spreader ring assembly. Tendons which are used to monitor the tensioned load must have a stressing section which remains unbonded and yet has corrosion protection. Grease or grease and plastic sheathing are normally used for this purpose.

Emplacement/Primary Grouting. Rod units are inserted into the hole after it has been filled with grout. They are then allowed to cure 3 - 7 days depending on anchor load and the cement type used. Cable anchors with 6 or less strands may be inserted by hand in the cleaned hole using several people to simply push the unit to the hole bottom. Tendons containing 12 or more strands usually have to be placed with the aid of a tugger hoist. The tugger is used in combination with several snatch blocks which allow force to be exerted on the tendon at the point where it enters the hole. Following emplacement, the anchorage section of the tendon is grouted. Typically, a water-cement mixture ratio of approximately 1L per 2.25Kg of Portland Type II cement is used. An expanding grout aid is usually added during grout mixing in the amount of 0.5-0.75 percent by weight. The grout is pumped to the hole bottom with a screw-type pump and allowed to flow back over and around the spreader rings and cable. In a 115mm diameter hole containing a 16-strand tendon, a 12m anchorage section may be achieved with 60L of water and 130Kg of cement.

Blockout Construction. Owing to the fact that rod units are usually of much less load capacity than cable units, special concrete blockouts are not required. Rather, a special steel anchor plate of sufficient size to provide the proper bearing surface is placed between the anchor nut-washer assembly and the rock. With 6-strand or smaller units a similar procedure could be followed if the rock can provide sufficient bearing. For larger units, a steel reinforced concrete blockout, which is cast about the hole collar, is generally required. A steel bearing plate with a connecting alignment tube allows the tendon to pass through the blockout.

Tensioning. Rod anchors are stressed with an electrically powered hydraulic jack. The lockoff stress is 70 percent ultimate and
the working or design level is 60 percent ultimate. Cable units are
tensioned after a suitable curing time for the tendon anchorage and
the blockout. A center hole jack is used which grips all the cables
simultaneously and stresses them. A head with a large conical wedge
or a head with individual conical wedges for each strand is normally
used to lock the forces on the cables. A jacking force of 80 percent
ultimate is applied to the cables. As the wedge or wedges are seated,
load loss occurs leaving an initial force of about 70 percent ultimate
on the cables. Long term creep in the cables and additional seating
losses result in a working force generally not less than 60 percent
ultimate which is used as a design level. Anchors which are to serve
as long term load monitors have a load cell placed between the steel
bearing plate and the head.

Corrosion Protection. Long term protection of the anchor can be
accomplished by several means. One method is to grease and plastic
wrap the stressing section of the unit during fabrication. Cable
units may alternatively be secondary grouted after tensioning by pumping a thin grout down the hole. Some form of corrosion protection is
considered a must in mining applications, even for short-life units.

Test Anchors

For moderate to large size projects involving more than about 50
anchors, the use of test anchors at the project start is recommended.
Test anchors allow variables such as ease of emplacement for a particular hole size, anchorage length and blockout size to be evaluated.
Modification in these variables may then be made so that optimum values can be selected for full-scale usage.

Displacement Monitoring

Anchor Forces. As previously mentioned, the anchor force in
selected units is usually monitored with load cells. Decreasing anchor loads may be indicative of a failing anchorage, a failing anchor
head bearing, or of anchor corrosion. Increasing anchor loads can
only be indicative of elongation in the anchor itself. Anchor elongation will only occur if the slope is failing and the anchors do not
resist the movement.

Slope Displacement. Anchored slopes should be monitored with
greater intensity than unsupported slopes to see if the support is effective. A survey net is usually the least expensive method for displacement monitoring, but extensometers and other devices may also be
used.

ARTIFICIAL SUPPORT COSTS - 1981

ROCK ANCHORS

Rod Variety. The hardware for a solid rod anchor is generally sold by weight. For a typical 340kN unit, 20m long, the cost would be approximately US$175. Hole drilling would add about US$200 to the cost. Sheathing, emplacement, grouting, tensioning and supervision costs would probably add 20 percent to the total anchor costs for a grand total cost of approximately US$450.

Cable Variety. The cable variety units are more complex in terms of hardware and emplacement, but larger forces can be obtained. Consequently, the total costs for an individual unit are much higher. A typical 50m long, 12-strand unit with a 1.32MN working force would cost approximately US$2,250.

Cost Comparison. For smaller support projects which do not require high capacity anchors, the rod variety may be substantially cheaper. Because the rod units are fairly straightforward to install, a mining company could do their own field work and thus save money on labor. Larger projects or those requiring high capacity anchors may have to use cable units. Owing to their more involved installation requirements, a high-cost specialty contractor may be required.

AUXILIARY SUPPORT SYSTEMS

Horizontal Stringers. Costs for this support item will directly depend on the size of the stringer. Based on past research costs (Barron, 1970), The present cost for 0.5m X 0.5m X 15m reinforced horizontal stringer would be approximately US$1,200.

Wire Mesh/Chain Link Fencing. Wire mesh or welded wire fabric would cost approximately US$6 per m^2 installed for No. 9 gauge mesh with 150mm X 150mm openings. Chain link fencing which uses No. 11 gauge wire with 25mm X 25mm openings would cost approximately US$4.50 per m^2 installed.

Rock Bolts. Costs for these units will depend on their type, diameter and length. Approximate cost for a single, 4m long, #7 bar, resin unit would be US$17 installed.

Shotcrete. The cost for shotcrete will vary considerably from one mine location to another depending on a variety of factors. If a mining operation already has shotcreting equipment and experienced personnel on-site, the cost for a 25mm to 50mm thick covering would be on the order of US$1.25 per m^2. Higher costs should be expected under less favorable conditions.

BUTTRESSES

The use of waste rock buttresses at the toe of a potential slope failure may have a wide cost range. Two important cost factors include the method and length of transport for the waste rock. If waste rock is already being removed from the mine and it could alternatively be placed at the toe of a nearby pit slope, a net cost savings may actually result. In most cases, waste rock buttresses will probably only be a viable possibility for small slope instabilities or for increasing the slope safety factor in abandoned portions of a mine. For estimating purposes, the cost of a typical rock buttress will be at least $3 per m^3.

CASE HISTORIES

ROCK ANCHORING

De Beers Mine. In 1970, a system(Anon,1971) was used at De Beers Mine in Kimberly, South Africa to ensure the safety of the main Witwatersrand-Cape Town railway line, residential areas and roads in the vicinity of the pit. Three 10m high benches approximately 325m long were supported. First, all loose rock was scaled from the faces and then the entire area was shotcreted. Bitumen was sprayed on the shotcrete to a thickness of 3mm, a reinforcing mesh was installed and then shotcreting was repeated. A total thickness of support cover was about 100mm. A single row of anchor bolts was then installed on each bench on 2.5m centers. Actual length and tension magnitudes are not known, but it is believed that they were on the order of 12m and 225kN sizes, respectively. Drain holes were placed in the support area to prevent groundwater pressures from building up behind the face.

Hilton Mine. The Canadian government, through its Mines Branch, undertook a support research project (Barron, 1971) at the Hilton Mine Quebec in 1969. Basically, the study involved the application of 4 rock anchors (three were cable anchors and one was a rod anchor), horizontal support stringers and wire mesh. The purpose of the study was to ascertain procedures, costs and effectiveness of the various techniques employed, not to prevent a potential slope failure. A great deal was learned about artificial support in open mines from this project and in many ways it has been the basis for stabilization projects which have followed.

Twin Buttes Mine. Following three years of investigation of support systems, the Anaconda Company, in its Mining Research Department, undertook a field trial(Seegmiller,1974;1979) of artificial stabilization in 1972 at its Twin Buttes Mine in Arizona. The primary objective of the trial was to determine cost and field performance of a support system placed in an active mine under operating conditions. The support system was comprised of a total of 40 cable anchor units which were

placed in an area measuring 60m high (4 benches) and 60m long. The
units varied in tension magnitudes, from 660kN to 1.76MN. Lengths
were either 30 or 45m. The test area was immediately adjacent to a
major slope failure zone and as mining proceeded on the opposite adjacent side, the effectiveness was evaluated. In essence, the test zone
was the only area which remained stable along the entire slope, as
shown in Figure 17, thus providing the first direct evidence that artificial stabilization techniques will work in open pit mining.

FIGURE 17

ANCHOR SUPPORTED SLOPE *(CIRCLED)* - TWIN BUTTES

Nacimiento Mine. This example(Seegmiller,1976) is the largest
single application of artificial stabilization yet undertaken in the
mining industry. In 1973 the mine operators of this sedimentary copper mine in New Mexico were faced with a situation of either abandon-
ing an entire portion of the pit or using some type of artificial stabilization. The decision was made to attempt slope support using cable anchors ranging in tension magnitude from approximately 900kN to
2.1MN. A total of 360 units were placed over a period of 14 months
under a variety of weather conditions. Lengths of the units varied
from 20m to 80m. An area measuring 200m in length and 200m in height
was supported. Owing to the urgency of the needed support, the cost
of the project, the variety of weather conditions encountered and the
fact the installation had to be coordinated with mining, a number of
unforeseen problems manifested themselves during the project. The
single worst problem was that of stress corrosion which eventually
caused tension losses in over 100 of the anchors. Compounding the
stress corrosion problem was the fact that corporate management

elected to close the mine, due to a depressed copper market, four months prior to the project's scheduled completion. Following further studies into the stress corrosion problem, all remaining anchors were flushed with a lime solution and secondary grouted. Although the stress corrosion problem allowed portions of the slope to fail, the remaining anchors did provide support and have demonstrated the effectiveness of artificial stabilization. A profile of the supported slope is shown in Figure 18.

FIGURE 18

ANCHOR SUPPORTED SLOPE - NACIMIENTO

Pipe Mine. An open pit nickel mine in Manitoba undertook in 1975, the research and development of artificial stabilization (Sage, 1977). A total of 22 cable anchors each 445kN and 45m long were installed over three 150m-long benches. The major project difficulty was access which had to be provided with suspended and mobile platforms. Consequently, the cost of the project was relatively high, but after one year the anchors were still functioning satisfactorily.

SHOTCRETE

The only documented example (Sage, 1977) of pit slope support using only shotcrete took place as a CANMET research project at the Ruttan Mine in Manitoba. The face of a single 15m bench was shotcreted with a 100mm thickness for a distance of approximately 30m. The objectives of the program were to determine application procedures and costs. Of significance is the fact that part of the bench had reinforced shotcrete and part had regular shotcrete. The reinforced

shotcrete had a shear strength approximately 80 percent greater than
the regular shotcrete, but cost more than three times as much.

Rock Buttress

Many case histories of rock buttressing exist, but few have involved detailed slope stability analysis and design. In 1980, the
highwall of a uranium mine in the western U.S.A. was to be abandoned.
The regulatory authorities desired that a "safe" slope be left in the
mine owing to the fact that the uranium bearing rock was classified as
a potentially hazardous material. The existing slope was approximately 175m high and had an overall angle of approximately 51 . The existing safety factor against slope failure was computed to be about
1.10. The decision was made to use a rock buttress at the toe of the
slope of such size that the safety factor against slope failure would
be increased to approximately 1.50. The result was a multi-terraced
buttress with a total height of 65m. The structure projected 115m into the pit in front of the slope toe and contained approximately
5300m^3 of waste rock per meter of slope length.

PRESENT KNOWLEDGE STATUS

Some twelve years have passed since the first rock anchors were
placed in a surface mining operation. The question arises as to what
has been learned in that period of time and what factors should be
given particular attention in any contemplated future usage. A summary of some of the most important of these factors is as follows:

1 - The most promising applications on a large scale, involving
    an entire pit slope, would be for well-defined planar shear
    failures in strong, hard rock. On a smaller scale, other
    near-surface failure modes could be supported. Major perma-
    nent installations such as an in-pit crusher may benefit by
    having their stability increased with mechanical support.

2 - A very detailed feasibility study is a must prior to the
    start of any field activity. Too many costly mistakes may
    result if installation is begun prior to completing the in-
    vestigation of the stability and support variables.

3 - It is an absolute must to have very close coordination bet-
    ween installation and mining. A sufficient working space,
    approximately 15m wide in the direction normal to the slope
    is important on the bench where drilling, emplacement, block-
    out construction and tensioning must take place. If mining
    takes place at such a fast rate that the main anchor working
    bench is cut to its final size, difficulties in completing
    all units will probably occur. If all units are not properly
    completed, as assumed in the stability design, the safety
    factor will be reduced and the effective support system will
    have to be increased.

4 - The problem of stress corrosion is greater in mining applications than was initially believed. Anchor units should be stressed and then secondarily grouted as soon as blockout and anchorage curing has occurred. An even better, but more costly alternative, would be to grease and plastic wrap all cable anchors when they are fabricated.

5 - The costs for an anchor support system generally end up being higher than planned. Therefore, if possible, all outside contractors should bid on a total cost basis or on a completed unit basis. Penalties for ineffective anchors or anchors not completed on schedule should be part of the contract.

REFERENCES

Albritton, J. A., 1974, "Report of Rockbolt Field Tests, Clarence Cannon Project, Missouri", U.S. Army Corps of Engineers, St. Louis District, 62 pp.

Anon., 1971, "Sidewall Support at De Beers Mine", South African Min. Eng. J., Jan.

Barron, K., et. al., 1970, "Support for Pit Slopes", CANMET Internal Report MR 70/46-LD, Mar.

Barron, K., et. al., 1970, "Artificial Support of Rock Slopes", CANMET Research Report, R228, Jul.

Barron, L., et. al., 1971, "Support for Pit Slopes", CIM Bull., Mar., pp 113-120.

Burman, B. C., 1969, "Theory and Practice of Anchored Rock Constructions", University College of Townsville, Department of Engineering, Vacation School in Rock Mechanics, May.

Coates, D. F., and T. S. Cochrane, 1970, "Development of Design Specifications for Rock Bolting from Research in Canadian Mines", CANMET Research Report R224, Jul.

Coates, D. F., and Y. S. Yu, 1971, "Rock Anchor Design Mechanics" CANMET Research Report R223, Jan.

De Freitas, et. al., 1973, "Some Field Examples of Toppling Failure", Geotechnique, Vol. 23, No. 4, pp 495-514.

Goodman, R. E., and J. W. Bray, 1976, "Toppling of Rock Slopes, Proc. Specialty Conference on Rock Engineering for Foundations and Slopes, Boulder, Colorado, ASCE, Vol. 2.

Heslop, F. G., 1974, "Failure by Overturning in Ground Adjacent to Cave Mining", Havelock Mine, Swaziland, Proc. Third Cong. Intn'l Soc. Rock Mechanics, Denver, Vol. 2B, pp 1085-1089.

Hoek, E., and J. W. Bray, 1977, Rock Slope Engineering, 2nd Ed., IMM, London, 402 pp.

Lang, T. A., 1966, "Theory and Practice of Rock Reinforcement", Proc. 45th Annual Meeting, Highway Research Board, Washington, D.C., Jan.

Littlejohn, G. S., and D. A. Bruce, 1977, "Rock Anchors—Design and Quality Control", Proc. 16th Symp. on Rock Mech., ASCE, New York, pp 77-88.

Londe, P., 1965, "Une Méthode d'Analyse à Trois Dimensions de la Stabilité d'une Rive Rocheuse, Annales des Ports et Chaussées, Paris, pp 37-60.

Londe, P., et. al., 1969, "The Stability of Rock Slopes, a Three-Dimensional Study", J. Soil Mech, and Foundation Div., ASCE, Vol. 95, No. SM 1, pp 235-262.

Londe, P., et. al., 1970, "Stability of Slopes - Graphical Methods", J. Soil Mech. and Foundation Div., ASCE, Vol. 96, No. SM 4, pp 1411-1434.

Major, G., et. al., 1977, Pit Slope Manual Supplement S-1—Plane Shear Analyses, CANMET Report 77-16, 307 pp, Aug.

Rawlings, G. E., 1968, "Stabilization of Potential Rock Slides in Folded Quartzite in Northwestern Tasmania", Eng. Geol. Vol. 2, No. 5, pp 283-292.

Reed, J. J., 1974, "Rock Reinforcement and Stabilization—Bolting Alternatives", Min. Cong. J., Vol. 60, No. 12.

Richards, D., and B. Stimpson, 1977, Pit Slope Manual Supplement 6-1—Buttresses and Retaining Walls, CANMET Report 77-4, 79 pp.

Sage, R., 1977, Pit Slope Manual Chapter 6—Mechanical Support, CANMET Report 77-3, 111 pp.

Seegmiller, B. L., 1974, "Rock Anchor Design Manual", Seegmiller International Internal Report 74-1, Feb.

Seegmiller, B. L., 1974, "How Cable Bolt Stabilization May Benefit Open Pit Operations", Min. Eng., Vol. 26, No. 12, pp 29-34

Seegmiller, B. L., 1975, "Cable Bolts Stabilize Pit Slopes, Steepen Walls to Strip Less Waste", WORLD MINING, Vol. 28, No. 8, p. 36.

Seegmiller, B. L., 1976, "A Case History of Support at Nacimiento Mine", CANMET Report 76-27, Dec.

Seegmiller, B. L., 1976, "How to Cut Risk of Slope Failure in Designing Optimum Pit Slopes", Eng. and Min. J., Vol. 177, No. 12, pp 53-59.

Seegmiller B. L., 1979, "Pit Limit Slope Design, Open Pit Mine Planning and Design, Crawford, Hustrulid, ed., SME-AIME, New York, pp 147-159.

Seegmiller, B. L., 1979, "Twin Buttes Pit Slope Failure, Arizona, USA", Rockslides and Avalanches, 2: Engineering Sites, Voight, B., ed., Elsevier, Amsterdam, pp 651-666.

Question

Could you comment more on potential for cost savings.

Answer

The potential for cost savings is enormous when one considers the possible reduction in required stripping using artificial stabilization. At some mines the steepening of slopes by only 1 or 2 degrees would allow millions of tons of rock to remain in place. At costs varying from 50¢ to $1.00 or more per ton of stripped material, we are talking about reducing mining costs by millions of dollars. Further, if a particular slope has a very high probability of failure and rock anchor support could prevent the failure, a major savings could be realized through the prevention of the failure and its associated disruption. To arrive at the actual cost savings for supporting any pit slope, a detailed stability study must be undertaken. The savings from reduced stripping plus the failure prevention minus the cost of the support system will provide an estimate of the true cost savings.

Question

How do you determine the number of units (anchors) required for a particular area. What is the cost of anchoring.

Answer

The number of anchors for the planar failure case, for example, may be determined as demonstrated in the example presented in the text of the artificial stabilization paper. Basically, one must choose a safety factor and then compute the total anchor force T required to achieve that safety factor. Then an anchor size and number of anchors is selected to produce the total anchor force necessary.

The cost of the anchors depends on their size in terms of load

capacity and length. Typically, the costs may range from $400.00 to
$2500.00 per unit.

Question

Please elaborate on describing problems of stress corrosion causing
loss of rock anchors.

Answer

The problem of either stress corrosion or hydrogen embrittlement is
perhaps the biggest single problem that one may encounter when rock
anchors are used in mining applications. The very nature of mining
operations, particularly base metal mines, exposes the rock anchors
to environments where certain ions (i.e., sulphide, arsenide) act as
surface catalysts or poisons on the steel strand and failure is
induced. Other solutions such as hydrogen sulphide and chlorides can
also affect the strand causing its rapid deterioration under any
significant stress level. Protection of the strand may be made by
keeping a physical barrier between the strand and the potentially
adverse agents or surrounding the strand with a high pH environment.
For the Nacimiento Mine a protective environment should have been
provided immediately upon emplacement and stressing. Future applications should investigate the geochemical environment of the potential
application area and immediate remedial action should be taken if
required.

Question

How could the New Mexico Mine have prevented stress corrosion.
What would be an approximate unit cost of a cable anchor.

Answer

Stress corrosion could have been prevented by greasing and plastic
wrapping each cable strand or by increasing the pH of the anchor
borehole and then fully grouting the anchors immediately after
stressing.

An approximate unit cost for a cable anchor would range from about
$400.00 for a short, low capacity unit to about $2500.00 for a long
high capacity unit

Question

For corrosion resistance, why not grout the steel to the rock over
its full length without greasing.

Answer

Either procedure may be used. That is, the anchors may be greased
and plastic wrapped or they may be fully grouted after tensioning.
In any case, it is important to have a barrier between the strand and
the surrounding rock.

Chapter 11

STABILIZATION OF ROCK SLOPES

C. O. Brawner

Dept. of Mining and Mineral Process Engineering
University of British Columbia, Vancouver
and
Brawner Engineering Limited

FACTORS THAT INFLUENCE STABILITY AND STABILIZATION METHOD

The most effective stabilization method and design can only be developed if the cause(s) and mechanics of the slide are known. Therefore, it is important that the stability be evaluated by an engineer who has had extensive rock stability experience.

The most important factors that influence stability are structural geology, ground water, blasting and the mining method or methods.

Structural Geology - The majority of failures in rock slopes occur along discontinuities in the rock (joints, bedding planes, shear or fault zones, etc.). Therefore, on all surface mining projects, the development and maintenance of a comprehensive structural geologic mapping program should be a first priority program that is considered a part of the operations function.

This program should include mapping all major features - faults, shear zones and through going joints. All structural features which could influence stability should be mapped and presented in a structural domain format for the entire mine (Canmet, 1977). Where planes, wedges or blocks will be undercut by the mining excavation, shear strength and strain to failure tests should be performed on the joints, fault gouge and gouge-rock contact surfaces. These tests must be performed on the samples in the direction of potential slide movement and must determine the range of peak and residual shear strength values. A preliminary assessment of friction angle can be obtained from measurements of failed surfaces in the field (Figure 1) or from core sliding tests (Barton, 1981).

An assessment must be made of the past geologic history of

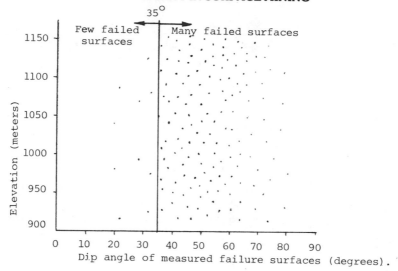

Figure 1. Field measurement of dip angles of failure surfaces
in one structural domain to provide estimate of friction angle.

discontinuities to determine whether peak, residual or an intermediate value shear strength is applicable. The unidirectional
direct shear test for residual strength is preferred to the reversal
procedure since the latter does not occur in practice.

Where very weak or weathered rock exists (more common in tropical
regions), the failure may occur through the "intact" material. For
this case, triaxial tests should be performed to determine rock
strength.

The orientation of structural geologic features and the range of
shear strength values can be evaluated using statistical methods.
However, the selection of the values to use in the stability
assessment requires experience and careful judgement.

Ground Water - In most large slides in rock (in excess of about
100,000 m^3) ground water has an important influence on stability.
The most important influence is the reduction in shear strength due
to buoyancy. The reduction in stability due to this cause (compared
to no water pressures) commonly exceeds 20 percent in practice. The
pore water pressure which influences stability is that in the
failure zone and considerable effort may be required to determine the
pressures in this zone. Measurement of pore pressures outside this
zone is usually erroneous.

If tension cracks have developed, the safety factor of the rock
slope is low and water filling the tension cracks can trigger the
failure. Where it has been decided to remove slide material,
filling the tension cracks with water is a low cost procedure that

may precipitate failure under controlled conditions.

It is important to recognize that the water pressure in the crack increases as the square of the depth and is independent of the width of the crack.

It is recommended that hubs be installed to monitor the movement and the cracks be filled as soon as possible with low permeability material.

Seepage flow creates pressures that usually act in the direction of movement to reduce stability. The pattern of this flow and determination of the continuity of pressure head with depth is important.   Stability can be improved by changing the direction of the seepage flow to be near perpendicular to the failure zone or into the slope. Methods to do this will be discussed later.

In northern climates the rock faces freeze in the winter and restrict the flow of water through the face. As a result pore water pressures increase in the slope. Many rock slides have occurred in Canada and Northern U.S.A. during extended cold periods. Measurement of piezometric pressures in the winter is just as important as during heavy rainfall or snow melt periods.

Blasting - Until recently the important influence of blasting on the stability of rock slopes in mining had not been adequately recognized (Brawner, 1968). The influence of excessive seimic acceleration forces, gas pressures and hydrodynamic shock is to develop fractures and open discontinuities which increase bench ravelling and reduce the overall stable slope angle.

The use of reduced blasting energy and controlled blasting near the final face usually justifies the additional cost by providing more stable slopes, steeper slopes and a reduced stripping ratio. In particular, closer spacing of large diameter drill holes along the final wall with reduced explosive charges and a line of buffer holes has usually been beneficial.

In heavily jointed rock or weathered rock, final rock slope development by ripping with a large bulldozer may be effective and allow steeper slopes. This technique has been successful at the Bougainville Mine in Papua-New Guinea.

Mining Method - The maintenance of stability or the institution of a stabilization program must be compatible with the mining method.

Where the slide is curvilinear, unloading material from the top of the slide will usually improve stability. However, the cost of this procedure may be excessive due to the cost of the additional excavation and loss of ore production for the equipment used in the stabilization program.

Truck and shovel operations are normally used in deep open pit mining where large slides usually develop slowly. This equipment can be moved quickly. The major concern is commonly, large rock fall from the face.

Draglines operate on top of slopes most commonly in sedimentary sequences, with weak clay, claystones, shales or slates. Failures in these materials may be rapid. Since the dragline cannot move rapidly, greater than normal attention must be paid to locating potential failure zones and to monitoring and detecting the initial movement.

Special attention must be paid to stability near haul roads, in-pit conveyors, skip ways or mine structures near the top of open pit slopes.

Where structures may be sensitive to settlement and the natural soil or rock is slightly to moderately compressible, changes in the subsurface pore water pressures may cause differential vertical settlement and stress relief due to mining and may cause lateral movement. During the early period of mining, ground elevations around the pit may undergo a period of rising due to stress relief followed by lowering as the mine gets deeper and rock stress increases.

MECHANICS OF THE MOVEMENT

The method of stabilization is influenced by the type of failure, size of slide, rate of movement and potential disruption in mine productivity.

Circular failures may be stabilized by removing material from the top of the slide, a procedure that is not effective for wedge or planar failures. The stability of block failures is usually improved by steepening the slope, a procedure that would reduce the stability of a circular slide.

Reduction in pore water pressures will improve the stability of all types of slides.

For small slides complete removal or changing the geometry of the slope may be practical. For very large slides the cost of these methods will usually be prohibitive. The larger the slide, the greater is the likelihood that some form of dewatering and pore water pressure reduction will be the most cost effective procedure to improve stability. Even in such dry areas as Arizona and northern Chile, the writer has not been involved in any deep mine where water pressures have not had a detrimental effect on stability. Very low precipitation does not mean no water table or pore pressures will be encountered.

The rate of movement will often dictate potential methods of stabilization. For very large slides, movements generally develop more slowly (except during severe earthquake conditions) and more time is usually available to develop the stabilization program. For small slides, particularly those which contain clayey gouge, clays or clayey rocks, failure can take place with so little warning that a stable slope must be maintained with assurance during mining. For example, failure of a dragline bench involving the dragline would be extremely serious on coal or tar sand projects. Special procedures to locate unstable areas and maintain stability must be developed in these instances.

INITIAL CONSIDERATIONS

When it first becomes apparent that failure of a rock slope is developing, the first considerations will usually include: (a) the potential danger to men, equipment and mine facilities and (b) the potential for disruption to the mining program.

If the assessment of danger or mine disruption results in the conclusion that the slide must be stabilized, a threefold program should be implemented immediately.

Develop a monitoring program to establish the mechanics, direction, extent and rate of the movement. The use of electronic distance measuring units is the most cost effective method for this purpose. A measurement accuracy to ± 5mm is adequate. If the mechanics of the movement are obvious, only measurement in the direction of movement is required (Brawner, Stacey and Stark, 1975). If the mechanics are not known, movement data should be used to establish the mechanism of failure and both angles and movement distance must be obtained. Since the surface movement will generally reflect the direction of the failure surface or zone, analysis of the movement pattern will usually be adequate to define the type of failure. Figure 2 illustrates how movement data indicates the type of failure.

The initial monitoring program should be concentrated down the general centerline area of the slide and readings should be taken initially at least once a day, until the mechanics are established.

Commence a geotechnical investigation to determine the geologic structure, pore water pressure profile, depth to the failure zone and likely shear strength parameters of material in the failure zone.

Most important of all is the performance of a "back anlysis" of the slide. It is assumed that the factor of safety is 1.0. The surface geometry down the centerline is determined. The top of the slide will be obvious from tension cracks and the toe of the slide can be observed or determined from an analysis of the surface movement data. The unit weight of the soil and rock is estimated or determined from drill core. The cohesion portion of the shear strength is assumed to be zero. The pore water pressure in the

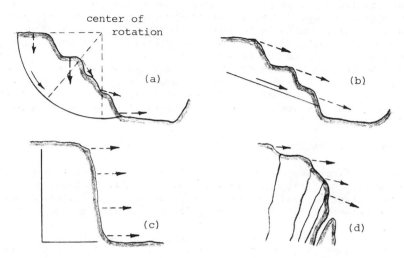

Figure 2. Movement measurements down the centerline of the
slide will indicate the type of failure,
(a) circular, (b) planar, (c) block and (d) toppling.

failure zone is measured or if this is not possible, it is assumed
for at least three elevations and the effective angle of friction is
calculated for each. This provides information for a sensitivity
analysis.

Procedures to commence stabilization of the slide should be
instituted immediately. As failure develops and movement occurs the
peak or maximum strength of the rock or discontinuity will usually be
reached in a short period of time. If stabilization procedures are
delayed, the continued strain will result in a shear strength less
than peak which will reduce eventually to the lowest or residual
value. If movement continues to a point where the available shear
strength is less than the peak value, the artificial stabilization
program must make up for this strength loss. In large slides this
can amount to millions of dollars. It must be concluded that a
stabilization program should be commenced as soon as possible after
it is recognized that a slide is developing. The stabilization
procedures will depend on the site conditions but it should be
recognized that pore water pressures so often contribute to
instability that an urgent program should be instituted to reduce
their influence. Procedures to do this are discussed later.

All tension cracks should be filled and all surface drainage
directed around the slide as an initial program.

GEOTECHNICAL INVESTIGATION

A geotechnical investigation should be instituted as soon as possible. The program should include the following:

(a) Surface monitoring of movements with emphasis on a cross section down the general center line of the slide and definition of the boundaries of the slide. Movement hubs should be installed across the tension cracks with the vertical and horizontal components determined. Movement readings should be obtained once a day until the pattern of movement is established.

(b) A borehole program with at least three drill holes located on the slide center line. A drilling and sampling procedure should be specified to obtain as close to 100% core recovery as possible. Any core that is not obtained usually represents the weakest and critical material. Triple tube core barrels are strongly recommended. If possible the program should include core orientation to provide the basis for an assessment of the frequency and orientation of discontinuities. The Christensen, Craelius and Call (Call, Savely and Pakalnis 1981) techniques are recommended for preferential consideration.

(c) A subsurface movement program which is based on the rate of movement. If very slow movement is occurring, the Slope Indicator, (Sinco, 1980) or equivalent technique is recommended. If rapid movement is occurring, the simple Sonde technique is recommended. A 5 mm diameter reinforcing rod, one meter long, is lowered on a steel fishing line down inside a pipe (which should also serve as a piezometer) to the piezometer tip located below the potential failure surface. This rod should be raised daily and lowered daily until eventually the slide movement bends the pipe so the rod cannot be raised. This point defines the base of the movement. Since the failure may be a zone, a second rod should be lowered from the surface to define the top of the failure. In some instances the slide may be failing in more than one zone.

(d) Establishment of a piezometric profile above, in and below the failure zone as soon as possible down the slide center line. For large slides the piezometer profiler developed by Patton should be considered (Patton, 1980)

As soon as the mechanics and depth of the movement are established, a back analysis should be performed. It is usual to assume the safety factor equal to 1.0 and the cohesion equal to zero after movement has commenced. The unit weight of the rock can be determined by measuring the unit weight of the core. If the pore water pressures are believed to have changed since the failure developed (the movement may open cracks and lower the pore water pressure), the back analysis should consider several pore pressure profiles. The average effective angle of friction can then be calculated for failure conditions assumed. Any stabilization procedures that are considered feasible can then be evaluated using the back analysis data and a comparative safety factor can be calculated. This method is the most accurate technique presently available to quantitatively assess the relative benefit of stabilization.

The next step is to assess the implications of the potential stabilization program on mining operations and finally to develop cost comparisons of alternative procedures.

The geotechnical investigation is best served by only performing that program which will yield information required to solve the problem and to keep the investigation program as simple as practical.

STABILIZATION METHODS

There are a number of stabilization procedures available to the mining engineer. Each is applicable to a limited variety of field conditions with the general exception of pore water pressure reduction, a procedure which will be beneficial in the majority of slides.

For large slides 'active' stabilization procedures include
(a) pore pressure reduction
(b) geometry modification
(c) cable anchoring
(d) vertical reinforcement
(e) blasting thin weak layers

Passive stabilization procedures include
(a) modification of the blasting program
(b) utilization of the mine and monitor program

Active Stabilization Procedures

(a) Pore Pressure Reduction

     1. Horizontal Drains. The most economical and rapid means of slope dewatering in developed countries has generally proven to be the horizontal drain technique (Figure 3). The technique, first developed to stabilize soil slides on highways, has direct application to rock slopes in mining.

Normally there will be insufficient time to determine the permeability of the in situ rock to design the most cost effective layout of the drains. Once the failure has commenced, urgent action should be taken to reduce the pore water pressures. As the program is developing at least two piezometers should be installed in the slide area with one near the crest on center line and one about mid height also on center line. The purpose is to monitor the effectiveness of the horizontal drain program (rate and extent of pore water reduction).

Figure 3. Horizontal drains installed to reduce the
pore water pressures in the slope. Multi-drains installed
from one location will reduce set up and moving costs.

Based on experience the guides for a horizontal drain program are
summarized:
- If the holes do no collapse, drain pipes are not required.
- If the holes even partially collapse, perforated drain pipes will
  be required. The pipe material should be selected for the
  strength, corrosion resistance, compatibility with milling
  (if the slice will be mined later), ease of installation and
  cost. The outer one third, up to 20 meters, should be left
  unperforated to ensure water which enters the pipe remains in
  the pipe.
- The length of drains should generally be equal to one half the
  height of slope with a minimum length of 15 meters and a maximum
  length of 100 to 125 meters.
- The drain holes should be installed at a plus 3-5 degree
  gradient to remain self cleaning.
- The gradient should be surveyed for the full length to ensure
  reasonable gradient control. If the drilling is forced too
  fast, the alignment may commence to raise rapidly above the
  phreatic line and the drains become ineffective. If the
  drilling is too slow the alignment will usually drop below the
  horizontal. The pressure will still cause flow and drainage
  but the adverse gradient will result in the pipe becoming
  plugged at an early date.
- Drains should be flushed once during each of the first and
  second years.

To reduce set-up and move time and to minimize the number of
locations were drainage must be controlled, four or five drain holes
should be fanned out from one set-up.

In cold climates the drain outlets must be protected from freezing. They can be covered, buried or more frequently, wrapped with heating cable. To minimize glaciation away from the drain, the rock for one bench depth below the base of the pit can be blasted so that the void volume developed will store the winter's drainage.

As the open pit deepens a second elevation of horizontal drains will ultimately be required.

2. Drainage Adits. In large pits where considerable water exists in the highwall, where fine infilled faults exist and act as a dam in the slope and in pits developed on a side hill topography, it may be economical to construct a drainage adit in the slope or under the pit (Figure 4). To make the adit more effective percussion drain holes should be drilled about 15 meters out from the adit in a fanned configuration. Mapping of the joints should be performed so that the drain holes can be angled to intersect as many joints as practical.

The drainage adit need be only large enough to allow efficient excavation, generally 1.5 x 2.2 meters. Drainage adits have been used at Marcopper and Atlas Consolidated in the Philippines, Canadian Johns-Manville in Canada Anamax in Arizona, U.S.A., and the Deye iron mine in China.

In underdeveloped countries where wages are low, the cost of drainage tunnels frequently makes them more attractive than horizontal drains.

Figure 4. Adit constructed to intercept subsurface water. To increase the effectiveness of the drainage tunnel, drain holes should be developed from the tunnel. To further increase the efficiency, the tunnel can be placed under a vacuum.

3. Wells. In some instances vertical wells will be feasible for rock slope drainage. Initial installations are often made within the pit to reduce the depth of wells required. Experience indicates however, that damage to the pumps, screens and pipe and continued pipe movement required, renders this more expensive than wells located external to the pit (Figure 5). The latter procedure allows permanent insulation to be installed for winter operation.

In the past, one or more pumping test have usually been performed to design the number, location, size and depth of pumping wells. This procedure is very slow and expensive. It is now more common to perform water packer permeability tests in holes that are used for other geotechnical purposes. This test is much less expensive and time consuming than the pumping test, more tests can be performed around the pit and a more accurate depth-permeability profile can be obtained to determine well locations for maximum efficiency.

Wells can be located near the edge of the pit and horizontal drains can be drilled to intersect the wells to develop a gravity flow system. The wells should be 30 - 45 cm diameter and filled with clean gravelly sand. If the horizontal drain misses the well during drilling, a 10 kg explosive charge in the hole near the well will usually develop hydraulic connection.

Figure 5. Dewatering wells constructed around the periphery of the open pit. At this location the wells are pumping up to 75,000 l. per minute. The drainage ditches must be watertight to prevent recirculation.

To assist in locating wet areas and seepage zones in the pit, all blast holes that are drilled should be mapped and the approximate depth of water in every hole should be noted. Seepage zones will be apparent near the pit wall where drainage is concentrated (Figure 6).

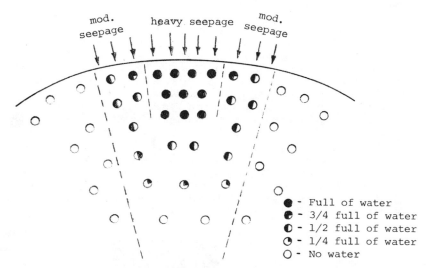

Figure 6. Blast hole plan with depth of water in all blast holes shown to demark seepage areas.

4. Vacuum Depressurization. Vacuum dewatering has been used for decades to dewater excavations in civil engineering. The author has proposed that the technique has application to slope stabilization in mining. (Brawner, 1977)

A recent field experiment at Gibraltar Mines near Williams Lake, Canada was performed where horizontal drains were installed. The outer 20 feet of the drains were grouted to the rock and the drains were subjected to a vacuum. Piezometers registered a dramatic reduction in water level soon after the vacuum was applied. With continued application of the vacuum the piezometer levels dropped to about 12 feet below the drains.

In addition to the very rapid drawdown that develops, the flow lines are directed nearly perpendicular to the drain. If the drains are below the failure surface, the seepage forces that develop create an additional downward force on the failure zone to increase stability. This is particularly important in attempting to improve stability and stop the slide quickly.

The most effective use of vacuum drainage will be where drainage tunnels are installed. The procedure involves construction of double bulkheads near the entrance of the adit beyond the zone of surface fracturing, installation of an air tight door in each bulkhead, and installation of a vacuum pump between the bulkheads to depressurize the tunnel. The vacuum line extends through the second bulkhead.

For active slides this system should be located below the failure surface to ensure that the tunnel is not damaged by movement and drain holes should be drilled up from the tunnel to intersect the failure zone.

5. Negative pore pressure development due to excavation unloading. In most soil/rock with low permeability, negative pore water pressures will develop as a result of stress relief induced by excavation (Brown, 1981). If the mining is very rapid, such as by dragline, the negative pore pressure may be sufficient to significantly increase stability initially. However, the negative pore pressures will induce the flow of water toward this zone because of the pressure gradient and depending on the permeability of the rock, the negative pore pressure will gradually reduce, thus reducing the stability. Piezometers must be installed to assess the pore pressure change with time and to assess the stability change with time.

This concept is important in determining the dragline cycle time in strip mines.

Where the permeability of the rock is high, where blasting increases the permeability, or where gas exists in the slope (i.e.coal or tar sands), there is limited possibility of the development of negative pore water pressures to practically benefit stability.

(b) Changing the Geometry

Changing the geometry of a slope is frequently used to improve stability. Since excavation is likely to continue below the toe on mining projects, the placement of a counterbalancing berm is not usually practical. For circular slides material may be removed from the top of the slide or the slope angle may be flattened. Such slides will likely occur in overburden soils, weathered rock (common in unglaciated regions), heavily fractured rock, mine waste or talus. However, geometric slope changes have little or no effect on planar or wedge type failures.

The other type of slide where changing the geometry will improve stability is the block slide, most common in sedimentary formations. The steeper the face angle, providing the material will stand steeply, the more stable the block. Weak interbed shales, claystone and mudstones often occur in uranium and coal deposits. In these materials, prone to block slides, the face angle and control of water in tension cracks are the most important factors to consider for stability. Where the angle of friction of weak interbed materials is low (less than 12°-15°), water pressure in tension cracks can actually cause blocks to move up dip.

(c) Cable Stabilization

A number of rock slopes have been stabilized with anchor cables
(Figure 7). (Seegmiller, 1981) The concept is to increase the
normal force on the failure surface to mobilize increased frictional
resistance.

Figure 7. Anchor cable installation to increase stability of
a major rock slope.

The technique has greatest potential where the ore body contact is
sharp, so that once stabilized, it is unlikely a further slice will
be mined from the slope.

The cables are easiest to install when no movement has as yet
commenced. If movement is occurring, the initial cables should
not be stressed until a sufficient number have been installed and all
can be tightened in a short time interval to stop the slide. If the
cables are stressed as each is installed, they are likely to be
broken by the movement.

The cables should be grouted full length to retain the stress in
the system and to reduce the potential of corrosion. If they are not
grouted, they must be retensioned periodically.

If mining is to be carried out below the cable stabilized slope,
it is possible the stabilization may have to be continued to lower
elevations, otherwise, the failure zone may extend below the anchors
or the slope may become overstressed below the cables so that the
failure undermines the stabilized area.

(d) Vertical Reinforcement.

Small to moderate sized slides can be stabilized or benches can be strengthened by installing unstressed cables (usually cables), long reinforcing steel rods or railway rails in vertical holes drilled with standard mine drills and concreted in the holes. For a one square inch (6.45 sq cm) cross section of cable wire, a shear resistance of about 40,000 psi (2770 kg/cm^2) is available.

If a slide is developing the number and location of cable or rod reinforcement bars and holes can be determined by performing a relative stability analysis.

This procedure (Figure 8) has been used at Utah Mines in Western Canada and Deye Mine in China and its influence on increasing shear strength has been described by Heuze, 1981.

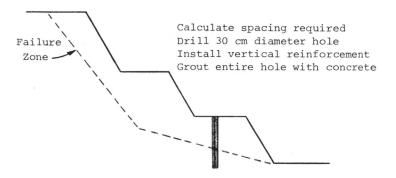

Figure 8. Use of vertical reinforcement in slide toe area to increase stability.

In areas where instability may be very serious, such as near haul roads, in pit conveyors, etc., this type of reinforcement can be installed to improve stability before any failure occurs.

(d) Blasting
There are numerous instances, such as on uranium, coal or tar sand projects where thin weak layers exist within the highwall slope in which failure may occur. The effective angle of friction may range from as low as 6° in high montmorillonite clayey layers to some 20°. Where failure may occur in these layers, aggravated by pore water pressures, blasting and disruption of these layers may improve stability. With proper blast design, the selective

disruption of the layer can increase the effective angle of friction and the increase in volume will result in a reduction of pore pressure. The long term benefit of the latter factor will depend on the potential water recharge of the zone from subsurface or surface sources prior to mining.

A program using holes about 10 meters apart was successful in improving stability of the highwall slope on the Syncrude tar sands project in Alberta, Canada. Test trenches excavated after the blast indicated that the clay layers were only disrupted in the vicinity of the blast. This indicates that a reduced hole spacing to about 3-5 meters should be used to substantially increase breakage.

The major increase in stability resulted from the sudden reduction in pore water pressure as a result of the increased rock volume.

The blast must be designed carefully so it does not precipitate the slide.

Passive Stabilization

Normally stabilization is considered to involve only some physical action taken to improve the stability of a rock slope which may or may be moving. However modification of final wall blasting and using the mine and monitor system in slide areas, offer a passive approach to dealing with stability.

(a) Improvement in stability can also be obtained by reducing seismic acceleration forces due to blasting (Brawner, 1968; Bauer and Calder, 1971). The author has inspected numerous mines where the amount of explosive per delay has ranged from 140,000 kg to 450,000 kg. At all these locations, serious instability developed as a result of very heavy blasting. Figure 9 shows a graph which illustrates the importance of reducing the amount of explosive per delay near the final wall. Figure 10 shows a final stable mine wall that has been developed using controlled blasting.

Controlled blasting involves using closer spaced blast holes along the final wall and intermediate buffer holes between the final wall and production holes. This procedure will usually result in allowable steeper overall slopes and safer working conditions. The one major exception will be where the slope angle is controlled by geologic structure. Numerous papers at this conference have illus-trated the benefit of final wall controlled blasting.

(b) The mine and monitor program developed so successfully for the Chuquicamata failure in Chile (Kennedy, B.A., et al, 1970) has application where a slide is developing and where economics do not justify the development of a stabilization program. A major application of the procedure is where the ore below the sliding area is nearly mined out and only a few more months of mining is required.

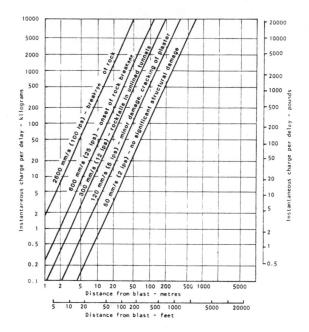

Figure 9. Relationship between maximum charge per delay, distance and particle velocity (Langefors and Kihlstrom, 1973; Ashby, 1981).

Figure 10. Controlled blasting used to develop stable mine wall.

Very large slides tend to develop slowly except those precipitated by earthquakes. Movement readings should be taken periodically to relate cumulative movement against time, precipitation and blasting occurrence. As stability reduces,acceleration will increase and the time of failure can be estimated. Mining can continue until some limiting rate of movement develops. Based on experience with large slides their complete failure will not occur within 24 hours providing the rate of movement is less than about 7.5 cm per day.

Figure 11 shows sequential photographs of a failure that occurred at Steep Rock Mines, Canada (Brawner, Stacey and Stark, 1975). Mining continued for three months after the slide condition was identified. A monitoring program using three systems, (a) measuring across the back tension crack, (b) measuring across all tension cracks and (c) using a remote Electronic Distance measuring unit,provided a good understanding of the movement and the potential danger. The example is particularly interesting in that the failure mechanism was toppling.

A modification of this technique is to artificially induce a slide to occur under controlled conditions. The most common procedure is to fill the tension cracks with water or introduce water under pressure into the general zone of sliding.

REFERENCES

Canmet, 1977. "Pit Slope Manual," Chapter 2, Supply and
          Services, Canada, Ottawa.

Barton, N., 1981. "Shear Strength Investigations for Surface
          Mining," Third International Conference on Surface
          Mining, SME of AIME.

Brawner, C.O., 1968. "The Three Major Problems in Rock Slope
          Stability in Canada," Second International Conference
          in Surface Mining, SME of AIME.

Brawner, C.O., Stacey, P. and Stark, R., 1975. "A Successful
          Application of Mining with Pitwall Movement,"
          Canadian Institute of Mining and Metalurgical
          Bulletin.

Call, R., Savely, J.P., and Pakalnis, R., 1981. "A Simple Core
          Orientation Technique," Third International
          Conference on Stability in Surface Mining, SME of
          AIME.

Sinco, 1980. "Geotechnical and Geophysical Instrumentation,"
          Slope Indicator Company, Seattle.

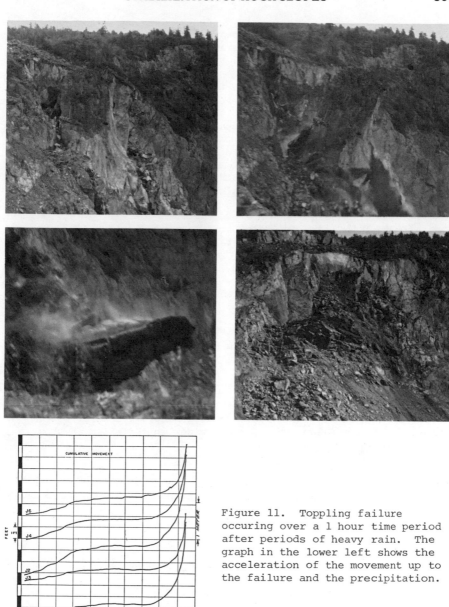

Figure 11. Toppling failure
occuring over a 1 hour time period
after periods of heavy rain. The
graph in the lower left shows the
acceleration of the movement up to
the failure and the precipitation.

Patton, F.D., 1979. "Groundwater Instrumentation for Mining Projects," Mine Drainage, Miller Freeman Publications Inc., San Francisco.

Brawner, C.O., 1977. "Open Pit Slope Stability Around the World," Journal of Canadian Petroleum Technology.

Brown, Adrian, 1981. "Influence and Control of Groundwater in Large Slopes," Third International Conference on Stability in Surface Mining, SME of AIME.

Seegmiller, Ben, 1981. "Artificial Support of Rock Slopes," Third International Conference on Stability in Surface Mining, SME of AIME.

Heuze, F.E., 1981. "Analysis of Bolt Reinforcement in Rock Slopes," Third International Conference on Stability in Surface Mining, SME of AIME.

Bauer, Alan and Calder, P.N., 1971. "The Influence of Blasting on Stability," Stability in Open Pit Mining, SME of AIME.

Kennedy, B.A., et al, 1970. "A Case Study of Slope Stability at the Chuquicamata Mine, Chile, SME of AIME.

Question

As to your statement, "large mass failing slowly", beware! Don't forget the Vaiont Slide (2 to 3 miles wide) which failed at about 60 mph.

Answer

The Vaiont Slide started slowly and gave plenty of warning of the impending failure. In my opinion, if drainage tunnels had been installed below the failure zone and placed under a vacuum, the slide would have stopped and there would have been no loss of life.

Question

What do you feel are the limits of permeability for the successful use of vacuum depressurization.

Answer

Permeability is only part of the question. The more important implication is to change the direction of the seepage forces to act approximately perpendicular to the failure zone thus increasing the normal load in the failure zone and thereby increasing stability. This will develop with low permeabilities.

Question

How do you assess the effectiveness of blasting discontinuities to improve rock slope stability. A limited experiment at Kennecott, within an active failure zone, left this issue unresolved.

Answer

The best way is to excavate a test trench to inspect the planes. Piezometers should be installed prior to the blasting to observe the influence on pore water pressure reduction which should be substantial.

Question

When you recommend a stabilization program for a typical open pit mining operation, do you base the magnitude of the stabilization program on a factor of safety or a probability of failure (or instability). What magnitude of factor of safety and what magnitude of probability of instability do you consider to be typically prudent for a general guideline.

Answer

I usually use the safety factor approach and assume a Safety Factor of 1.0 when I perform the standard back analysis. This is the most accurate shear test available since it incorporates all field and environmental site conditions.

For the design of stabilization I will generally use a safety factor of about 1.10 for large slides and about 1.20 for small slides. If the slide is endangering major facilities, I will increase these to about 1.2 and 1.3 respectively.

Question

Could you expand on the rock condition at your vacuum depressurization test site (i.e. fracture intensity; fracture filling, if any; clay content of rock; etc.).

Answer

The rock was jointed and fractured with a fracture frequency of about 6 per meter near the surface. Clay content in the joints at the site was negligible. The fractures were either reduced in frequency with depth or were moderately tight with depth since it was possible to develop the vacuum with about 20 feet of grout. For heavily blasted slopes this would not likely be possible.

Question

Concerning piezometic monitoring of (a) general pit area and (b) specific failures, how can one get a quick idea of order of depth and spacing of installations.

Answer

It is always recommended that piezometers be installed, prior to installing drains, to monitor the change in pore water pressures. Drains should be installed at the lowest pit elevation or in the toe area of the slide. Four to five drains should be developed from one set up with the ends of the drains 10 to 15 meters apart. Length of the drains should be about one half total slope height, up to a maximum length of 125 meters.

Question

Would you consider initiating a detailed stabilization program for short and long range stability in the same way or with what diffences. Any slope stabilization program must take into account the mine design and the operational life of the wall. Please comment.

Answer

If theoretical studies and stability analysis indicate that slope drainage will allow the use of steeper slopes, I will specify that blast plans show water levels in the blast holes. If water begins to be encountered at the final wall, as shown in Figure 6, I will immediately recommend horizontal drains.

Question

Monitoring pore water pressure as a stability check would seem to call for a monitoring frequency of at least once daily. Is this reasonable, and if so, would remote monitoring techniques be cost effective.

Answer

The frequency of monitoring pore pressures can be reduced when there is a gradual reduction occurring in the levels or when the pressures are lower than they have been in the past. More frequent readings should be taken when the pressures are increasing, and, during and following heavy precipitation or spring snow melt.

Question

Both in your paper and in others presented to date, rocks are described as having initially a peak shear strength followed by a reduced shear strength. Some people would call this a strain softening behavior. However, there are other rocks which have a strain hardening behavior, i.e. the strength increases with strain. Does this affect influence the general theme of your paper.

Answer

The so called strain softening (weakening) is by far the most common condition of the two that is encountered. I prefer to take the conservative approach, assuming this condition until field evidence suggests the opposite may be occuring.

Chapter 12

RESEARCH REQUIREMENTS IN SURFACE MINE STABILITY
AND PLANNING

G. Herget and O. Garg

Research Scientist, Elliot Lake Laboratory,
CANMET, Department of Energy, Mines and Resources,
Elliot Lake, Ontario

Chief, Long Range Planning,
Iron Ore Company of Canada Ltd.,
Labrador City, Newfoundland

ABSTRACT

Trends will continue towards more automated, and sometimes larger
mining and haulage equipment to reduce pit development and haulage
costs. To save labour costs, larger capitalization of open pits is
required and pit design and stripping ratios have to be firmed up at a
very early stage of mine planning.

Bench failures will assume a greater importance with larger equipment and at present we have only a limited capability of predicting
stable slopes in complex structural geological settings. Deep open
pits will pose particular problems in regard to high toe and floor
stresses.

This will require the development of more efficient systems for
rock mass exploration and an improvement of our predictive capability
in regard to pit slope stability and appropriate management of groundwater and materials.

To meet the challenge, we have to improve:

a) Analytical tools to assess slope stability in structurally complex
    settings as found in large open pits;

b) Core orientation methods during diamond drilling operations;

c) Geophysical logging methods to obtain geotechnical and structural
    geological information for large volumes of rock;   and

d) Geotechnical data handling systems to define mechanical properties of rock masses and their parameter uncertainties.

Probabilistic procedures will provide a powerful tool for organizing and evaluating site investigation data and engineering practice.

INTRODUCTION

In the last decade considerable advances have been made in regard to pit slope design for hard rock mines from a geotechnical point of view. The significance of structural geology on pit slope angles has been realized, non-uniform slope angles around open pits have been adopted, and the control of groundwater in open pits has made considerable progress.

There are many workers in the mining, geotechnical and earth sciences who contributed significantly to a better understanding of stability problems in open pits and comprehensive publications like the book on "Rock Slope Engineering" by E. Hoek and J.W. Bray (1974) and the "Pit Slope Manual" by CANMET, E.M.R. (1977) have placed expertise and know how into the hands of many resident open pit engineers. Cases are becoming less frequent where geotechnical considerations are entered into pit design only when something has gone wrong.

This achievement is certainly commendable, but many unanswered questions remain. They need urgent attention if we want to fulfil requirements which are going to be asked in the near future of the geotechnical community. To understand these needs, a brief description of the present concerns in open pit operations is helpful.

CONCERNS AND CHOICES IN OPEN PIT OPERATIONS

During the last fifteen years, significant productivity improvements have been made by open pit mine operators based on:

a) Bigger and better equipment;

b) Better utilization of manpower and supplies; and

c) Application of advanced open pit technology (Hogan and Kennedy, 1980; Rance, 1979).

However, since 1973 and particularly during the last five years, open pit mine operators have been faced with a significant increase in operating costs. These costs have increased mainly due to:

a) Substantial increases in energy and other material costs; and

b) Increased costs of manpower.

In addition to increased operating costs, fluctuating and depressed market conditions for metals and minerals and the shortage of skilled manpower have forced management to review all aspects of their operation.

Based on a review of published data and discussions with operators and planners, it appears that the current stagnant productivity trends in open pit mining and the rising costs cannot be changed if the industry stays with its present or conventional technology (Alberts and Dippenaar, 1980; Chadwick, 1981; Cranford and Hustrulid, 1979; Pelzer, 1979; Laswell and Laswell, 1980).

The following four areas offer the greatest potential for productivity improvements in open pit mining operations during the 1980's:

1) Development of more economical systems to haul ore and waste material;

2) Improvements in drilling and blasting technology;

3) Reduction of stripping ratio; and

4) Optimization of mine planning using computers.

Haulage Systems

For large open pit mining operations the cost of hauling ore and waste varies between 25% and 45% of the direct operating cost depending on the equipment used and the distances involved in hauling ore and waste material.

A more disturbing consideration is the rate of increase of haulage costs which amounted to about 20% per year during the last two years for a typical large open pit mine using trucks. These more than normal cost increases are due to the increased cost of diesel fuel, tires, and labour.

Considering that energy and labour costs will continue to rise and that haulage distances will be longer as mining advances with time to deeper levels, the haulage costs could easily double in the next five years (Chadwick, 1981).

Alternatives to conventional haulage using diesel-electric trucks are (Almond, 1980; Faulkner, 1980): a) belt conveying, b) railroads, and c) trolley assisted haulage systems.

Belt Conveying. Handling of ore and waste material using belt conveying offers high capacity and flexibility. In some cases both ore and waste material may have to be crushed before these can be conveyed out of the active pit area to the dumping point.

Although belt conveying systems are capital intensive, the savings in operating costs, including maintenance, amount to about 30% as compared

to truck haulage systems. Fig. 1 shows a typical belt conveying system
for handling waste material (Pelzer, 1979).

Fig. 1: Conveyor belt set up for handling ore and waste material.

Railroad Transport. In some open pit operations, the use of in-pit
and out-of-pit railroad systems (unit trains) for hauling ore and waste
material may be more economical than belt conveying and haulage trucks
if long haulage distances exist and the topography is favourable.

Trolley-Assist Haulage. As an interim measure, haulage of ore and
waste using trolley-assisted trucks could reduce the unit cost by:

a) Reduction in fuel consumption (up to 50%);

b) Reduction in cycle time due to increased speed of trucks (from 12
    km/hour to 20 km/hour at Shishen in South Africa); and

c) Reduced maintenance cost on engines (Gagnon and Rochfort, 1972).

    Figure 2 shows a diesel electric haulage truck equipped with panto-
graphs for a trolley-assisted haulage truck system (Chadwick, 1981).

Improvements in Drilling and Blasting Technology

    Overburden preparation is one of the most expensive operations in a
surface coal mine. It is basic to good stripping machine performance.
A large bucket or dipper requires good fragmentation of strata for
optimum production. Bucket wheel excavators have high continuous output, low operating costs, uniform and low power demands and excellent
conveyor belt loading and discharge. Bucket wheel excavators dig soft

Fig. 2: Trolley-assisted diesel-electric haulage truck.

upper strata, but will not handle harder or large fragments.

Advances in drilling and blasting are not going to be in the form of great technical break-throughs, but in the determination of an optimum choice for the given technology.

<u>Drilling</u>. During the last ten years the productivity from the blast and drill cycle has improved, mainly because of the availability of larger and heavier drilling rigs and better drill bits. However, the recent trend towards using large (440 cm.) diameter drill holes for production drilling has slowed down due to the following reasons:

a) High per foot cost and longer delivery times for large diameter drilling bits;

b) More than normal damage to final pit walls;

c) Excessive capital cost to replace the existing drills with machines capable of drilling large diameter drill holes.

Now, more emphasis is being placed on:

a) Drilling more feet per shift by monitoring the performance of drills as measured by automatic recording instruments; and

b) Obtaining rock strength data and geological information during the drilling stage for use in determining the type and quantity of explosives to be loaded into holes.

Blasting. Blasting is used to break strata too hard for ripping and future improvements related to blasting operations are likely to be based on the following aspects;

1) Selection of the quantity and type of explosives to be loaded according to rock conditions encountered during drilling;

2) Reduction of the use of more expensive slurry type of explosives for wet holes and improved blasting caps with more precise delay times;

3) Placing more emphasis on blast designs when approaching final pit walls. This aspect will be more significant in mines where substantial savings can be realized by steepening the final pit wall slopes; and

4) Simulation of blast performance by computers to select types of explosives, loadings and patterns based on rock properties, geological conditions and planning requirements. This will minimize the experimentation in the field based on trial-and-error (Borquez, 1981).

Reduction of Stripping Ratio

A significant part of the direct mining cost of any open pit mining operation is the cost of removing waste (stripping) material. Therefore, one important way to improve productivity and thereby reducing the unit cost of mining is to reduce the overall stripping ratio to a minimum. One of the more important factors influencing the stripping ratio is the slope angle at which the final pit wall will be designed.

Table 1 outlines the calculated reduction in stripping volume in relation to final pit wall slope angles for a pit wall which is 300 m long and 300 m deep.

Applications of recently developed pit slope engineering techniques have clearly demonstrated that some and probably most, open pit final wall slopes may be steepened consistent with safety considerations.

Slope angles were steepened from 45° to 60° at the Kimbley Pit and from 40° up to 55° at the mines in the Schefferville area. These successful cases show how R & D efforts towards advanced pit slope design techniques can result in a significant decrease in the stripping

TABLE 1.   Relationship Between Steeper Slope Angles
and Reduction in Volume of Stripping.

| Final Pit Wall Slope Angle | | Reduction in Volume of Stripping |
From	To	1000 m³
45°	46°	470
50°	51°	402
55°	56°	351

ratio (Ross-Brown, 1973; Garg, 1980; Garg, 1981).

Optimization of Mine Planning

In the planning and design of open pit mines the culmination of the work is the cost analysis. Mine management generally requires several alternate mining plans before an optimum mining plan is selected and approved. These plans are developed in order to answer "what if" type of questions. Also, the financial analyses of ultimate pit designs, including the effects of various cut-off grades and stripping ratios, have been particularly significant during the last 3-5 years because of the fluctuating market conditions.

Perhaps the only practical way to develop several alternate mining plans in a relatively short time is by using computer based mine planning systems. The most promising areas where computers will be used to improve planning and productivity at large open pit mines are (Dziubek, 1980; O'Hara, 1980):

a) Design of ultimate pit limits based on geological and financial
data banks;

b) Development of short and long range mine plans including scheduling
of stripping;

c) Studies related to improvements in equipment productivity, such as
automatic truck dispatching systems.

Within this optimization process an output by a geotechnical engineer of pit slope angles against probability of failure or reliability provides a better base for optimizing financial returns than an output recommending one fixed slope angle only.

RESEARCH REQUIREMENTS FOR SLOPES IN SURFACE MINES

The previous section on the operating concerns in open pits has shown the engineering developments and system choices which are likely to be made. Generally, we will be dealing with more specialized and sometimes larger equipment which has special operating requirements in regard to stability of intermediate slopes and permissible crest loadings. To make use of economics of scale, operators will have to choose less flexible and more capital intensive ore loading and haulage systems. This requires, however, reasonably assured ore/waste ratios which translates for the geotechnical engineer, into a need to more accurate predictions of slope performance. This will not only require an answer to the question "what are stable slope angles?", but also "what reliability can we attach to our stability predictions?".

The slope design process commences in most cases with a review of past open pit mining and design experience for the particular region. The location of the pit site can suggest typical problems, such as permafrost, typical zones of weathering, peculiar rock types, major zones of shearing, regional trends of rock formations or topography, and special groundwater conditions.

This should be followed by a collection of information on material properties and structural geology for the specific site with main emphasis on the description of the shear strength parameters and the groundwater conditions in the rock mass.

Having obtained a data base, slope stability calculations can be carried out and a pit can be planned based on operational requirements.

When mining commences, monitoring equipment is installed to check on slope performance and to verify design assumptions.

The slope design process can therefore be broken down into four stages:

1)   Site data collection;

2)   Parameter quantification;

3)   Slope design;   and

4)   Performance monitoring.

Failure Mechanisms in Large Open Pits

Most of the learning process occurs during the monitoring of slope performance because our computational concepts of rock mass behaviour are not always reflected in reality.

One often overlooked point is the scale of open pit mine slopes which involves huge volumes of inhomogeneous rock containing a variety of geological structural features and rock types. Classical theories of slope failure can, at best, only achieve crude approximations.

Slope failure is considered here as the state of a slope where final collapse occurs. Failure surfaces in large slopes rarely follow a pure plane or a circular failure mode.

Transition Zones. As shown in Fig. 3 (Kvapil & Clews, 1979) failure surfaces

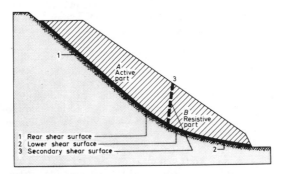

Fig. 3:   General scheme of slope failure with transition zone
          simulated by secondary shear surface (Kvapil & Clews 1979)

have a typically steep portion at the top (A) and a shallow section at
the bottom (B).   (A) can be called the active part and (B) the resistive part.   Major failures above the slope toe can only occur under the
condition that a transition zone exists in the slope between (A) and
(B).   In this transition zone thorough shearing and crushing of the
rock mass is necessary if movement is to proceed.

   Typically, development of failure occurs over long periods of time,
often many years and slope deformation can exhibit many regressive and
progressive phases (Zavodni and Broadbent, 1978).   The regressive phase
or slow down can be explained by the energy input required for the
breakdown of the rock mass, especially in the transition zone, see Fig. 4.

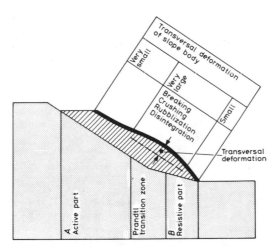

Fig. 4:   Principle of transversal deformation of slope body
          (Kvapil and Clews, 1979).

In addition, groundwater pressures can reduce during this process and the coarse granular material can have a greater angle of repose than the internal friction angle of the original rock mass. This means the material can be stable at a steeper angle than the original slope before the start of deformation.

The effect of the transition zone on slope stability has not been fully explored and the Prandtl prism mechanism has been suggested (Kvapil and Clews, 1979) as a possible model. The Prandtl prism has been applied successfully to foundation assessments and models the wedge type transition zone with development of shear surfaces between vertical compression of the zone below a foundation and the passive resistance of the soil in horizontal direction.

Progressive Failure. Another concern is progressive failure in slopes. A review of case histories of instability of natural slopes in plastic Champlain clays has shown quite clearly that during sliding, peak shear resistance is not mobilized along the total failure surface. Most cases of failure plot along the post peak strength envelope, suggesting that we are dealing with progressive failure and anisotropy of deformation, see Fig. 5 (Lo and Lee, 1974).

For rock slopes very little quantitative information is available in regard to time/shear strength relationships, but considering that already small shear movements along rough surfaces will lead to high stresses at points of contact, progressive failure needs to be considered as a significant process.

Therefore, high peak shear resistance values which are suggested from measurements of surface roughness and waviness on rock surfaces have to be used with caution if time effects are not considered.

Various case histories of slope failures in rock corroborate this case because back analysis indicated strength properties for the given failure surfaces to be closer to residual values than to peak values (Hoek and Bray, 1974).

Computational Concepts

So far only two areas have been pointed out where we require R & D, and where we have to be cautious at present when it comes to the design of slopes for large open pits. But a discussion on research requirements for open pit slopes is incomplete if we do not discuss the computational techniques which are used at present to predict the stability of slopes.

The most common way of expressing the stability of a slope is with the aid of the safety factor. One can find quite a variety of definitions in literature, but the most general is probably that the safety factor represents the ratio of the resisting forces to the driving forces along a potential failure surface (Gedney and Weber, 1978). The factor of safety is a useful concept, but we know that a

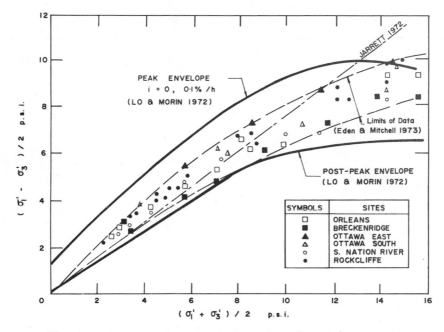

Fig. 5: Summary of strength data for Champlain sea clays (Lo and Lee, 1974).

factor of safety larger than one cannot guarantee safety if the variations of rock strength properties and stresses are not taken into account. Numerous examples exist in literature where failure has occurred despite a safety factor larger than one. It is not difficult to show that the safety factor itself is a variable.

Raising permissible safety factors can help to adjust a computational model so that a particular slope can be accepted as shown in Table 2 (Walton and Atkinson, 1978).

Adjusting safety factors does, however, not address the source of the problem and probabilistic procedures offer a better way of combining objective information with subjective judgement.

As people from the real world we could reject probability approaches by following the statement in Van Nostrand's Scientific Dictionary (Van Nostrand, 1958), "that application of probability theory to the real world encounters formidable logical difficulties". This is true, but if we are satisfied to deal with subjective probabilities as a set of weights which express an individual measure of the relative likelihood of the outcome of events, then a world of possibilities is opened up which allows incorporation of judgement factors into slope stability analysis in a numerical way. Many engineering disciplines have made this step.

TABLE 2. Safety Factor Selection

		Case 1: risk to external persons and property	Case 2: high risk to mine personnel	Case 3: low risk to mine personnel
A	Design based on peak or maximum shear strength	1.5	1.3	1.1
B	Design based on residual/ultimate shear strength parameters	1.3	1.2	1.1
C	Fluid- or slurry-retaining embankments in seismic area, incorporating both residual strength and seismic loading	1.2	1.15	1.1

This statement taken out of context could imply that deterministic slope designs are wrong. This is far from the truth. Classical methods are completely sufficient for the majority of cases. Imperfect "theoretical" analyses and "imperfect" material sampling can lead to sound engineering selections if calibrated against previous experience (Hoeg, 1979).

The attractiveness of probabilistic approaches is that research is directed into the data base of slope stability design. This leads to more meaningful communication between the various disciplines contributing to a slope stability analysis because by necessity hydrologists and geologists present most of their site investigation data in the form of frequency distributions. One argument which has been advanced against using probabilistic approaches has been the implicit admission that there is no ultimate safety, but a definable risk and that in human terms risk is not acceptable. This attitude is, however, unrealistic and it is better to identify the risks and learn to minimize them.

The one advantage we have is that a measurable risk of failure in a large slope does not necessarily imply a measurable risk of failure for personnel and machinery. Accidents are most often caused, not by large failures, but by small slides resulting in rock falls and small wedges which are admittedly difficult to appraise beforehand, but for which the safety factor approach does not provide a better answer.

Subjective probability analysis, however, can provide insight into mechanisms and effects of various parameters on slope stability which are stifled by a deterministic safety factor approach.

Let us look at some examples.

Parameter Uncertainty. One item which is addressed in a probabilistic analysis is the variation of input parameters. Handbooks which list experience parameters rarely report variations of properties.

Brawner, 1978, lists eight subject areas essential for the assessment of slope stability. These are listed below:
1) Geology,
2) Shear strength of the materials in the slopes and the base of the pit,
3) Topography,
4) Climate,
5) Hydrology,
6) Mining method and the rate of mining,
7) Need for, and program of blasting, and
8) Mine economics.

Apart from the last item, all others contribute to the limit equilibrium equation for slope stability as given below:

$$W \sin i = C A + W \cos i \cdot \tan \phi$$

with: W = weight of material above potential failure plane, modified by expressions for effective stress, seismicity and blast accelerations
  i = slope angle
  C = cohesion
  tan ϕ = coefficient of friction
  A = area of potential failure surface.

Just looking at the values of the residual angle of friction as determined during laboratory testing, we find considerable variation. An example is given below:

Material	No. of Samples	Residual Coefficient of friction	Standard Deviation	Coefficient of Variation	Reference
calcareous silt	60	0.533	±0.091	0.17	Novosad, 1978
calcareous quartzite	63	0.760	±0.140	0.18	Gyenge & Herget, 1977

Here we are dealing with a finite laboratory example and have not considered yet the vagaries of geological structure, groundwater distribution and change in material properties over a large body of rock. A good example is given in the literature for soil embankments where a study on the probability of failure of embankments showed an "unexpected high value" of 0.15 and 0.2 of the probability of failure for an ordinary range of safety factors between 1.1 and 1.5 (Matsuo and Kuroda, 1974, quoted by Athanasiou-Grivas, 1980). One can see that slope stability calculations without considering parameter uncertainty can be good advice, but can misrepresent accuracy and reliability. We urgently need to introduce procedures which require an a priori assessment of parameter uncertainty and tailor investigations accordingly.

However, it is difficult to assess parameter uncertainty in a field situation if no basic information is available on the various populations concerning individual parameters. An attempt has been made to list the distribution functions used by a variety of authors (Table 3). This topic still requires considerable study. At the moment when distribution functions are known, we can use statistics to identify the size of the sampling required.

Another area where probabilistic approaches have assisted us is in the determination of the overall shear resistance of the rock mass, such as prediction of stepped failure surfaces or the occurrences of wedges from two joint sets.

Step Path Failure Geometry. This case (Fig. 6, 7 and 9, Call and Nicholas, 1978) was studied with the aid of probability by using a 2-dimensional probabilistic model formed by the intersection of two joint sets that have similar strikes. The master joint set is gently dipping (20-50°) while the cross joint set is steeply dipping (50-90°).

Most significant inputs for the model are distributions of fracture dip, length, spacing, overlap, slope height and maximum length of intact rock. The output is in the form of a cumulative % plot of the step path angle (Fig. 8). The resultant angle of the step geometry and the percentage of intact rock are obtained.

Occurrence of Wedge Failure. Another example comes from an open pit coal mine (Hutchings and Gaulton, 1978) where joints with a dip of 40-80° have a significant effect on the stability of the faces which are mined with bucket wheel excavators. The method used for predicting stability problems is based on the probability of occurrence of a critical wedge or plane shear failure from joint surveys.

Each joint is combined with every other joint and from n joints surveyed, n(n-1)/2 wedges are generated by this procedure. This implies that all joints are a true sample of the overall joint population and location has no effect on joint characteristics. However, there are variations of joints with depth and along the face. Therefore, limitations are placed on combining joints from various areas. The analysis showed that the most significant parameter influencing the probability of instability was the orientation of the face. The probabilistic approach allowed judgements to be made on slope height, slope angle and slope orientation (Fig. 10).

Table 3. Distribution Functions of Slope Properties

Type of Parameter	Distribution Function	Coefficient of Variation $(\frac{\text{Mean}}{\text{Std. Dev.}})$	Reference*
Density	normal	0.01-0.1	1
Deformation modulus	log normal	0.3	2
Cohesion	normal	0.1-0.5	1,3
	log normal		2
Residual angle of friction (degrees)	normal	0.05-0.2 0.16	1,3,4,5,8
Internal angle of friction (degrees)	normal		2
Orientation of fracture (dip)	normal	0.1-0.2	3,6,8
Spacing of fractures	negative exponential		6,7
Length of fractures	negative exponential		6
Roughness (degrees)	negative exponential		6

* 1) **Favre**, 1979   2) Mikheer, 1979
  3) Förster and Hagen, 1979   4) Athanasiou-Grivas, 1980
  5) Seycek, 1978   6) Call, Savely, Nicholas, 1977
  7) Priest and Hudson, 1976   8) Herget, 1977

The examples mentioned have illustrated only a limited aspect of a probabilistic slope stability analysis. It generally requires the geotechnical engineer to come up with a probability schedule where reliability or probability of failure is plotted against slope angle as shown in Fig. 11 (Major, Ross-Brown and Kim, 1978). These schedules can then be used for a cost/benefit analysis.

Site Data Collection

Probabilistic approaches allow us to rationalize or quantify judgements, but this is only meaningful if we have good case histories and an adequate data base from site investigations.

Ideally, an agreement should be reached at the outset of site investigations for a new project as to what data are required and with what precision. A value has to be put on the reduction of uncertainty if this problem is to be resolved.

Generally, we have no short-comings in regard to procedures for outcrop mapping, surface sampling, and computer based storage, retrieval and analysis systems (e.g., DISCODAT, Pit Slope Manual (Cruden et al. 1977)), but we still have difficulties with basic site investigation tools, such as 100% core recovery in drilling, core orientation and geophysical data collection.

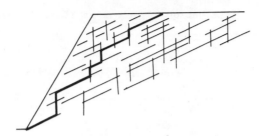

Fig. 6: Continuous step path
(Call & Nicholas, 1978)

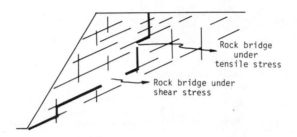

Fig. 7: Step path separated by rock bridges
(Call and Nicholas, 1978)

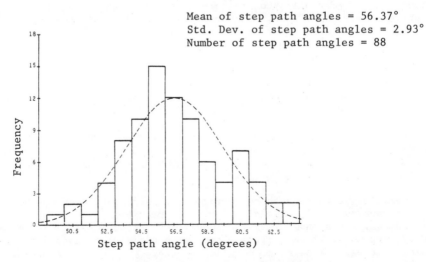

Mean of step path angles = 56.37°
Std. Dev. of step path angles = 2.93°
Number of step path angles = 88

Fig. 8: Step path angle distributions
(Call and Nicholas, 1978)

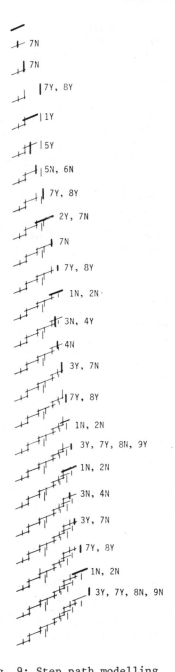

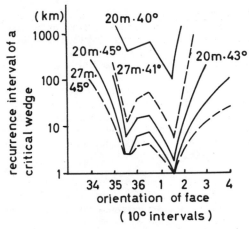

Fig. 10: Statistical analysis of coal
face stability (Hutchings and
Gaulton, 1978)

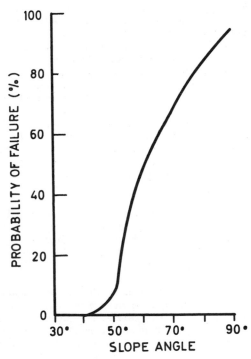

Fig. 11: Probability of slope failure
(Mayor, Ross-Brown and Kim,
1978).

Fig. 9: Step path modelling
(Call & Nicholas, 1978)

Core Orientation. Drilling is the only means to expose rock at depth and drilling for geotechnical purposes has rather stringent requirements compared to exploration drilling. We not only need a sample of the rock material, but need to recover a sample of the rock mass complete with discontinuities in their known orientation.

This indicates a need for 100% core recovery and core orientation. In softer formations, special core barrels are necessary and under no circumstances can we afford to sacrifice quality (recovery) for production. Especially the layers which are easily washed away during drilling (clay, coal, bentonite) can have far reaching consequences for slope stability.

Reviews of core orientation devices are available (Hoek and Bray, 1974; CANMET, 1977), but many of the procedures and the equipment are not very popular because they almost double drilling time, require an engineer on site and are very delicate. Accuracy of core orientation can be better than 3°, but if core recovery is not 100%, the data may be erroneous as it is not always possible to tell from what section the core is lost.

Geophysical Methods. Geophysical methods can often fill important gaps in geological information by means of relatively quick surveys of the subsurface, especially in cases of horizontal stratification. For pit site investigations the most useful geophysical methods are those which can distinguish between fractured and solid rock, soft and hard materials, and water bearing and dry strata. Favoured for such capabilities are seismic or sonic methods, electrical resistivity or radioactive methods, and fluid temperature measurements.

Geophysical methods have seen considerable progress in the last decade which improved the sensitivity and reliability of information significantly. Today, for instance, coal quality assessments and distribution are not considered complete without gamma ray, resistivity and density logging.

In regard to structural geological information there is the drawback that borehole exposures are required to prove interpreted geologic structures from geophysical logs to remove interpreter bias. Further development of computing techniques and log interpretation are needed to obtain reliable structural information from borehole logs and surface geophysical methods.

Deep seated sliding surfaces have been detected in landslides with borehole seismic measurements. The interpretation was based on a comparison of the physical properties of the deformed part of the slope with those of the undeformed part below the failure surface. This was located at a depth of about 150 m and affected a rather large area of slope.

Geo-acoustic methods have shown their potential to allow very early detection of slope movement, although only qualitatively. They depend

on registration of sub-audible noises which can be determined meaning-
fully even if slope movements occur only at a rate of 5 to 10 mm/year
(Novosad, 1978).

CONCLUSIONS

On the basis of concerns expressed by mine operators an examination
of useful avenues of research for surface mine slopes has been made and
these are listed below:

1) Documentation of case histories of large slope failures with
   description of failure surface topography, slope deformation and
   back-analysis of shear strength parameters.

2) Determination of progressive failure in rock slopes.

3) Application of computational methods which consider parameter
   uncertainty.

4) General studies of frequency distributions of input parameters.

5) Development of better tools and methods for site investigations and
   a priori determination of site investigation requirements.

REFERENCES

Alberts, B.C. and Dippenaar, A.P., 1980, "An In-pit Crusher Overburden
Stripping System for Grooteqeluk Coal Mine," Symposium on Materials
Handling in Opencast Mining held in Pretoria, South Africa, Sept.

Almond, R.M., 1980, "Overland Conveyors", Engineering Mining Journal,
vol 181, August.

Athanasiou-Grivas, D., 1980, "Probabilistic Seismic Stability Analysis
- A Case Study", Canadian Geotechnical Journal, vol 17, May,
pp 352-360.

Borquez, G.V., 1981, "Estimating Drilling and Blasting Costs - An Analysis
and Prediction Model", Engineering Mining Journal, vol 182, January.

Brawner, C.O., 1978, "Stability in Open Pit and Strip Mining Coal
Projects", Proceedings 1st International Symposium on Stability in
Coal Mining, Vancouver, pp 13-28.

Call, R.D., and Nicholas, D.E., 1978, "Prediction of Step Path Failure
Geometry for Slope Stability Analysis", Proceedings 19th U.S. Rock
Mechanics Symposium, Nevada, May.

Call, R.D., Savely, J.P., and Nicholas, D.E., 1977, "Estimation of Joint
Set Characteristics from Surface Mapping Data", Monograph on Rock
Mechanics Application in Mining, W.S. Brown, S.H. Green and
W.A. Hustrulid, Ed., AIME, New York, Chapter 9, pp 65-73.

CANMET, 1977, "Pit Slope Manual", Energy, Mines and Resources, Ottawa, 10 Chapters with 16 Supplements.

Chadwick, J.R., 1981, "Materials Handling in Opencast Mining", World Mining, vol 34, no. 1, January.

Chadwick, J.R., 1981, "Trolley Assisted Trucks at Sishen Iron Ore Mine", World Mining, vol 34, no. 3, March.

Cranford III, J.T. and Hustrulid, W.A., 1979, "Open Pit Mine Planning and Design", Society of Mining Engineers of AIME, New York.

Cruden, D., Ramsden, J. and Herget, G., 1977, "DISCODAT Program Package", Pit Slope Manual, Supplement 2-1.

Dziubek, J.A., 1980, "Computer-Age Mine Planning is Here", Coal Age, July.

Faulkner, I.A., 1980, "In-Pit Crushing: A Review of Techniques and Practices", Quarry Management and Products, June.

Favre, J.L., 1979, Contribution 3.5, Proceedings 7th European Conference on Soil Mechanics and Foundation Engineering, Brighton, United Kingdom, Chapter 3 Statistics, Reliability Theory and Safety Factors.

Förster, W. and Hagen, C., 1979, "Zur Genauigkeit bodenmechanischer Aussagen bei der Untersuchung von Böschungen und zur Verantwortung des Sachverständigen", Zeitschrift für Angewandte Geologie, Bd 25, Heft 12, pp 624-631.

Gagnon, R. and Rochefort, F. 1972, "Diesel-Electric Truck Haulage Improved Through Trolley Assist", CIM Bulletin, vol 65, December.

Garg, O.P., 1980, "Report of the Sub-Committee on Rock Slopes Submitted to the Canadian National Committee on Rock Mechanics", Annual Meeting in May.

Garg, O.P., 1981, "Successful Implementation of Steeper Slope Angles in Open Pit Mines in Labrador", Proceedings 3rd International Conference on Stability in Surface Mining, Vancouver, June.

Gedney, D.S. and Weber, W.G., 1978, Chapter 8, Landslides, Analysis and Control, Transportation Research Board, National Academy of Sciences, Special Report 176, R.C. Schuster and R.J. Krizek, eds., Washington, D.C.

Gyenge, G. and Herget, G., 1977, "Mechanical Properties", Pit Slope Manual, Chapter 3, Energy, Mines and Resources, Ottawa.

Herget, G., 1977, "Structural Geology", Pit Slope Manual, Chapter 2, Energy, Mines and Resources, Ottawa.

Hoeg, K., 1979, Contribution 3.16, Proceedings 7th European Conference on Soil Mechanics and Foundation Engineering, Brighton, United Kingdom, Chapter 3, Statistics, Reliability Theory and Safety Factors.

Hoek, E. and Bray, J.W., 1974, "Rock Slope Engineering", Institution of Mining and Metallurgy, London, 309 p.

Hogan, R.O. and Kennedy, B.A., 1980, "Open Pit and Underground Mining - Trends, Equipment", World Mining, Yearbook.

Hutchings, R. and Gaulton, R.J., 1978, "Slope Stability in Australian Brown Coal Open Cuts", Proceedings 1st International Symposium on Stability in Coal Mining, Vancouver, B.C. p 65-74.

Kvapil, R. and Clews, K.M., 1979, "An Examination of the Prandtl Mechanism in Large-Dimension Slope Failures", Transactions, Section A (Mining Industry) of the Institute of Mining and Metallurgy, London, vol 88, p A1-A5.

Laswell, B.J. and Laswell, G.W., 1980, "New Surface Mining Equipment for the 80's", Canadian Mining Journal, vol 101, no. 12, December.

Lo, K.Y. and Lee, C.F., 1974, "An Evaluation of the Stability of Natural Slopes in Plastic Champlain Clays", Canadian Geotechnical Journal, vol 11, no. 1, p 165-178.

Major, G., Ross-Brown, D., and Kim, H.S., 1978, "A General Probabilistic Analysis for Three-Dimensional Wedge Failures", Proceedings 19th U.S. Rock Mechanics Symposium, Nevada, vol 2, p 45-56, May.

Mikheer, V.V., 1979, Contribution 3.7, Proceedings 7th European Conference on Soil Mechanics and Foundation Engineering, Brighton, United Kingdom, Chapter 3, Statistics, Reliability Theory and Safety Factors.

Novosad, S., 1978, "The Use of Modern Methods in Investigating Slope Deformation", Bulletin of the International Association of Engineering Geology, no. 17, p 71-73, June.

O'Hara, T.A., 1980, "Quick Guides to the Evaluation of Ore Bodies", CIM Bulletin, vol 73, February.

Pelzer, H.K., 1979, "Long Distance Conveyors: Economics and Operating Experience with the Sahara Phosphate Conveyor", Proceedings 3rd International Symposium on Transport and Handling of Minerals, Vancouver, October.

Priest, S.D. and Hudson, J.A., 1976, "Discontinuity Spacings in Rock", International Journal Rock Mechanics and Mining Science and Geomechanic Abstracts, vol 13, p 135-148.

Rance, D.C., 1979, Seminar on Forecasts of Mining and Metallurgy - Open Pit Mining, Ottawa, November.

Ross-Brown, D.M., 1973, "Design Considerations for Excavated Mine Slopes in Hard Rock", Engineering Geology, Quarterly Journal of, vol 6.

Seycek, J., 1978, "Residual Shear Strength of Soils", Bulletin of the International Association of Engineering Geology, no. 17, p 73-75, June.

Van Nostrand, 1958, Scientific Encyclopedia, 3rd edition, D. Van Nostrand Company, Toronto, 1839 p.,

Walton, G., and Atkinson, T., 1978, "Discussion of Some Geotechnical Considerations in the Planning of Surface Coal Mines", Transactions Section A (Mining Industry) of Institute of Mining and Metallurgy, London, vol 87, p 147-171.

Zavodni, Z.M. and Broadbent, C.D., 1978, "Slope Failure Kinematics", Proceedings 19th U.S. Rock Mechanics Symposium, May, vol 2, p 86-94.

2
Investigation, Research, and Design

Chapter 13

SEDIMENTOLOGICAL CONTROL OF MINING CONDITIONS IN THE PERMIAN COAL
MEASURES OF THE BOWEN BASIN, AUSTRALIA

C.W. Mallett

Senior Research Scientist, CSIRO Division of Applied Geomechanics,
Victoria, Australia

ABSTRACT

The distribution and properties of interseam and overburden rocks
in coal mines is largely controlled by conditions at the time of
their original deposition. Within the developed areas of the Bowen
Basin, depositional environments range from fluvial through upper and
lower detaic, to marginal marine and intertidal facies. Each
environment has a characteristic association of lithologies with
specific range of properties, and each lithology has a characteristic
geometry.

Fragmentation problems are associated with ancient sand channel
systems, where the difficulties encountered depend on the dimension
and shape of channels, their continuity, orientation to mining face
and associated sediment bodies formed as levees or interdistributary
muds. Stability problems in spoil, pit floor and highwalls are
usually associated with mudstones which contain swelling clays derived
from the original Permian volcanic source. This clay problem is
aggravated by deep Tertiary weathering.

Case studies are presented of detailed analyses and prediction of
geomechanical properties on a mine site scale. These are used in
investigating or forecasting fragmentation and stability hazards. The
studies incorporate detailed field mapping, bore log data and computer
data processing. They allow the development of detailed depositional
models for specific interseam intervals within mines. As areas of
depositional environment are identified they allow the construction of
maps of similar mining conditions and the determination of the most
suitable mining techniques.

INTRODUCTION

Permian Basins in eastern Australia are the source of most of the Australian hard coal presently mined.

Mines are located in the Sydney and Bowen Basins, which are separated by an area of younger sediments. The Sydney Basin has been most extensively developed, predominately by underground mining, with surface workings concentrated at the north west margin of the Basin.

The Bowen Basin has been developed mainly by surface mines which have concentrated on coking coal production in the last twenty years. Mines are located along the western margin of the Basin and in the south eastern corner (Figure 1). The mines are long strips dug with draglines, and there is currently approximately 200 km of strip length exposed. This is expected to increase by hundreds of kilometres in the next decade.

COAL MEASURE SEDIMENTS

Coal accumulates in a very specialised environment, requiring moist conditions to encourage plant growth, high water table to preserve organic material and commence coalification, the complete exclusion of terrigineous sediment, and most importantly, that this occurs in a setting stable enough to accumulate large thicknesses of peat which can be buried by subsequent sediments and preserved. The limited number of places where all these conditions combine, mean that the depositional environments and sediments associated with coals are restricted.

The patterns of sediments found with coals allow the development of depositional models for ancient coal measures. By looking at the properties of the coal, the continuity of seams, associated sediment type and internal structures and geometry, it is possible to draw analogies with modern depositional processes. This approach was first developed as a tool in coal exploration as a way to predict where the major coals may have accumulated in a prospective sedimentary basin. A newer and equally important use is in the mine development phase. Impetus to this is provided by a more rigorous approach to mine design which requires better geotechnical data of the sequence to be mined, and the increase in the percentage of surface workings which allow the study of large sections of coal measure sequences.

An example of a depositional model for coal measures is that proposed by Horne *et al*. (1978) Figure 2. This has met with considerable acceptance particularly in North America. It was first applied to the Appalachian coalfields where many similarities to the modern Mississippi Delta were recognised in the coal measures. The model outlines typical patterns in the number and thickness of seams, sulphur content, and lithology and structures in the interseam sediments. Marginal marine or barrier sands are associated with back-

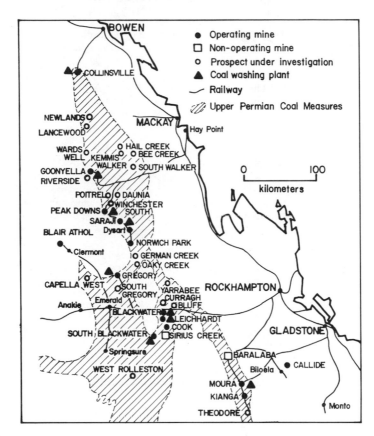

Fig. 1.   Locality map of coal mine sites in the Permian Bowen Basin, eastern Australia.

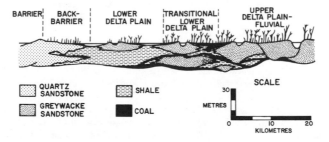

Fig. 2.   Cross-section of depositional model for coal accumulation by Horne *et al.* 1978. The model shows coal accumulation sites on a flat coastal area ranging from marine to fluvial conditions with associated lithologies.

barrier coals, and distinction made between a lower and upper delta environments. Further 'inland' the model has coals associated with fluvial river channels and flood plain deposits.

The scale of these areas is large, and any particular mine lease could expect to be within only one part of this large system. Thus the lithologies present would be peculiar to the processes operating in that subsystem only.

SURFACE MINE STABILITY

In the Bowen Basin draglines are used to dig the strip mines. Stability studies have therefore been directed to the particular mining configuration imposed by this system.

Spoil Failures

Most large spoil failures have been caused by failure on a weak surface at the base of the spoil which dips gently into the pit.

The origin of this weak plane varies, it can be due to:

(a) A weak zone in the spoil where material is weakened by reworking and water uptake.

(b) Weak spoil/floor boundary which is usually related to deterioration of the floor when exposed during mining.

(c) Weak layers in the floor which cannot maintain the load of the spoil pile on the dipping floor of the pit.

Linear shear planes at the base and steeply dipping within the spoil pile give a failure geometry approximated by two wedges. The lower wedge moves into the pit on a weak basal plane and the upper wedge moves vertically down on steeply dipping shear planes or shear zones. This failure mechanism has been described by Gonano (1980) and Boyd *et al.* (1978).

Highwall Failures

Highwalls are less than 60 metres high and often include significant thicknesses of Tertiary sands, clays, and volcanics overlying the Permian Coal Measures (Hagan *et al.* 1979). These are associated with a weathering profile in the Coal Measures and result in the upper portion of the highwall being considerably weaker than the lower section. Circular slip failures are rare, but occur in this upper section. Prestripping necessary in deeper pits removes much of this material and it becomes less significant during the mine's life.

Slips also occur on the unconformity surface between the Permian and the Tertiary. The surface has weathered clays associated with it, and as the Coal Measures are impervious, groundwater is channelled along the unconformity which wet any movement occurring. These failures are relatively uncommon and small.

The largest highwall failures are related to the bedding planes and joints within the Coal Measures. Mass movement occurs on weak bedding planes which dip gently into the pit. The features and processes which are considered responsible for these failures are as follows:

(a) The highwall has weak black clay bands dipping into the pit and is traversed by near vertical joints.

(b) Blasting damages the highwall by inducing movement on joints and bedding planes.

(c) On excavation stress relief results in moveout into the pit on clay layers and further opening of joints.

(d) Water ingress along joints in association with the reworking caused by stress relief movement, raises the water content of clays and seriously reduces the strength along discontinuities.

(e) Hydrostatic pressure may build up in joints during periods of rain.

(f) Significant moveout on a gently dipping plane leads to the formation of a more steeply dipping secondary shear plane at the back of the failure which joins to the vertical joint system. Mass movement of the block defined by these surfaces occurs, which can be sustained by energy from the down and outward movement of a large mass bounded by the secondary shear plane and joints at the back of the failure.

The geometry of such failures depends on the orientation and extent of potential failure surfaces. The configurations formed were reported by Gonano (1980) and are indicated in Figure 3. The simplest case is illustrated by Figure 3(a) where a bedding plane and joints proscribe the failure block. This is commonest in smaller failures. Large failures involve a secondary failure plane which dips into the pit at a much higher angle than the basal slip surface (Figure 3(b)). It connects the weak clay bands with vertical joints, which do not penetrate the whole height of the highwall. Gonano (1980) postulated that this secondary plane was through intact coal measures. Cox (1981) analysed stress patterns which develop in highwalls with progressive stress relief on excavation. He showed that as stress relief progressed the greatest potential for shear surfaces was along planes dipping into the pit at approximately 54°, commencing from positions located beneath the crest of the highwall and the edge of the bench at the basal slip plane. He noted that discontinuities such as bedding at lower angles could preferentially fail in an

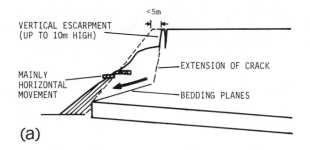

(a)

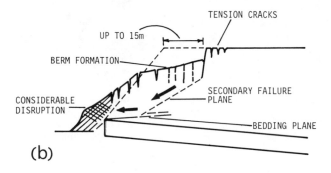

(b)

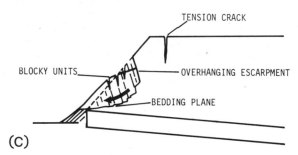

(c)

Fig. 3. Highwall failure configurations generated by shearing on
discontinuities within the Permian Coal Measures, Bowen Basin,
after Gonano 1980.

inhomogeneous highwall. Failures with overhanging scarps as shown in
Figure 3(c) generate when the joint penetration is limited, the basal
weak bedding plane is not continuous, or the generation of the
secondary failure plane is inhibited.

Barton and McKean (1978) showed that failures occurred when the weak clay bands dipped toward the pit, although the angle of dip did not appear critical. The angle of the failure surfaces was mathematically analysed by Toh (in Fuller 1981) and Cox (1981) although both authors differed in their methods and conclusions, they both emphasised that the variations in material properties of the shear planes were considerably more significant than the angle of the planes.

A long term study of inclinometer strings inserted in a highwall was noted in Mallett and Wooltorton (1981). This study illustrated that over several years the different layers in the highwall moved as independent layers. The boundaries of units were bedding planes separating lithologies and unconformities. Incremental movement of one unit was followed by a quiet period while the others caught up. In this way the total moveout of the wall was fairly uniform. Appreciable movement was usually associated with an adjacent coal production blast.

BEDDING ORIENTATION

The orientation of bedding in the Coal Measures is related to the original depositional environments and compactional history. As material fails by sliding into the pit on bedding surfaces which dipped into the pit, rather than away from it, like the coal, a survey was made of the bedding orientations throughout surface mines in the Bowen Basin. Recurring patterns of bedding geometry and lithological associations were found.

(a) Beds parallel to coal which maintain the same orientation as the coal seam throughout the interseam interval. They are usually claystones and mudstones.

(b) Lenticular sandstone beds which form high angular dips of up to 25° to the coal seams.

(c) Intermediate beds of interbedded sandstones, silts and claystones with moderate dips to coal.

Beds Parallel to Coal

The claystone beds parallel to coal accumulate by the settling of suspended sediment in lakes, bays and on flood plains. Considerable variation in composition and properties of the thinly bedded sediments is possible as the initial supply of suspended material varies, and the chemical conditions in which it accumulates is capable of great variation. This controls the type of clays and other minerals which settle or are precipitated from groundwater. The percentage of organic material is also highly variable. Smectites and expansive clays are common. Despite this, the beds cause little problem in highwalls as they follow the dip of the coal away from the pit. They are significant in spoil stability when they occur immediately below

a coal seam and form the floor on which spoil piles are built. When
spoil is dumped on these dipping materials they may not be able to
support the load, and in-floor failures lead to major spoil instability.

Failures of this type have occurred throughout the life of No. 1
Pit in B Seam at Moura in the Southern part of the Bowen Basin
(O'Regan *et al.* 1981). At this locality successive spoils have failed
with associated buckling of the coal and pit floor. The coal and
immediate floor dips at approximately 8°. The interseam sediments
above and below B Seam show high dips to the coal, but this has not
influenced stability (Figure 4). Immediately underlying the coal and
parallelling it is 4 metres of carbonaceous mudstone with some thin
sandy interbeds. This unit accumulated in a marshy lake which
preceded the establishment of B Seam coal accumulation. Muds and sand
were mixed with organic matter in sufficient quantity to form mudstone
rather than coal. A failure plane is located about the centre of the
unit where more clay indicates drowning of vegetation or increased
terrigineous input at the time of deposition.

Beds Dipping to Coal

The most unfavourable geometry in highwalls is formed by steeply
dipping sandstones, which however do not have weak clays associated
with them. These steeply dipping units developed in peat environments
where channels depositing sand impinge upon the peat, compressing it
and allowing for the migration of the channel into the compressed
marginal zone. Down warping of the peat as the channel migrates over
it results in the channel sand deposits being bent into a curve,
giving the configuration illustrated in Figure 5. The compactional
mechanism needed to generate the high angle dips is only associated
with rapid strip loading of the peat margin characteristic of high
energy channels. These channels drop only coarse sandy bedload
material and are rarely associated with claystones.

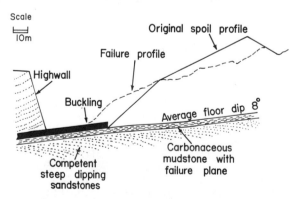

Fig. 4.   Cross-section of Pit 1B, Moura, Southern Bowen Basin, showing
          weak carbonaceous mudstone floor in which shear planes
          develop in association with spoil failure (after O'Regan *et
          al.*, 1981).

MOURA   MINE

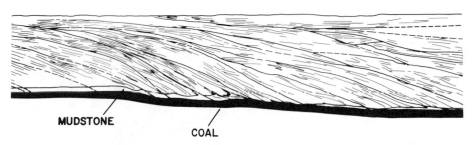

SCALE $\frac{V}{H}$ = 1

PIT   1C

Fig. 5.   Sandstone beds dipping to coal at a high angle in a highwall
at Moura, Southern Bowen Basin. A thin layer of mudstone
overlies the coal beneath the sandstone.

Many highwall failures are associated with lower angle beds where
sandstones, siltstones and claystones are interbedded.   These can be
dipping in one direction as with the sandstones above, or dipping
symmetrically around a channel (Figure 6).   Unidirectional units are
formed in the same way as the sandstones mentioned above by compaction
of the peat.   They differ from the sandstones in that the finer
sediments form more extensive sheets and thinner units over a longer
time, resulting in a less dramatic compaction of the peat and a lower
angular difference in the bedding generated.   Symmetrical units are
formed around sandy channels where beds thin and become more clayey
away from the sediment source.   Subsequent compaction of the clayey
beds further emphasises the antiform geometry.

These geometries are most susceptible to failures in highwalls, as
although the dips may be low they are commonly associated with weak
clayey material interbedded with stronger siltstone and sandstone.

One of the mines in which highwall failures have been studied is
Goonyella.   During 1977 a failure 450 metres in length occurred
adjacent to a line of instruments installed in the highwall (Figure 7).
An analysis and the highwall kinematics was reported by Cox (1981).
Movement into the pit occurred near the base of the highwall along
weak carbonaceous mudstone layers, which dipped into the pit at angles
of 3 to 8°.   Cox (1981) postulated the development of a steeper
secondary shear plane through intact coal measure overburden at the
back of the failure.   However, when the geological cross-section of
the highwall is plotted (Figure 8) it can be seen that bedding dips so
steeply into the pit that shearing most likely occurred along bedding

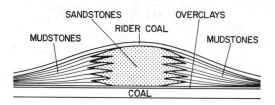

Fig. 6. Symmetrical antiforms or drape structures developed over
         channel sands and marginal sediments.

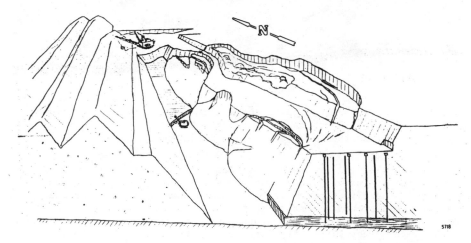

Fig. 7. Sketch of highwall failure at Goonyella Mine, 1977, from
         Cox, 1981.

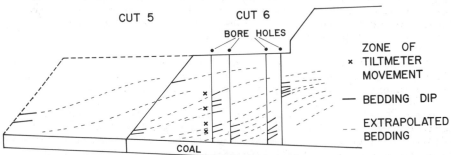

Fig. 8. Cross-section of highwall at the failure site at Goonyella
         Mine, Bowen Basin. Bedding values were measured in the
         highwalls, and boreholes. Dips from bore core were oriented
         to the same direction as that found in the two previous
         highwalls, as mapping of successive cuts verifies the general
         continuity of dip direction.

plane surfaces. Dips in excess of 30° were recorded in boreholes. The position of movement in the highwall indicated by inclinometer strings are indicated by asterisks. The movements were related to adjacent coal production blasting. The sites of movement correspond to clayey horizons in the highwall. At this site, the presence of weak carbonaceous mudstone bands dipping into the pit provided an initial failure plane. Extension of the failure boundaries well back into the highwall and along the pit was facilitated by steeper dipping beds higher in the wall. These bedding planes were oriented in a similar direction to the plane of greatest shear stress, which developed as stress relief occurred in the vicinity of the highwall face. In addition they parallel the highwall along the cut. Thus when failure initiated low in the highwall at one locality, conditions existed for its propogation back into and along the wall, generating a large mass movement. The dipping secondary shear plane extended back and up to the open vertical joints, which formed the edge of the failure where it intersected the surface. A vertical scarp formed as the mass slid down along these joints.

CRITICAL LITHOLOGIES

Although similar lithologies and geometries occur at a number of surface mines in Queensland, highwall failures of this type have been restricted to one mine. The difference in performance is caused by subtle lithological differences in the source of sediments, diagenesis, depositional conditions and weathering. In some areas these factors result in a more clayey highwall where even the individual grains of rock fragments in the sandstones are now mineralogically clays. The prevalence of clay affects the intact rock strength and the rock mass, as joints and discontinuities are infilled with clay. Depositional conditions have also produced some clay beds which have moderate strength in the virgin state, but whose strength can be significantly reduced by conditions imposed by mining. The most important of these are carbonaceous mudstones referred to locally as the Black Clay Bands. The presence of suspect lithologies has been tested for by slaking experiments and applied on a routine basis in mine surveying, (Godfrey, 1978; Boyd et al. 1978). Improved methods of classification and identification of suspect clays have been recommended by Seedsman. and Emerson (1981) who studied the factors affecting the strength of bonding in the clays. They identified the relative contribution of initial organic content, water take up, mechanical reworking, and exchangeable cations to strength reduction. This now allows the identification of problem lithologies from small borehole samples prior to development.

CONCLUSION

In the Bowen Basin weak strain softened lithologies cause instability when associated with particular bedding geometries and positions in the mine sequence. These lithologies are mainly

carbonaceous mudstones originally deposited in lakes and swamps.
Spoil pile failure is initiated by movements on layers within the
immediate floor, which dip into the pit parallelling the coal seams.
Unusual depositional processes in the Permian Coal Measures have
resulted in the beds overlying the coal seams often dipping at
significant angles to the coal. This has created a geometric
distribution of beds which in certain orientations to the strip
encourage the initiation and extension of large highwall failures on
shear surfaces developed in the weakest bedding planes.

REFERENCES

Barton, C.M. and McKean, R.M., 1978, "Review of Engineering Geology
    Aspects of an Investigation of Slope Stability in Strip Coal
    Mining at Goonyella Queensland," CSIRO Aust. Division of Applied
    Geomechanics, Technical Report No. 48.

Boyd, G.L., Komdeur, W. and Richards, B.G., 1978, "Open Strip Pitwall
    Instability at Goonyella Mine - Causes and Effects. *Proc. Ann.
    Aust. I.M.M. Conf. North Qld.*, pp. 139-157.

Brawner, C.O. and Dorling, P.F. (Eds.) 1979, *Stability in Coal Mining,*
    Millar Freeman, San Francisco, pp. 1-496.

Cox, R.H.T. 1981, "Analysis of the Stability of a Highwall in an
    Opencut Coal Mine Queensland", CSIRO Aust. Division of Applied
    Geomechanics, Geomechanics of Coal Mining Report No. 34.

Fuller, P.G. (Compiler), 1981, "Stability Problems in Open Strip
    Excavations and Spoil Piles - Final Report", CSIRO Aust. Division
    of Applied Geomechanics, Geomechanics of Coal Mining Report No. 25.

Godfrey, N.H.H., 1978, "Highwall Stability in Goonyella Open Pit Mine
    with Particular Reference to Clay Types", Msc. Thesis James Cook
    Univ., Townsville.

Gonano, L.P., 1980, "An integrated Report on Slope Failure Mechanisms
    at Goonyella Mine - November 1976", CSIRO Aust. Division of Applied
    Geomechanics, Technical Report No. 114.

Hagan, T.N., McIntyre, J.S. and Boyd, G.L., 1979, "The Influence of
    Blasting in Mine Stability", In *Brawner and Dorling,* pp. 95-122.

Horne, J.C., *et al.,* 1978, "Depositional Models in Coal Exploration
    and Mine Planning in Appalachian Region", *Bull. Am. Assoc. Petrol.
    Geol.* Vol. 62(12), pp. 2379-2411.

Mallett, C.W., and Wooltorton, B., 1981, "Characteristics and
    Mechanisms of Highwall Failures in Surface Coal Mines, Bowen
    Basin Queensland", *Proc. Aust. I.M.M. Sym., Strip Mining 45 Metres
    and Beyond,* In Press.

O'Regan, G.J., Dunbavan, M., and Mallett, C.W., 1981, "An Investigation into Spoil Stability at Moura", Proc. Aust. I.M.M. Sym. Strip Mining 45 Metres and Beyond. In Press.

Seedsman, R.W., and Emerson, W.W., 1981, "Dispersion Tests and Spoil Pile Stability", CSIRO Aust. Division of Applied Geomechanics, Geomechanics of Coal Mining Report No. 33.

Question

What are the economic effects of the spoil and highwall failures.

Answer

Costs are very significant. To date no major equipment such as a dragline has been lost, although there have been dangerous situations where the dragline has had to be walked out through rapidly failing ground. Coal is frequently lost in spoil failures. Up to 25% of coal is being left in some pits. The major economic impact is in the cost of extra rehandling of material necessitated by failures. This is seen in the production cost per tonne produced, and in capital outlay on additional equipment required to meet production schedules.

Question

What methods of control and stabilization are being employed.

Answer

Methods applied are selected for particular mining situations.

Highwalls: As highwall failures relate to inherent properties of the mined sequence and the effect of the mining process, and it is difficult to modify the primary strength of the rocks, remedial measures concentrate on the effects of mining. Water and movement are critical to strain softening within the highwall, and are minimised by reducing as far as possible surface water ingress and blasting damage (Hagan et al. 1979). The most encouraging results have been achieved by modifications to blasting procedures.

Spoils: Spoil failures develop either in the floor at the spoil/floor interface or in the base of the spoil. Most mines in the Bowen Basin have only a few metres of weak material below the floor in which failures can propagate. In the example discussed at Moura the floor is now being dug and removed prior to spoiling. A discussion of remedial measures applicable to the site is given in O'Regan et al. (1981). Prior fracturing of the floor to disrupt continuous planes of discontinuity has not been attempted. Given the strain softening characteristics of the clays present, it would most probably be counter productive.

Spoil/floor interface failures are commonly related to deterioration of the floor in the time between successive cuts, which may be longer than a year. Care is needed prior to spoiling, with soft weathered rock dug out and removed. As the strength of the base of the spoil is critical in in-spoil failures of the two wedge type, stability is improved by selective placement of spoil. Highwall rock is indexed to obtain the best material to place at the base and dragline operators notified accordingly. This has been the only way to effectively stabilise some spoil piles even though stockpiling and several rehandlings of the best material was required. Re-enforcement of spoil has not been attempted. Experiments indicate that chemical strengthening of the spoil could be achieved by calcium exchange in the sodium rich clays (Seedsman and Emerson 1981). Cross-pit buttressing may be necessary in some situations and initial experiments at South Blackwater Mine are encouraging (G. Boyd, Personal Communication).

Chapter 14

GEOLOGY AND ROCK SLOPE STABILITY--APPLICATION OF
A "KEYBLOCK" CONCEPT FOR ROCK SLOPES

Richard E. Goodman and Gen-Hua Shi

Professor of Geological Engineering
Department of Civil Engineering
University of California, Berkeley

Research Engineer, California Mining and Mineral Resources
Research Institute; on leave from the Research Institute for
Water Conservation and Hydroelectric Power, Peking

ABSTRACT

    In hard, discontinuous rocks, failure modes and stability are
controlled to a great extent by the intersection of discontinuities
with the excavated surface. We have solved the general problem of
finding the shape and size of all blocks formed through the intersection of any number of joints and excavation surfaces and will
demonstrate a simple graphical method for determining which of these
is the "keyblock." An example will explain this methodology applied
to a simple open pit. The effects of slope curvature in plan or
section, will be demonstrated. Finally, we will show how to determine
the optimum orientations for the cut slopes. The methodology is
based upon a powerful topologic theorem concerning the finiteness
of spatial intersections. This paper is a companion to another
dealing with underground openings in the Proceedings of the 22nd
Symposium on Rock Mechanics (MIT), July 1981.

    This paper discusses the rock blocks created by the intersection
of discontinuities and excavation surfaces in a rock mass. It
establishes principles and procedures by means of which those blocks
that are finite and critically located can be identified and
described. The main assumptions are the following:

1. The rock itself is assumed to be strong so that the only
    concern for stability is the movement of rock blocks them-
    selves.

2. The orientations of the joint sets can be determined as
    constants. In future research it should be possible to
    introduce statistical distributions for orientations.

The fundamental proposition underlying this work is that some
joint blocks are more important than others and that certain combinations of joints and excavation surfaces liberate blocks that are
keystones for the excavation stability. In any joint and excavation
system, certain combinations of surfaces will be found to define
only infinite blocks that cause no worry if the rock cannot fracture.
In contrast, it will be found that other combinations of joint and
excavation surfaces create finite blocks and some of these may rest
in orientations that cannot be maintained unless support is provided.
Should such a block be located where other finite blocks rely upon
it for support, a larger rock movement will follow the loss of the
first block. For example, Figure la shows four types of joint
blocks defined by joint and excavation surfaces. Block A is the
most critical. Block B can move only by virtue of the movement of
block A and blocks C can move only after movement of block B. Since
the survival of block A is the key to the stability of a larger mass
within the slope, and possibly to overall stability, it is appropriately denoted a keyblock. Another example is shown in Figure lb,
where A is a potential keyblock for the failure of B. In this
example, however, B cannot move even if A moves. If the toe of the
slope were lengthened or steepened as shown, a new keyblock would
be liberated, this one being of type B. Its movement would then
undermine the entire slope. It is not generally obvious which
combinations of blocks will be potentially dangerous and therefore
we rely on a formal analysis.

The analysis is developed on the basis of a theorem introduced by
Dr. Shi using methods of mathematical topology. This theorem is
stated and elaborated in a companion paper (Shi and Goodman, 1981)
but will be reiterated here for clarity. *A physical rock block
represents the intersection of n half spaces determined by n planes
of given orientation and position. Let each plane be shifted without
rotation until it passes through a common origin. If the intersection of half spaces on this shifted diagram contains any points
other than the origin itself, the block defined by the intersection
of these half spaces is infinite in extent. On the other hand, if
the rock block is finite, the intersection of half spaces in the
shifted diagram will contain only one point in common, namely the
origin.* The theorem is rigorous, and encompasses blocks formed by
the intersection of any number of non-parallel surfaces.

Figure 2a presents a simple two-dimensional example with a finite
block formed by two joint planes and a free surface. Each surface
separates the whole plane into two half planes, which are denoted
U and L. The particular block in question is the intersection of
U_1, U_2, and L_3. In Figure 2b, each plane passes through a common
origin. The only region common to U_1, U_2, and L_3 is the single

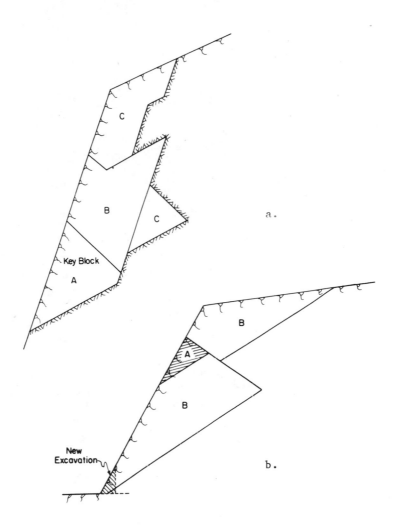

Fig. 1.   Examples of keyblocks in rock slopes

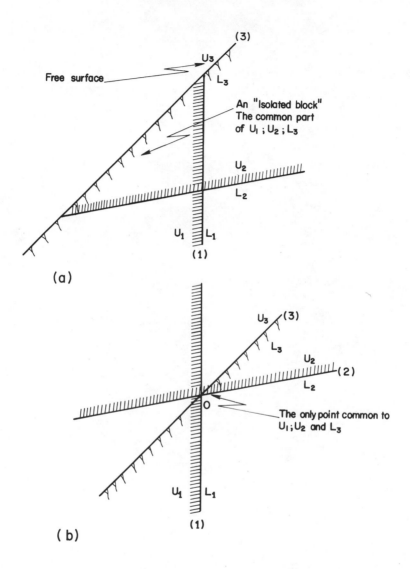

Fig 2a. A two dimensional, finite, isolated block; Fig. 2b. Application of the theorem for this example.

point representing the origin itself. By the theorem, the block is finite.

Figure 3a offers a contrary example, with an infinite block formed by the intersection of U_1, L_2, and L_3. We form the shifted diagram in Figure 3b, where it is seen that there is a sector common to U_1, L_2, and L_3. Therefore, by the theorem, the block must be infinite.

In order for a rock block to be a keyblock, it must be free to move into the excavated space. If the block meets this condition, we call it isolated. In the first example, the finite block was in fact isolated. But in Figure 4 we have an example of a finite block, B, that is not isolated. Since it has a tapered shape, block B cannot move into the excavated space. A second theorem establishes whether or not a block will be tapered in this way. *If a block is finite by the intersection of n non-parallel joint surfaces without any free surfaces, it will be tapered when cut by a free surface.* Figure 4 illustrates this theorem. Block A, formed by U_1, U_2, U_4, and L_5 is seen to be finite and isolated. To test the theory, we shift all the joint planes in question through a common origin but omit the free surface (5) (Figure 4b). Since there is a region common to U_1, U_2, and U_4, the block formed by the joint planes alone is infinite. Therefore it will not be tapered when made finite by its intersection with one or more free surfaces, in particular with L_5. In contrast, block B, formed by the intersection of L_1, U_2, U_3, and L_5 is finite and tapered. In the shifted diagram, Figure 4c, the joint half-spaces L_1, U_2, U_3 are seen to possess no region in common except for the origin and therefore define a finite block. By the theorem, this block will be tapered and not-isolated when intersected by L_5.

These two-dimensional examples are instructive because the finiteness and tapering conditions are identifiable by inspection. In a three-dimensional case, drawings and sections are not so lucid. All cases of interest to rock mechanics are three dimensional, however. The easy solution is to adopt the stereographic projection (see Goodman, 1976). Figure 5 shows the features of this technique when used to represent joint or excavation planes. In the stereographic projection, we adopt as a working convention that all planes and lines will pass through a common origin at the center of a reference sphere of arbitrary radius R. The reference sphere is then a three dimensional analogue to the shifted diagram of half spaces shown in Figures 2b, 3b, 4b and 4c. The points of intersection of the planes with the surface of the reference sphere are projected to a horizontal plane through the center of the reference sphere by means of a focus at the bottom of the reference sphere, from which all rays emenate. In this manner, the horizontal plane is established as a circle of radius R about 0 (shown by the broken line in Figure 5). Choosing north arbitrarily on the circumference of this circle, we can project an inclined plane with dip α (below horizontal) by another circle with center O_1 and radius R_1. The distance $0-0_1$ is equal to $R \tan \alpha$; and the radius R_1 has magnitude $R/\cos\alpha$. (The derivation of these

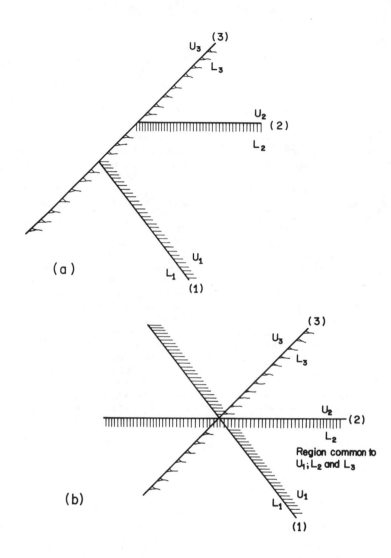

Fig. 3a. A two-dimensional, infinite, non-isolated block.
Fig. 3b. Application of the theorem for this example

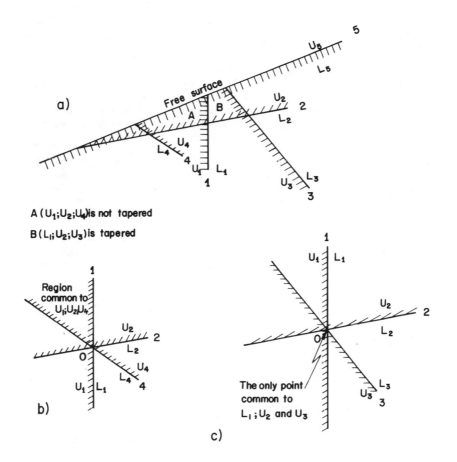

Fig. 4a. Tapered and non-tapered blocks; Fig. 4b. application of the tapering theorem for block A; Fig. 4c. application of the theorem for block B.

formulas is immediate from a vertical section through the reference
sphere along the line of dip; see Figure 5 of Shi and Goodman, 1981.)
Thus, in this upper hemisphere stereographic projection, the inclined
plane with dip 50° in direction 240° (S 60° W) is represented by the
circle drawn with a solid line. The region inside this circle is the
half space above the plane and the region outside the circle is the
half space below the plane.

It is important to appreciate that the stereographic projection
maps only point on the <u>surface</u> of the reference sphere. Therefore,
the origin itself is not contained on the stereographic projection.
Application of the first theorem then consists of finding regions
that have <u>no</u> intersection with the half spaces formed by a set of
excavation surfaces.

In the context of a surface excavation, the blocks that are
critical might be formed by the intersection of one or more excavation
surfaces. The simplest excavation is formed by a single planar rock
cut; more complicated excavations include benches and faces, and may
be curved in plan. To illustrate the method for handling such
excavations we consider a compound excavation formed by planes 5, 6,
and 7 on Figure 6a. The face of the cut, B, is formed by plane 5
dipping 60°. The crest of the slope, A, is formed by the intersection
of plane 5 with 6 dipping 20°. The toe of the slope, C, is created
by the combination of plane 5 with a horizontal surface (plane 7).

The kinematics of blocks intersecting the slope face are
represented by Figure 6b, which shows a stereographic projection of
plane 5 alone. The theorem stated previously can be applied in this
diagram by plotting a circle for each joint plane and examining the
spherical polygons defined by their intersections in the stereographic
projection. If a spherical polygon of joint half-spaces plots
entirely within the region U_5 of Figure 6b, it is therefore completely
contained in the region <u>above</u> plane 5 and consequently has no intersection with the half space <u>below</u> plane 5. Such a joint block is
finite and is a potential keyblock in the slope face.

A similar analysis can be made for the crest of the slope (region
A in Figure 6a). By the first theorem, if a block formed by the
intersection of joint sets and excavation surfaces is finite, it
must have no intersection with the region below both planes 5 and 6.
The polygonal region on the stereographic projection defined by the
intersections of joint half spaces must therefore plot entirely
within the shaded area of Figure 6c constructed by the union of the
circles representing planes 5 and 6. Since the union of two circles
is larger than either alone, the crest region has more potential
keyblocks than the slope face alone and is therefore more likely to
degrade.

In the case of the toe region (C of Figure 6a), the first theorem
means that a joint block must have no intersection with the half

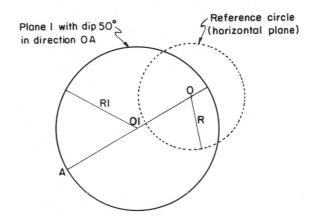

Fig. 5. Stereographic projection (upper hemisphere, conformal) of
a horizontal plane and an inclined plane.

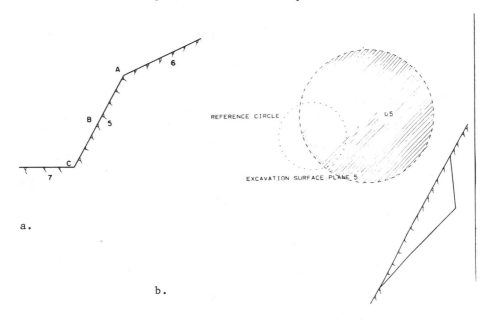

a.

b.

Fig. 6a. Definition of crest region (A), slope face (B), and
toe region (C) of a rock slope; Fig. 6b. to be isolated
in the face, a rock block must plot entirely in the
shaded region.

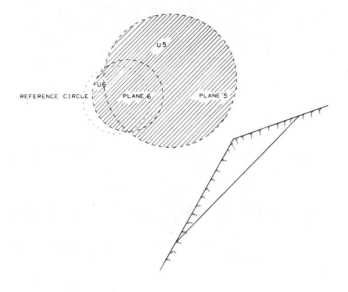

c.

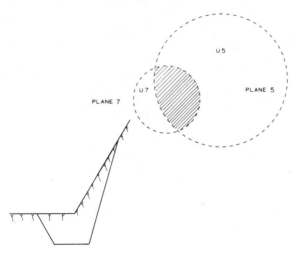

d.

Fig. 6c. In order to be isolated in the slope crest, a rock block
must plot entirely in the shaded region; Fig. 6d. to be
isolated in the toe of the slope, a rock block must plot
in the smaller shaded region.

spaces below both planes 5 and 7. On the stereographic projection, Figure 6d, the toe of the excavation must therefore be represented by the planes 5 and 7; in other words, the polygonal region of intersection of the joint set half spaces must plot entirely within the shaded region of Figure 6d that is generated by finding the area common to the circles of planes 5 and 7. Since the shaded region is smaller than the region within either circle alone, the toe region will have fewer keyblocks than the crest region, or than the face region.

We are now in a position to examine a three dimensional example. Consider the rock mass divided by four sets of joints with dip and dip direction given in Table 1. The orientations of the cut slopes are also stated.

TABLE 1

Plane	Dip (°)	Dip Direction (°)
Joint Set 1	70	160
Joint Set 2	25	160
Joint Set 3	45	110
Joint Set 4	50	220
Slope Face, 5	60	70
Slope Crest, 6	20	70
Slope Base, 7	0	

In Figure 7 are shown the circles for each of the joint sets. The various regions delimited by sections of the arcs of different joint circles are labelled with four digits, e.g., 0110. The digits are given in the order of the joint set numbers; the number 0 denotes the half space above the joint (previously called U); the number 1 denotes the half space below the joint (previously called L). Thus the number 0110 represents block $L_1 U_2 U_3 L_4$. Such a block is formed by sections of arcs of each of the four joint sets and is inside circles 1 and 4 and outside circles 2 and 3.

Figure 8 is a tree drawn to show all the possible combinations of half spaces created by the four joint sets, considering one half space per joint (i.e., omitting blocks formed with one or more sets of parallel faces). Since there are four joint sets and two half spaces per joint set, there are sixteen possible blocks formed by their intersection. However, examination of Figure 7 reveals that only fourteen blocks have an intersection with the reference sphere. The missing regions are 1100 and 0011. These blocks are therefore finite even without any excavation surfaces. By the second theorem,

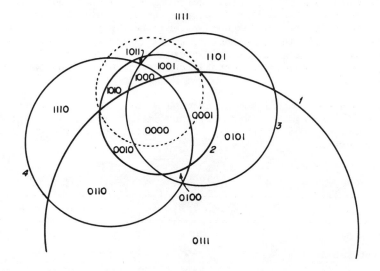

Fig. 7.   Rock blocks are represented by the spherical polygons
created by the intersections of the four joint planes
of Table 1 that are projected on this Figure.

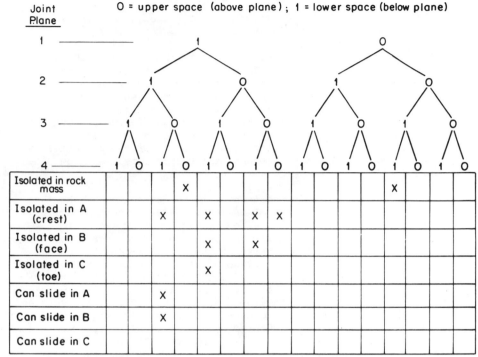

O = upper space (above plane) ; 1 = lower space (below plane)

Joint Plane																
Isolated in rock mass			X									X				
Isolated in A (crest)		X		X		X	X									
Isolated in B (face)				X		X										
Isolated in C (toe)				X												
Can slide in A		X														
Can slide in B		X														
Can slide in C																

Fig. 8.   A tree describing all possible combinations of half spaces,
created by the four joint planes of Table 1, that delimit
rock blocks formed of one joint plane each together with
a free surface or set of free surfaces.   X indicates the
results of the analysis described in later Figures.

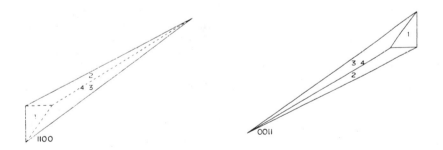

Fig. 9.   The tapered blocks of the system of joint planes.

they will be tapered in the excavation and cannot be keyblocks. The blocks are shown in Figure 9a and 9b.

It is interesting to observe that the blocks 0011 and 1100 are centro-symmetric. In general, forming a new block by interchanging all the half spaces defining another block will create a centro-symmetric sister block. If one block is finite when excavated by a half space, the image block will be finite when excavated by the opposite half space. This is a practical result for underground excavations, wherein each excavated half space can have its opposite in another part of the excavation. It is not a practical matter in a surface excavation and only one of the sister blocks will be realized.

In Figure 10a, we apply the theorem to determine the blocks that will be isolated in the crest of the slope. Recall from Figure 6c that the crest region is represented by the union of excavation planes for the crest and the face of the slope. The excavation at the crest is therefore represented by the region within the dashed arcs. (Note that the lightly dotted circle is the reference circle, i.e., the horizontal plane.) There are four regions on the stereographic projection of joint sets that are contained entirely within the dashed arcs of Figure 10a; these are blocks 1101, 1001, 1000, and 1011. Other blocks can be formed by deleting one or two of the planes. A deleted plane is denoted by the symbol 2. Thus, additional blocks isolated by the crest of the slope are 1201, 1002, 1021, 2011, 1202, 1102, and 1200. Since there are a number of isolated blocks, additional analysis is required to choose the keyblocks.

One possible additional analysis is to consider the sliding directions caused by a specific set of forces comprising the combined action of self weight, inertia, water pressure, rock supports, etc. In another paper we intend to explain how to perform such an analysis. Here we simply show the results when the only loads are those due to self weight. Figure 10b presents the directions of potential sliding under gravity for all the blocks. The symbol 13 on the shaded spherical triangle means that the block will tend to slide along the intersection of planes 1 and 3. The other sliding directions are attached to blocks that are not isolated and therefore are not potential keyblocks. By this figure, we determine that the keyblocks of the slope crest are 1101, 1102, 1201, and 1202.

Another approach to reducing the number of isolated blocks is to draw all the candidate keyblocks. Using vector analysis we have developed a BASIC program that does this on a minicomputer with graphics display (Tektronix 4051). The various blocks considered are shown in Figures 10c through 10m. By mere inspection of these figures, it is possible to appreciate whether any of these blocks hold potential for movement. The excavation surfaces are denoted by 5 and 6 on these figures. All figures are drawn for an observer looking from the Northeast, i.e., slightly off the direction of dip of the slope, which is to the N 70 E.

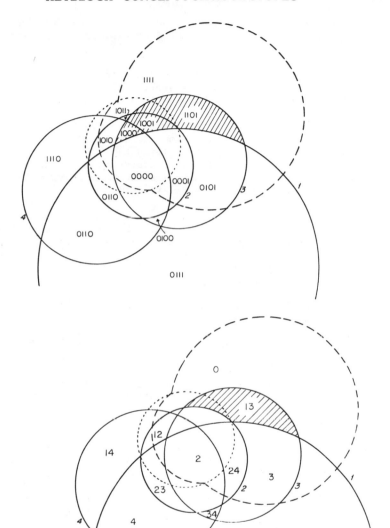

a.

b.

Fig. 10. Isolated blocks of the crest of the rock slope; Fig. 10a.
        stereographic projection of the joints and application of
        the theorem to determine the isolated blocks (shaded); Fig.
        10b. results of an analysis to determine the sliding direc-
        tions under gravity alone. 13 means block slides on the
        intersection of planes 1 and 3 if it is isolated.

- Figure 10c shows the block 1100, which we already know to be
  tapered, when it is intersected by the excavation surfaces (5
  and 6).

- Figure 10d shows block 1011. Note that the potentially sliding
  edge along the line of intersection of planes 1 and 2 is horizontal. Therefore the block will not slide under gravity alone.

- Figure 10e shows block 2011, i.e., the block formed with planes
  2, 3, and 4 and the excavation surfaces. The lines of intersection 23, 34, and 24 controlling the shape of the block do
  not daylight downward in the space created by planes 5 and 6.
  Therefore this block cannot slide on intersection lines. All
  the joint planes daylight upward into the free space. This
  block is safe.

- Figure 10f shows block 1021. Like block 1011, the line of
  intersection 12 is horizontal and poses no threat.

- Figure 10g shows block 1001. Again, intersection 12 is horizontal so the block is safe.

- Figure 10h shows block 1000. All the possible intersections and
  faces daylight upward into the excavation.

- Figure 10i shows block 1200. The potentially troublesome
  intersection 14 does not daylight downward into the excavation.

- Figures 10j, 10k, 10l, and 10m show blocks 1101, 1102, 1201, and
  1202. All are similar in presenting a potential sliding mode
  along the intersection of planes 1 and 3 that does daylight
  into the excavation. These are the keyblocks of the crest.

Now we turn to the face of the slope, formed by excavation plane 5
alone. Figure 11a shows the stereographic projection of the joints
with the single half space of the excavation indicated by the dashed
line. (The lightly dotted line is the reference circle.) The
isolated blocks formed of four joint planes are 1101, 1001, and 1011.
Deleting some planes produces additional candidate keyblocks 1201,
1021, 2011, and 1102. These are drawn to reveal their potential
stability.

- Figure 11b shows the tapered block 1100 as intersected by
  plane 5.

- Figure 11c shows the very slender block 2011. The intersections
  and faces daylight only upward, so it is not a keyblock.

- Figure 11d shows block 1011. The daylighting intersection 12
  is horizontal so the block cannot slide.

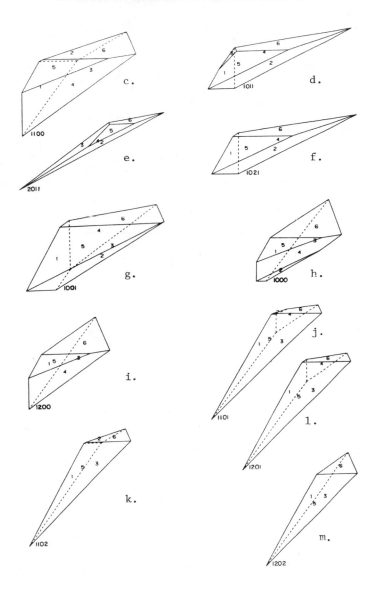

Fig. 10.   Isolated blocks in the crest region: Fig. 10c. to 10m.
drawings of potential keyblocks. The slope faces are
planes 5 and 6 in these drawings.

- Figure 11e, block 1021, and Figure 11f, block 1001, are similar to Figure 11d and cannot slide.

- Figures 11g, 11h, and 11i show blocks 1102, 1101, and 1201. All are isolated and all can slide along intersection 13. These are the keyblocks of the slope face.

Table 2 summarizes the results of the analysis for the keyblocks of the crest and the face presented in Figures 10 and 11. It also presents the keyblocks for the toe of the slope, the figures for which are not presented here. In this case, there are two isolated blocks, 1011 and 2011, but neither can slide. The toe of the slope is therefore stable. It has no keyblocks.

TABLE 2--SUMMARY OF KINEMATIC ANALYSIS

Region of slope	Tapered Blocks	Isolated Blocks that cannot slide under gravity	Isolated blocks that could potentially slide
A: Edge of bench (convex upward) (Figure 10)	1100	1011 2011 1021 1001 1002 1000 1200	1101 1102 1201 } slide along intersection 1, 3 1202
B: Face of slope (Figure 11)	1100 0011	2011 1011 1021 1001	1101 1102 } slide along 1201 intersection 1,3
C: Back of bench, concave upward (No figure)	1100 0011	1011 2011	None

The stereographic projection lends itself to analysis of the influence of the direction of the excavation in a rock mass with defined joint orientations. For the joint sets and slope angles given in the previous examples, trials in various slope directions established that slopes with dip directions between 221° and 33° (221° < dip direction < 33°) have no keyblocks. This would be true even if the joints had great extent, had no shear strength, and were

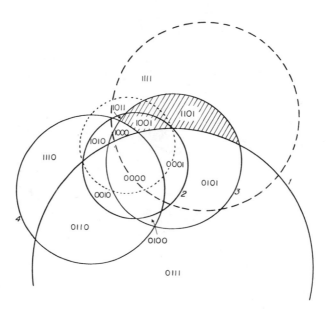

Fig. 11. Isolated blocks of the slope face. Fig. 11a. stereographic
projection of the joint sets and application of the theorem
to determine the isolated blocks (shaded).

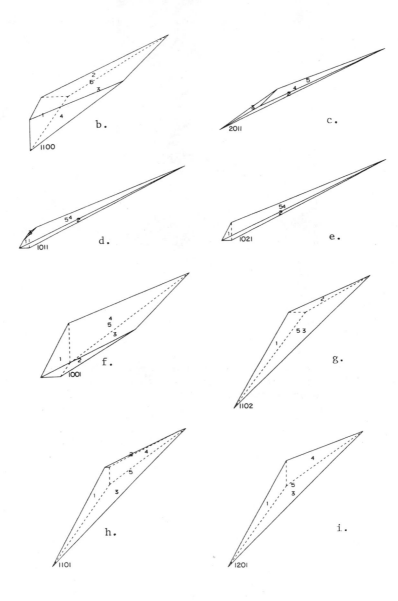

Fig. 11. Isolated blocks of the slope face. Fig. 11b to 11i. are
drawings of potential keyblocks in the slope face;
the plane of the slope face is represented by 5.

closely spaced; (except that if the latter two conditions were taken to an extreme, the first assumption of this paper would be violated). Space does not permit a full elaboration of the full set of slope directions but illustrations are given in good and bad directions respectively in Figures 12a and 12b.

The analysis of a cut with dip direction 33°, and face and crest angles of the cut respectively 60 and 20 degrees, is shown in Figure 12a. While there are isolated blocks within the dashed arcs that represent the crest of the slope, analysis of the sliding directions under gravity alone shows that there are no isolated blocks that can slide. This is because the line of intersection 13 is actually contained in the face of the slope and can never daylight. In contrast, in the case of a slope directed towards the South East (160°), the condition is potentially unstable. The large shaded region in Figure 12b with label 2 indicates that this large isolated block can slide on plane 2. Figure 12c is a drawing of the block in question (0000).

CONCLUSIONS

In this article we have introduced a novel and rigorous method for performing shape analysis for potential rock block movements. The idea behind this is that the entire rock excavation will be safe if the keyblocks are discovered and supported or if the excavation is oriented and sloped, if possible, so that there are no keyblocks. The entire analysis can be performed by hand using the stereographic projection. However, the method is amenable to automatic computation so that it is feasible to perform a kinematic analysis for virtually any problem. The analysis will yield the locations and design requirements for support when the actual locations of the rock joints are known. A hypothetical example is shown in Figure 13.

ACKNOWLEDGMENT

This paper was prepared with partial support from the Office of Surface Mining, U.S. Department of the Interior, through a grant to the California Mining and Mineral Resources Research Institute.

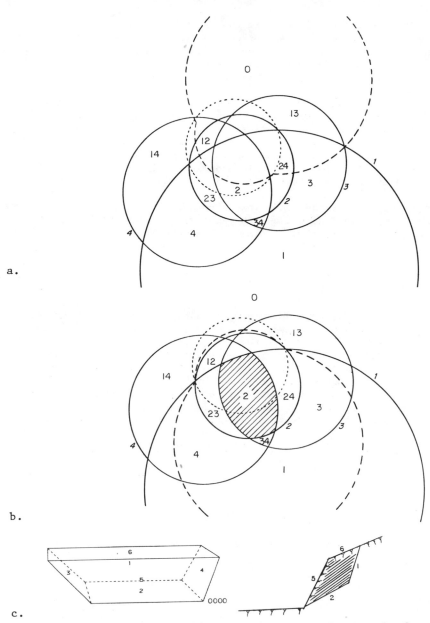

a.

b.

c.

Fig. 12. Influence of the direction of the slope face on the formation
of keyblocks. Fig. 12a is a stereographic projection of blocks
formed in a cut dipping N 33° E; there are no keyblocks.
Fig. 12b is a stereographic projection to find the keyblocks
in a slope dipping S 20° E. Fig. 12c shows block 0000 of 12b.

REFERENCES

Shi, Gen-Hua and Goodman, R. E., 1981, "A New Concept for Support of Underground and Surface Excavations in Discontinuous Rocks Based on a Keystone Principle," July, Proceedings 22nd Symposium on Rock Mechanics, MIT.

Goodman, R. E., 1976, <u>Methods of Geological Engineering in Discontinuous Rocks</u>, West Publishing Co., St. Paul, Chap. 3 "Principles of Stereographic Projection and Joint Surveys," pp. 58-90.

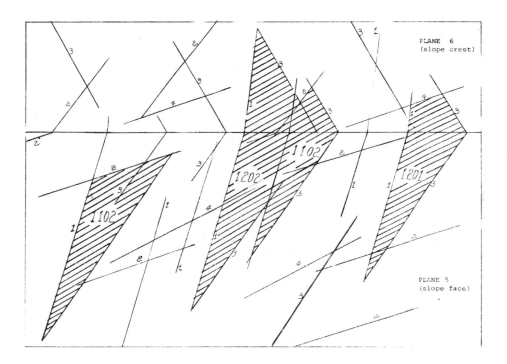

Fig. 13. Application of the kinematic analysis to identify specific keyblocks in the slope crest. The map is a plan of the crest region with projected geology and with slope dips as given for planes 5 and 6 in Table 1. Keyblocks are ruled.

Question

Is there an existing publication or one planned.

Answer

The lecture closely followed the text of the paper published in the Proceedings of this Conference. A companion paper is in the Proceedings of the 22nd Symposium on Rock Mechanics, held in July 1981. These two papers cover respectively applications in surface and underground excavations. A fundamental discussion of the underlying theory is currently being prepared. We are also contemplating a short book, to be published by John Wiley. This will contain computer programs.

Question

Can the analysis you introduced give an answer to the interactive processes and interlocking of the individual blocks if the key block is removed.

Answer

The program, examined either graphically or computationally, will give the keyblocks for any defined set of excavation surfaces. If the excavation shape is changed by removal of a keyblock, it is possible then to redefine the excavation shape accordingly. Thus, in an interactive computer setup, it will be possible to ascertain through use of the program whether or not the removal of the keyblock creates a worsening situation. This would lead to a progressive deterioration of the excavation. It may be that the progression will achieve a final, still stable configuration after removal of some keyblocks. Thus a new tunnel or surface excavation shape will be attained. If this is the case, it should be possible to find the optimum excavation shape through use of an interactive program. We are actually now examining this kind of analysis for tunnels.

It should be noted, however, that the keyblock concept is an attempt to find the important blocks that merit support to prevent overall rock loosening. The spirit of the method is that keyblocks should not be permitted to move. If the joints open, after keyblocks move, it will be difficult to address the hazards created.

Question

How often are tapered blocks associated with keyblocks and is there a definitive relationship. Secondly, is there benefit derived from bolting into tapered blocks.

Answer

The number of blocks formed of one plane each from n joint sets, is equal to $(2)^n$. The number of regions formed on the sphere by these blocks is equal to the quantity $(n)(n - 1) + 2$. By the second theorem, therefore the number of tapered blocks is the difference between these expressions evaluated for the particular number of joints (n). For four joint sets, there are thus two tapered block types, only one of which will be realized on a single side of the excavation. If this block could be definitely established in the wall of the excavation or at some slight depth, it could be used to site anchors to sustain light reinforcement, for example in defense of ravelling during blasting.

Question

It was pointed out that Mr. Riching's paper stressed the need for an evaluation of the probability of failure for different slopes. Where in the analysis would you recommend including the uncertainties introduced by variations in the orientation, spacing, irregularity, etc., of the discontinuities analyzed.

Answer

Statistical distributions for spacing, orientation, extent, surface roughness, and other parameters of discontinuous rocks can be considered but they have not been introduced in our work as yet. As we have no results to offer, the following is merely conjectural.

If joint orientations are deterministic, then we normally would not consider blocks formed with two faces per joint set. Such blocks would have trouble sliding because the merest irregularity would prevent movement without rupture through virgin rock material. This was pointed out by Terzaghi in his paper on tunnel geology, in Proctor and White, Rock Tunneling with Steel Supports (1946). If the joint orientations are dispersed about central tendencies, two faces of one block that are presumably members of the same joint set might not in fact be parallel. Through statistics, it should be possible to derive a probabilistic estimate of the departure from parallelism given specific distributions for joint orientations. For a given roughness, it would then be possible to predict whether or not a specified block could actually slide. It might be necessary to expand the list of keyblocks in the probabilistic analysis.

Consideration of variability of joint extent leads to a distribution of block sizes when integrated in three dimensions. Our analysis can lead to a distribution of block shapes, given orientation and spacing distributions. These are distinct concepts but could be coupled to yield a probabilistic weighing of the potential importance of various keyblocks.

If a specific geological map is not available, a statistical simulation could be developed to provide a generic map, or a series of probable maps, when the statistics of jointing are introduced. The keyblock concept can then be applied to yield sample distributions of keyblocks on the generic maps. In this manner, a pattern and length of rock bolts could be selected rationally to achieve a certain level of effectiveness.

Question

Does the concept of keyblocks take into account cohesion and friction along joints. How is the keyblock selected if joints comprising the keyblock have different friction characteristics.

Answer

The keyblock concept is a lower bound condition. If keyblocks are actually present, they may or may not slide by virtue of their shear strength characteristics in relation to the forces present. The analysis locates the keyblocks. Then, well developed existing methods of limit equilibrium analysis can consider the stability of these blocks, with variable friction on the different faces if desired. For example, the stereographic projection wedge analysis proposed by Klaus John or methods published in Rock Slope Engineering by Hoek and Bray can be used to examine the stability of wedge shaped keyblocks after they have been identified through the keyblock analysis. Potential rotations of keyblocks can be analysed using methods published by Wittke or by Goodman (Methods of Geological Engineering in Discontinuous Rock)

Question

How do you find keyblocks in heavily fractured rock.

Answer

If the jointing is closely spaced in only one set, for example parallel to bedding planes, the analyis is unaffected in method. If there is close jointing on all sets, the numbers of blocks become so large that the rock mass might resemble a soil and failure through the body of rock, or on complex paths embracing steps here and there along the different joint sets, would nullify the strict use of a keyblock concept. However, a block shape analysis could be made using the methods of this paper even for such rock masses. This could be of interest in rock breaking, block caving, and other technologies. In closely jointed rock, a sliding surface may wander from one to another member of a specific joint set, in this way, lengthening individual joints. The result is that the sizes of keyblocks might grow larger than in rocks with widely spaced joints. Also, in view of the previous statement, the keyblock list would have to be expanded to embrace blocks with parallel faces formed of two individuals from a given joint set. (compare with the answer to previous question, "It was pointed out......")

Question

It is noted that the keyblock concept implies that if individual joints are mapped, only the keyblocks need be supported (if necessary). This means that pattern bolts would not be necessary. However, how would support be designed if joint sets are recognized but individual joints are not mapped.

Answer

This is an important question. Some pattern bolts are usually necessary regardless of keyblocks to retain the thin zone of blast-loosened rock from further loosening. If specific geological maps are not obtainable, it will be prudent to design for the largest probable joint blocks occurring in any locality. Thus a pattern of joints will be required that assures support for the largest probable block and all smaller subblocks.This will be expensive and cost savings will be possible if specific geological information and rock mechanics data can be introduced. (see also last paragraph in answer to question, "It was pointed out......")

Chapter 15

ANALYSIS OF SLOPE STABILITY IN VERY HEAVILY JOINTED OR
WEATHERED ROCK MASSES

Evert Hoek, Principal

Golder Associates, 224 West 8th Avenue
Vancouver, Canada   V5Y 1N5

SUMMARY

An empirical criterion is presented for use in estimating the strength of heavily jointed or weathered rock masses. This criterion incorporates the intact strength of the rock material and introduces two dimensionless parameters m and s which characterize the behaviour of the interlocking particles in a jointed rock mass. The use of rock mass classification systems for estimating the values of these parameters is discussed.

The application of this non-linear failure criterion to the analysis of slopes is illustrated by means of a worked example.

INTRODUCTION

A large proportion of rock mechanics literature deals with the behaviour of intact rock or of rock containing one or two families of discontinuities in regular patterns. On the other hand, most soil mechanics literature deals with residual materials in which structural patterns have been destroyed by weathering and/or movement and where the resulting material can be treated as a homogeneous mass.

In the no-man's land between these two disciplines lie some important engineering materials which have received scant attention. Typical of such materials is heavily jointed weathered rock in which particle interlocking has not been destroyed by movement of the rock mass. The behaviour of such materials is poorly understood and, because of the difficulty of applying conventional testing techniques to such materials, there is a serious lack of reliable data for use in engineering design.

This paper presents an empirical failure criterion which, when used in conjunction with a rock mass classification system, provides a basis for estimating the strength reduction associated with an increase in the frequency of discontinuities and the weathering of a rock mass. The application of the criterion to the design of slopes in heavily jointed weathered rock masses is illustrated by means of a worked example.

DERIVATION OF FAILURE CRITERION

Figures 1 and 2 illustrate the influence of scale upon the behaviour of rock masses. Intact rock can generally be treated as a homogeneous isotropic elastic material. The introduction of one or two discontinuities results in strongly arisotropic behaviour in which the failure process is dominated by sliding on the discontinuity surfaces. When four or more sets of closely spaced (relative to the size of the structure) discontinuities are present, the rock mass behaviour is controlled by sliding, rotation, crushing and splitting of the individual pieces of rock within the mass.

The overall behaviour of such a rock mass approximates to that of a homogeneous isotropic mass of interlocking angular particles.

The difference between the behaviour of an intact rock specimen and that of a jointed rock mass is the result of the greater mobility of the individual rock pieces within the interlocking matrix. As a first approximation, the intact sample may be regarded as a small scale model of the jointed rock mass and a destruction of the tensile bond between individual grains or crystals results in the greater mobility which is characteristic of a jointed rock mass. This analogy was used in an elegant series of experiments in which marble specimens were heated to destroy the tensile bond between grains and the resulting material showed many of the characteristics of a heavily jointed rock mass (ROSENGREN and JAEGER, 1968).

For a failure criterion to be useful for both laboratory scale studies and full scale engineering design, it must be capable of predicting the strength reduction associated with the increase in jointing illustrated in Figure 1. It must also be capable of modelling the highly non-linear relationship between major and minor principal stresses or between shear stress and normal stress at failure.

A complete discussion on the derivation of the empirical failure criterion presented below exceeds the scope of this paper. Briefly, the criterion was developed by a trial and error process based upon experience of both theoretical and experimental studies of rock failure (HOEK and BROWN, 1980).

The resulting failure criterion can be expressed in the following forms:

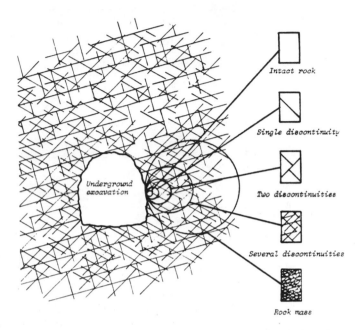

Figure 1 : Transition from intact to heavily jointed
rock mass with increase in sample size.

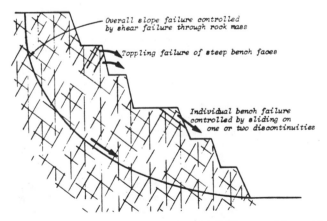

Figure 2 : Relationship between size of slope, discontinuity
spacing and mechanism of slope failure.

$$\sigma_1 = \sigma_3 + \sqrt{m\sigma_c\sigma_3 + s\sigma_c^2} \tag{1}$$

$$\tau = A\sigma_c \left(\frac{\sigma}{\sigma_c} + \frac{\sigma_t}{\sigma_c} \right)^B \tag{2}$$

where: σ_1 = major principal stress at failure,

σ_3 = minor principal stress at failure,

σ_c = uniaxial compressive strength of intact rock pieces,

m = material constant which controls the curvature of the σ_1 vs. σ_3 curve,

s = material constant which controls the location of this curve in space,

τ = shear strength,

σ = normal stress at failure,

$\left. \begin{array}{c} A \\ B \end{array} \right)$ = material constants for Mohr envelope,

σ_t = $1/2 \, \sigma_c \, (m - \sqrt{m^2 + 4s})$ is the apparent tensile strength of the rock mass.

Typical curves defined by equations 1 and 2 are illustrated in Figure 3. A complete discussion on the determination of the values of σ_c, m, s, A, B, and σ_t from triaxial test data is given in Appendix 2 at the end of this paper.

Note that the strongly anisotropic behaviour resulting from the presence of one or two discontinuities in the rock sample has not been included in this discussion but is dealt with elsewhere (HOEK and BROWN, 1980).

FAILURE OF INTACT ROCK

In order to demonstrate the application of the empirical failure criterion to the analysis of intact rock failure, the results of a study on granite are presented in Figure 4. In order to compare the strength data for the seven samples of granite included in this study, each data set has been normalized by dividing individual strength values by the uniaxial compressive strength of the rock sample under consideration. This process makes it possible to plot all of the data for granite on a single graph.

Figure 4 shows that all of the granites included in this study exhibit very similar strength characteristics and that there is considerable justification for the derivation of a single value of the material constant m to represent these materials.

Similar trends have been found in other rock types and in groups of rock types and the following list represents an attempt to relate average values of the constant m to different categories of rock.

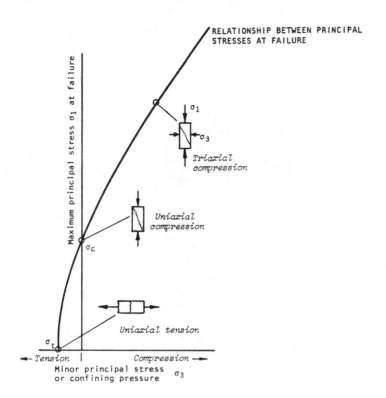

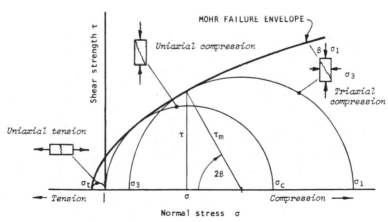

Figure 3 :   Typical relationship between major and minor principal stresses and between shear strength and normal stress for rock failure.

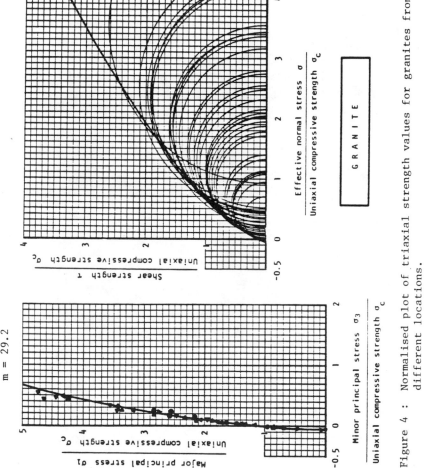

Figure 4 : Normalised plot of triaxial strength values for granites from different locations.

Carbonate rocks with well developed crystal cleavage $m = 7$
(dolomite, limestone and marble),

Lithified argillaceous rocks (mudstone, siltstone, $m = 10$
shale and slate (normal to cleavage)),

Arenaceous rocks with strong crystals and poorly de- $m = 15$
veloped crystal cleavage (sandstone and quartzite),

Fine grained polyminerallic igneous crystalline rocks $m = 17$
(andesite, dolerite, diabase and rhyolite),

Coarse grained polyminerallic igneous and metamorphic $m = 25$
crystalline rocks (amphibolite, gabbro, gneiss, granite,
norite and quartzdiorite)

Note that $s = 1$ for all these samples of intact rock.

USE OF ROCK MASS CLASSIFICATIONS FOR ESTIMATING CONSTANTS FOR JOINTED ROCK

In order to provide a rational basis for estimating the values of the material constants m and s for jointed and weathered rock masses, use is made of the rock mass classifications proposed by Bieniawski (BIENIAWSKI, 1974, 1976) and by Barton et al (BARTON, LIEN and LUNDE, 1974). In the interests of space, only the CSIR Geomechanics Classification proposed by Bieniawski will be discussed in this paper.

Table 1 shows that the following five parameters are included in the CSIR Geomechanics Classification.

(1)   Strength of intact rock material,

(2)   Rock Quality Designation (RQD) defined as the percentage of pieces of intact core of length >100 mm in the total length of core drilled,

(3)   Spacing of joints,

(4)   Condition of joints,

(5)   Ground water conditions.

An additional adjustment for the orientation of the major joint system is made on the basis of the information listed in Table 1B.

The following example illustrates the use of the classification system.

TABLE 1 — CSIR GEOMECHANICS CLASSIFICATION OF JOINTED ROCK MASSES

A. CLASSIFICATION PARAMETERS AND THEIR RATINGS

	PARAMETER		RANGES OF VALUES						
1	Strength of intact rock material	Point load strength index	> 8 MPa	4 - 8 MPa	2 - 4 MPa	1 - 2 MPa	For this low range - uniaxial compressive test is preferred		
		Uniaxial compressive strength	> 200 MPa	100 - 200 MPa	50 - 100 MPa	25 - 50 MPa	10-25 MPa	3-10 MPa	1-3 MPa
		Rating	15	12	7	4	2	1	0
2	Drill core quality RQD		90% - 100%	75% - 90%	50% - 75%	25% - 50%	< 25%		
	Rating		20	17	13	8	3		
3	Spacing of joints		> 3 m	1 - 3 m	0.3 - 1 m	50 - 300 mm	< 50 mm		
	Rating		30	25	20	10	5		
4	Condition of joints		Very rough surfaces Not continuous No separation Hard joint wall rock	Slightly rough surfaces Separation < 1 mm Hard joint wall rock	Slightly rough surfaces Separation < 1 mm Soft joint wall rock	Slickensided surfaces or Gouge < 5 mm thick or Joints open 1-5mm Continuous joints	Soft gouge > 5mm thick or Joints open > 5mm Continuous joints		
	Rating		25	20	12	6	0		
5	Ground water	Inflow per 10m tunnel length	None		< 25 litres/min.	25 - 125 litres/min.	> 125 litres/min.		
		Ratio joint water pressure / major principal stress	0		0.0 - 0.2	0.2 - 0.5	> 0.5		
		General conditions	Completely dry		Moist only (interstitial water)	Water under moderate pressure	Severe water problems		
	Rating		10		7	4	0		

B. RATING ADJUSTMENT FOR JOINT ORIENTATIONS

Strike and dip orientations of joints		Very favourable	Favourable	Fair	Unfavourable	Very unfavourable
Ratings	Tunnels	0	-2	-5	-10	-12
	Foundations	0	-2	-7	-15	-25
	Slopes	0	-5	-25	-50	-60

C. ROCK MASS CLASSES DETERMINED FROM TOTAL RATINGS

Rating	100 — 81	80 — 61	60 — 41	40 — 21	< 20
Class No	I	II	III	IV	V
Description	Very good rock	Good rock	Fair rock	Poor rock	Very poor rock

D. MEANING OF ROCK MASS CLASSES

Class No.	I	II	III	IV	V
Average stand-up time	10 years for 5m span	6 months for 4 m span	1 week for 3 m span	5 hours for 1.5m span	10 min. for 0.5m span
Cohesion of the rock mass	> 300 kPa	200-300 kPa	150-200 kPa	100-150 kPa	< 100 kPa
Friction angle of the rock mass	> 45°	40° - 45°	35° - 40°	30° - 35°	< 30°

Consider the case of a rock mass in which a slope is to be excavated. The classification is carried out as follows:

Classification Parameter	Value or Description	Rating
(1) Intact rock strength	150 MPa	12
(2) RQD	40%	8
(3) Joint spacing	150 mm	10
(4) Condition of joints	slightly rough surfaces separation <1 mm, soft joint wall contact	12
(5) Ground water	water under moderate pressure	4
		46

The orientation of the slope is such that very few joints dip out of the slope at angles in excess of 30 degrees. This situation is classed as favourable and Table 1B gives a rating adjustment of −5. Hence, the final rating for the rock mass under consideration is 46 − 5 = 41 which falls into the category of "fair rock".

On the basis of what little experimental evidence is available and the use of a substantial amount of engineering judgement, a table relating rock mass classification to the rock mass failure characteristics (defined by m, s, A, B, and σ_t) has been compiled and is reproduced as Table 2.

DESIGN OF SLOPES IN HEAVILY JOINTED WEATHERED ROCK MASSES

Most rock slope failures occur as a result of sliding and/or rotation of blocks or wedges defined by intersecting structural discontinuities. However, when the rock mass contains a number of discontinuity sets and the spacing of the discontinuities is small relative to the size of the slope, failure can occur as a result of sliding along a shear surface similar to that which occurs in soil slopes. In analyzing this type of failure, the highly non-linear failure characteristics of the interlocking granular rock mass must be taken into account and an example of such an analysis, using the simplified Bishop method, is presented in Appendix 3.

CONCLUSIONS

Heavily jointed weathered rock masses exhibit failure characteristics midway between those of rock and those of soil. Relationships between principal stresses and between shear and normal stress are strongly non-linear as a result of the interlocking of pieces of rock within the jointed or weathered matrix.

An empircial failure criterion has been developed and is presented in this paper. The use of rock mass classification systems for esti-

TABLE 2 - APPROXIMATE RELATIONSHIP BETWEEN ROCK MASS QUALITY AND EMPIRICAL CONSTANTS.

Empirical failure criterion $\sigma_1 = \sigma_3 + \sqrt{m\sigma_c\sigma_3 + s\sigma_c^2}$ $\tau = A\sigma_c\left(\sigma/\sigma_c - \sigma_t/\sigma_c\right)^B$	CARBONATE ROCKS WITH WELL DEVELOPED CRYSTAL CLEAVAGE *dolomite, limestone and marble*	LITHIFIED ARGILLACEOUS ROCKS *mudstone, siltstone, shale and slate (normal to cleavage)*	ARENACEOUS ROCKS WITH STRONG CRYSTALS AND POORLY DEVELOPED CRYSTAL CLEAVAGE *sandstone and quartzite*	FINE GRAINED POLYMINERALLIC IGNEOUS CRYSTALLINE ROCKS *andesite, dolerite, diabase and rhyolite*	COARSE GRAINED POLYMINERALLIC IGNEOUS AND METAMORPHIC CRYSTALLINE ROCKS *amphibolite, gabbro, gneiss, granite, norite and quartz-diorite*
INTACT ROCK SAMPLES *Laboratory size specimens free from joints.* CSIR rating 100 NGI rating 500	m = 7.0 s = 1.0 A = 0.816 B = 0.658 $\frac{\sigma_t}{\sigma_c} = -0.140$	m = 10.0 s = 1.0 A = 0.918 B = 0.677 $\frac{\sigma_t}{\sigma_c} = -0.099$	m = 15.0 s = 1.0 A = 1.044 B = 0.692 $\frac{\sigma_t}{\sigma_c} = -0.067$	m = 17.0 s = 1.0 A = 1.086 B = 0.696 $\frac{\sigma_t}{\sigma_c} = -0.059$	m = 25.0 s = 1.0 A = 1.220 B = 0.705 $\frac{\sigma_t}{\sigma_c} = -0.040$
VERY GOOD QUALITY ROCK MASS *Tightly interlocking undisturbed rock with unweathered joints at ± 3m.* CSIR rating 85 NGI rating 100	m = 3.5 s = 0.1 A = 0.651 B = 0.679 $\frac{\sigma_t}{\sigma_c} = -0.028$	m = 5.0 s = 0.1 A = 0.739 B = 0.692 $\frac{\sigma_t}{\sigma_c} = -0.020$	m = 7.5 s = 0.1 A = 0.848 B = 0.702 $\frac{\sigma_t}{\sigma_c} = -0.013$	m = 8.5 s = 0.1 A = 0.883 B = 0.705 $\frac{\sigma_t}{\sigma_c} = -0.012$	m = 12.5 s = 0.1 A = 0.998 B = 0.712 $\frac{\sigma_t}{\sigma_c} = -0.008$
GOOD QUALITY ROCK MASS *Fresh to slightly weathered rock, slightly disturbed with joints at 1 to 3m.* CSIR rating 65 NGI rating 10	m = 0.7 s = 0.004 A = 0.369 B = 0.669 $\frac{\sigma_t}{\sigma_c} = -0.006$	m = 1.0 s = 0.004 A = 0.427 B = 0.683 $\frac{\sigma_t}{\sigma_c} = -0.004$	m = 1.5 s = 0.004 A = 0.501 B = 0.695 $\frac{\sigma_t}{\sigma_c} = -0.003$	m = 1.7 s = 0.004 A = 0.525 B = 0.698 $\frac{\sigma_t}{\sigma_c} = -0.002$	m = 2.5 s = 0.004 A = 0.603 B = 0.707 $\frac{\sigma_t}{\sigma_c} = -0.002$
FAIR QUALITY ROCK MASS *Several sets of moderately weathered joints spaced at 0.3 to 1m.* CSIR rating 44 NGI rating 1.0	m = 0.14 s = 0.0001 A = 0.198 B = 0.662 $\frac{\sigma_t}{\sigma_c} = -0.0007$	m = 0.20 s = 0.0001 A = 0.234 B = 0.675 $\frac{\sigma_t}{\sigma_c} = -0.0005$	m = 0.30 s = 0.0001 A = 0.280 B = 0.688 $\frac{\sigma_t}{\sigma_c} = -0.0003$	m = 0.34 s = 0.0001 A = 0.295 B = 0.691 $\frac{\sigma_t}{\sigma_c} = -0.0003$	m = 0.50 s = 0.0001 A = 0.346 B = 0.700 $\frac{\sigma_t}{\sigma_c} = -0.0002$
POOR QUALITY ROCK MASS *Numerous weathered joints at 30 to 500mm with some gouge / clean waste rock.* CSIR rating 23 NGI rating 0.1	m = 0.04 s = 0.00001 A = 0.115 B = 0.646 $\frac{\sigma_t}{\sigma_c} = -0.0002$	m = 0.05 s = 0.00001 A = 0.129 B = 0.655 $\frac{\sigma_t}{\sigma_c} = -0.0002$	m = 0.08 s = 0.00001 A = 0.162 B = 0.672 $\frac{\sigma_t}{\sigma_c} = -0.0001$	m = 0.09 s = 0.00001 A = 0.172 B = 0.676 $\frac{\sigma_t}{\sigma_c} = -0.0001$	m = 0.13 s = 0.00001 A = 0.203 B = 0.686 $\frac{\sigma_t}{\sigma_c} = -0.0001$
VERY POOR QUALITY ROCK MASS *Numerous heavily weathered joints spaced < 50mm with gouge / waste with fines.* CSIR rating 3 NGI rating 0.01	m = 0.007 s = 0 A = 0.042 B = 0.534 $\frac{\sigma_t}{\sigma_c} = 0$	m = 0.010 s = 0 A = 0.050 B = 0.539 $\frac{\sigma_t}{\sigma_c} = 0$	m = 0.015 s = 0 A = 0.061 B = 0.546 $\frac{\sigma_t}{\sigma_c} = 0$	m = 0.017 s = 0 A = 0.065 B = 0.548 $\frac{\sigma_t}{\sigma_c} = 0$	m = 0.025 s = 0 A = 0.078 B = 0.556 $\frac{\sigma_t}{\sigma_c} = 0$

mating the parameters incorporated in this empirical criterion is described.

An example of the application of this failure criterion to the design of slopes is given in an appendix to this paper. All the calculations included in the appendices were carried out on a Hewlett Packard model 41C programmable calculator. Copies of the programs may be obtained by writing directly to the author.

ACKNOWLEDGEMENTS

Most of the information contained in this paper has been extracted from a textbook on underground excavations in rock recently published by the author in conjunction with Professor E.T. Brown of Imperial College in London (HOEK and BROWN, 1980). The contributions of Professor Brown to the ideas presented in this paper are gratefully acknowledged.

REFERENCES

BALMER, G. (1952), "A general analytical solution for Mohr's envelope", American Society for Testing Materials, Vol. 52, pages 1260-1271.

BARTON, N., LIEN, R. and LUNDE, J. (1974), "Engineering classification of rock masses for the design of tunnel support", Rock Mechanics, Vol. 6, No. 4, pages 189-236.

BIENIAWSKI, Z.T. (1974), "Geomechanics classification of rock masses and its application in tunnelling", Proc. 3rd Intnl. Cong. Rock Mech., Denver, Vol. 11A, pages 27-32.

BIENIAWSKI, Z.T. (1976), "Rock mass classification in rock engineering", Proc. Symposium on Exploration for Rock Engineering, Johannesburg, Vol. 1, pages 97-106.

BISHOP, A.W. (1955), "The use of the slip circle in the stability analysis of slopes", Geotechnique, Vol. 5, pages 7-17.

HOEK, E. and BRAY, J.W. (1980), Rock Slope Engineering, Third Edition, The Institution of Mining and Metallurgy, London.

HOEK, E. and BROWN, E.T. (1980), Underground Excavations in Rock, The Institution of Mining and Metallurgy, London.

JAEGER, J.C. (1970), "The behaviour of closely jointed rock", Proc. 11th Sump. Rock Mech., Berkeley, AIME, New York, pages 57-68.

JAEGER, J.C. and COOK, N.G.W. (1976), Fundamentals of Rock Mechanics, Chapman and Hall, London.

JANBU, N. (1954), "Application of composite slip circles for stability analysis", Proc. European Conference on Stability of Earth Slopes, Stockholm, Vol. 3, pages 43-49.

JANBU, N. (1973), "Slope stability computations", in Embankment Dam Engineering, Hirschfeld, R.C. and Poulos, S.J., editors, Wiley, New York, pages 47-86.

MORGENSTERN, N.R. and PRICE, V.E. (1965), "The analysis o the stability of general slip surfaces", Geotechnique, Vol. 15, No. 1, pages 79-93.

MORGENSTERN, N.R. and SANGREY, D.A. (1978), "Methods of stability analysis", in Landslides, Analysis and Control, Transportation Research Board Spec. Rep. 176, National Academy of Sciences, Washington, pages 155-171.

NONVEILLER, E. (1965), "The stability analysis of slopes with a slip surface of general shape", Proc. 6th Intnl. Conf. Soil Mech. Found. Eng., Montreal, Vol. ll, pages 522-525.

ROSENGREN, K.J. and JAEGER, J.C. (1968), "The mechanical properties of a low porosity interlocking aggregate", Geotechnique, Vol. 18, pages 317-326.

SARMA, S.K. (1973), "Stability analysis of embankments and slopes", Geotechnique, Vol. 23, No. 3, pages 423-433.

WHITMAN, R.V. and BAILEY, W.A. (1967), "use of computers for slope stability analysis", Jnl. Soil Mech. Found. Div. ASCE., Vol. 93, No. SM 4, pages 475-498.

APPENDIX 1 - NOTATION

A   -   material constant for Mohr envelope

a -   moment arm for water force in tension crack

B   -   material constant for Mohr envelope

b   -   distance of tension crack behind slope face

c_i -   instantaneous cohesive strength

c_i' -   instantaneous cohesive strength for effective stress

H   -   overall slope height

h   -   height of individual slice within slope

h_w -   height of phreatic surface above slice base

m   -   material constant for rock mass

n -   number of pairs in data set

R   -   radius of circular failure surface in slopes

r^2 -   coefficient of determination in regression analysis

s   -   material constant for rock mass

X   -   horizontal distance of slip circle center from slope toe

Y   -   vertical distance of slip circle cneter from slope toe

z   -   depth of tension crack in slope

α   -   inclination of slice base in slip circle analysis

Δx -   width of slice in slip circle analysis

γ   -   unit weight of rock

γ_w -   unit weight of water

ϕ_i -   instantaneous angle of friction

ϕ_i' -   instantaneous angle of friction for effective stress

σ   -   normal stress

σ' —   effective normal stress

σ_1 —   major principal stress

σ_3 —   minor principal stress

σ_{3m} —   maximum value of σ_3 in data set

σ_c —   uniaxial compressive strength of intact rock

σ_t —   apparent tensile strength of rock mass

τ —   shear stress

τ_f —   available shear strength

τ_m —   maximum shear stress

APPENDIX 2 - DETERMINATION OF MATERIAL CONSTANTS FROM
TRIAXIAL TEST DATA

Intact Rock

The empirical criterion given by equation 1:

$$\sigma_1 = \sigma_3 + \sqrt{m\sigma_c\sigma_3 + s\sigma_c^2} \tag{1}$$

may be rewritten as:

$$y = m\sigma_c.x + s\sigma_c^2 \tag{A.1}$$

Where: $y = (\sigma_1 - \sigma_3)^2$ and $x = \sigma_3$.

For intact rock s = 1 and the uniaxial compressive strength σ_c and the material constant m are given by:

$$\sigma_c^2 = \frac{\Sigma y_i}{n} - \left[\frac{\Sigma x_i y_i - \frac{\Sigma x_i \Sigma y_i}{n}}{\Sigma x_i^2 - \frac{(\Sigma x_i)^2}{n}} \right] \frac{\Sigma x_i}{n} \tag{A.2}$$

$$m = \frac{1}{\sigma_c} \left[\frac{\Sigma x_i y_i - \frac{\Sigma x_i \Sigma y_i}{n}}{\Sigma x_i^2 - \frac{(\Sigma x_i)^2}{n}} \right] \tag{A.3}$$

Where x_i and y_i are successive data pairs and n is the total number of such data pairs.

The coefficient of determination r^2 is given by:

$$r^2 = \frac{\left[\Sigma x_i y_i - \frac{\Sigma x_i \Sigma y_i}{n} \right]^2}{\left[\Sigma x_i^2 - \frac{(\Sigma x_i)^2}{n} \right]\left[\Sigma y_i^2 - \frac{(\Sigma y_i)^2}{n} \right]} \tag{A.4}$$

The closer the value of r^2 is to 1.00, the better the fit of the empirical equation to the triaxial test data.

Broken or Heavily Jointed Rock

For broken or heavily jointed rock, the strength of the intact rock pieces, σ_c , is determined from the analysis presented above. The value of m for the broken or heavily jointed rock is found from equation A.3 and the value of the constant s is given by:

$$s = \frac{1}{\sigma_c^2} \left[\frac{\Sigma y_i}{n} - m\sigma_c \frac{\Sigma x_i}{n} \right] \qquad (A.5)$$

The coefficient of determination r^2 is found from equation A.4.

When the value of the constant s is very close to zero, equation A.5 will sometimes give a negative value. In such a case, put s = 0 and calculate m as follows:

$$m = \frac{\Sigma y_i}{\sigma_c \Sigma x_i} \qquad (A.6)$$

Note that equation A.4 cannot be used to calculate the coefficient of determination r^2 when s is negative.

Mohr Envelope

The relationships between the shear stress τ and the normal stress σ and the principal stresses σ_1 and σ_3 are defined by the following equations (BALMER, 1952):

$$\sigma = \sigma_3 + \frac{\tau_m^2}{\tau_m + m\sigma_c/8} \qquad (A.7)$$

$$\tau = (\sigma - \sigma_3)\sqrt{1 + m\sigma_c/4\tau_m} \qquad (A.8)$$

Where: $\tau_m = \frac{1}{2}(\sigma_1 - \sigma_3)$.

By substituting successive pairs of σ_1 and σ_3 values into equations A.7 and A.8, a complete Mohr envelope can be generated. While this process is convenient for some applications, it is inconvenient for slope stability calculations in which the shear strength of a failure surface is required for a specified normal stress value. Consequently, a more useful expression for the Mohr envelope is:

$$\tau/\sigma_c = A\left(\sigma/\sigma_c - \sigma_t/\sigma_c \right)^B \qquad (A.9)$$

Where A and B are empirical constants which are determined from the values of σ and τ given by equations A.7 and A.8 by the following analysis:

Rewrite equation A.9 as follows:

$$y = ax + b \qquad\qquad (A.10)$$

Where:

$$y = \log{^\tau/_{\sigma_c}}$$
$$x = \log({^\sigma/_{\sigma_c}} - {^{\sigma_t}/_{\sigma_c}})$$
$$a = B$$
$$b = \log A$$
$${^{\sigma_t}/_{\sigma_c}} = \tfrac{1}{2}(m - \sqrt{m^2 + 4s})$$

The values of the constants A and B are found by linear regression analysis of equation A.10 for a range of values of σ and τ calculated from equations A.7 and A.8.

$$B = \frac{\Sigma x_i y_i - \dfrac{\Sigma x_i \Sigma y_i}{n}}{\Sigma x_i^2 - \dfrac{(\Sigma x_i)^2}{n}} \qquad\qquad (A.11)$$

$$\log A = \frac{\Sigma y_i}{n} - B.\frac{\Sigma x_i}{n} \qquad\qquad (A.12)$$

The instantaneous friction angle ϕ_i and the instantaneous cohesion for a given value of σ are defined by:

$$\mathrm{Tan}\,\phi_i = AB({^\sigma/_{\sigma_c}} - {^{\sigma_t}/_{\sigma_c}})^{B-1} \qquad\qquad (A.13)$$

$$\qquad\qquad (A.14)$$

$$c_i = \tau - \sigma\,\mathrm{Tan}\,\phi_i$$

Calculation Sequence

The analysis presented in this appendix can be carried out with the aid of a programmable calculator. The following calculation sequence can be used:

Intact Rock

1. Enter triaxial test data in the form $x = \sigma_3$, $y = (\sigma_1 - \sigma_3)^2$
2. Calculate and accumulate:
$$\Sigma x_i, \quad \Sigma x_i^2, \quad \Sigma y_i, \quad \Sigma y_i^2, \quad \Sigma x_i y_i$$
3. Calculate σ_c from equation A.2
4. Calculate m from equation A.3
5. Note that s = 1 for intact rock
6. Calculate r^2 from equation A.4

Broken or Heavily Jointed Rock

1. Enter value of σ_c for intact rock
2. Enter triaxial test data in the form $x = \sigma_3$, $y = (\sigma_1 - \sigma_3)^2$
3. Calculate and accumulate:
$$\Sigma x_i, \quad \Sigma x_i^2, \quad \Sigma y_i, \quad \Sigma y_i^2, \quad \Sigma x_i y_i$$
4. Calculate m from equation A.3
5. Calculate s from equation A.5
6. Calculate r^2 from equation A.4
7. Where s < 0 in step 5, put s = 0 and calculate m from equation A.6
8. Note that equation A.4 is not valid when s < 0.

Generation of Mohr Envelope

1. Enter values of σ_c, m and s
2. Calculate σ_1 for a chosen value of σ_3 from equation 1
3. Calculate $\tau_m = \frac{1}{2}(\sigma_1 - \sigma_3)$
4. Calculate σ from equation A.7
5. Calculate τ from equation A.8
6. Repeat for new values of σ_3 until the curve has been defined.

Calculation of Constants A and B

1. Input values of σ_c, m and s; calculate $\sigma_t = \frac{1}{2}\sigma_c (m - \sqrt{m^2 + 4s})$
2. Calculate values of σ and τ from equations A.7 and A.8 for values of σ_3 equal to:
σ_{3m}, $\sigma_{3}m/2$, $\sigma_{3}m/4$, $\sigma_{3}m/8$, $\sigma_{3}m/16$, $\sigma_{3}m/32$, $\sigma_{3}m/64$, $\sigma_{3}m/128$, $\sigma_{3}m/256$, $\sigma_t/4$, $\sigma_t/2$, $3\sigma_t/4$ and σ_t.
   where σ_{3m} is the maximum value of σ_3 in the problem under consideration
3. Enter each of these calculated values in the form:
$$x = \log(\sigma/\sigma_c - \sigma_t/\sigma_c) \quad \text{and} \quad y = \log \tau/\sigma_c,$$
4. Calculate and accumulate:
$$\Sigma x_i, \quad \Sigma x_i^2, \quad \Sigma y_i, \quad \Sigma y_i^2 \text{ and } \Sigma x_i y_i,$$
5. Calculate the constants A and B from equations A.11 and A.12.

Determination of Instantaneous Friction Angle and Cohesion

1. Input A, B, σ_c and σ_t from previous calculations. Note that σ_t is always negative
2. For a chosen value of σ calculate τ from equation A.9
3. Calculate ϕ_i and c_i from equations A.13 and A.14.

Practical Example

The following resuts were obtained from a series of triaxial tests on intact samples of Andesite:

σ_1 (MPa)	269.0	206.7	503.5	586.5	683.3
σ_3 (MPa)	0	6.9	27.6	31.0	69.0

Analysis of these data gave the following results:

σ_c = 265.5 MPa, m = 18.84, s = 1, r^2 = 0.85, A = 1.115, B = 0.698

Note that, in this case, σ_{3m} = 69 MPa.

Very carefully drilled 152 mm (6 inch) diameter cores of very heavily jointed Andesite were tested by Jaeger (JAEGER, 1970) who obtained the following results at low confining pressures:

| σ_1 (MPa) | 1.24 | 6.07 | 8.96 | 12.07 | 12.82 | 19.31 | 20.00 |
| σ_3 (MPa) | 0 | 0.35 | 0.69 | 1.24 | 1.38 | 3.45 | 3.45 |

Analysis of these data, using σ_{3m} = 3.45 MPa, gave the following results:

σ_c = 265.5 MPa, m = 0.277, s = 0.0002, σ_t = -0.191 MPa, r^2 = 0.99, A = 0.316 and B = 0.700.

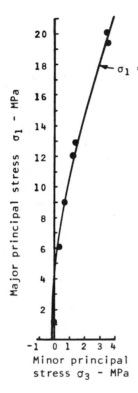

$$\sigma_1 = \sigma_3 + \sqrt{m\sigma_c\sigma_3 + s\sigma_c^2}$$

$\sigma_c = 265.5$ MPa

$m = 0.277$

$s = 0.0002$

Figure A.2.1 - Relationship between major and minor principal stresses from triaxial tests on heavily jointed Andesite tested by Jaeger (JAEGER, 1970).

Figure A.2.2 - Mohr envelope for triaxial tests on heavily jointed Andesite (JAEGER, 1970).

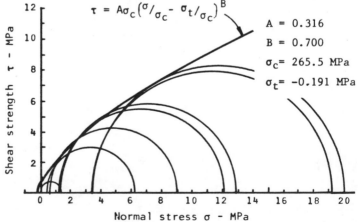

$$\tau = A\sigma_c\left(\frac{\sigma}{\sigma_c} - \frac{\sigma_t}{\sigma_c}\right)^B$$

$A = 0.316$

$B = 0.700$

$\sigma_c = 265.5$ MPa

$\sigma_t = -0.191$ MPa

APPENDIX 3 - SLOPE STABILITY ANALYSIS FOR MATERIALS WITH
NON-LINEAR FAILURE CHARACTERISTICS

Introduction

In heavily jointed and weathered rock masses in which failure is
not controlled by one or two dominant discontinuities, slope failure
can occur on an approximately circular failure surface such as that
illustrated in Figure A.3.1. Under these circumstances, conventional
stability analyses developed for soil slopes (BISHOP, 1955, JANBU,
1954, 1973, MORGENSTERN and PRICE, 1965, NONVEILLER, 1965, SARMA,
1973) can be used for slope design, provided that the non-linear characteristics of the Mohr failure envelope are taken into account.

Simplified Bishop Circular Failure Analysis

In order to demonstrate the method of incorporating a non-linear
failure criterion into a stability analysis, the simplified Bishop
method will be considered. The geometry of a circular failure surface
passing through the toe of the slope is defined in Figure A.3.1. The
influence of a partially water-filled tension crack will be included
in this analysis.

The factor of safety of the failure surface considered is defined
as:

$$F = \frac{\text{available shear strength}}{\text{shear strength mobilized for equilibrium}} = \frac{\tau_f}{\tau} \qquad (A.15)$$

Considering overall moment equilibrium of the failing mass:

$$\Sigma \gamma h \Delta x \, \text{Sin}\, \alpha . R \; + \; \tfrac{1}{2}\gamma_w z^2 . a \; = \; \Sigma \frac{\tau_f \, \Delta x . R}{F \, \text{Cos}\, \alpha} \qquad (A.16)$$

The available shear strength is:

$$\tau_f = (c_i' \; + \; \sigma' \, \text{Tan}\, \phi_i') \qquad (A.17)$$

where σ' is the effective normal stress acting on the base of the
slice and, from equations A.13 and A.14 in Appendix 2:

$$\text{Tan}\, \phi_i' \; = \; AB(\sigma'/\sigma_c - \sigma t/\sigma_c)^{B-1} \qquad (A.18)$$

$$c_i' = A\sigma_c (\sigma'/\sigma_c - \sigma t/\sigma_c)^{B} - \sigma' \, \text{Tan}\, \phi_i' \qquad (A.19)$$

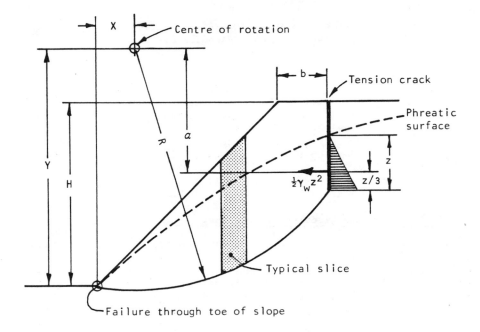

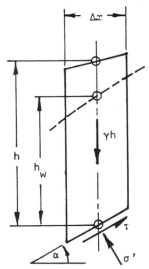

Note : Angle α is negative when sliding is uphill.

Figure A.3.1 : Failure surface and typical slice geometry for simplified Bishop circular failure analysis.

Substitution of equation A.17 into A.16 and rearrangement gives:

$$F = \frac{\Sigma(c_i' + \sigma' \tan\phi_i') \dfrac{\Delta x}{\cos\alpha}}{\Sigma \gamma h \Delta x \sin\alpha + \frac{1}{2}\gamma_w z^2 \cdot {}^a/_R} \qquad (A.20)$$

In order to solve equations A.18, A.19 and A.20, a value is required for the effective normal stress σ' acting on the base of each slice. The simplest solution to this problem is to ignore all inter-slice forces and to assume that the effective normal stress is:

$$\sigma' = \gamma h \cos^2\alpha - \gamma_w h_w \qquad (A.21)$$

Substitution of this expression for σ' into equation A.20 yields an equation for the factor of safety which is identical to that given by the method known as the Fellenius method, the U.S. Bureau of Reclamation method, the common method of slices or the ordinary method (MORGENSTERN and SANGREY, 1978). This factor of safety is known to be inaccurate (WHITMAN and BAILEY, 1967) but it does provide a useful starting point for the more accurate Bishop method described below.

A more realistic determination of the effective normal stress is obtained by following the method used by Bishop (BISHOP, 1955). This involves summing the vertical components of the forces acting on each slice, assuming zero shear stress between slices. The resulting equation for σ' is:

$$\sigma' = \frac{\gamma h - \gamma_w h_w - \dfrac{c_i' \tan\alpha}{F}}{1 + \dfrac{\tan\phi_i' \tan\alpha}{F}} \qquad (A.22)$$

Substitution of equation A.22 into equation A.20 results in the following well known equation for the factor of safety:

$$F = \frac{\Sigma\left((c_i' + (\gamma h - \gamma_w h_w)\tan\phi_i'\right) \dfrac{\Delta x}{\cos\alpha\,(1 + \tan\alpha\tan\phi_i'/F)}}{\Sigma\gamma h \Delta x \sin\alpha + \frac{1}{2}\gamma_w z^2 \cdot {}^a/_R} \qquad (A.23)$$

Since F appears on both sides of this equation, it is necessary to solve it by an iterative process. In order to incorporate the non-linear failure characteristics discussed in Appendix 2, the following calcultion sequence can be used:

1. Calculate the effective normal stress σ' acting on the base of each slice using equation A.21.

2. Using this value of σ', calculate values for $\tan\phi_i'$ and c_i' for each slice from equations A.18 and A.19.

3.   Substitute these values for σ', $\text{Tan } \phi_i'$ and c_i' into equation A.20 to obtain a first estimate for the factor of safety F.

4.   Using this value of F and the values of $\text{Tan}\phi_i'$ and c_i' calculated in step 2, calculate new values for the effective normal stress σ' on the base of each slice from equation A.22.

5.   Substitute these values of σ' into equations A.18 and A.19 to obtain new values of $\text{Tan } \phi_i'$ and c_i'.

6.   Use these values of σ', $\text{Tan}\phi_i'$ and c_i' to calculate a new value for the factor of safety F from equation A.20.

7.   Calculate a new value for the effective normal stress σ' for each slice from equation A.22.

8.   Return to step 5 and repeat the iteration process until the difference between successive factors of safety is less than 0.001. It will be found that convergence is rather slow and that approximately 10 iterations will be required for a typical problem.

9.   Repeat the entire analysis for different failure surfaces until the minimum factor of safety for the slope has been obtained.

In programming these steps on a computer or programmable calculator, error messages should be incorporated to warn the user if either of the following conditions occur:

a. $\text{Cos } \alpha \ (1 + \text{Tan } \alpha \ \text{Tan } \phi_i' / F) < 0.2$, which can occur near the toe of a slope with a deep failure surface (WHITMAN and BAILEY, 1967). In an extreme situation, the numerator in equation A.22 can become very small or negative and the computed effective normal stress will have an unreasonably large influence upon the factor of safety. If this occurs, the conditions at the toe should be examined in detail and, if no reasonable alternative is available, the analysis should be abandoned.

b. $\sigma' < 0$ in either equations A.21 or A.22 is unacceptable. This can occur near the top of a steep failure surface or when the water pressure acting on the failure surface is exceptionally high. If this occurs, the slice data should be examined for errors or for alternative conditions such as the incorporation of a tension crack. If no other solution is available, set $\sigma' = 0$ and note that this has been done when reporting the result.

Approximate Location of Centre of Critical Circle

In order to assist the reader in obtaining an approximate location
for the centre of the critical circle for a slope, two charts published by Hoek and Bray (HOEK and BRAY, 1977) are reproduced in Figures
A.3.2. and A.3.3. The dimensions X and Y are defined in Figure A.3.1.
and in order to find these dimensions, the following steps are followed:

a. From a knowledge of the ground water conditions in the slope,
    decide whether Figure A.3.2. or A.3.3. should be used.

b. From a plot of the Mohr failure envelope for the material,
    estimate the average angle of friction for the range of nor-
    mal stresses which apply to the problem under consideration.

c. The intersection of the curves defined by the average fric-
    tion angle and the average slope face angle, in Figures
    A.3.2. and A.3.3. gives the approximate location of the
    critical circle centre.

d. The distance b of the tension crack behind the slope crest
    (see Figure A.3.1.) can be estimated from the lower graphs in
    Figures A.3.2. and A.3.3.

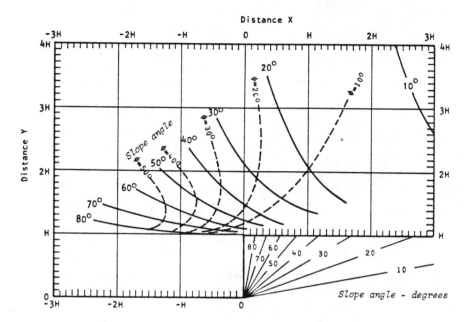

Location of centre of critical circle for failure through toe

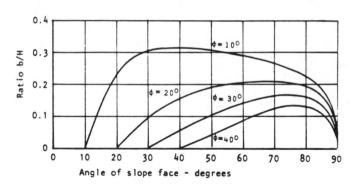

Location of critical tension crack position

Figure A.3.2: Approximate location of critical failure surface and critical tension crack location for fully drained slopes.

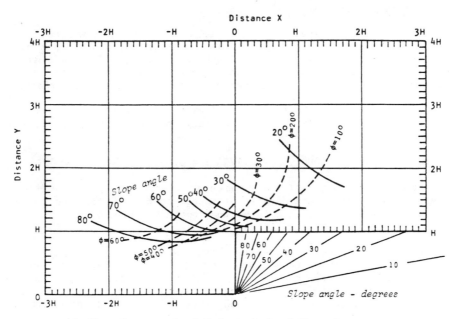

Location of centre of critical circle for failure through toe

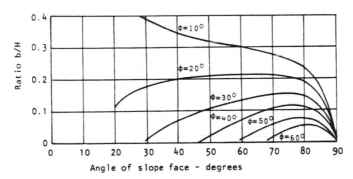

Location of critical tension crack position

Figure A.3.3: Approximate location of critical failure surface and critical tension crack location for slopes with groundwater conditions similar to those illustrated in figure A.2.1.

Example of Circular Failure Analysis

A slope is to be excavated in a highly weathered andesite. The slope, illustrated in Figure A.3.4, is to consist of three 15 m high benches with two 8 m wide berms. The bench faces are inclined at 75 degrees to the horizontal and the top of the slope is cut to 45 degrees from the top of the third bench to the natural ground surface.

Classification of the rock mass gives a CSIR rating of about 20. The material properties for a poor quality andesite rock mass are estimated, from Table 2, to be:

m = 0.09, s = 0.00001, A = 0.172, B = 0.676, σ_t/σ_c = -0.0001

The uniaxial compressive strength of the intact pieces of rock within the weathered rock mass is estimated to be 150 MPa and the in situ unit weight of the rock is assumed to be γ = 0.025 MN/m^3 for both dry and saturated conditions.

From a plot of the Mohr envelope, the average friction angle for the normal stress range from 0 to 1 MPa is estimated at 40 degrees and, from Figure A.3.3, the critical failure circle center and the location of the tension crack are determined. The phreatic surface illustrated in Figure A.3.4 has been estimated from local ground water conditions.

Figure A.3.4 shows the typical layout of a stability analysis and includes the detailed results of the 7th and final iteration to determine the factor of safety. The successive factors of safety calculated are as follows:

1.160, 1.234, 1.272, 1.287, 1.293, 1.295, 1.296

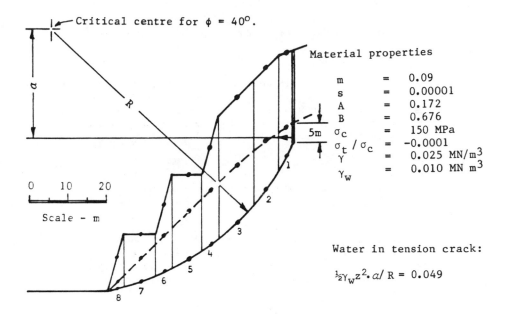

Critical centre for $\phi = 40°$.

a

R

5m

Material properties

m	=	0.09
s	=	0.00001
A	=	0.172
B	=	0.676
σ_c	=	150 MPa
σ_t / σ_c	=	-0.0001
γ	=	0.025 MN/m^3
γ_w	=	0.010 MN m^3

0   10   20

Scale - m

Water in tension crack:

$$\tfrac{1}{2}\gamma_w z^2 \cdot a/ R = 0.049$$

7th iteration

Slice	α (°)	h (m)	h_w (m)	Δx (m)	ϕ_i' (°)	c_i' (MPa)
1	62	26.5	7.5	4	45.27	0.107
2	55	30.8	12.5	6.5	42.14	0.129
3	44	31.8	15.0	8.5	39.80	0.150
4	36	25.0	14.0	4	41.40	0.135
5	31	21.8	12.5	8	42.07	0.130
6	25	17.5	9.5	4	43.17	0.121
7	20	12.5	6.0	8	45.36	0.107
8	12	7.5	0.2	4	46.92	0.098

Factor of safety = 1.296

Figure A.3.4:   Example of circular failure analysis

Question

Would you elaborate on the type of rock ripped at Bougainville.

Answer

The rock mass is a heavily jointed Andesite. The unconfined compressive
strength of the rock is 1900 kg/sq cm2 and it is fresh and unweathered
The joints are oriented in a multitude of directions and spaced at
about 5 cm. apart. Consequently, the rock mass can be regarded as a
tightly packed array of hard angular particles.

Question

What was the gain in slope angle achieved by ripping.

Answer

The bench face angle in the trials conducted to date has been
increased by about 10 degrees. It must be emphasised that this cannot be regarded as a general rule because it will depend upon the
particular types of failures which control the stability of
individual benches.

The figure of 10 degrees may be reasonable for tighty interlocked
heavily jointed rock masses which are particularly sensitive to
blast damage. In the case of more massive rocks with wide joint
spacing there may be no benefit in ripping as compared with blasting.

The process of bench face excavation, whether by ripping or by
blasting, influences near surface rock to a depth of perhaps 5 metres.
If overall slope stability is controlled by deep-seated structures
or failure surfaces, the method used to excavate the benches will
have no impact upon the stability of the overall slope. Consequently,
whatever method of bench excavation is chosen, there remains a need
to investigate the overall stability of the slope.

Question

What is the significance, if any, of progressive failure related to
circular failures in view of the brittle nature of rock (in contrast
to soil), on the assumption that part of the strength of heavily
jointed rock masses derives from the intact rock blocks.

Answer

This is a question of stress level compared with the strength of the
intact rock. In the case of a heavily jointed mass of very strong
rock particles, the stress levels which operate in a slope are only
small fraction of the intact rock strength. Consequently, the process
of slope failure is controlled by sliding and rotation of the

interlocking rock pieces and not by failure of the intact rock. Under these circumstances the rock mass behaviour tends to be plastic rather than brittle and the slope failure process resembles that of a soil.

When the stress levels are high (as in the case of underground excavations) a significant amount of intact rock failure can be induced and the overall rock mass behaviour reflects the brittle nature of the failure process of hard rock. In such cases, popping and spalling of the rock may constitute a significant hazard and sudden failures may be more frequent than "squeezing" failures.

Question

Have you considered using limit analysis methods incorporating parabolic-type failure conditions. As an example, similar work has been reported by Reinike and Ralston in 1977 or 1978.

Answer

I am not familiar with the paper referred to but there is no in-herent difficulty in using a parabolic relationship to represent the failure criterion.

In the analysis presented in this paper the failure criterion is based upon a parabolic relationship between maximum and minimum principal stresses at failure. This relationship is useful in the analysis of underground excavation failure but is inconvenient when applied to a slope stability problem in which the shear strength for a given normal stress level is required. Consequently, the original relationship has been transformed into one that is more convenient for slope stability calculations.

Note that the process is empirical and does not depend upon the nature of the equation used to define the relationship between shear strength and normal stress--provided that this equation gives a good fit to available experimental data. In fact in the computer program used by Golder Associates for slope stability analysis, several options are available for the relationship between shear strength and normal stress. Which relationship is used depends upon the preference of the user and the adequacy of the relationship in defining the experimental data behaviour.

Question

Some years ago at a symposium on natural slopes, Ralph Peck said that "We are unable to predict the stability of natural slopes". As a specialist in this field, has your experience led you to contradict Peck's statement. In particular, I refer to your statement on the difficulty of estimating shear strength in the case of large rock slopes.

Answer

The problem with predicting the stability of natural slopes is
knowing where to look. If someone decides that the stability of a
particular natural slope requires investigation, the process which
would be followed would be very similar to that used to evaluate the
stability of an open pit mine slope. In both cases the mechanics of
slope failure is controlled by structural discontinuities in the rock
mass and by the ground water conditions in this mass. Methods for
investigating these problems are relatively well established and the
difficulties of interpreting the data are the same. Given that
enough money was spent on the problem, an approximate assessment of
the stability of a natural slope could be made but it must be
realised that the accuracy may not be adequate for the problem
under consideration.

Presumably, the reason why anyone would want to assess the stability
of a natural slope would be that this slope had given some indication of instability. One would need to know how close this slope was
to the condition of limiting equilibrium and whether there was a
significant risk of failure.

Methods of slope stability assessment, taking into the account the
uncertainty of input data on rock strength and ground water
conditions for large slopes, may only be accurate to about +10 per
cent. Hence the factor of safety would be assessed as between
0.9 and 1.1. This degree of accuracy would not be accurate enough to
permit a realistic assessment of risk of failure.

In the case of an open pit mine slope, the same degree of uncertainty would not be a problem in the case of a slope designed for a
factor of safety of 1.3. Factors of safety between 1.2 and 1.4 are
both acceptably safe.

Hence, in spite of advances in methods of rock slope stability
assessment, the problem of accurately predicting the stability of
natural slopes which have shown signs of instability is no closer to
a solution than when Peck made his comments.

Chapter 16

THE APPLICATION OF STOCHASTIC MEDIUM THEORY
TO THE PROBLEM OF SURFACE MOVEMENTS DUE TO OPEN PIT MINING

Liu Baoshen   and Lin Dezhang

Associate Professor and Doctor,
Research Institute of Mining and Metallurgy, Changsha, China

Postgraduate Student of Rock Mechanics,
Research Institute of Mining and Metallurgy, Changsha, China

ABSTRACT

Vertical and horizontal movements beyond the perimeter of the open pit have resulted from the development of deep and extensive open pits; the surface point always moves towards the centre of the open pit. The surface movements are not only great in relation to the stability of the open pit slope and the safety of mining operations but also are of significance to the stability of structures and services on the surface beyond the perimeter of the open pit.

Recently, some studies on factors contributing to the surface movements have been carried out. Particular attention has been paid to prediction of vertical and horizontal movements, differential subsidence and horizontal strain and careful assessment of their influences on existing structures in the adjacent area.

In the application of the theory of stochastic medium, it is assumed that ground subsidence process has the property of Markov Process. Based on the superposition principle, surface movements due to open pit mining are derived from the Kolmogorov Equation, and calculated in a digital computer.

Quantitative calculation of surface movements is made according to the method employed in this paper and compared with the results of measurement in Morwell Open Pit, Victoria, Australia and Baiyin Open Pit, Gansu, China. The method of calculation gives reasonably satisfactory results.

STOCHASTIC PROCESS MODEL FOR THE ACTUAL PROBLEM

The results of observations of surface movements indicate that subsidence and horizontal displacements near open pits increases continuously with the development of mining operations and decreases gradually after stopping excavation.

In order to describe in some quantitative manner the nature of ground movements, we consider a simple random model for describing ground movements due to mining operations. In this analysis, the random model describing ground movements is given by the Markov Process. Our main concern is the determination of the statistical properties of the ground movements in terms of the combined statistical properties of surface movements and extraction.

In the study of the contribution of extraction to surface movements, the modelling of random movements by stochastic process is useful in many practical cases. A small amount of extraction leads to a small amount of movement on the surface, and the resultant movements of a point is given by the sum of movements caused by the extraction of different areas.

In the following analysis only the two-dimensional problem is considered, but the nature of the associated three-dimensional problem is similar. Let W be the subsidence on the surface, $W_f(x,z)$ represents its probability density, $W(x,z)$ represents the amount of probability of vertical movement. The stochastic process itself is not Markovian but its vertical component, i.e. subsidence, is a Markov Process. For this reason we would like to focus our attention on the subsidence process and clearly identify the role played by the probability density function, $W_f(x,z)$. The Markov Process is completely characterized by its density function. Therefore, if $W_f(x,z)$ is given, the knowledge of the transition probability density $W_f(x,z|x_0,z_0)$ completely specifies the stochastic process, that is, if the induction movement at (x,z) is specified, the transition probability describes the subsidence on the surface due to the induction movement.

From the assumption above, the subsidence movement process is referred to as a Markov Process. The transition probability density certainly satisfies Markov Process conditions, i.e. the Chapman-Kolmogorov Equation

$$W_f(x_3,z_3|x_1,z_1) = \int_{-\infty}^{+\infty} W_f(x_3,z_3|x_2,z_2) \cdot W_f(x_2,z_2|x_1,z_1)dx_2 \tag{1}$$

and the Kolmogorov Equation

$$\frac{\partial W_f(x,z|x_0,z_0)}{\partial z} = \frac{1}{2}\frac{\partial^2}{\partial x^2}[a(x,z)W_f(x,z|x_0,z_0)] - \frac{\partial}{\partial x}[b(x,z)W_f(x,z|x_0 z_0)] \tag{2}$$

with the boundary condition

$$W_f(x,z_0|x_0,z_0) = \delta(x - x_0) \tag{3}$$

where $a(x,z)$, $b(x,z)$ are parameters representing the characteristics of the rock medium, and function of position (x,z). $\delta(x-x_0)$ is Dirac function, i.e.

$$\delta(x-x_0) = o \qquad x \neq x_0$$

$$\delta(x-x_0) = \infty \qquad x = x_0$$

$$\int_{-\infty}^{+\infty} \delta(x-x_0) \, dx = 1$$

Let us first mention that, in the present work we consider only relatively small movements. The large displacements such as slope slip or slide or toppling are not considered here.

Introducing the symmetrization of the stochastic rock medium for a horizontally situated layer and superposed layers or not stratified. The stochastic medium is isotropic horizontally. It can be shown that a(x,z) is the function a(z) of only vertical coordinate, and b = o. Equation 2 can be simplified as follows:

$$\frac{\partial}{\partial} W_f(x,z|x_0,z_0) = \frac{1}{2}a(z)\frac{\partial^2}{\partial x^2} W_f(x,z|x_0,z_0) \tag{4}$$

DERIVATION OF FORMULAS

The subsidence distribution function and the horizontal movement distribution function can be derived by means of the transition probability density. Equation (4) can be rewritten as

$$\frac{\partial W_f}{\partial z} = \frac{1}{2} a(z)\frac{\partial^2 W_f}{\partial x^2} \tag{5}$$

Solving the parabolic partial differential equation we obtain

$$W_f = [2\pi A(z)]^{-\frac{1}{2}}\exp[-\frac{(x-x_0)^2}{2A(z)}] \tag{6}$$

where

$$A_z = \int_{z_0}^{z} a(t)dt$$

Introducing the notion of the range of the main influences β we obtain, correspondingly, the relation between the quantities $a(z_0)$ and r, namely

$$A(z_0) = r^2/4\pi \tag{7}$$

where r - denotes the range of main influences.

We introduce the function $h(z_0)$ which depends only on vertical coordinate. Assuming the following form

$$h(z_0) = \sqrt{\pi}/r \tag{8}$$

From Equation (7) and Equation (8) we obtain

$$A(z_0) = 1/4h^2(z_0) \tag{9}$$

Substitution of Equation (9) into Equation (6) yields the subsidence transition W_f in terms of the parameter $h(z_0)$

$$W_f = h(z_0)\exp[-h^2(z_0)(x-x_0)^2]/\sqrt{\pi} \qquad (10)$$

Fig. 1 shows the cross section through the open pit where H is the exploitation depth, the plan equation for the top level of the open pit is

$$z = o$$

the plan equation for the bed level is

$$z = -H$$

and the plan equation for the slope is

$$z = x.tg\alpha$$

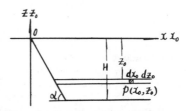

Fig. 1   Schematic section of open cut

Consider a point $P(x_0,z_0)$ in the extraction area where an induction subsidence has taken place in the horizontal level $z = z_0$. In terms of the Chapman-Kolmogorov Equation and Equation (10) we obtain

$$\Delta W(z_0) = \int_{-\infty}^{+\infty} \Delta z_0 \frac{h(z_0)}{\sqrt{\pi}} \exp[-h^2(z_0)(x-x_0)^2]dx_0 \qquad (11)$$

The small amount of increment of distribution function of subsidence $\Delta W(z_0)$ can be considered to be a result of the small amount of induction subsidence.

Bearing in mind the range of the main influence, Equation (11) may be presented in the following form

$$\Delta W(z_0) = \int_{-z_0 ctg\alpha}^{+\infty} \Delta z_0 \frac{h(z_0)}{\sqrt{\pi}} \exp[-h^2(z_0)(x-x_0)^2]dx_0 \qquad (12)$$

If we consider the resultant subsidence as superposition of these small amounts of subsidence, Equation (12) can be replaced by the following form

$$W(x) = \int_{-H}^{o} \int_{-z_0 ctg\alpha}^{+\infty} \frac{h(z_0)}{\sqrt{\pi}} \exp[-h^2(z_0)(x-x_0)^2 dx_0 dz_0 \qquad (13)$$

If we write $x\varepsilon (-\infty .o)$

$$h(z_0)(x-x_0) = \lambda \qquad dx_0 = -d\lambda/[h(z_0)]$$

Equation (13) becomes

$$W(x) = \frac{1}{2} \int_{-H}^{o} \int_{-h(z_0)(x+z_0 ctg\alpha)}^{\infty} \frac{2}{\sqrt{\pi}} e^{-\lambda^2} d\lambda dz_0 \qquad (14)$$

Using the notion of the range of the main influences we get

$$\Upsilon(z_o) = -z_o/tg\beta \tag{15}$$

From Equation (7) and Equation (8) we obtain

$$h(z_o) = -\sqrt{\pi}tg\beta/z_o \tag{16}$$

Substituting Equation 16 into Equation 14, we can obtain

$$W(x) = \frac{1}{2} \int_{-H}^{o} [1-erf[\frac{\sqrt{\pi}tg\beta}{z_o}(x+z_octg\alpha)]]dz \tag{17}$$

$$x\epsilon \ (-\infty,o)$$

Maximum subsidence is only attained at x = 0, it is meaningless when x >0, we have

$$W_{max}(x) = H[1-erf(\sqrt{\pi}tg\beta ctg\alpha)]/2 \tag{18}$$

Assuming that the stochastic medium is incompressible, i.e.

$$\epsilon_x + \epsilon_y = o$$

$$(\epsilon_y = o \text{ for the plain strain problem})$$

that is,

$$\frac{\partial U}{\partial x} + \frac{\partial W}{\partial z} = o$$

Solving the equation above we obtain

$$U = \int \frac{\partial W}{\partial z}dx + C$$

Thus, we can obtain the horizontal transition probability density U_f from the transition probability density W_f (1). (Liu, Liao and Yan, 1979)

$$U_f = h^1(z_o)(x-x_o)exp[-h^2(zo)(x-x_o)]/\sqrt{\pi} \tag{19}$$

Similarly, integration yields the horizontal displacement distribution function

$$U = \int_{-H}^{o} \int_{-Z_octg\alpha}^{\infty} \frac{h^1(zo)}{\sqrt{\pi}}(x-x_o) exp[-h^2(z_o)(x-x_o)^2]dx_odz_o \tag{20}$$

$$x\epsilon \ (-\infty,o)$$

Since $h(z_o) = \sqrt{\pi}tg\beta/z_o$, $h^1(z_o) = \sqrt{\pi}tg\beta(z_o^2)^{-1}$, we have

$$U = \frac{1}{2\pi tg\beta} \int_{-H}^{o} exp[-\frac{\pi tg^2\beta}{z_o^2}(z_octg\alpha x_o + x_o)^2]dz_o \tag{21}$$

$$x\epsilon \ (-\infty,o)$$

Maximum horizontal displacement is attained at x = o, it is meaningless when x > o

$$U_{max}(x) = U(o) = H\exp(-\pi ctg^2\alpha\, tg^2\beta)/(2\pi tg\beta) \tag{22}$$

The differential subsidence $T(x)$ can be obtained from Equation (17)

$$T(x) = tg\beta \int_{-H}^{o} \frac{1}{z_o} \exp\left[-\frac{\pi tg^2\beta}{z_o^2}(x_o + z_o ctg\alpha)^2\right]dz_o \tag{23}$$

$$x\epsilon\ (-\infty, o)$$

The horizontal strain $\epsilon\ (x)$ can be obtained by differentiating Equation (21) with respect to $\epsilon(x)$

$$\epsilon(x) = -tg\beta \int_{-H}^{o} \frac{(z_o ctg\alpha + x_o)}{z_o} \exp\left[-\frac{\pi tg^2\beta}{z_o^2}(z_o ctg\alpha + x_o)^2\right]dz_o \tag{24}$$

$$x\epsilon\ (-\infty, o)$$

where $T(x)$, $\epsilon\ (x)$ are singular points at $x = o$

As to other extraction cases, the subsidence W, horizontal displacement U, differential subsidence T and horizontal strain ϵ can also be obtained using Equations (13), (20), (23) and (24), only upper and lower limits must be rewritten according to the boundary conditions. Let us consider only the formulas for the subsidence.

The common extraction form of open pit is shown in Fig. 2 where the respective slope angles on the two sides are α_1 and α_2, the depth of open pit is H and the width at the bottom L. We have

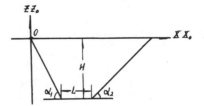

Fig. 2 Schematic section of open cut

$$W(x) = \int_{-H}^{o} \int_{-z_o ctg\alpha}^{\frac{z_o+H}{tg\alpha_2}+(Hctg\alpha_1+L)} \frac{h(z_o)}{\sqrt{\pi}} e^{-h^2(z_o)(x-x_o)} dx\, dz_o \tag{25}$$

$$X\epsilon(-\infty, o), [H(ctg\alpha_1, + ctg\alpha_2, +L, \infty]$$

It is clear that the influence of the far extracted area can be neglected, so that the formulas for Fig. 2 can be replaced by the formulas for Fig. 1.

In Fig. 3 the slope angle is α, the natural slope angle α^1 the depth of extraction H.

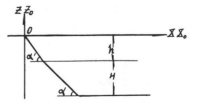

Fig. 3 Schematic section of open cut

$$W(x) = \int_{-(H+h)\ h\mathrm{ctg}\alpha^1}^{-h} \int_{-(z+h)\mathrm{ctg}\alpha}^{\infty} \frac{h(z_0)}{\sqrt{\pi}} e^{-h^2(z_0)(x-x_0)^2} dx_0 dz_0 \tag{26}$$

$$x \in (-\infty, 0)$$

The following example of practical application in Baiyin Open Pit is the case in Fig. 2.

Fig. 3 is a special case of Fig. 2 where L = 0. For the case L = 0, Equation (27) is reduced to

$$W(x) = \int_{-H}^{0} \int_{-\mathrm{ctg}\alpha z_0}^{\frac{z_0+H}{\mathrm{tg}\alpha_2} + H\mathrm{ctg}\alpha_1} \frac{h(z_0)}{\sqrt{\pi}} e^{-h^2(z_0)(x-x_0)^2} dx_0 dz_0 \tag{27}$$

$$x \in (-\infty, 0), \ [H(\mathrm{ctg}\alpha_1 + \mathrm{ctg}\alpha_2), \ \infty]$$

ANALYSIS OF PRACTICAL EXAMPLES

Morwell Open Pit

The maximum horizontal movement recorded is 1.9 m. at the edge of Morwell Open Pit Mining, Latrobe Valley, Victoria, Australia during the period 1961-1972. (Gloe, James and MacKenzie)

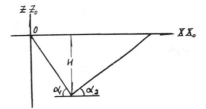

Fig. 4 Schematic section of open cut

Contours of horizontal displacement and subsidence in the area adjacent to Morwell Open Pit are shown in Fig. 5. The area of influence of horizontal displacement is approximately circular in shape. The area of influence of subsidence is considerably larger and less regular than that of horizontal displacement.

The C-line (Fig. 6) passes through the southern part of Morwell township at right angles to the northern edge of the open pit. On the C-line it is possible to evaluate the angle of the range of main influences by the method mentioned above using the results of observation in April 1972.

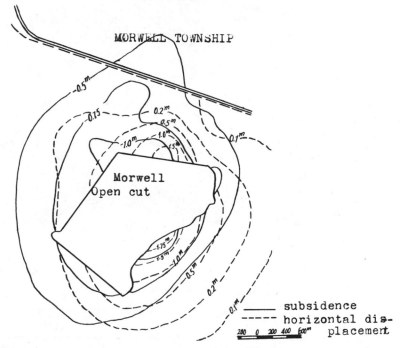

Fig.5 Movements adjacent to Morwell open cut

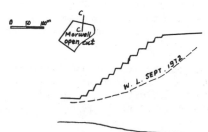

Fig .6 Section through the C-line, Morwell
        open cut

The value of the depth H attained 110 m., the slope angle $\propto$ was 39°. In addition to these, it was possible to determine with sufficient accuracy the value of W(0) = 1200 mm. subsidence, from the curve of subsidence. We have from Equation (18)

$$\beta = 36° \ 32' \ 10''$$

Inserting β into Equation (22), U(0) can be obtained

$$U(0) = 1700 \ mm.$$

A survey of surface movements around Morwell Open Pit shows that reduction in groundwater pressure follows the excavation of over- burden and coal as the open pit is progressively developed. The re- moval of material disturbs the stress equilibrium and a stress read- justment occurs, associated with movements around the open pit. This influence of removal extends for more than one kilometer.

We introduce groundwater influence coefficients K_w and K_u asso- ciated with subsidence and horizontal displacement respectively.

In Morwell Open Pit K_w = 75 and K_u = 30 are given.

A comparison between the calculated and measured values is represented in Fig. 7.

Baiyin Open Pit

The subsidence and horizontal displacement have been recorded of the edge of Baiyin Open Pit by sur- vey department of this open pit mine. The value of subsidence measured in June 1972 is 125mm., the estimated value β using Equation (18) is 39° 05' when $\propto$ = 34.3 , H = 88m. (Fig. 6). Upon substituting the values $\propto$, β and H into Equation (22), the calculated value of horizontal dis- placement is 200 mm., the measured value is about 175mm. W and H can also be obtained by inserting data in Fig. 8 into Equation (26)

It can be seen that the value of β may be obtained from the value of W or U at the edge of the open pit by direct measurement, then the movement of surface points can be estimated.

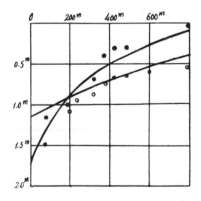

Fig. 7   Comparison between the theoretical and measured values

In mining design, it is essential to determine the characteristic parameter of rock mass, i.e. the angle of main influence. The movement of an arbitrary point on the surface can be estimated by the method mentioned above. The influences of surface mining on the adjacent area can be forecast.

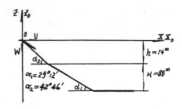

Fig. 8 Section through Baiyin Open Cut

CONCLUSION

A preliminary quantitative analysis of the movements and deformations due to open pit mining is given by means of the theory of stochastic media in this paper. To evaluate the validity of the method of calculation a comparison between the measurement in two open mines and the theoretical values showed the method developed here can be used to predict the character of the ground movements and of the deformations due to open pit mining, as well as to forecast their magnitude.

REFERENCES

Liu Baoshen, Liao Kuohua, and Yan Roungui, "Research in the Surface Ground Movement due to Mining", Proc. of 4th Congress I.S.R.M. 1979.

C.S. Gloc, J.P. James and R.J. McKenzie, "Earth Movements Resulting from Brown Coal Open Cut Mining - Latrobe Valley, Victoria". Subsidence in Mines, Editor A.J. Hargraves. The Australian Institute of Mining and Metallurgy, Illawarra Branch.

Chapter 17

ANALYTICAL ESTIMATION OF PARABOLIC
WATER TABLE DRAWDOWN TO A SLOPE FACE

by

Stanley M. Miller
Department of Civil Engineering

University of Wyoming, Laramie, Wyoming

ABSTRACT

For rock or soil materials assumed to be generally isotropic and
homogeneous, seepage theory and analytic geometry can be used to develop a mathematical technique for estimating the steady state, piezometric drawdown to a slope face. The resulting parabolic drawdown
curve is a function of slope angle and the height of the original
water table above the slope toe. When the slope is cut at a rate
faster than that which allows a steady state to be attained, a nonsteady state analysis approximates the transient drawdown. Monte
Carlo simulation provides realizations of possible drawdown curves
that can be included in slope stability risk analyses.

INTRODUCTION

Methods of predicting the water table drawdown that results from
a slope cut are often mathematically or graphically complicated.
Some complexity is involved even if the slope material can be assumed
to have generally isotropic and homogeneous hydrologic properties
with uniform seepage along the slope face. Because of intricate computations or flow-net constructions, these techniques are not practically suited for probabilistic slope stability studies which require
repeated simulations of possible drawdown curves dependent on the
natural variability and measurement uncertainty of hydrologic properties.

Therefore, an accurate, simplified procedure to predict water
table drawdown would be advantageous if it could meet the following
criteria:

1. The basic computations should be simple enough to allow the slope engineer in the field to perform them with a hand-held (portable) calculator.

2. The procedure should be amenable to Monte Carlo simulation, allowing for the natural variability and measurement uncertainty of hydrologic parameters to be included in the slope stability analysis.

Seepage theory and analytic geometry can be used to develop such a procedure for estimating the parabolic water table drawdown to a slope face.

SEEPAGE THEORY CONTRIBUTION

Seepage through earth dams and embankments has been a topic of much investigation and analysis in the geotechnical branch of civil engineering. The work of L. Casagrande (cited by Harr, 1962, p. 52) is especially suited for application to steady state, ground water seepage at a slope face. The hydrologic free surface (drawdown curve) is parabolic in shape.

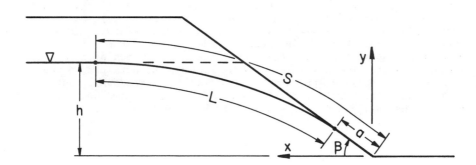

Figure 1. Parabolic water table drawdown to a slope face, based on seepage theory (steady state conditions)

Referring to Figure 1, the quantity of seepage at the slope face is given by:

$$q = ka(\sin^2\beta) \tag{1}$$

where: q = quantity of seepage in volume per unit time per unit
length of slope strike

k = seepage coefficient, or permeability, in distance per unit
time

a = upslope distance from the slope toe to the seepage point,
and

β = slope angle.

By using the hydraulic gradient, dy/ds (where s is measured along the
free surface), the seepage quantity can also be determined by the expression,

$$q = -ky \frac{dy}{ds} .$$ (2)

To solve for the upslope distance to the seepage point, equations
(1) and (2) are combined and then integrated over appropriate limits.

$$-y \, dy = a \sin^2\beta \, ds$$

$$- \int_{h}^{a \sin\beta} y \, dy = \int_{0}^{S-a} a \sin^2\beta \, ds$$

.
.
.

$$(S-a)^2 = S^2 - \frac{h^2}{\sin^2\beta}$$

$$a = S - \sqrt{S^2 - \left(\frac{h}{\sin\beta}\right)^2}$$ (3)

Thus, the upslope distance to the seepage point depends on the
height of the original water table above the slope toe, the slope
angle, and the length of the free surface. The first two parameters
can be measured or specified by the analyst. Length of the free surface can be determined by analytic geometry.

ANALYTIC GEOMETRY CONTRIBUTION

Parabolic drawdown curves are tangent to the water table and to
the slope face. For a specified slope angle, there is a family of
such parabolic curves, but only one will satisfy the steady state seepage analysis presented above. This unique parabola can be determined
by plane analytic geometry.

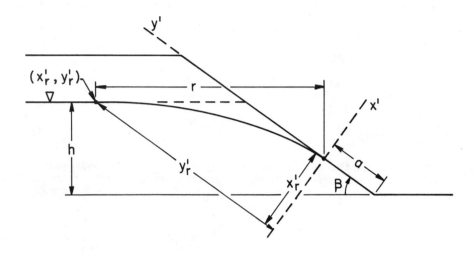

Figure 2.   Parabolic water table drawdown to a slope face, based on
            analytic geometry

    In Figure 2, the reference axes have been translated and rotated
so that the y´ axis coincides with the slope face and the new origin
coincides with the seepage point. Assuming that the parabolic drawdown curve is of the form

$$(y´)^2 = Cx´ \quad \text{(where C is a constant)},$$

geometric and algebraic manipulations can be used to quantify the relationships between the drawdown curve and the slope configuration.

    The radius to null drawdown, r, is the horizontal distance from
the seepage point to the point of null drawdown on the original water
table (Fig. 2). The x´ and y´ coordinates at the point of null drawdown $(x´_r , y´_r)$, are determined below:

$$x´_r = (r + a \cos\beta) \sin\beta - h \cos\beta$$

$$x´_r = r \sin\beta - h \cos\beta + a \cos\beta \sin\beta . \tag{4}$$

At the point (x_r', y_r'),

$$\frac{dy'}{dx'} = \frac{1}{2} \sqrt{\frac{C}{x_r'}} \tag{5a}$$

and,

$$\frac{dy'}{dx'} = \frac{x_r'/\tan\beta}{x_r'} = \frac{1}{\tan\beta} \ . \tag{5b}$$

Combining equations (5a) and (5b) provides an expression for the constant C,

$$C = \frac{4\ x_r'}{\tan^2\beta} = \frac{4}{\tan^2\beta} \ (r\ \sin\beta - h\ \cos\beta + a\ \cos\beta\ \sin\beta) \ . \tag{6}$$

Combining equations (4) and (6) provides an expression for y_r' ,

$$y_r' = \sqrt{C\ x_r'} = \sqrt{\frac{4(x_r')}{\tan^2\beta}} = \frac{2\ x_r'}{\tan\beta}$$

$$y_r' = \frac{2}{\tan\beta}\ (r\ \sin\beta - h\ \cos\beta + a\ \cos\beta\ \sin\beta) \ . \tag{7}$$

    The upslope distance from the slope toe to the seepage point can be expressed in terms of x_r' and y_r':

$$a = \frac{h}{\sin\beta} - (y_r' - \frac{x_r'}{\tan\beta})$$

$$a = \frac{h}{\sin\beta} - \frac{1}{\tan\beta}\ (r\ \sin\beta - h\ \cos\beta + a\ \cos\beta\ \sin\beta)$$

$$a = \frac{h}{\sin\beta} - \frac{r\ \sin\beta - h\ \cos\beta}{\tan\beta} - a\ \cos^2\beta$$

$$a(1 + \cos^2\beta) = \frac{h}{\sin\beta} - \frac{r\ \sin\beta - h\ \cos\beta}{\tan\beta}$$

$$a = \frac{1}{1 + \cos^2\beta} \left[\frac{h}{\sin\beta} - \frac{r\ \sin\beta - h\ \cos\beta}{\tan\beta} \right] \ . \tag{8}$$

Referring back to equation (3), the length of the free surface, S, can now be expressed in terms of the geometric parameters given above. This is accomplished by evaluating the length of the parabolic drawdown curve by integration. A differential element of the curve has a length given by:

$$dL = \sqrt{(dx')^2 + (dy')^2}$$

$$= \sqrt{\left(\frac{dx'}{dy'}\right)^2 + 1} \ dy'$$

$$= \sqrt{\left(\frac{2y'}{C}\right)^2 + 1} \ dy'.$$

The length of the parabolic curve is then determined by:

$$L = \int_L dL = \int_0^{y'_r} \sqrt{\frac{4}{C^2}(y')^2 + 1} \ dy'$$

$$L = \frac{y'_r}{2}\sqrt{\left(\frac{2y'_r}{C}\right)^2 + 1} + \frac{C}{4} \ell n \left[\frac{2y'_r}{C} + \sqrt{\left(\frac{2y'_r}{C}\right)^2 + 1}\right]. \tag{9}$$

The above integral was evaluated by using a formula given by Selby and Weast (1970).

Because the length of the free surface, S, is expressed as the sum, a + L, a substitution into equation (3) yields the following:

$$a = a + L - \sqrt{(a + L)^2 - \left(\frac{h}{\sin\beta}\right)^2}$$

$$L^2 = (a + L)^2 - \left(\frac{h}{\sin\beta}\right)^2$$

$$a = \sqrt{L^2 + \left(\frac{h}{\sin\beta}\right)^2} - L. \tag{10}$$

Thus, equation (10) represents an application of steady state seepage theory, and equation (8) represents a result from plane analytic geometry. Given that h and β are known, each equation has two unknowns, a and r. Solving for these two unknowns is not a trivial procedure because the L term in equation (10) is quite complicated. Therefore, a trial-and-error method will be used.

STEADY STATE ANALYSIS

The steady state, parabolic drawdown curve will be determined by solving for the upslope distance to the seepage point, a, and the null radius to drawdown, r. Procedures for a trial-and-error solution method are presented below.

Values for h and β are measured or specified by the analyst. A value for r is assumed; usually, 2.5(h) is a good first guess.

The following variables are calculated:

1. a_1, from equation (8)

2. C, from equation (6)

3. y_r', from equation (7)

4. L, from equation (9)

5. a_2, from equation (10).

If the correct value of r was used, then a_1 and a_2 will be equal; and the solutions would be r and a_1. If a_2 is less than a_1, then r was too small; r should be increased by approximately ten percent and the above calculations (steps 1 through 5) repeated. If a_2 is greater than a_1, then r was too large; r should be decreased by approximately ten percent and the above calculations (steps 1 through 5) repeated. Solutions for a and r will usually be determined within three to ten trials.

The calculations can be readily performed with a hand-held calculator that has scientific functions. However, to eliminate computational errors and to expedite the process, a programmable desktop or hand-held calculator is recommended. With efficient programming, the trial-and-error solutions for a and r are computed in a few seconds.

NONSTEADY STATE ANALYSIS

When slope excavation intersects the water table, nonsteady ground water flow to the slope face will occur until steady state conditions are attained. Therefore, an analysis is needed to estimate

the elapsed time from initial drawdown to final, or steady state,
drawdown; and to predict the shape of the transient drawdown curve
at any time during the period of nonsteady flow.

Adaptations from Polubarinova-Kochina (1962) and Glover (1974)
provide a basis for analyzing the transient drawdown to a slope face.
In Figure 3, the height, h_t, is given by:

$$h_t = H \sqrt{\frac{2}{\pi} \int_0^{\eta} e^{-w^2} dw} \qquad (11)$$

for: $\eta = \frac{x}{2} \sqrt{\frac{m}{kDt}}$

where: x = horizontal distance from seepage point to the location at
            which h_t is to be determined
        m = porosity, expressed as a decimal
        k = permeability, in distance (length) per unit time
        D = vertical distance from seepage point to base of geologic
            formation, and
        t = time, usually expressed in days.

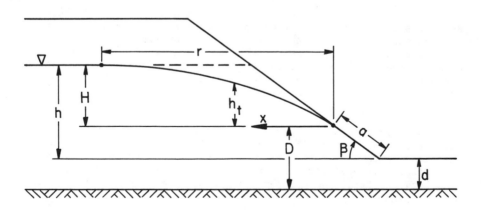

Figure 3.   Parabolic water table drawdown to a slope face for non-
            steady state analysis

When steady state is attained, x equals the null radius to drawdown, r, and h_t equals H. Consequently, for steady state conditions equation (11) becomes:

$$\frac{h_t}{H} = 1.0 = \sqrt{\frac{2}{\pi} \int_0^\eta e^{-w^2} dw} \quad . \tag{12}$$

The expression under the radical is known as the "probability integral", and it equals 1.0 only when η is infinitely large. For slope stability calculations that will use an estimate of the drawdown curve, let us set the ratio of h_t to H equal to 0.995, a sufficiently good approximation having a 0.5 percent accuracy. Equation (12) then becomes:

$$0.995 = \sqrt{\frac{2}{\pi} \int_0^\eta e^{-w^2} dw}$$

$$0.990 = \frac{2}{\pi} \int_0^\eta e^{-w^2} dw \quad . \tag{13}$$

By consulting a table of the calculated values of the probability integral, the value of η can be determined ($\eta = 1.822$) for which equation (13) is valid. The time at which steady state is attained can now be estimated:

$$\eta = 1.822 = \frac{r}{2} \sqrt{\frac{m}{kDt_s}}$$

$$t_s = \frac{m}{kD} \left(\frac{r}{3.644}\right)^2 \tag{14}$$

The null radius to drawdown at time t is predicted by:

$$r_t = 3.644 \sqrt{\frac{kDt}{m}} \quad . \tag{15}$$

However, before using equation (15) to solve for r_t, the value of D must be calculated. Algebraic manipulation of equation (8) yields the following expression:

$$a^2 - r^2 \left(\frac{\cos\beta}{1 + \cos^2\beta}\right)^2 - \frac{2ah}{\sin\beta} + \left(\frac{h}{\sin\beta}\right)^2 = 0 \ . \tag{16}$$

Let $\alpha = (3.644)^2 \left(\frac{kt}{m}\right)$,

then $r_t^2 = \alpha D$.

But $D = d + a_t \sin\beta$ (from Figure 3), $\tag{17}$

so $r_t^2 = \alpha d + \alpha a_t \sin\beta$. $\tag{18}$

Substitution of equation (18) into equation (16) produces the following quadratic equation:

$$a_t^2 - (\alpha d + \alpha a_t \sin\beta)\left(\frac{\cos\beta}{1 + \cos^2\beta}\right)^2 - \frac{2a_t h}{\sin\beta} + \left(\frac{h}{\sin\beta}\right)^2 = 0$$

$$a_t^2 + a_t \left[-\alpha\sin\beta\left(\frac{\cos\beta}{1 + \cos^2\beta}\right)^2 - \frac{2h}{\sin\beta}\right]$$

$$+ \left(\frac{h}{\sin\beta}\right)^2 - \alpha d\left(\frac{\cos\beta}{1 + \cos^2\beta}\right)^2 = 0 \ . \tag{19}$$

This quadratic equation can be solved for a_t by the following formula:

$$a_t = \frac{-b \pm \sqrt{b^2 - 4c}}{2} \quad \text{(negative sign gives correct answer)}$$

where: $b = -\alpha\sin\beta\left(\frac{\cos\beta}{1 + \cos^2\beta}\right)^2 - \frac{2h}{\sin\beta}$, and

$$c = \left(\frac{h}{\sin\beta}\right)^2 - \alpha d \left(\frac{\cos\beta}{1 + \cos^2\beta}\right)^2 .$$

Once the value of a_t has been determined, the value of D is calculated by using equation (17), and r_t is evaluated by using equation (15). As a result, the shape of the transient drawdown curve at time t can now be described by a_t and r_t.

In the above analysis, the entire slope of interest is assumed to be cut instantaneously. This assumption is valid for a slope that is cut progressively in phases, provided that steady state flow is not attained during any phase other than the final phase. If steady state flow is attained during an intermediate phase, a time correction will have to be included in subsequent phases. The correction would be the time interval from t_s to the beginning of the next phase of slope excavation. During this interval, no active (transient) drawdown of the water table occurs; so this length of time is subtracted from the time t estimated for excavating the entire slope of interest, and the new t is used to calculate a_t and r_t for the entire slope.

MONTE CARLO SIMULATION OF DRAWDOWN

The above analytical procedures and mathematical relationships are amenable to Monte Carlo simulation techniques. Simulation can provide a way of using the variability and uncertainty of hydrologic properties to generate a distribution of possible drawdown curves. This distribution can then be incorporated into probabilistic slope stability analyses.

Probability densities and cumulative distributions must be estimated for the height of the water table above the slope toe (h), the permeability (k), and the porosity (m). Usually, data histograms are used to approximate the distributions. If insufficient data is available, then engineering experience and subjective judgment must be used. If a dependency between permeability and porosity can be measured or conjectured, then it should be included in the simulation. The theory and computational procedures of Monte Carlo simulation, including a treatment of dependencies between variables, are presented by Newendorp (1975).

The drawdown simulation procedures can be easily programmed as a subroutine or subprogram of any slope stability computer program that simulates a distribution of safety factors (from which the risk of failure is calculated). Therefore, a different simulated drawdown curve is used in each stability simulation pass (iteration) that calculates one safety factor.

EXAMPLE PROBLEM

    To illustrate the drawdown calculations, the open pit slope geom-
etry shown in Figure 4 will be analyzed. The water table located at
the base of the andesite is known (from previous measurements) to
fluctuate less than 0.5 m in elevation; therefore, its elevation will
be treated as a constant. Field pump tests have indicated that the
permeability of the uranium-bearing arkosic sandstone is approximately
0.012 m/day with an accuracy estimated to be 25 percent. Porosity
tests of 30 drill core specimens yield a mean porosity of 0.254 with
a standard deviation of 0.041. Available data do not suggest a measurable dependency between permeability and porosity. Mine scheduling
dictates that the slope will be 30 m high at two years, 60 m high at
four years, and 75 m high at six years (ultimate slope). The estimated water table drawdown at two year increments is deterministically
calculated using the trial-and-error method for steady state analysis.
The results are summarized in Table 1.

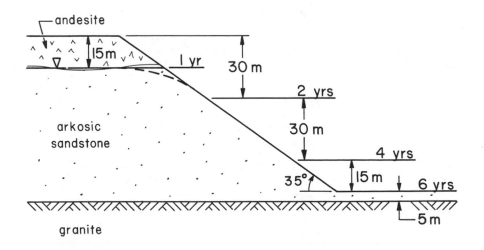

Figure 4.　Section of example pit slope, showing geology, water table,
and sequence of mining

Table 1. Results of steady state seepage analysis for 35-degree
slope

Mine Life (years)	Slope Height (m)	h (m)	First Guess for r (m)	Computed Values	
				a (m)	r (m)
2	30	15	37.5	7.8	37.5
4	60	45	112.5	23.7	111.6
6	75	60	150.0	31.6	148.8

Nonsteady State Analysis

The first step in the nonsteady state analysis is to calculate
the time required for the flow to reach steady state (t_s) during the
second year of mining, after the slope cut intersects the water table:

$$t_s = \frac{m}{kD} \left(\frac{r}{3.644} \right)^2$$

$$= \frac{0.254}{(0.012) \ (50 + 7.8 \sin 35°)} \left(\frac{37.5}{3.644} \right)^2$$

$$= 41 \text{ days},$$

which is much less than the 365 day interval in which the slope is
cut below the water table. Therefore, steady flow occurs for the last
324 days of the second mining year.

At the end of the fourth mining year, the time interval in which
nonsteady flow (active drawdown) occurs is 771 days, given by 1,095
days minus 324 days. The time to attain steady state conditions for
the 60 m high slope is

$$t_s = \frac{0.254}{(0.012) \ (20 + 23.7 \sin 35°)} \left(\frac{111.6}{3.644} \right)^2$$

$$= 591 \text{ days},$$

which does not exceed 771 days and indicates that steady flow occurs
for the last 180 days of the fourth mining year.

At the end of year six, nonsteady seepage has occurred for 1,321
days, given by 1,825 days minus 324 days minus 180 days. The time to
attain steady state conditions for the 75 m high slope is

$$t_s = \frac{0.254}{(0.012)(5 + 31.6 \sin 35°)} \left(\frac{148.8}{3.644}\right)^2$$

$$= 1,525 \text{ days},$$

which exceeds 1,321 days and indicates that steady state conditions are not attained for the ultimate slope at the end of the sixth mining year. Results from the transient seepage analysis are summarized in Table 2.

Table 2. Summarized results of transient seepage analysis for 35-degree slope

Mine Life (years)	Time Interval for Seepage (days)	Duration of Nonsteady Flow (days)	Duration of Steady Flow (days)
2	730 − 365 = 365	41	324
4	1460 − 730 = 730	591 − 41 = 550	771 − 591 = 180
6	2190−1460 = 730	1525−591 = 934*	1321−1525 = −204*

*Steady state conditions will not be attained at the end of the sixth year of mining, but nonsteady flow will persist until the 204-th day of the seventh mining year.

The drawdown curve corresponding to 1,321 days of nonsteady flow is described by r_t and a_t, which are calculated by equations (15) and (19), respectively:

$$r_{1321} = 143.2 \text{ m}$$

$$a_{1321} = 34.4 \text{ m} .$$

note the difference between these values and those in Table 1.

Monte Carlo Simulation

To illustrate a Monte Carlo simulation of the transient drawdown, let us assume that the ultimate slope is completed at the end of three years, and that a steady state flow condition is not attained at any prior time. Laboratory tests show the porosity of the arkosic sandstone to be normally distributed with a mean of 0.254 and a standard deviation of 0.041. For this example, the permeability is also assumed to be normally distributed with a mean of 0.012 m/day and a standard deviation of 0.001 m/day. This standard deviation was estimated by setting 0.003 (25 percent of 0.012) equal to three standard deviations, which represents a range that includes 0.9974 of the area under the normal density curve.

Typical results from a 100-iteration simulation are shown in Table 3. The estimated minimum time for reaching steady state is 981 days; the maximum is 2,208 days. Using average permeability and porosity values, the calculated t_S is 1,525 days. A relative-frequency histogram (Figure 5) indicates that the t_S values approximate a normal distribution. Results from each of the iterations could be incorporated into a safety-factor calculation for the slope, yielding a distribution (essential for risk analysis) of 100 safety factors, all of which are partially dependent on realizations of parabolic water table drawdown.

Table 3. Typical results from drawdown simulation

Iteration	Permeability (m/day)	Porosity	t_s (days)	Drawdown Parameters at 1,095 Days	
				a (m)	r (m)
Avg. Values	0.01200	0.2540	1525	38.1	135.7
1	0.01217	0.2720	1659	39.1	133.6
2	0.01078	0.1916	1319	34.7	142.7
3	0.01233	0.1644	990	31.1	150.0
4	0.01141	0.2960	1926	42.0	127.7
.					
.					
.					
98	0.01290	0.2671	1537	37.6	136.6
99	0.01312	0.2014	1140	31.8	148.4
100	0.01091	0.2550	1735	40.0	131.8

Seepage Quantities

The seepage quantity for steady state conditions is estimated (from seepage theory) to be:

$$q = ka \sin^2 \beta$$

$$= 0.012 \ (31.6) \ (\sin 35°)^2$$

$$= 0.125 \ (m^3/day)/m \ .$$

Seepage quantities for nonsteady state conditions are predicted by using the following equation from Glover (1974):

$$q = kHD \ / \ \sqrt{\frac{\pi kDt}{m}} \qquad (20)$$

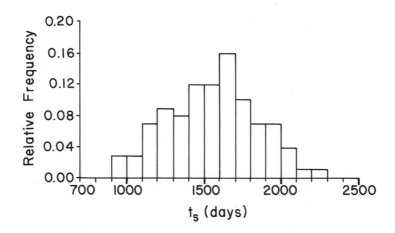

Figure 5.   Relative-frequency histogram of simulated values of t_s (time to attain steady state) for example problem

$$q = \frac{0.012\ (41.875)\ (23.125)}{\sqrt{3.1416\ (0.012)\ (23.125)\ (1525)}\Big/\ 0.254}$$

$$= 0.161\ (\text{m}^3/\text{day})/\text{m}\ .$$

This seepage quantity is higher than that predicted by seepage theory, probably due to the assumption in equation (13) where 0.995 approximates one.

At the end of three years (1,095 days), the average seepage at the face of the ultimate slope is estimated to be:

$$q = \frac{0.012\ (38.15)\ (26.85)}{\sqrt{3.1416\ (0.012)\ (26.85)\ (1095)}\Big/\ 0.254}$$

$$= 0.186\ (\text{m}^3/\text{day})/\text{m}\ .$$

Seepage quantities can be predicted as part of the simulation, which consequently provides a distribution of flow quantities that can be expected at any given time in the mine life. For this example, the simulated values approximate a normal distribution with a mean of

0.182 and a standard deviation of 0.015 $(m^3/day)/m$.

CONCLUSIONS

For a slope excavated in rock or soil material that can be considered hydrologically isotropic and homogeneous, an analytical method developed from seepage theory and plane analytic geometry can estimate the parabolic water table drawdown to the slope face. The drawdown curve is a function of slope angle and the height of the original water table above the slope toe. Its defining parameters are calculated by a trial-and-error method that usually converges after three to ten iterations.

When the slope is cut at a rate faster than that which allows steady state flow conditions to be attained, a nonsteady state analysis is used to approximate the transient drawdown. If the permeability is known, seepage quantities at the face can be estimated at any time during, or at the end of, the active drawdown period.

Calculations in the above procedures are readily performed with any hand-held calculator (with scientific functions), but programmable desk-top or hand-held calculators are recommended to expedite the computations and to minimize errors. The procedures are also amenable to Monte Carlo simulations, allowing for water table drawdown to be included in slope stability risk analyses.

An example problem, based on actual geologic and hydrologic data, was analyzed to illustrate this method of estimating the parabolic drawdown. The simulation in this example produced distributions of the drawdown curve parameters and the seepage quantity at the slope face.

REFERENCES

Glover, R.E., 1974, Transient Ground Water Hydraulics, Dept. of Civil Engineering, Colorado State University, Fort Collins, Colorado, 413 p.

Harr, M.E., 1962, Groundwater and Seepage, McGraw-Hill, Inc., New York, 315 p.

Newendorp, P.D., 1975, Decision Analysis for Petroleum Exploration, The Petroleum Publ. Co., Tulsa, Oklahoma, 668 p.

Polubarinova-Kochina, P.Ya., 1962, Theory of Ground Water Movement, translated by R. DeWiest, Princeton University Press, New Jersey, 613 p.

Selby, S.M., and Weast, R.C., eds., 1970, CRC Standard Mathematical Tables, 18th Edition, The Chemical Rubber Co., Cleveland, Ohio, p. 417.

Question

Are the slopes blasted. If so what adjustment did you make for permeability changes due to blasting.

Answer

This method of estimating the water table drawdown has only been applied to "soft-rock" slopes that are capable of being excavated by scraping or ripping. However, it would also be useful for "hard-rock", blasted slopes, provided that major geologic structures are absent and that joints are connected enough to allow overall drawdown of the water table. For high slopes, the bench geometries are usually neglected in most two-dimensional stability analyses, and thus, could also be neglected in the drawdown analysis. For slopes that only contain several benches (or mining levels), blast-induced permeability changes should be included in the drawdown analysis. I do not know at this time how such changes could be estimated, especially at the feasibility stage of a mine.

Question

Could you comment on the applicability of your method in view of your basic assumption of $K_V = K_H$. Have any case studies been done which show a correlation between actual and theoretical results.

Answer

In order for this drawdown analysis to be applied to an actual slope cut, the isotropy assumption (which includes $K_V = K_H$) is understood to be valid on the scale of the slope cut. Small sandy or clayey lenses are considered to be insignificant on the scale of the geologic formation that is being cut by the slope excavation. If field evidence indicates strongly preferred flow directions, then a more sophisticated drawdown analysis should be used. If field data are not available or are somewhat limited, then the simplified method presented in my paper is a useful tool for predicting possible drawdown curves; even though the slope material is assumed to have generally isotropic and homogeneous hydrologic properties.

At this time, no case studies have been conducted to compare analytical results with actual results.

Question

Why go to the trouble and expense of a Monte Carlo simulation at the feasibility stage of a mine.

Answer

It is at the feasibility stage of a mine when geotechnical data is limited, and the uncertainty in parameter values is high. Uncertainty implies variability, which is expressed as a distribution of possible values for a given parameter, or variable. The purpose of Monte Carlo simulation is to be able to account for variabilities in the slope stability analysis. Therefore, simulation techniques should be essential in feasibility slope studies. Distributions of the input variables can be estimated from available data or subjectively estimated by the geologist or engineer most qualified to do so. To use a "best" value (average) of a parameter in the analysis is to deny that it has any variability, and also implies that the analyst is capable of predicting the exact value of an unknown variable.

In my experience, the benefits of a Monte Carlo simulation far outweigh the "trouble" and expense of conducting it. The computer programming of a simulation analysis is relatively straight forward, and running the analysis is usually not exceptionally time consuming or expensive. The procedure is amenable to sensitivity analyses, and most importantly, it provides a distribution of safety factors that allows for the estimation of slope reliability.

Chapter 18

A COMPUTER PROGRAM FOR FOOTWALL
SLOPE STABILITY ANALYSIS IN STEEPLY
DIPPING BEDDED DEPOSITS

Brian Stimpson
and
Keith E. Robinson

Associate Professor of Mining Engineering, Department of
Mineral Engineering, Edmonton, Alberta, Canada

Principal, Robinson, Dames & Moore, Vancouver, B.C., Canada

INTRODUCTION

In inclined sedimentary strata slope failure may occur by sliding
along bedding and along a discontinuity or weak zone, as illustrated
in Fig. 1. This mode of failure may be called 'bi-linear', as the
failure surface is comprised of two linear or approximately linear
segments.

This type of failure has been suggested for the famous Frank Slide,
Alberta, Canada (Cruder and Krahn, 1973) in which an estimated 90 M
tons (8.97×10^5 MN) of rock separated from Turtle Mountain and
buried the town of Frank. Also, with the increasing depth of open
pits and the construction of highways through inclined sedimentary
strata this mode of failure is of major concern to the rock slope
stability engineer.

A computer program for calculating factors of safety for this
slope failure mode is provided and includes the ability to vary shear
strengths of potential slide surfaces, ground water elevations and
wedge geometry.

LIMITING EQUILIBRIUM MODEL

The simple limiting equilibrium model used in the analysis is
shown in Fig. 2. It is assumed that jointing is orthogonal to bedding and that the interface between the upper and lower wedges therefore corresponds to a joint plane or planes. No shear strength is
attributed to the interface so that the resultant force, P, from the
upper wedge acts normally across the interface. The normal loads

437

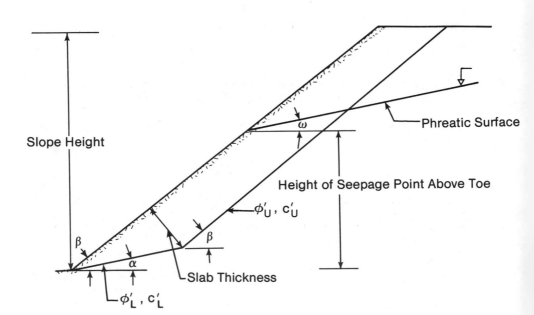

Slope Height

Phreatic Surface

ω

Height of Seepage Point Above Toe

ϕ'_U, c'_U

β

β

Slab Thickness

α

ϕ'_L, c'_L

Figure 1 - Failure by sliding along a discontinuity

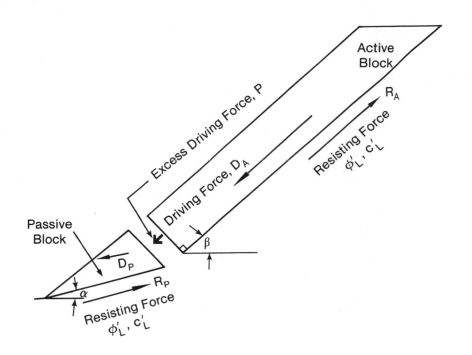

1. $P = D_A - R_A$
2. Apply P to Passive Block

Figure 2 - Simple limiting equilibrium model

across the upper and lower failure surfaces are assumed to be uniformly distributed. For the upper surface the effective normal load
is calculated from the effective weight of the upper wedge; for the
lower wedge the effective normal load is calculated from the effective weight of the lower wedge plus the contribution from the excess
force, P, from the upper wedge.

For simplicity, the slope face is parallel to the upper failure
surface and each surface can provide shear strengths according to
the linear relationship:

$s = c'L + N \tan \phi'$ where

c' the effective cohesion

L the length of the upper or lower failure surfaces

N the effective normal load

and ϕ' the effective angle of friction.

The program is written so that upper and lower failure surfaces
can have different shear strength properties.

Owing to the anisotropic nature of sedimentary strata the distribution of water pressure in slopes is particularly adverse when
compared to the distribution in an isotropic slope. Fig. 3 shows
calculated equipotentials for isotropic and bedded slopes and indicates the build up of high water pressures immediately behind the
slope in an isotropic strata.

For the stability analysis a linear water table is assumed and
groundwater pressures are based on a simple hydrostatic distribution.
Fig. 3 contracts the hydrostatic case with the pressures calculated
from equipotentials in the isotropic and anisotropic cases. The
higher groundwater pressures in the anisotropic slope are clearly
illustrated in this figure. In some cases the anisotropic situation
may not be as markedly different from the isotropic case as a result
of weathering and opening of cross-joints near surface and cross
drainage created by the presence of the lower potential failure plane
(Fig. 1). For the purpose of this analysis, the hydrostatic distribution was used to model water pressures though it probably under-
estimates the magnitude of these pressures.

The assumed distributions of water pressure along the potential
failure surfaces are illustrated in Fig. 4.

Calculations are carried out in two stages. Firstly, the resisting and driving forces for the upper or active wedge are computed. Any excess 'driving' force from this wedge is then applied

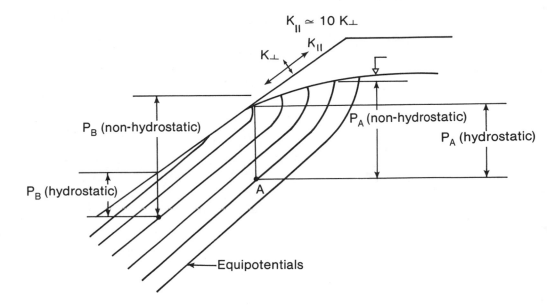

$K_{||} \simeq 10\,K_{\perp}$

$K_{||}$

$K_{\perp}$

P_B (non-hydrostatic)

P_A (non-hydrostatic)

P_A (hydrostatic)

P_B (hydrostatic)

A

Equipotentials

Due to weathering and/or blasting the degree
of anisotropy near the face may be
reduced so that the isotropic situation is
approached:-

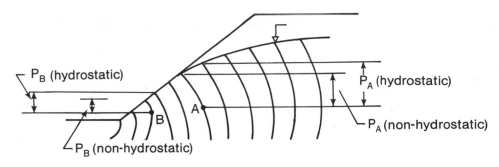

P_B (hydrostatic)

P_A (hydrostatic)

B   A

P_A (non-hydrostatic)

P_B (non-hydrostatic)

Figure 3 - The hydrostatic case with the pressures
        calculated from equipotentials in the
        isotropic and anisotropic cases

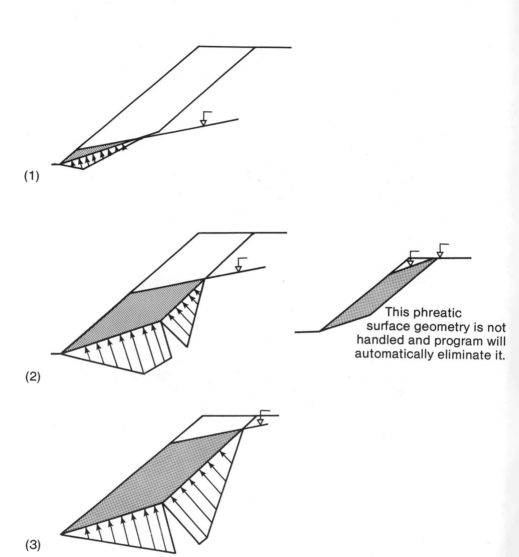

This phreatic surface geometry is not handled and program will automatically eliminate it.

Figure 4 - Assumed distributions of water pressure along potential failure surfaces

to the lower or passive wedge. Results are computed in terms of either Factors of Safety (ratio of resisting to driving force for upper and lower slide surfaces) or as the effective angle of friction required along the lower surface in order to give a Factor of Safety along that surface of 1.0.

Do-loops are employed to enable a range of parameters to be used for each run. Documentation is provided in Appendix I.

The program was written in for use with British units -ft, lbs/sq ft, lbs/cu ft.

REFERENCES

1. Cruden, D.M. & Krahn, J. A reexamination of the Geology of the Frank Slide. Canadian Geotechnical Journal, 10, 581-591, 1973.

APPENDIX I

Program Documentation

To limit size of paper, a listing of all variables and their usage is not given but may be obtained from the authors.

1. Program Identification

Title: "SLOPE STABILITY LIMITING EQUILIBRIUM SLAB
                    ANALYSIS" WITH BI-LINEAR FAILURE SURFACE

Authors: Dr. Brian Stimpson and Mr. Keith E. Robinson

Version: 0

Language: MTS Fortran IV - G compiler

Availability: The source program and additional system documen-
                    tation is available from the authors.

Abstract: Program computes the factors of safety, using a standard limiting equilibrium approach, for slopes in which the potential failure surface is formed by two planar features - one parallel to the slope face (e.g. bedding) and a second dipping out of the slope and daylighting on the slope face (Fig. 1). A phreatic surface may be included. Alternately, for given strength properties along the planar surface parallel to the slope, the program will compute the average normal and shear stresses along the lower surface and the effective friction angle required to give a Factor of Safety of 1.0 along this surface.

The program has the facility to vary parameters by amounts specified by the user so that sensitivity analysis can be carried out. It
has been developed primarily for analysis of slopes in bedded deposits where bedding is parallel to the slopes.

II. Engineering Documentation DESCRIPTION: Program TSLAB – SLOPE
STABILITY LIMITING EQUILIBRIUM SLAB ANALYSIS BI-LINEAR FAILURE SUR-
FACE – Calculate Factors of Safety or required effective friction
angles for a Factor of Safety of 1.0 on the lower potential failure
plane.

The slope angle, the dip of the upper potential failure surface,
and the inclination to the horizontal of the phreatic surface (which
is approximated to a straight line) are fixed for each run. Slope
height, slab thickness, position of phreatic surface, dip of lower
potential failure surface, and the effective cohesion and friction
angle of both surfaces can be varied during a single run. (Fig. 2).

The Factors of Safety are computed from the ratio of resisting to
driving forces. A linear shear strength relationship is as assumed
for each surface.

METHOD OF SOLUTION: For analysis the wedge bounded by the two potential slide surfaces is divided into an upper, or active block and a
lower, or passive block (Fig. 2). The interface between the active
and passive blocks is at right angles to the upper surface and
corresponds in attitude to the joints commonly found in bedded
strata, i.e., perpendicular to bedding. The shear resistance along
the interface is not included in the analysis so that the results are
slightly conservative. For a long thin slab the effect of omitting
the interface resistance is negligible.

The resisting and driving forces parallel to the upper surface are
calculated from simple statics for the upper block considered as an
isolated wedge, taking into account the uplife forces for sections
below the phreatic surface. The ratio of the two forces is calculated. If this ratio or 'Factor of Safety' is equal to or less than
1.0 the calculation is stopped at this stage since no excess driving
forces from the upper block can be applied to the lower block (the
stability of the lower block could then be considered as a simple
case of planar sliding). If the ratio is less than 1.0 the excess
force is applied to the passive wedge and the resisting and driving
forces parallel to the lower surface computed. The ratio of these
two forces gives the Factor of Safety of the lower surface. The
ratio of the sum of resisting to the sum of driving forces for both
surfaces provides the total Factor of Safety.

Instead of calculating Factors of Safety the program provides the
option of calculating the average normal and shear stresses on the
lower surface and the effective angle of friction required for a
Factor of Safety of 1.0 on the lower failure surface.

A hydrostatic distribution of pressure is assumed along the potential failure surfaces.

The various uplift pressure distributions assumed for the different possible locations of the phreatic surface are shown in Fig. 4. The program automatically identifies these different situations and carries out the appropriate calculations.

If the uplift forces exceed the component of weight normal to the surface, a negative effective force is obtained. To avoid this situation the program automatically sets the effective force to zero if it becomes negative. With zero effective force the frictional resistance is zero so that if additionally no cohesion is present the total resisting force is zero. For this case the Factor of Safety is set to 0.0.

If height of the phreatic surface above the toe (Fig. 2) is equal to or exceeds the slope height the program will do no further calculation and transfers to the next problem. Similarly, the situation in which the phreatic surface intersects the slope crest within the wedge boundaries (Fig. 4) is not considered and transfer to the next problem is automatically made. This phreatic surface geometry was considered unlikely to occur in practice.

Slope geometry is also analysed by the program and certain anomalous situations are automatically eliminated. For this particular analysis the interface between active and passive wedges must intersect the slope face but not the slope crest.

It is assumed for all practical purposes that the density of material below the phreatic surface is the same as that above the phreatic surface. The unit weight of water (62.4 pcf) is built into the program.

For each run of the program the following values are fixed:

a. Slope angle and thereby the dip of the upper potential failure surface.

b. Inclination to the horizontal of the phreatic surface.

c. Unit weight of rock mass.

For each run on the computer the following variables may be varied from an initial value by an increment specified by the user: -

a. Effective cohesion of the lower surface.

b. Effective friction angle of lower surface.

c. Effective cohesion of the upper surface.

d. Effective friction angle of upper surface.

e. Dip of lower surface.

f. Height above toe of seepage point on face.

g. Slab thickness.

h. Slope height.

    The increment takes one value for each variable, e.g. vary slope
height from an initial value of 100 ft. to 300 ft. in 20 ft. increments.

    The variables (a-h) above are nested together in a series of DO
loops with the innermost variable being the effective cohesion of the
lower failure surface (a) and the outermost variable being the slope
height (h). Thus, the program will first of all vary the lower surface cohesion through all the specified values while keeping variables (b-h) set at their initial values. Then the effective friction
angle of the lower surface is increased by one increment and the
lower surface cohesion varied through all the chosen values again.
This process continues until all the iterations are completed. It
may be appreciated that with eight variables the number of problems
increases rapidly with the number of increments required. For
example, if all variables are required to take 4 different values the
total number of problem is 65,536 and for 5 different values 390,625.
Each problem requires one line of print-out so that the amount of
output can be overwhelming. Therefore, it is necessary to consider
very carefully the number of increments for each variable.

    FEATURES, CAPABILITIES & LIMITS: Though there are numberical
limitations to the values of variables that can be used, the FORMAT
statements are such that most likely practical situations are covered.
Effective cohesion up to 99,999 psf (694 psi), slab thickness up to
999 ft., and slope height up to 9,999 ft. can be used and each variable may be incremented up to 998 times.

    The program contains checks on various situations which could
produce unexplained program failures, e.g. if the level of phreatic
surface increase to a value greater than the slope height or if the
dip of the lower surface equals or exceeds the slope angle.

    Angles may take any value from 0.0 to 89.0 degrees.

    INPUT/OUTPUT: The input data include a project title card, an
initial slope geometry card, a slope geometry unit increment card, a
slope geometry number of calculations control card, an initial slope
material properties card, a slope material unit increment card, a
slope material number of calculations control card, and finally an

option control card.

Program Options: By changing one data card the user may either compute Factors of Safety for each problem or, given all values except the strength parameters of the lower surface, compute the average normal and shear stresses along the lower surface and the effective friction angle required for a Factor of Safety of 1.0 along this plane.

The card must be formatted as follows:

1. PROJECT TITLE I.D. CARD (10A8)

   Field

   　　i) title up to 80 char. describing data

2. INITIAL SLOPE GEOMETRY DATA CARD (2F5.0,F4.0,3F2.0,F3.0) (See Fig. 2)

   Field

   　　i)　initial slope height in feet, H.

   　　ii)　initial height above toe of point of seepage on slope face in feet, Hw

   　iii)　initial slab thickness in feet, D

   　　iv)　slope angle in degrees (constant), θ

   　　v)　initial dip of lower face in degrees, α

   　　vi)　angle of phreatic surface to horizontal in degrees (constant), w

   　vii)　unit weight of rock mass in pcf, γ

3. SLOPE GEOMETRY UNIT INCREMENT CARD (4F3.0,F2.0)

   Field

   　　i)　increment in slope height in feet

   　　ii)　increment in height of phreatic surface in feet

   　iii)　increment in slab thickness in feet

   　　iv)　increment in dip of lower surface in degrees

4. SLOPE GEOMETRY NUMBER OF CALCULATIONS CONTROL CARD (413)

Field

    i)   number of calculation in slope height

    ii) number of calculations in height of phreatic surface

   iii) number of calculations in slab thickness

    iv) number of calculations in the dip of lower surface

5. INITIAL SLOPE MATERIAL PROPERTIES CARD (2F2.0,2P5.0)

Field

    i)   initial effective angle of friction of upper surface in degrees, FU

    ii) initial effective angle of friction of lower surface in degrees, FL

   iii) initial effective cohesion of upper surface in psf, CV

    iv) initial effective cohesion of lower surface in psf, CL

6. SLOPE MATERIAL UNIT INCREMENT CARD (2F2.0,2F4.0)

Field

    i)   increment in upper surface friction angle in degrees

    ii) increment in lower surface friction angle in degrees

   iii) increment in upper surface cohesion in psf

    iv) increment in lower surface cohesion in psf

7. SLOPE MATERIAL NUMBER OF CALCULATIONS CONTROL CARD (413)

Field

    i)   number of calculations in upper surface friction angle

    ii) number of calculations in lower surface friction angle

   iii) number of calculations in upper surface cohesion

iv)   number of calculations in lower surface cohesion

8. OPTION CONTROL CARD (II)

N=1   program calculates Factor of Safety

N=0   program calculates average normal and shear stresses along
lower surface and friction angle required for FS=1.0

All results are output via LUN 6. The following items are reported:

i)   all current slope geometry and material constants

ii)   Factor of Safety of upper surface

iii)   Factor of Safety of lower surface

iv)   Total Factor of Safety

v)   all geometric and physical constants except the cohesion and
friction angle of the lower surface

vi)   average normal stress

vii)   average shear stress

viii)   required friction angle to achieve a Factor of Safety of 1.0

All the parameters for each problem are printed under appropriate
headings and units. Typical outputs for the two types of problem
which can be run with this program are shown in Table 1.

OTHER OUTPUTS: When inadmissible geometries are encountered the
program 'side-steps' these situations but does not produce any
printed output signifying this.

If the upper block is found to be stable the print-out will state
'UPPER BLOCK STABLE' together with the input data for that problem.

The computer program listing follows.

TABLE I

SENSITIVITY TO CU AND FU

SLOPE ANGLE = 40. ANGLE OF PHREATIC SURFACE = 0.

CL PSF	FL DEG	CU PSF	FU DEG	ALPHA DEG	HW FT	D FT	H FT	FS UPPER SURFACE
5000.	35.	0.	20.	0.	0.	50.	1500.	0.43
5000.	35.	600.	20.	0.	0.	50.	1500.	0.55
5000.	35.	1200.	20.	0.	0.	50.	1500.	0.67
5000.	35.	1800.	20.	0.	0.	50.	1500.	0.79
5000.	35.	2400.	20.	0.	0.	50.	1500.	0.91
5000.	35.	3000.	20.	0.	0.	50.	1500.	1.02
5000.	35.	0.	25.	0.	0.	50.	1500.	0.56
5000.	35.	600.	25.	0.	0.	50.	1500.	0.67
5000.	35.	1200.	25.	0.	0.	50.	1500.	0.79
5000.	35.	1800.	25.	0.	0.	50.	1500.	0.91
5000.	35.	2400.	25.	0.	0.	50.	1500.	1.03
5000.	35.	3000.	25.	0.	0.	50.	1500.	1.14
5000.	35.	0.	30.	0.	0.	50.	1500.	0.69
5000.	35.	600.	30.	0.	0.	50.	1500.	0.81
5000.	35.	1200.	30.	0.	0.	50.	1500.	0.92
5000.	35.	1800.	30.	0.	0.	50.	1500.	1.04
5000.	35.	2400.	30.	0.	0.	50.	1500.	1.16
5000.	35.	3000.	30.	0.	0.	50.	1500.	1.28

FS LOWER SURFACE	FS TOTAL
0.70	0.51
0.73	0.60
0.78	0.69
0.88	0.80
1.24	0.93

UPPER BLOCK STABLE	
0.73	0.60
0.78	0.69
0.89	0.80
1.27	0.93

FS LOWER SURFACE	FS TOTAL
UPPER BLOCK STABLE	0.71
UPPER BLOCK STABLE	0.82
0.79	0.95
1.39	

UPPER BLOCK STABLE
UPPER BLOCK STABLE
UPPER STABLE

```
 1      C*                  BI LINEAR LIMITING EQUILIBRIUM ANALYSIS
 2      C       -----THIN SLAB ANALYSIS - BI-LINEAR FAILURE SURFACE
 3      C       PROGRAM COMPUTES FACTOR OF SAFETY OF EACH SURFACE AND OF THE TOTAL
 4      C       SLOPE
 5      C       OR
 6      C       AVERAGE SHEAR AND NORMAL STRESS ALONG LOWER FAILURE SURFACE AND
 7      C       FRICTION ANGLE REQUIRED FOR FACTOR OF SAFETY OF ONE ALONG LOWER SU
 8      C*****INPUT DATA SLOPE HEIGHT(H) FT AND INCREMENTS
 9      C       SLOPE ANGLE(BETA)
10      C       SLAB THICKNESS (D) FT AND INCREMENTS
11      C       HEIGHT OF SEEPAGE POINT ON FACE ABOVE TOE(HW) AND INCREMENTS
12      C       DIP ANGLE(ALPHA) OF LOWER SURFACE AND INCREMENTS
13      C       FRICTION ANGLE OF UPPER SURFACE(FU) AND INCREMENTS
14      C       FRICTION ANGLE OF LOWER SURFACE (FL) AND INCREMENTS
15      C       COHESION OF UPPER SURFACE(CU) AND INCREMENTS
16      C       COHESION OF LOWER SURFACE (CL) AND INCREMENTS
17      C*****OUTPUT PRODUCES AS NOTED ABOVE
18      C
19      C
20      C***
21              REAL*8 A,B,C,D,E,F,G,H,III,JJJ
22      C*READ AND PRINT TITLE
23              READ (5,47) A,B,C,D,E,F,G,H,III,JJJ
24              WRITE (6,48) A,B,C,D,E,F,G,H,III,JJJ
25      C*INPUT INITIAL GEOMETRY AND DENSITY
26              READ (5,56) H,HW,D,BETA,ALPHA,OMEGA,GAMMA
27              WRITE (6,54) BETA,OMEGA
28      C*INPUT INCREMENTS IN GEOMETRY
29              READ (5,57) HI,HIW,DI,AIALP
30      C*INPUT NUMBER OF VALUES
31              READ (5,58) NIH,NIHW,ND,NIALP
32      C*INPUT INITIAL STRENGTHS
33              READ (5,59) FU,FL,CU,CL
34      C*INPUT INCREMENTS IN STRENGTH
35              READ (5,60) FUI,FLI,CUI,CLI
36      C*INPUT NUMBER OF VALUES
37              READ (5,61) IFU,IFL,ICU,ICL
38      C*INPUT OPTION TO CALCULATE FACTOR OF SAFETY OR FRICTION REQUIRED
39              READ (5,62) N
40              IF (N.EQ.O) GO TO 1
41              WRITE (6,49)
42              GO TO 2
43      1       WRITE (6,51)
44      C*SET UP DO LOOPS
45      2       SBMO = SIN(0.0175*(BETA-OMEGA))
46              DO 46 I=1,NIH
47              IF (I.EQ.1) GO TO 3
48              CH = CH+HI
49              GO TO 4
50      3       CH = H
51      4       DO 46 J=1,ND
52              IF (J.EQ.1) GO TO 5
53              CD = CD+DI
54              GO TO 6
55      5       CD = D
56      6       XZ = CD/SBMO
57              DO 46 K=1,NIHW
58              IF (K.EQ.1) GO TO 7
59              CHW = CHW+HIW
60      C*CHECK WHETHER POINT OF INTERSECTION OF PHREATIC SURFACE WOULD BE
61      C       ABOVE SLOPE
62              IF (CHW.GE.CH) GO TO 46
63      C*CHECK WHETHER PHREATIC SURFACE INTERSECTS CREST AT TOP OF WEDGE
64              EKR = XZ*SIN(O.O175*OMEGA)
65              HWKR = EKR+CHW
66              IF (HWKR.GT.CH) GO TO 46
67              GO TO 8
68      7       CHW = HW
```

```
 69        8    DO 45 L=1,NIALP
 70             IF (L.EQ.1) GO TO 9
 71             CALP = CALP+AIALP
 72             GO TO 10
 73        9    CALP = ALPHA
 74   C*CHECK THAT DIP OF LOWER SURFACE LESS THAN SLOPE ANGLE OR UPPER SURFACE
 75       10    IF (CALP.GE.BETA) GO TO 45
 76             BMA = BETA-CALP
 77   C***
 78   C*COMPUTE WEIGHT OF UPPER BLOCK
 79   C*CHECK WHETHER LOWER SURFACE INTERSECTS SLOPE CREST
 80             SBMA = SIN(0.0175*BMA)
 81             AE = CD/SBMA
 82             HLS = AE*SIN(0.0175*CALP)
 83             IF (HLS.GE.CH) GO TO 45
 84             AB = CD/TAN(0.0175*BMA)
 85             SIBET = SIN(0.0175*BETA)
 86             HD = AB*SIBET
 87             HDD = CH-HD
 88   C*CHECK WHETHER INTERFACE INTERSECTS SLOPE CREST
 89             IF (HDD.LE.0.) GO TO 45
 90             BC = HDD/SIBET
 91             VOL1 = BC*CD
 92             DD = CD/TAN(0.0175*BETA)
 93             ED = BC+DD
 94             VOL2 = 0.5*DD*CD
 95             W1 = (VOL1+VOL2)*GAMMA
 96   C*COMPUTE WEIGHT OF LOWER BLOCK
 97             W2 = 0.5*AB*CD*GAMMA
 98   C*COMPUTE BASIC INTERNAL DATA
 99             SAMO = SIN(0.0175*ABS(CALP-OMEGA))
100             COALP = COS(0.0175*CALP)
101   C*CHECK WHETHER SLOPE IS DRY
102             IF (CHW.EQ.0.) GO TO 21
103             AX = CHW/SIBET
104             COSOM = COS(0.0175*OMEGA)
105   C*COMPUTE UPLIFT FORCES
106   C*CHECK WHETHER PHREATIC SURFACE INTERSECTS LOWER FAILURE SURFACE
107             IF (CALP.LE.OMEGA) GO TO 11
108             AC = AX*SBMO/SAMO
109             IF (AC.GT.AE) GO TO 11
110             H2 = AX*SBMA/COALP
111             UL = 31.2*AC*H2
112             UU = 0.
113             GO TO 22
114   C*CHECK WHETHER PHREATIC SURFACE INTERSECTS ON SLOPE SURFACE ABOVE LOWER
115   C*SLIP SURFACE OR ABOVE UPPER SLIP SURFACE
116       11    D2 = CHW/TAN(0.0175*BETA)
117             D1 = AE*COALP
118             IF (D1.LT.D2) GO TO 16
119   C*COMPUTE UPLIFT FORCES WHERE INTERSECTION ABOVE LOWER SLIP SURFACE
120             AB = D2/COALP
121             BE = AE-AB
122             H2 = AX*SBMA/COALP
123             IF (SAMO.GT.0.) GO TO 12
124             DH1 = 0.
125             GO TO 13
126       12    DH1 = BE*SAMO/COSOM
127             IF (CALP.GT.OMEGA) GO TO 14
128       13    H1 = H2+DH1
129             GO TO 15
130       14    H1 = H2-DH1
131       15    U2 = (H1+H2)*BE
132             U3 = H2*AB
133             UL = (U2+U3)*31.2
134             EZ = H1*COSOM/SBMO
135             UU = 31.2*H1*EZ
136             GO TO 22
137   C*COMPUTE UPLIFT FORCES WHERE INTERSECTION ABOVE UPPER SLIP SURFACE
```

```
138    16    H3 = CD/COS(0.0175*BETA)
139          YZ = H3*COSOM/SBMO
140          IF (SAMO.GT.O.) GO TO 17
141          ZZ = O.
142          GO TO 18
143    17    ZZ = XZ*SAMO/SBMA
144          IF (CALP.GT.OMEGA) GO TO 19
145    18    EZ = AX+ZZ
146          GO TO 20
147    19    EZ = AX-ZZ
148    20    EY = EZ-YZ
149          U3 = H3*YZ*0.5
150          U2 = H3*EY
151          UU = (U2+U3)*62.4
152          UL = 31.2*AE*H3
153          GO TO 22
154    21    UU = O.
155          UL = O.
156    C*COMPUTE RESISTING AND DRIVING FORCES ALONG UPPER SLIDE SURFACE
157    C*WEIGHT COMPONENT DOWN DIP
158    22    WCDD = W1*SIN(0.0175*BETA)
159    C*EFFECTIVE NORMAL FORCE ACROSS UPPER SLIP SURFACE
160          ENFU = W1*COS(0.0175*BETA)-UU
161    C***
162    C*CHECK WHETHER EFFECTIVE NORMAL FORCE IS POSITIVE OR NEGATIVE
163          IF (ENFU.GE.O.) GO TO 23
164          ENFU = O.
165    C*FRICTIONAL AND COHESIVE RESISTANCE ALONG UPPER SLIP SURFACE
166    23    DO 44 II=1,IFU
167          IF (II.EQ.1) GO TO 24
168          CFU = CFU+FUI
169          GO TO 25
170    24    CFU = FU
171    25    DO 44 JJ=1,ICU
172          IF (JJ.EQ.1) GO TO 26
173          CCU = CCU+CUI
174          GO TO 27
175    26    CCU = CU
176    27    DO 44 KK=1,IFL
177          IF (KK.EQ.1) GO TO 28
178          CFL = CFL+FLI
179          GO TO 29
180    28    CFL = FL
181    29    DO 44 LL=1,ICL
182          IF (LL.EQ.1) GO TO 30
183          CCL = CCL+CLI
184          GO TO 31
185    30    CCL = CL
186    31    TRFU = ENFU*TAN(0.0175*CFU)+ED*CCU
187    C*CHECK WHETHER RESISTING GREATER THAN OR EQUAL TO DRIVING FORCES
188          IF (TRFU.GE.WCDD) GO TO 38
189    C*COMPUTE NORMAL AND SHEAR FORCES ACTING ALONG LOWER SLIP SURFACE
190          P = WCDD-TRFU
191          FCDD = W2*SIN(0.0175*CALP)+P*COS(0.0175*(BETA-CALP))
192          FCAD = W2*COS(0.0175*CALP)+P*SIN(0.0175*(BETA-CALP))-UL
193    C*CHECK WHETHER EFFECTIVE NORMAL FORCE IS POSITIVE OR NEGATIVE
194          IF (FCAD.GE.O.) GO TO 32
195          FCAD = O.
196    32    IF (N.EQ.O) GO TO 40
197    C*CHECK WHETHER CALCULATION OF FACTOR OF SAFETY ALONG LOWER SLIP SURFACE
198          TRFL = FCAD*TAN(0.0175*CFL)+AE*CCL
199    C*CHECK WHETHER RESISTING FORCE ALONG UPPER SURFACE IS ZERO
200          IF (TRFU.EQ.O.) GO TO 33
201          FSU = TRFU/WCDD
202          GO TO 34
203    33    FSU = O.
204    C*CHECK WHETHER RESISTING FORCE ALONG LOWER SURFACE IS ZERO
205    34    IF (TRFL.EQ.O.) GO TO 35
206          FSL = TRFL/FCDD
```

```
207              GO TO 36
208       35     FSL = O.
209     C*CHECK WHETHER FACTORS OF SAFETY OF BOTH SURFACES ARE ZERO
210              IF (FSU.EQ.O..AND.FSL.EQ.O.) GO TO 37
211       36     FSUL = (TRFU+TRFL)/(WCDD+FCDD)
212              GO TO 43
213       37     FSUL = O.
214              GO TO 43
215       38     FSU = TRFU/WCDD
216              IF (N.EQ.1) GO TO 39
217              WRITE (6,55) CCU,CFU,CALP,CHW,CD,CH
218              GO TO 44
219       39     WRITE (6,50) CCL,CFL,CCU,CFU,CALP,CHW,CD,CH,FSU
220              GO TO 44
221       40     IF (FCAD.EQ.O.) GO TO 41
222              AVNS = FCAD/AE
223              AVSS = FCDD/AE
224              FRIC = ATAN(AVSS/AVNS)*57.28
225              GO TO 42
226       41     AVNS = O.
227              AVSS = FCDD/AE
228              FRIC = 90.
229       42     WRITE (6,52) CCU,CFU,CALP,CHW,CD,CH,AVNS,AVSS,FRIC
230              GO TO 44
231       43     WRITE (6,53) CCL,CFL,CCU,CFU,CALP,CHW,CD,CH,FSU,FSL,FSUL
232       44     CONTINUE
233       45     CONTINUE
234       46     CONTINUE
235              STOP
236     C
237       47     FORMAT (10A8)
238       48     FORMAT (1H ,20X,10A8)
239       49     FORMAT (1H ,127HCL PSF    FL DEG    CU PSF    FU DEG    ALPHA DEG
240            $HW FT      D FT      H FT      FS UPPER SURFACE  FS LOWER SURFACE
241            $   FS TOTAL)
242       50     FORMAT (1H ,8(F6.O,4X),7X,F5.2,10X,18HUPPER BLOCK STABLE)
243       51     FORMAT (1H ,112HCU PSF    FU DEG    ALPHA DEG HW FT      D FT
244            $H FT      AV NORMAL STRESS  AV SHEAR STRESS  REQUIRED FRIC ANG)
245       52     FORMAT (1H ,6(F6.O,4X),7X,F10.2,7X,F10.2,7X,F5.1)
246       53     FORMAT (1H ,8(F6.O,4X),7X,F5.2,12X,F5.2,10X,F5.2)
247       54     FORMAT (1H ,14HSLOPE ANGLE = F5.O,28HANGLE OF PHREATIC SURFACE = F
248            $5.0)
249       55     FORMAT (1H ,6(F6.O,4X),30X,18HUPPER BLOCK STABLE)
250       56     FORMAT (2F5.O,F4.O,3F2.O,F3.O)
251       57     FORMAT (3F3.O,F2.O)
252       58     FORMAT (4I3)
253       59     FORMAT (2F2.O,2F5.O)
254       60     FORMAT (2F2.O,2F4.O)
255       61     FORMAT (4I3)
256       62     FORMAT (I1)
257              END
```

END OF FILE

Question

What is the typical range of ϕ', c' values you have used along the bedding planes.

Answer

We hesitate to give values of properties as if they applied universally when they are, in reality, site specific. However, in the Canadian Rocky Mountains in sequences of flexurally folded mudstones and sandstones we would suggest ϕ' s of 20°-25° for bedding planes in mudstones along which flexural slip has clearly occurred and 30°-35° for bedding planes in sandstones under the same conditions. In all cases we would assume c' = 0.

Question

Are you faced with clay mylonite problems ($\phi_r' \simeq 11^{\circ}$) in these steeply dipping beds.

Answer

We presume you are referring to the narrow, sheared, high moisture content, low shear strength zones reported in coal-bearing strata by various authors (cf. "Clay Mylonites in English Coal Measures - Their Significance in Opencast Slope Stability" by B. Stimpson and G. Walton, 1st Int. Congress of the International Association of Engineering Geology, Paris, September 8-11, 1970). We have not observed these narrow clay bands in the Rockies but have seen some evidence in the Foothills and Plains of Canada. However, high quality drilling and centimetre by centimetre geotechnical logging is required (1-4 cms. typically).

Chapter 19

ANALYSIS OF BOLT REINFORCEMENT IN ROCK SLOPES

Francois E. Heuze

Leader, Rock Mechanics Program
Lawrence Livermore National Laboratory

INDEX

INTRODUCTION

Rock slope stability typically is governed by the geological discontinuities. This stability can be improved by drainage, unloading, adjustment in slope orientation, adjustment in slope angle, and also by reinforcement such as bolts or tendons. This paper addresses the modeling of bolt reinforcement in rock slopes.

Most geological discontinuities such as joints have rough surfaces; they dilate during shear displacement. Their shear strength is characterized by two envelopes (Fig. 1); the peak envelope is distinct from the residual envelope when the joints can dilate. If the dilation is restrained transversely, such as with bolts, the normal stress will increase and the joint strength also will increase. It has been demonstrated that this increase may be considerable (Obert et al., 1976). The new normal stress at peak strength is a composite function of the properties of the rock mass and of the steel (Heuze, 1979). Hence, the dilation and its resultant effects on joint shear strength must be included in the analysis and design of the reinforcement. We present a model which can be used to perform such an analysis.

Early studies of rock reinforcement in mining and civil applications (Lang, 1957 - Panek, 1962 - Osen and Parsons, 1966 - Lang, 1972), did not consider the specific effects of bolts on joint shear strength. In 1973, Heuze and Goodman reported direct shear tests on rough sandstone joints with bolts perpendicular to the joint plane (Fig. 2). The peak shear strength of such joints was greatly in excess of the shear strength of the unbolted joints. The excess

457

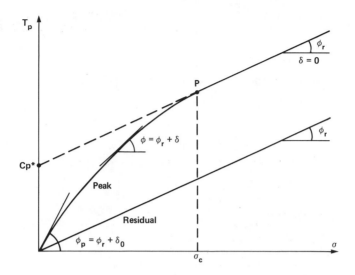

Figure 1: Shear Strength Envelopes for Natural Rock Joints

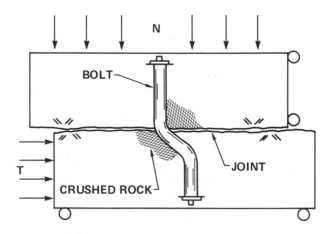

Figure 2: Direct Shear Test on a Bolted Joint

strength can be explained in terms of three contributions:
- the shear strength of the steel dowel
- the crushing strength of solid rock surrounding the bolt, in the vicinity of the shear plane
- the dilatant joint effect: the stretched bolt reacts by increasing the normal stress on the joint, thus increasing its strength.

Further studies (Bjurstrom, 1974 - Gerdeen et al., 1977 - Haas et al., 1978 - Kwitowski and Wade, 1980) confirmed that the contribution of a steel dowel to joint shear strength goes well beyond its own shear capacity.

However, current methods for analysis of bolt reinforcement, such as the stereographic projection (Heuze and Goodman, 1972 - Hoek and Bray, 1974 - Goodman, 1976), only accommodate the bolt tension and its shear capacity, in a limit equilibrium approach. The new model developed by the author does include the pertinent dilation effects.

THE DILATANT EFFECTS MODEL

The new model is based on the simple concept represented by Fig. 3. The joint is shearing and is subjected to lateral restraint from adjacent springs which have a stiffness KNEFF. The springs represent the adjacent rock, and/or the reinforcement. Under a small shear displacement, Δu, the joint will open a small amount, Δv. The Δv represents a compromise between the opening tendency, due to the instantaneous dilation angle δ, and the reclosing tendency due to the increased normal stress acting on the normal stiffness of the joint itself, KN. The complete model derivation is given in a previous paper (Heuze, 1979). In summary, the net increase in normal stress on the joint is obtained as:

$$\Delta\sigma = \tan\delta \cdot \frac{KN \cdot KNEFF}{KN+KNEFF} \Delta u$$

This model has been implemented in the finite element code JPLAXD of the author, and several applications have been reported (Heuze and Barbour, 1980). A particular example relevant to rock slope stability is given next.

APPLICATION: ANALYSIS OF A BOLTED SLOPE

A simple geometry, amenable to two-dimensional analysis, was chosen for illustration purposes (Fig. 4). A pervasive rough joint is assumed to be present at the site; when the cut is fully excavated, failure could take place along the daylighting joint. A small cut is made at the top and a bolt reinforcement is installed. If the bolt is emplaced before full excavation, it will restrain the dilatant shear and increase the normal stress on the joint, according to the mechanism explained previously. Thus the joint shear strength will increase significantly and the slope safety

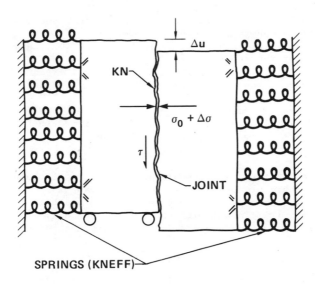

Figure 3: Conceptual Model of a Transversely Restrained Joint in Shear

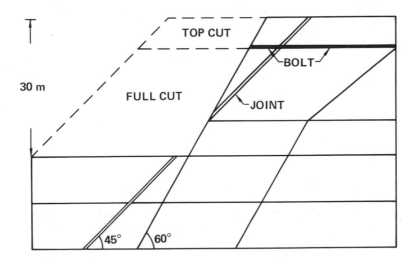

Figure 4: Model of a Pre-Reinforced Rock Slope, With a Master Joint

factor will increase. The analysis was performed in a sequential fashion, following the construction steps: top cut excavation, bolt installation, and full cut excavation. The material properties for the rock, the steel and the rock joints are summarized in Tables 1 and 2. The steel thickness in the plane of the model is 5 cm. As represented, the bolt is considered fully grouted.

Table 1: Initial Rock Properties for JPLAXD Calculations

	Intact Rock	Steel	Excavated[1] Rock
Mass density	2.73	8.	0.
Modulus (GPa)	20.	200.	10^{-2}
Poisson's ratio	0.25	0.10	0.25
Peak cohesion (MPa)	10.	400.	400.
Residual cohesion (MPa)	0.	0.	0.
Peak friction (°)	45.	0.	0.
Residual friction (°)	25.	0.	0.
Tensile strength (MPa)	5.	400.	400.

[1]The excavated rock is made very light and soft, but very strong so as to bypass the failure criterion routines.

Table 2: Initial Joint Properties for JPLAXD Calculations

	Joint in Intact rock	Steel joint	Excavated Rock Joint
Normal stiffness (MPa/m)	10^4	10^6	10^{-3}
Shear stiffness (MPa/m)	10^3	10^5	10^{-4}
Maximum closure (m)	$-1.27.10^{-2}$	$-2.5.10^{-3}$	$-2.5.10^{-3}$
Peak cohesion intercept (MPa)[1]	1.5	0.	0.
Peak cohesion[2]	0.	400.	200.
Initial dilation angle (°)[3]	10.	0.	0.
Critical normal stress (MPa)[4]	10.	10^4	10^4
Tensile strength (MPa)	0.	400.	100.
Residual friction (°)[5]	35.	0.	0.

[1] C_p on Fig. 1.
[2] Zero, because both envelopes go through the origin.
[3] δ_o on Fig. 1; zero for non-dilatant joints.
[4] σ_c on Fig. 1; zero for non-dilatant joints.
[5] Peak friction is ϕ_p on Fig. 1; residual friction is ϕ_r. The instantaneous dilation angle δ varies continuously with the normal joint stress σ.

If failure of joints or solids were to take place, the program automatically recalculates material properties to conform to the given
constitutive relations.

In the four cases studied, the total friction angle of the joint was
taken as 45°. However in cases A it was modeled as a flat $\phi_r = 45°$
(Fig. 1), whereas in cases B it was modeled as $\phi = \phi_r + \delta$ where
$\phi_r = 35°$ and δ is $= 10°$ at zero normal stress, and decreases
when σ increases. Using conventional definition, the shear factor
of safety for the joint plane below the bolt is compared in Table 3
for the A and B models. Both tensioned and untensioned bolts are
considered. Bolt tension was 70 MPa, applied upon installation.
The explicit dilation model calculates significantly higher values
of F.S. From a practical point of view, this means that a less
sophisticated analysis may indicate an unsatisfactory reinforcement,
when in fact the reinforcement is adequate. Hence, the new model is
apt to produce more economical designs. Bolt tensioning is providing a large increase in F.S. In practice, the bolts should be
regularly distributed over the shear area, to improve the uniformity
of the normal stress increase on the joint.

Table 3: Comparison of Shear Factors of Safety (F.S.)

Case	Condition	Dilation (°)	F.S.	% Increase
A–U	Untensioned bolt	$\phi = 45 + 0\ (\delta)$	1.05	---
B–U	Untensioned bolt	$\phi = 35 + 10\ (\delta)$	1.29	23%
A–T	Tensioned bolt	$\phi = 45 + 0\ (\delta)$	3.52	---
B–T	Tensioned Bolt	$\phi = 35 + 10\ (\delta)$	4.88	38%

THREE-DIMENSIONAL CASES

Our example of a block or slab on a plane could reasonably be
analyzed with a two-dimensional model. However, many rock slope
problems involve wedges, and require a three-dimensional approach.
Three-dimensional finite element programs with joints are very
scarce, and none includes dilatant models. Until such a refined
three-dimensional numerical analysis is available, it should be
possible to combine the use of the stereographic projection with the
above two-dimensional models. Whether plane slides or intersection
slides are considered, the system of driving and resisting forces on
the slide plane(s) can be determined for the non-dilatant case. In
turn, this can be used as a starting condition for an individual
dilatant analysis of the sliding surface(s). This will necessitate
that additional joint parameters be determined, which are not

required for stereographic analysis: the normal and shear stiffness, and the variation of the dilation angle with normal stress. However, those parameters are natural by-products of the type of shear testing which should be done, in any case, to obtain the cohesion and friction angle values required for stereographic analysis.

SUMMARY

This simple application has emphasized the need to explicitly take dilation into account, in order to estimate the strength of joints in laterally restrained shear. The results indicate that pre-bolting and pre-tensioning can be very beneficial; the rock movement could be kept below the value where joint peak strength is exceeded, and where the benefit of dilatant effects is lost. When the joint is properly restrained in shear displacement, the normal stress due to tensioning is further enhanced as described. Also, a smaller shear displacement will allow less joint opening which, in turn, will minimize fluid percolation through the joint. Even with an untensioned fully grouted reinforcement, appreciable dilatant strength increase can also be mobilized.

REFERENCES

1. Bjurstrom, S. (1974) "Shear Strength of Hard Rock Joints Reinforced by Grouted Untensioned Bolts", _Proc. 3rd Congress Int. Soc. for Rock Mechanics_, Denver, Colorado, Sept., v II-B, pp 1194-1199.

2. Gerdeen, J. C., et al. (1977) "Design Criteria for Roof Bolting Plans Using Fully Resin-Grouted Non Tensioned Bolts to Reinforce Bedded Mine Roofs. Volume 1: Executive Summary and Literature Review", Michigan Tech. University, Houghton for U.S. Bureau of Mines, July, (PB80-180052).

3. Goodman, R. E. (1976) "_Methods of Geological Engineering_", West Publishing Co., 472 p.

4. Haas, C. J. et al. (1978) "An Investigation of the Interaction of Rock and Types of Rock Bolts for Selected Loading Conditions", University of Missouri, Rolla for U.S. Bureau of Mines, (PB-293988).

5. Heuze, F. E. and Goodman, R. E. (1972) "Three-Dimensional Approach for Design of Cuts in Jointed Rock", _Proc. 13th U.S. Symposium on Rock Mechanics_, Urbana, Ill., ASCE Ed., pp 397-441.

6.   Heuze, F. E. and Goodman, R. E. (1973) "Finite Element and Physical Model Studies of Tunnel Reinforcement in Rock", <u>Proc. 15th U.S. Symposium on Rock Mechanics</u>, Custer State Park, S.D., Sept. 17-19, ASCE Ed., pp 37-67.

7.   Heuze, F. E. (1979) "Dilatant Effects of Rock Joints", <u>Proc. 4th Congress Int. Soc. for Rock Mechanics</u>, Montreux, Switzerland, Sept., v. 1, pp 169-175.

8.   Heuze, F. E. and Barbour, T. G. (1980) "New Models for Rock Joints and Interfaces", Lawrence Livermore National Laboratory Report <u>UCRL-85222</u>, Dec., 28 p.

9.   Hoek, E. and Bray, J. (1974) "<u>Rock Slope Engineering</u>", Institute of Mining and Metallurgy, London, 306 p.

10. Kwitowski, A. J. and Wade, L. V. (1980) "Reinforcement Mechanisms of Untensioned Full-Column Resin Bolts", <u>U.S. Bureau of Mines RI 8439</u>.

11. Lang, T. A. (1957) "Rock Behavior and Rock Support in Large Excavations", Proc. Symp. on Underground Power Stations, ASCE Power Division, New York, N.Y., Oct.

12. Lang, T. A. (1972) "Rock Reinforcement", <u>Bulletin Assoc. Eng. Geologists</u>, v. IX, n. 3, pp 215-239.

13. Obert, L. et al. (1976) "The Effect of Normal Stiffness on the Shear Resistance of Rock", <u>Rock Mechanics</u>, v. 8, pp 47-72.

14. Osen, L. and Parsons, E. W. (1966) "Yield and Ultimate Strength of Rock Bolts Under Combined Loading", <u>U.S. Bureau of Mines RI 6842</u>.

15. Panek, L. (1962) "The Combined Effects of Friction and Suspension in Bolting Bedded Mine Roof", <u>U.S. Bureau of Mines RI 6139</u>.

ACKNOWLEDGMENT

This paper was prepared under support from the Department of Energy, on contract W-7405-ENG-48.

Chapter 20

A SIMPLE CORE ORIENTATION TECHNIQUE

R. D. Call, J. P. Savely, and R. Pakalnis

President, Call & Nicholas, Inc., Consulting Engineers
Tucson, Arizona

Chief Geological Engineer, Inspiration Consolidated Copper Company
Inspiration, Arizona

Department of Mineral Engineering
The University of British Columbia, Vancouver, B.C.

ABSTRACT

A simple and inexpensive clay imprint core orienting device has
been developed by Dr. R. D. Call. It has a minimum of moving parts,
is durable and easily used by drillers, and adds only 15 minutes to
a core run. Conventional wire-line drilling and a drill hole inclined 30° to 60° from the horizontal plane are required. The
device will not operate in a vertical hole. The orientor is an
eccentrically weighted downhole device which can be constructed in
a mine shop from an old inner core barrel and lead. Modeling clay
is packed in the core lifter case, which is screwed on the downhole
end of the orientor.

A clay imprint of the bottom of the hole is taken after each coring run. This imprint is matched to the top of the succeeding core
run, the drill core is pieced together, and a reference line representing the drill hole is scribed on each core. Fracture attitudes
are measured relative to the reference line and the core axis.

On recent drilling projects, tests for imprint reproducibility
were conducted. The results of these tests show an average variation
of ±3° in location of the scribed reference line between imprints.
Comparisons of surface mapping data with oriented core data are good.
It is rare that the entire length of drill core in intensely fractured zones can be oriented because it is impossible to piece the
core together. However, if a triple tube set-up is used, a modification in the scribing procedure can result in a good statistical sampling of the fracture population. Attitudes of major structures have
been oriented with this technique.

INTRODUCTION

Geotechnical studies for mine design require a knowledge of rock fabric, as well as general geology. Adits or drill holes must be planned for subsurface investigation if sufficient information cannot be obtained by surface mapping. Even if surface mapping were possible, there is no assurance that fracturing at the surface (the sampled population) would correspond to that at depth (the target population).

Stability analysis for ultimate mine designs requires knowledge of the rock substance and rock fabric beyond the ore boundaries. Oriented drill core provides a means of obtaining this knowledge.

When drilling geotechnical holes for slope design, it is preferable to drill an angle hole to maximize intercepting structural fractures in the critical orientation dipping into the pit. If the direction and plunge of an inclined hole are obtained by a conventional drill hole survey, which side of the core is up is the only additional information needed to orient core.

In 1970, a clay imprint orientor was used on the Tazadit Pit Slope Study in Mauritania (Call, 1972). This device used an Eastman multishot to determine the top of the core. Subsequently, the device described below was developed; it has been used on a number of projects, and has proven to be operational.

ORIENTOR PROCEDURE

The orientor is an eccentrically weighted, downhole device, which consists of a 1 m length of inner core barrel half full of lead (Figure 1). Drilling fluid in the drill pipe surrounds the orientor during the operation and acts as a liquid bearing. In an inclined hole, the eccentricity produced by the weighted bottom of the orientor always causes it to rotate to the same position as it is dropped down the drill pipe. A line etched on the side of the orientor opposite the weighted side will be in the same direction as the drill hole and will represent the top of the hole. Conventional wire-line drilling and a drill hole inclined 30° to 60° from the horizontal plane are used. The device will not operate in a vertical hole.

Modeling clay is packed in the core lifter case, which is screwed on the downhole end of the orientor. The clay-packed lifter case is used for imprinting the bottom of the drill hole before each core run. Prior to imprinting, the drill string is raised about 6 in. (15 cm). The orientor is dropped inside the drill pipe and allowed to free-fall through a full mud column. When the orientor reaches the bit, the mud pump is engaged to hold the orientor in place at the bottom of the hole. In deep holes, it may be necessary to engage the mud pump and to pump the orientor to the bottom of the

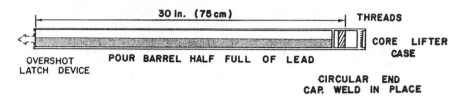

Figure 1: Clay Core Orientor

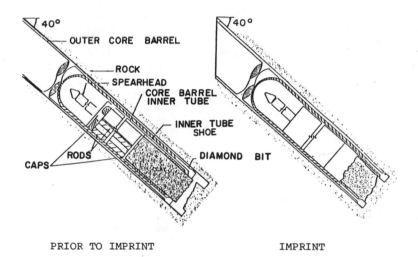

PRIOR TO IMPRINT   IMPRINT

Figure 2: Imprinting Core Stub

hole. The drill string is lowered, without rotation, to take the imprint (Figure 2), and the overshot is then used to retrieve the orientor. The core lifter case containing the clay imprint of the bottom of the hole is unscrewed and then laid in the tray with the top of the core run next to the imprint. The imprint is matched to the top of the core run, the core pieces are fitted together, and a reference line, which represents the hole orientation, is scribed on the core. This reference line is an extension of the etched line on the unweighted side of the orientor.

As the drill core is logged, angles are measured for each structural feature present. A drill-hole survey provides the orientation of the drill hole and, thus, the orientation of the reference line on the core for various depth intervals. This survey is interpolated to specific depths in order to calculate the true orientations of each structural feature.

It is rare that the entire length of the drill core in intensely fractured zones can be oriented because it is impossible to piece the core together. However, if an attempt to imprint is made for every run, a good statistical sample of the fracturing can usually be obtained. In massive rock, it may not be necessary to imprint every run if the core can be reliably pieced together from one run to the next.

Triple tube or split tube core barrels greatly decrease the disturbance of the core and should be used if possible.

Reproducibility of the Orientor

On a recent quartz monzonite drilling project, two tests for imprint reproducibility were conducted. Test 1 was a 58° inclined hole, drilled to a depth of 97 ft. The orienting device was dropped six times. The first drop was considered the initial imprint, from which the variation in rotation of the five succeeding drops could be measured. Test 2 was a 52° inclined hole drilled to a depth of 513 ft. Table 1 summarizes the reproducibility of imprinting.

From the results of these two tests, it is apparent that the orientor averages about ±3° variation in location of the scribed reference line between imprints.

Table 1

Number of Drops	Test 1 Hole Depth = 97 ft Variation from First Dump	Test 2 Hole Depth = 513 ft Variation from First Dump
1	0.0°	0.0°
2	+3.0°	+1.1°
3	-2.2°	+2.1°
4	-1.1°	+5.3°
5	-2.2°	+1.1°
6	-4.3°	-4.8°

+ = clockwise rotation of orientor
- = counterclockwise rotation of orientor

MEASURING AND RECORDING FRACTURE DATA

Orientations of fractures in the drill core are measured relative to the core axis and the reference line, using a plexiglas goniometer (Figure 3). The measurements made with the goniometer are converted to dip direction and dip of the fractures by vector mathematics and the drill hole survey data.

A standard data sheet, prepared in key-punching format, is used to record the goniometer measurements (Figure 4). For each fracture, the following are recorded:

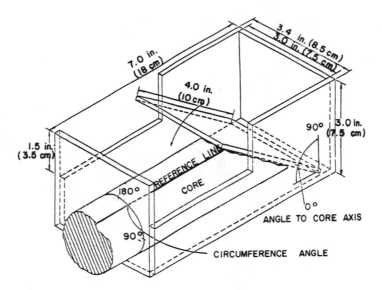

Figure 3: Plexiglas Core Orienting Goniometer

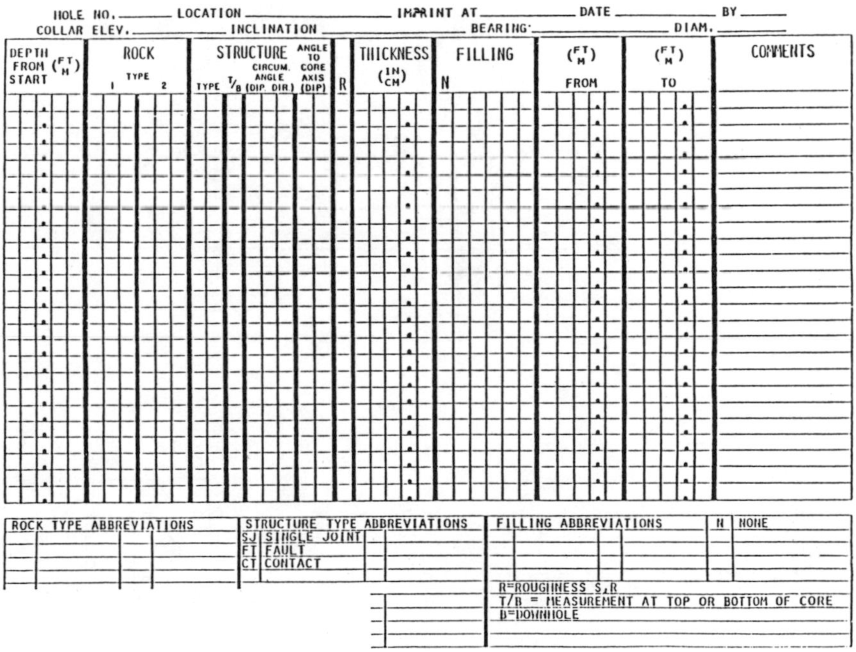

HOLE NO. ____ LOCATION ____ IMPRINT AT ____ DATE ____ BY ____
COLLAR ELEV. ____ INCLINATION ____ BEARING ____ DIAM. ____

DEPTH FROM START (FT/M)	ROCK TYPE 1 2	STRUCTURE TYPE T/B CIRCUM. ANGLE (DIP. DIR.)	ANGLE TO CORE AXIS (DIP)	R	THICKNESS (IN/CM)	FILLING N	(FT/M) FROM	(FT/M) TO	COMMENTS

ROCK TYPE ABBREVIATIONS

STRUCTURE TYPE ABBREVIATIONS
SJ SINGLE JOINT
FT FAULT
CT CONTACT

FILLING ABBREVIATIONS
N NONE

R=ROUGHNESS S,R
T/B = MEASUREMENT AT TOP OR BOTTOM OF CORE
D=DOWNHOLE

Figure 4: Data Sheet for Oriented Core

Depth from Start of Drill Run. The distance from the top of the drill run to the point where the reference line intercepts the fracture.

Rock Type. A three-character alphanumeric code to describe the rock type, such as SCH for schist.

Structure Type. A two-character code identification of the genetic nature of the fracture, such as SJ for single joint.

Bottom (or Top). "B" if the goniometer measurement is taken from the bottom of the top core stick, or "T" if the measurement is taken from the top of the bottom core stick. Either side of a fracture surface can be measured, but it is better to measure the bottom, or "B" end, of the top core stick (Figure 5).

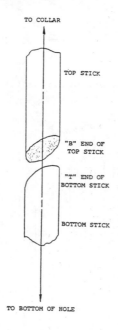

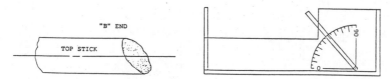

NOTE: THIS MEASUREMENT WOULD HAVE A "B" DESIGNATION

Figure 5: "B" or "T" Designations for Recording Data

<u>Circumference Angle (β)</u>. An angle measurement of the dip direction relative to the reference line.

<u>Angle to Core Axis (α)</u>. An angle measurement that is the complement of the dip angle relative to the core axis.

<u>Roughness</u>. An alphanumeric character that qualitatively describes the nature of the fracture surface: "S" = smooth; "R" = rough surface.

<u>Thickness</u>. Width of the measured fracture opening.

<u>Filling</u>. Six columns for acknowledging the presence of fracture-filling materials.

<u>From - To</u>. Depth from collar to top and bottom of core run.

VECTOR AND STEREOGRAPHIC PROJECTION SOLUTIONS TO DETERMINE FRACTURE ORIENTATION

The following sections present both vector and stereographic projection solutions to determine the true fracture orientations from the fracture measurements which have been measured relative to the core axis.

Determination of Dip Direction and Dip by Vector Solution

The input orientation data are

1) angle to core axis (α),
2) circumference angle (β),
3) bearing of drill hole (B), and
4) plunge of drill hole (P).

The circumference angle (β) would be the dip direction, if the hole were vertical and the top of the hole were north when the bottom end (B) of the core is inserted into the measuring box. If the top end (T) has been inserted, the circumference angle (β) must be subtracted from 360° before the conversions. The angle to the core axis (α) would be the complement of the fracture dip if the hole were vertical.

1) Conversion to lower hemisphere pole:

     For bottom reading, S = 180° + β
     For top reading,   S = 360° - β
                     D =   90° - α

2) Calculation of directional cosines:

    X = cos S sin D
    Y = sin S sin D
    Z = cos D

NOTE: The top of the hole is +X, east +Y, and down +Z.

3) Axis rotation for hole inclination:

    ϕ = 90° - plunge of hole
    X' = X cos ϕ + Z sin ϕ
    Y' = Y
    Z' = Z cos ϕ - X sin ϕ

4) Axis rotation for bearing of hole (θ):

    X" = X' cos θ - Y' sin θ
    Y" = Y' cos θ + X' sin θ
    Z" = Z'

5) Conversion to dip direction and dip:

$$A = \left| \text{Tan}^{-1} \left(\frac{Y"}{X"} \right) \right|$$

    Y" > 0 and X" $\geq$ 0 DDR = A + 180
    Y" $\geq$ 0   X" $\leq$ 0 DDR = 360 - A
    Y" $\leq$ 0   X" $\geq$ 0 DDR = 180 - A
    Y" < 0   X" $\leq$ 0 DDR = A

$$\text{Dip} = \cos^{-1} Z"$$

Determination of Dip Direction and Dip by
  Stereographic Projection

   The lower hemisphere equal angle net can be used to determine the
fracture orientation. The usual procedure is to use a transparent
overlay for plotting fracture poles. Figure 6 is an example of the
four steps that are to be followed in determining the dip direction
and dip from the recorded oriented core data.

CASE STUDY

   A geotechnical study was undertaken on a large porphyry copper
operation in Northern British Columbia (1979). This operation is
one of the largest operating copper pits in Canada, with ultimate
dimensions on the order of 8000 ft x 4000 ft x 1200 ft. The stability study required the mapping of all exposed bench faces within
the mine area (Figure 7). A drill program for evaluating the ore/
waste contact that would be encountered on the north wall of the
pit below the 720 ft level was initiated in March, 1980. The area

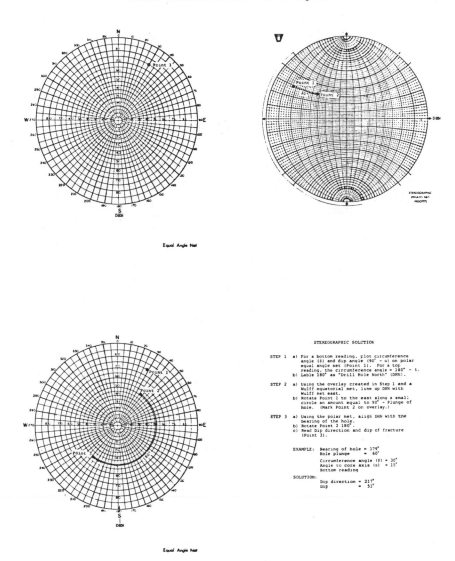

Figure 6: Stereographic Solution

involved was at ultimate limit above the 720 level, and it was at
this point that the findings of the geotechnical study were to be
implemented. The majority of the proposed drill holes were to be
inclined between 50° and 57° and dip into the pit area. A core
orienting technique, where discontinuity information could be obtained through the orientation of the drill core was applied.

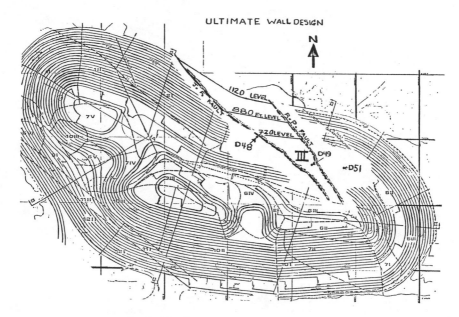

Figure 7:   Planned Final Pit

Various orienting techniques were analyzed (Table 1); most proved to
be too technically involved and/or greatly impeded the normal drill
rate. The device designed by Dr. Call was employed for the following reasons:

1) it was the least expensive method of orientation (10 percent
   increase in drill cost at mine site);
2) it was technically simple (procedure was conducted by
   drillers);
3) it was accurate (the ±3° deviation per drop was verified at
   the mine site through back/fore orienting the core); and
4) it had a minimal effect on drill rate (10 percent slower at
   mine site).

Procedure

   The procedure followed was that outlined by Dr. Call, where the
orienting device was dropped approximately every 10 ft. A modification of the method was that a reference line was drawn along the
entire length of the retrieved core, with the only restriction being
that the line be parallel to the core axis. It was found that an
edge of the splits from the triple tube barrel facilitated drawing
of the line. The core stub, whose impression had been taken, was
additionally referenced with an extension from the unweighted scribe
line from the clay pot orientor.

Therefore, the stub has two parallel scribe lines, and the deviation from the true scribe is recorded on the goniometer, the true scribe being the line corresponding to the inclination of the drill hole.

This recorded deviation is then applied to all oriented data obtained for that particular core run.

Three diamond drill holes, D48, D49, and D51, a total of 2486 ft, were drilled simultaneously with the clay pot orienting technique. The drill core was NQ-3 size with triple tube wireline procedures employed throughout.

A typical drill hole, D48, gave the following information:

1) 65 percent of core was oriented 170/275 m ;
2) 993 structures were noted;
3) RQD was less than 60 percent; and
4) geology was that of a fresh andesite.

Drill rate was as follows: 20 m of core per ten hour shift were drilled and oriented, with one hour out of ten employed for orienting. Normal operations would yield 23 m/ton per ten hour shift.

The geotechnical study of 1979 suggested that the location of D48 would fall in Domain III, which is separated from adjacent areas by the JF and RP faults (Figure 7). Domain III was designed using structural data obtained from mapping exposed bench faces; at the time of the 1979 report, this involved the area between 1102 and 880 ft level. Structural data representing the 1979 mapping and the oriented data for D48 are shown in Figures 8 and 9, respectively. The two stereo-nets revealed the strong S1 concentration, but, as expected, the oriented hole would result in a blind band that would neglect structures parallel to the trend of the drill hole. Therefore, S1, S2, and S3, which were present in Domain III, were absent in D48. Initially, that this concentration would not be detected with holes drilled parallel to the strike of S1 and S3 was obvious; however, because of poor continuity, they did not appear to dominate the stability of Domain III.

A concentration, S5, was detected in D48, but not in the 1979 mapping (Figure 9). Further mapping of exposed benches at the 880 to 720 level resulted in the plot shown in Figure 10, which has the same concentrations as depicted in 1979, with the addition of S5. The conclusion was that the same concentrations can be expected to elevations of 40 m, which is the bottom of D48.

The technique was further used on an exploration property in Northern British Columbia. The following observations were made:

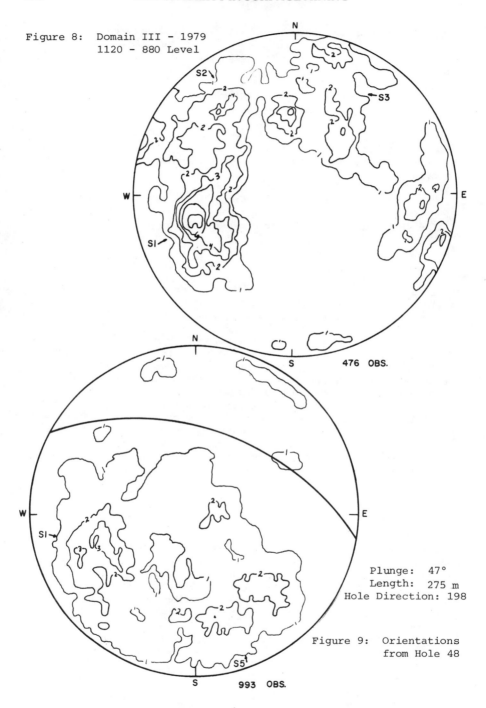

Figure 8:   Domain III - 1979
            1120 - 880 Level

476 OBS.

Plunge: 47°
Length: 275 m
Hole Direction: 198

Figure 9:   Orientations
            from Hole 48

993 OBS.

Figure 10: Orientations from 800 to 720 Level - Domain III

1) 75 percent of the core was oriented 221/300 meters;
2) RQD was 80 percent; and
3) geology was that of a feldspar porphyry.

Drill rates were as follows: 45 m of core per ten hour shift were drilled and oriented. Normal operations would yield 65 m per ten hour shift.

Due to the high drill rate in this type of rock, drill rates were impeded by 30 percent. The orientor, however, was dropped every 6.4 m, and, due to the intact nature of the core, it could be back/fore oriented by matching up the impression stub (fore-running core) and the mirror-image stub (back-running core).

Salient Points

In addition to the blind band, a gap was evident in D48, D49, and D51. Upon closer examination, it became apparent that the gap had an orientation that reflected all discontinuities perpendicular to the core axis (orientations in Figures 9, 11, and 12). At the mine site, these perpendicular structures to the core were assumed to be due to the drilling procedure and not natural breaks, and, therefore, they were not measured. It was also found that recording Tropari readings at the bottom, middle, and top of the hole was sufficient.

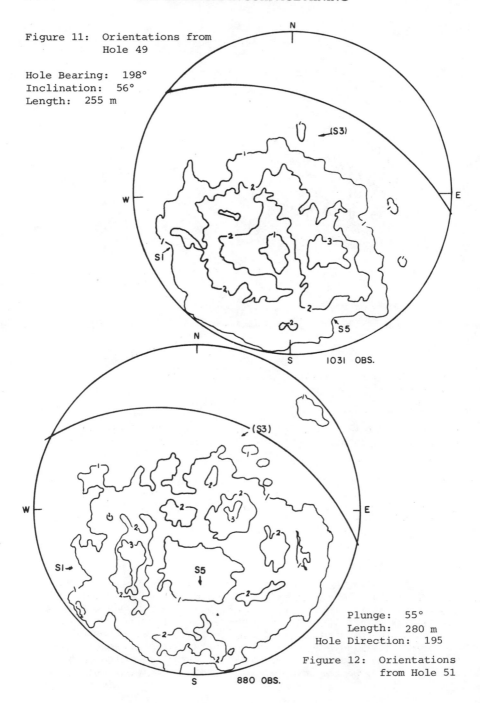

Figure 11:   Orientations from
             Hole 49

Hole Bearing: 198°
Inclination: 56°
Length: 255 m

1031 OBS.

Plunge: 55°
Length: 280 m
Hole Direction: 195

Figure 12:   Orientations
             from Hole 51

880 OBS.

Bits with the water course on the inside helped drillers to know when the core orientor had reached its destination. This occurred because the orientor blocked passage of water through the bit annulus; as a consequence, water pressure increased.

The device was used by exploration geologists at Utah Ltd. to orient particular faults, enabling the geologists to obtain a better understanding of the genesis of the deposit.

Summary of Borehole Structural Logging Techniques

Method	Borehole Size	Principal of Operation	Restrictions	Cost
Paint Marker	no limit	gravity	Inclined hole, difficult to apply paint mark under water	Double (50% slower)
Craelius Core Orientor	wide range	mechanical profiling	inclined hole, requires expert on site	$2200/unit (buy) Slows drilling Can only obtain 8½ft core/10 ft run
Borehole Camera	60 mm	photography of borehole wall	restricted to clear water, low head	$5000/month (rent) (+$300/day/expert)
Borehole Television	63 mm	video of wall	water head is bad, require clear water	$70/day(rent) (+$300/day/expert)
Christensen	no limit	mechanical scribe plus photo of scribe superimposed on a compass	slow	$10,000/mo (rent) (double cost of drilling - 50% slower)

Costs are obtained from Roctest Stockholm, Sweden; CANMET; B.C. Hydro; Christensen, Salt Lake City, Utah; 1980 figures.

CONCLUSIONS

Drill core can be reliably and inexpensively oriented, using a clay imprint device. Tests have shown that fracture orientations can be accurately obtained from core drilled in an inclined drill hole, using conventional wire-line coring procedures. The orientor is simple, with a minimum of moving parts; mechanical failure is, therefore, nearly eliminated. The clay imprint on the bottom of the hole produces an imprint that can be reliably matched to the extracted drill core. Since the drill core must be fitted together in the core tray, some drill runs in intensely fractured rock cannot be oriented. However, in most holes, a sufficient number of fractures can be oriented to give a good statistical sample of the fracture population. Major structure, as well as rock fabric, can be oriented.

ACKNOWLEDGMENTS

The authors thank Messieurs Pickering, Robertson, Janes, Lamb, Flemming, and Thornton of Island Copper Mines, Utah Mines Ltd. for their input into the core orientation project that was conducted at the mine site.

REFERENCES

Call, R. D., 1972, Analysis of Geologic Structure for Open Pit Slope Design: unpublished Ph.D. dissertation, University of Arizona, Tucson.

Question

What is validity of applying line bias correction to oriented core to enhance joint sets subparallel to the core. Does line bias connection give reliable results for oriented core.

Answer

We prefer to use the line bias correction for computing the spacing of a fracture set rather than enhancing the stereographic projection plots. Generation of an artificial fracture count for the attitude of a single fracture gives the impression of a tight orientation distribution, which is false. When only one fracture of a set is intersected by a hole, the true spacing is not known. Applying the line bias correction of $1/\sin\alpha$, where α is the angle between the fracture orientation and the core axis, assumes the fracture spacing is the minimum, when in fact the true spacing could be anything greater. Thus, unreal concentrations can be generated by applying the line bias correction.

The blind zone should be located on the stereographic projection so that those orientations can be identified where fracture sets could be present but would not appear on the plot. When computing the spacing of fractures, the line bias must of course be used to determine true spacing.

Question

Although your concern was to orient core from inclined holes, do you have any comments on the usefulness of the core orienting devices for use in vertical holes.

Answer

The Christiansen-Hugel system can be used in a vertical hole. Other than the cost and the limitations of a magnetic orientation device, it is an effective system.

Question

Is there a practical limit (RQD or fracture spacing) for proper orientation employing core orientor device. RQD's mentioned were above 50%.

Answer

The amount of orientable core decreases with the intensity of fracturing. Our experience has been that below about 20% RQD less than 10% of the core can be oriented. There is a dependence on the nature of the fracturing. In regularly bedded or foliated rock the core may separate along the bedding plane or foliation, giving a low RQD, but it can still be oriented because the pieces fit together.

Comment

The claypot method for orienting drill cores is the first new method of core orientation to appear in over ten years. It is a method that many individuals may use because they can fabricate their own hardware. Owing to the fact that the method has mostly been used only by the inventor it would be interesting to obtain data from other users to substantiate the ± 3 degrees precision that was claimed. The claypot method conceptually works in much the same way that the Craelius Core Orientator (CCO) works. That is, an inclined hole is required and only the first piece of core from the next run is orientated. The CCO method will, however, work in a hole inclined from 0 to 75 degrees from horizontal which is a greater range than the claypot method. One problem that some users have had with the CCO is that they do not fully understand how to use the device in wireline equipment. At Seegmiller International we have to date successfully oriented slightly more than 5090m (16,700 ft) of core by the CCO method under a variety of conditions and in a variety of mines (copper, iron, uranium, etc.). We have had an over-all success ratio (number of meters oriented divided by number of meters attempted) of more than 87%. One common misconception that was stated during the claypot paper presentation is that when using the CCO, less than a full core barrel is obtained. Present methods of using the CCO dictate that after the CCO has been releas-ed in the core barrel, it is retrieved from the hole and then a full core barrel may be obtained as with other methods. Our ex-perience would further indicate that a CCO operator can be success-fully trained in as little time as a half-day, thus allowing site personnel to perform all their own orientation.

Chapter 21

MONITORING THE BEHAVIOR OF HIGH ROCK SLOPES

W.B. Tijmann

Senior Development Engineer, Slope Indicator Company
Seattle, Washington U.S.A.

ABSTRACT

Maintaining safe, yet economical, slope geometries in a mining
operation is paramount. When design analysis and engineering judgement have dictated conservative and usually more expensive problem
solutions, monitoring has often proven to be a valuable method in
dealing with potential slope instability. A well planned and executed
monitoring program can allow operation at a lower theoretical margin
of safety. Monitoring provides a valuable check on design parameters
and can measure effects of remedial improvements. Warning of impending failure can often be established well in advance.

The monitoring problem situations in civil engineering works and
open pit mining are much alike. Instruments used for monitoring in
both industries are often the same. In this paper, geotechnical instruments for measuring deformation and groundwater pressure in rock
and overburden are discussed with a broad-brush approach. Results
obtained over extended time periods are presented in a graphical presentation for a case history. Descriptions of various types of instruments, mostly used to monitor stability of rock slopes, are discussed, and data analysis and approximate costs are presented.

INTRODUCTION

Instruments used for long-term monitoring should be as simple as
possible. Instruments should be uncomplicated to read, preferrably
mechanical, pneumatic rather than electrical, reliable, robust and
portable. Preferrably, data reduction should be possible in the
field. Electrical instruments are ideally suited for remote monitoring and can be used in otherwise inaccessible and hazardous locations.

It has become possible to quickly reduce data in the field with the
steadily increasing application of microprocessors in the indicators.
Careful visual observations by a trained eye cannot easily be replaced by the electronics technology. Instrumentation is just another available tool and should be used wisely by trained personnel.

The engineer must bear in mind, when planning a monitoring program,
that the overall costs of the slope monitoring program go far beyond
the initial costs of hardware and installation. Obtaining the measurements, reducing and plotting the data for a long-term monitoring program can be significantly more costly than the initial costs of the
instruments installed. Contingency plans are usually formulated at
the outset for dealing with data indicating potential failure by providing some redundancy for critical measurements to increase certainty. The best approach is usually to combine several monitoring techniques to facilitate interpretation of slope behavior. Cost benefit
of monitoring must be evaluated against the costs and risks of failure. This is often as difficult as evaluating slope stability itself.

PLANNING

The planning of instrumentation programs and measurement methods
should be selected early in the development of the project. The value
of a cost effective program to the overall cost of the project should
be estimated. It may sometimes be less expensive to "do nothing"
since excavations of slopes of a particular area have been purposely
designed with a comfortable margin of safety. On the other hand,
areas within the same project may require a very sophisticated observational and instrumentation plan.

A procedure checklist could include the following: Purpose and use
of the instrumentation, for example, may require an exploration program to define groundwater conditions, subsoil and rock stratigraphy
and engineering properties.

From the above, the purpose and type of instruments, and extent of
a monitoring program can be determined. Points to be considered are:
Safety, construction control and assuring the design adequacy of conditions anticipated. Predictions of slope behavior, parameters to be
measured, and ranges for which instruments are available should be
determined. Select where to measure and in what direction for least
interference during progress of excavations and production of the
project. Predict, if possible, the required accuracy and sophistication of the instruments desired.

SELECTION OF INSTRUMENTS

Investigate what type of instruments and components are available
to measure a particular variable needed for slope design. Inquire

from specialist and manufacturer of the instruments, not only the
price and quality of the components, but also availability, durability, simplicity and repeatability of sensors and readout components
during long-term use. Warranty of equipment performance, maintenance
cost for repairs, and availability of parts should be compared for the
instrument system selected. The indicator readout can usually be
rented first and purchased later, which has the advantage of trying
out the unit first. Compatibility with other equipment, particularly
interfacing with "standard", well-known computer facilities or in-
house software facilities and computers are essential to process and
interpret the data quickly.

Do not hesitate to ask for references of clients familiar with
similar equipment, and check performance of instruments with them.
Remember, someone said: "The bitterness of poor quality remains long
after the sweetness of a low price is forgotten."

Consider not only the cost of hardware, but also the difficulty of
installation of components, routing of leads, etc. Usually the readout indicator is the most expensive component of a system offered,
and determines the accuracy with which the whole program can be monitored. Determine if a single backup instrument and redundancy of
measurements to monitor the movements can be used. Consider protection of the components so that damage by excavation equipment and
blasting procedures can be kept to a minimum. Note whether environmental conditions, for example, frost, rain, flooding, etc., may cause
damage to sensors and other parts of the system. It is suggested
that the systems be kept simple, but that components be selected carefully for the best available quality and use. Determine who is to
decide on the implementation of the information obtained and action
required. Ascertain who has overall responsibility when things go
wrong. The cost of these considerations may outweigh the cost of the
hardware many times over for long-term projects of moderate size.

Another very important consideration may be the available skills of
the personnel who will have the responsibility to obtain the data. Do
they need special training, by whom, and how much time is involved to
monitor the instruments? Any instrumentation program should ideally
have the flexibility in its design to allow for modification of alternate programs, so that timely results can be obtained when quickly
needed.

OBJECTIVE AND IMPLEMENTATION OF INSTRUMENTS

Settlement Surveying

Surface settlement and vertical movements of slopes to check stability are measured with conventional surveying equipment. Level
circuits using permanent measuring points and stable benchmarks with
self-leveling features of survey equipment are preferred to speed up

monitoring. Tape and transit-lines can be used for short distances; however, theodolite, laser and/or precision automatic level and elec-tro-optic distance measuring methods can be used for longer distances and higher slopes. Order of accuracies have been proposed to be on the order of 1.5 x (L/2) mm, where L is length of level line in meters.

Photographs taken from easily accessible points at the ground sur-face at certain time intervals would be helpful in mapping original conditions and direction of movements during the progress of the pro-ject under study. (Figure   1.) Sudden anticipated movements, as a result of blasting, can be monitored successfully from a stationary movie camera station with a manual or trigger starter placed across the area. Also aerial photographs are useful to complement movement data records, particularly when certain points and coordinates can be established on the ground surface for future identification. They also make good monthly records.

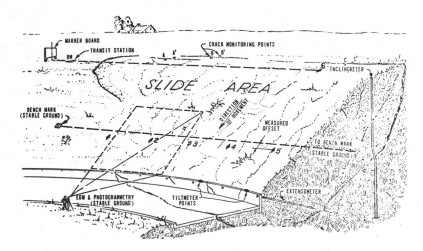

Figure 1 - Movement Measurements in a Typical Slide Area

Surface and Sub-surface Extensometers

To measure surface movements, reference points, i.e., stakes and grouted-in pegs or pipes, placed securely in the soil or rock surface between cracks, can be used. As an example, settlement and heave can be monitored that may have developed because of the instability of slopes on the top and bottom of a failure. Movements between cracks can be measured using conventional survey chains and anchors located from benchmark points behind the cracks unaffected by the movements. Measurements should be periodically repeated and compared with pre-viously recorded threshold base data. More sophisticated installa-tions can connect grouted peg or pipe anchors to (invar) rods or

cables with mechanical, universal swivel joints on the surface of the
ground. Measurements can be obtained very accurately with mechanical
depth micrometers and dial indicators between reference points attached to the rods or cables; however, temperature corrections for rods
and cable length should apply. Rods can also be buried in shallow
trenches protected by, for example, enclosures of polyvinylchloride
(PVC) pipe installed across the cracks and anchored on one end in the
rock or soil. The movement can be measured remotely on the opposite
free-end of the rod with a rectilinear potentiometer or other type of
electrical sensor connected by (electrical) cable to a portable indicator. Movement of slopes should always be monitored in more than
one direction and recorded as accurately as possible, noting the time,
date, etc., to be of value.

Deep sub-surface settlement instruments and extensometer arrays can
be installed in vertical, horizontal or inclined drilled boreholes.
Multiple extensometers can be designed for attaching anchors at different levels by expanding prongs or springs inside the borehole, or
grouted anchors may be used to fasten the anchors permanently. Measurements are usually obtained from the collar-end of the borehole,
using a common reference plate or reference point accessible at the
surface. Stainless steel or invar rods are employed, protected with
(PVC) pipe to connect the collar-end to the anchor-end of the installation placed into a stable reference area. Steel cables are less
popular because of the stretching effect within the cable under tension. Once extensometers have been installed, initial readings are
taken and become the basis for measuring the net changes of movements,
as the project progresses. Readings are taken periodically at pre-
selected time intervals to obtain the changes of the instrument measurements. By comparing movements between adjacent anchors of multi-
ple-position extensometers, it is possible to locate slide planes,
bed separations, failure zones and creep as a result of stress. Particularly when rates of movements of extensometers can be graphically
illustrated and plotted against time, time rates of movements can be
compared and often predicted.

Another special portable tool to accurately measure horizontal and
vertical movements up to 30 m is a tape extensometer, commonly used in
mine openings. It usually consists of an engineer's chain, a dial
indicator, a compression or tension spring or proving ring, and swivel
bearings. The tension can be adjusted to take readings with the same
tension applied to the chain for each reading, e.g., $\pm$20 kg (40 lbs.).
The installation of the system consists of placing opposing reference
anchors of special design across a failing plane. The tape extensometer is attached to one anchor and the chain-end attached to the
other anchor. Spring tension is applied and the reading taken from
the dial indicator. An accuracy of 0.13 mm (0.005 inches) can be
obtained.

Slope Indicators

Horizontal movements and tilt are monitored frequently by installing vertical Slope Indicator Casing in the unstable area. The casings
are placed in vertically drilled boreholes. The annulus between the
casing and the hole can be back-filled with sand, gravel or grouted
with a cement grout. One should use a grout which modulus duplicates
closely the elastic modulus of the material in which it is placed.

The casing is permanently installed in the ground to a sufficient
depth so that the bottom can function as a stable reference point.
Three sizes of casing are available from 49 mm (1.9-inch) O.D. to 85
mm (3.34-inch) O.D. Casings are made from aluminum or plastic, and
are provided with four internal longitudinal grooves for guiding and
orienting the inclinometer sensors used. Couplings are provided to
connect the casings for butt joints or telescoping joints. The latter
are used if movements parallel to the casings are expected (down-drag).
One set of grooves are normally installed in the direction of anticipated movement.

Inclinometer monitoring of the casing is done with a portable sensor, control cable and indicator. (Figure 2.) The sensor is guided
to the bottom of the grooved casing by special spring-loaded wheels.
Readings are taken at regular depth intervals by raising the sensor
with the marked control cable and recording the slope displayed on the
portable indicator. Repeating these measurements periodically provides data on the elevation, magnitude, direction and rate of movements of the casing installed within the soil or rock.

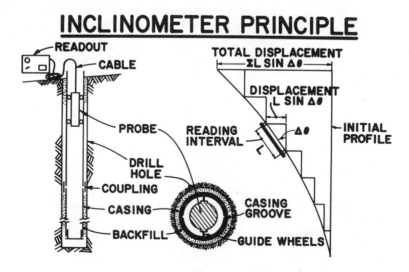

Figure 2 - Inclinometer Principle.

A uniaxial sensor using a pendulum-actuated potentiometer circuit sensor has been used to measure the angle of the sensor axis with its vertical axis. This type of sensor has been successfully used for over twenty years. Other types of bonded wire and vibrating wire circuit principles are used in similar types of sensors to measure angles. The pendulum-type sensors have a sensitivity of one part in 1,000 over the instrument range of $\pm12^{\circ}$ from vertical with a wheel spacing of 12 inches. About twelve years ago, servo-accelerometer tilt sensors were introduced and employed, for example, in the Digitilt line of slope indicators to measure angles of the sensor with a sensitivity of one part in 10,000 over a range of $\pm30^{\circ}$ from vertical. Standard sensor units are available with biaxial readout capabilities, whereby the axes are situated at 90° from each other to measure two directions simultaneously. Special units can measure angle ranges from 0° to 90° with less sensitivity. From monitoring the two opposite sets of casing grooves, the magnitude of horizontal movement and direction can be calculated from vector summing the data. Sensors, cables and indicators are now available in metric or English units. Indicators are portable or stationary units, and display the output of the two accelerometers of the sensor. Readings can show inclination of the sensor in 2 sine $\emptyset$, or displacement in feet or millimeters, at any level inside the casing to be monitored. The data obtained can be recorded manually and the results processed with a small calculator, or the data can be recorded on an integral magnetic tape cassette recorder and the data interfaced through a compatible data terminal modem to a computer. A computer can be programmed to process the data, which is printed out and graphically plotted to display the result of changes from a base reading. (Figures 3 and 4.)

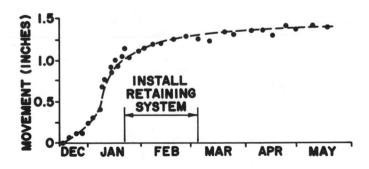

Figure 3 - Movement vs. Time

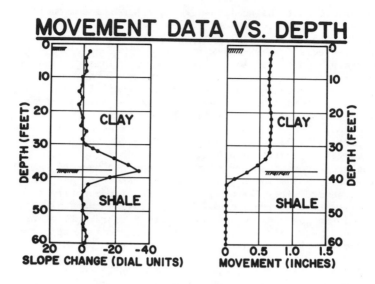

Figure 4 - Movement Data vs. Depth.

The latest improvement in portable indicators is the Digitilt RPP (Recorder-Processor-Printer). This indicator has computer capabilities, and a keyboard to enter commands and data using micro-chip circuits. This data can now be reduced quickly, checked and analyzed in the field. A hard copy printout of the data is provided and a graph can be made with the electrostatic printer in the field, while data is stored automatically on cassette tape. This is a considerable time saving improvement in data reduction.

An alarm-type system can consist of a standard casing installation, which allows several sensors to be placed permanently, interconnected with pipe and universal joints, above each other in the casing. Each sensor measures changes in slope or deformation over its pipe length via cable to a portable readout or an automatic monitoring console. Options include remote readout via telephone lines and an automatic alarm system for immediate warning when a sensor exceeds a preset threshold. A long-term project in the Northwest of the United States (Packwood Hydraulic Project) was monitored with this system. (Figure 5.)

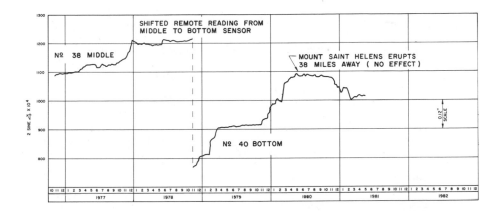

Figure 5 - Remote In-Place Inclinometer Readings (Typical)

Tiltmeters

Tiltmeters are portable uniaxial sensors, which can measure the rotation of an accessible permanent surface point, for example, a rock
outcrop of a moving earth mass, or the side of a building. Special
temperature-stable ceramic reference plates are epoxied to the rock or
structural building surface to form the base plate and measuring
point on which the sensor is manually placed. The sensor utilizes the
same servo-accelerometers and provides the same sensitivity as the
Digitilt system and hence, the same readout indicators can be used.

Water Pressure Measurements

Groundwater and/or pore-water measurements are most important to
determine their driving force, which influences the stability of rock
slopes. The monitoring of water pressures (piezometric pressures) and
flow can provide this information. A piezometer can measure the head
of water or the pore-pressure in the soil or rock at the level of the
sensor at which it is placed.

Standpipe Piezometers

The measurements can be a vertical distance measurement to the
water level in a standpipe. The standpipe is provided with a filter
stone or perforated pipe on the bottom, installed in a drillhole. The
sensing tip is usually isolated from waterbearing stratification above
it with bentonite seals and surrounded by permeable clean sand which
functions as filter material and increases the sensing area.

Several designs of standpipe tips and so-called "Casagrande" type piezometers are on the market, some provided with two tubes or pipes to allow flushing of the tips periodically when placed in fine grained, silty and/or clayey type materials. These piezometers are monitored with a "dip stick" or probe. They consist of an insulated, marked, electrical cable connected to an indicator provided with a voltmeter, light source and/or buzzer, and a DC (battery) power source. The other end is connected to a sensor, which is lowered in the standpipe. When the probe reaches the water surface, contact is made at the tip and current is transmitted through the cable to the meter, and gives a visual or audible signal. The marked, electrical cable and tip will give the distance from the top of the standpipe to the water surface.

Pneumatic-Electrical Transducer Piezometers

Other more sophisticated remote-type piezometers are available. They use an electrical or pneumatic sensor or transducer component with special filters, a readout indicator, a terminal point and electrical cable or pneumatic tubing to connect a terminal at the surface to the sensor installed in the soil or rock. Two distinct methods can be used. The electrical-type piezometers and the pneumatic-type piezometers both have advantages and disadvantages. Cost of pneumatic piezometers is much less than electrical piezometers. Cost of cable and jacketed multi-tubing is about the same. This holds true for the cost of indicators also.

The pneumatic sensors use a balancing diaphragm membrane against the pressure of the water at the tip, which includes a valve-type arrangement. The valve is activated from a gas source with toggle valves from the portable indicator at the terminal location. When the balance of the gas pressure is achieved, the sensor valve is closed. An in-line dial pressure gauge on the indicator can show the balanced pressure and hence the water pressure of the sensor on the dial in kg/cm^2, pascals or pounds per square inch. In some systems, the sensor valve closes when the gas pressure is balanced against the water pressure, but the principle is the same.

The indicator accuracy determines the sensitivity with which the system can be monitored. For long-term monitoring, this is the best and most stable remote system available to date; however, distances between sensors and terminal points in excess of 600 m (2,000 ft.) become more time consuming to read. Flowmeters that measure flowrate of the gas, automatic flow controllers, and the quality and range of the gauge of the indicator are all points that can influence the selection of a pneumatic system. Automatic readout with this system is possible, but expensive.

The electrical piezometer transducer can be of many different electrical design circuits. The most stable, but expensive sensors employ the vibrating wire principle. The basic system of electrical sensor, cable, terminal and electrical indicator is the same; however, readings can be obtained much more quickly than with pneumatic sensors. Also, pore-pressures subject to dynamic impulses or vibrations from blasting, for example, can be monitored. The system is more adaptable to automatic readout at a moderate expense. Disadvantages are hysterisis of the electrical components with time, and sensitivity to temperature changes. Also, protection of the system against lightning is essential. Short circuiting of leads due to a moist or wet environment of connectors and terminals is also more troublesome for the electrical systems.

Installation costs for piezometers, including drilling, mobilization and demobilization costs, can be as high as five to eight times the cost of the hardware for the piezometer system. This comparison depends, of course, on drilling conditions and the quality, type, lengths and depths of the piezometer installation selected.

Acoustic Emission

Geo-Monitors employ sensitive detectors to detect acoustic emission (sounds) from rock and soil. Strain of soil and rock, in an impending failure, for example, produces sound waves, which can be detected with very sensitive transducers and amplified with a readout monitor. A count of events per minute can be a measure of the strain emitted.

Several monitor components with built-in filters to filter out background noise, and event counters, are available. Event counters record each occurrence of an event above an adjustable threshold level. These events can be recorded per one minute or ten minute sample period intervals. Available accessories are tape recorders, storage oscilloscopes and oscillographs. The measuring sensor transducer attached to a cable is lowered in a borehole or crack to measure events. This system also has been used successfully to detect leaks in pipes, and locations of major water piping and seepage in dams. Some background noises are difficult to filter out completely. This system was used on the Thornton Bluffs project in the United States. (Figure 6.)

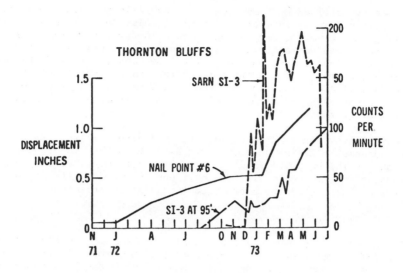

Figure 6 - Event of Acoustic Emission at Thornton Bluffs.

Vibrating Wire Strain Gages

An electric vibrating wire strain gage is basically a steel music
wire strung between two fixed points of the gage components, enclosed
within a small tube. During the reading, the gage is surrounded by a
set of magnets from the pickup sensor. One is used to excite the wire
continuously during the readout period. The other magnet reads the
harmonic pitch or frequency of the wire during the same readout period.
The strain gages can be spot welded to a steel bar or structural steel
member to be measured. The higher the tension of the member, the
higher the frequency that will be recorded. A special indicator records the oscillating frequency, which is displayed in microstrain or
micro-inches per inch. An initial reading of zero strain is needed as
a base reading with which readings are compared during loading or
straining of the unit. The strain of a tieback anchor bar used, for
example, in borehole grouting or bolting to decrease movements of a
ground or rock mass, can be monitored with this type of instrument.
Automatic scanning and reading options for the indicator recorder are
available.

CONCLUSIONS

The instrumentation discussed here is intended only as a brief description of the hardware available today. An instrumentation program
can result many times in a decrease of construction and operating
costs of a project. Stability of slopes can be checked against the

design and movements monitored. It can often give enough warning to protect lives and damage to structures and roadways under the worst conditions of slope failure. Cost of failures, which were not planned, are usually much more costly than the expenses for instrumentation programs. Data acquisition, processing and interpretation are all factors to be considered for estimating the total cost of an instrumentation program. An instrumentation program; however, should not be substituted for human observation and simple measurements. Those persons involved in monitoring instruments will normally give their full attention to the matters at hand, which will benefit the human factor of a project considerably.

Since proper installation is as important as the quality of the instrumentation selected, it is suggested that intensive training of personnel be given by specialists during initial installation of first components, i.e., piezometers in boreholes. At that time, training of data monitoring acquisition, processing and interpretation can be taught also. More details of the instruments discussed can be obtained through specialty consultants and/or instrumentation manufacturers, who will be pleased to be of service.

Question

There have been several efforts made to develop equipment which can measure the twist or rotation of the Slope Indicator Casing once installed. Could you comment on the present state of development of this tool and what success it has had so far.

Answer

We manufacture a Spiral Checking Device, which can be rented or purchased, to check the spiral twist of casing along the length of the installation. This probe is used with the same cable and readout indicator that is used with the Digitilt probe.

Chapter 22

BLASTING TO ACHIEVE SLOPE STABILITY IN WEAK ROCK

G. Harries

Senior Blasting Physicist, ICI Australia Operations Proprietary
Limited, Melbourne, Australia.

ABSTRACT

The mechanism of blasting and the effect that blasting has on
rock properties including the generation of new cracks and the
opening of existing joints is discussed and compared with changes
in seismic velocities. Knowing the extent of blast induced
damage and the overall stable slope angle it is shown that it
is possible to design blasting so as not to induce any damage
behind the toe of the blast area. Practical methods of blasting
to meet this criterion and the results of these blasts are then
discussed.

INTRODUCTION

This paper discusses methods of blasting in order to achieve
the maximum angle at which it has been determined previously
the slope will remain stable. The various ways in which blasting
can affect slope stability and the ways in which blasting can be
modified to reduce these effects is discussed in some detail
with particular reference to blasting in or near a weak shale
which forms the footwall in two large Australian iron ore mines.

The mechanism of blasting and methods of measuring and
calculating changes in the properties of the rock in the final
limits area are discussed. A criterion to minimise damage to
the rock and to enable slopes to be blasted to the final designed
angle is proposed. Methods of blasting to meet this criterion
in practice and some results are then discussed.

497

MECHANISM OF BLASTING

Following the detonation of an explosive the rock around the blasthole is strained. The strains produced in rock by blasting can be calculated by assuming that the blasthole is a thick-walled cylinder of rock subjected suddenly to an internal pressure equal to the pressure the explosive would have if it reacted in its own volume. This pressure is the explosion pressure not the detonation pressure. As the wall of the blasthole is expanded the volume of the explosion gases increases and the pressure falls until equilibrium is reached.

In order to calculate the strain, Young's modulus and Poisson's ratio of the rock have to be known. The explosion pressure and the isentropic path along which the explosive product gases expand have also to be known. A comparison of calculated and measured strains is given by (Harries 1973).

As rocks are strong in compression and weak in tension the tangential component of the strain will create a pattern of radial cracks all around the blasthole; the number and extent of the cracks being determined by the strain induced by the explosive and the tensile breaking strain of the rock. These cracks develop uniformly all around the hole until the hole becomes aware of the existence of a free face. The radial compressive strain wave produced by the explosion is reflected at the free face as a radial tensile wave and tends to open up those cracks to which it is tangential which are the forward facing cracks. The explosion product gases stream into the forward facing cracks extending them until they reach the free face when the burden starts to move.

In practice the explosive induced radial cracks will propagate until they meet natural joints in the rock. Depending on the orientation of these joints they too can be opened up. If the strain is high enough new cracks can be propagated beyond the joints as well as opening and extending them. The maximum distance to which one explosive induced crack can be propagated is considered below.

Criterion Used for Assessing the Effects of Blasting

In order to be able to calculate the effects of blasting on slope stablility, a method which will relate the effects of blasting to some criterion of slope stability is necessary.

A survey of literature has shown that a reduction of cracking, usually by overbreak control blasting, is significant in improving slope stability. Hoek (1977) believes that by overbreak control blasting it may be possible to steepen slopes by 10°.

Dimock and Clayton (1977) have studied the effects of particle velocity but only as an indication of the extent to which rock may be fractured. Vibrations from blasting do not seem to have been considered as directly provoking slope failure – no instance appears to have been recorded. Certainly reducing the extent of cracking will also mean reduced vibration.

It is assumed, therefore, that the distance to which blasting can affect the stability of a slope is the distance to which a blast can produce one crack or extend an existing crack. The creation of new cracks or extension of existing cracks will make the slope less stable. Beyond this distance blasting will not produce any new cracks and the slope will not be weakened any further.

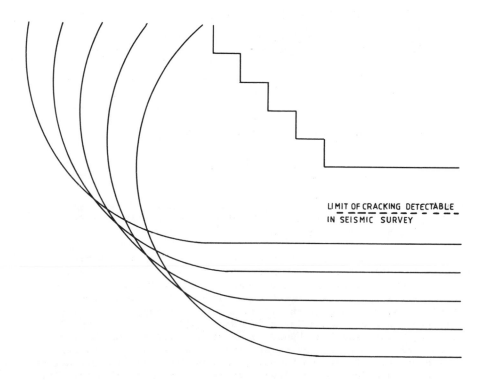

LIMIT OF CRACKING DETECTABLE
IN SEISMIC SURVEY

Fig. 1.   Calculated limit of cracking for multiple hole firings in shale using ANFO density 0.8 t.m^{-3}.

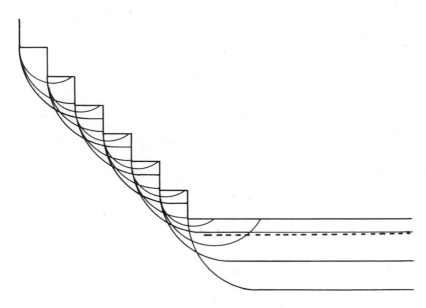

- - - - - - CALCULATED LIMIT OF CRACKING DETECTABLE
IN SEISMIC SURVEY

Fig. 2.   Calculated limit of cracking for single hole firings
with low bulk power ANFO charges between 18-37 m from
the foot of the slope.

Blasting on a particular bench will produce cracking behind
and to the side of the bench and will also deepen the excavation
by the bench height. It can be seen from Figs. 1 and 2 that the
angle of the interface between the cracked and uncracked rock is
dictated by the angle given by the bench width and height. The
thickness of the cracked layer is controlled by the blasting
practice. The cracking should not extend behind the toe of the
final limit. In practice the cracking is not allowed to extend
behind the toe of the bench above, that is, any subsequent blasting
will not weaken any bench already formed on the final limits. This
does not eliminate all damage to the final limits because of the
cracking from the bottom of the charge from the bench above.
This can be reduced by supra-grade drilling.

Given the configuration of the final slope, blasting should be engineered so as not to further damage any benches or berms already created and form subsequent benches or berms with the minimum damage. This can be achieved by restricting the extent of blast induced cracking to the toe of the slope.

To solve the problem, therefore, it is necessary to know the extent of cracking and how this is influenced by various methods of blasting.

The Number and Length of Blast Induced Cracks

The tangential strain ϵ generated by an explosive charge around a blasthole radius b at a distance R from the blasthole has been shown by Harries (1973) to be

$$\epsilon = (Kb/R) \exp(-\alpha R/b) \qquad\qquad -(1)$$

where K is the strain at the blasthole wall and α is the strain absorption co-efficient.

If the rock has a tensile breaking strain T the number of radial cracks N which will be found on the circumference of a circle radius R whose centre is the blasthole is

$$N = (N_o \, b/R) \exp(-\alpha R/b) \qquad\qquad -(2)$$

where $No = K/T$ = number of cracks at the blasthole wall.

The strain at the blasthole wall K can be found from the elastic rock properties (Young's modulus, Poisson's ratio) and the explosive properties (explosion pressure and isentropic path along which the explosive gases expand). The strain absorption co-efficient has been found experimentally to vary from 0.002 to 0.008 (Harries, 1973). A typical graph of crack length versus number of cracks of that length is shown in Fig. 3.

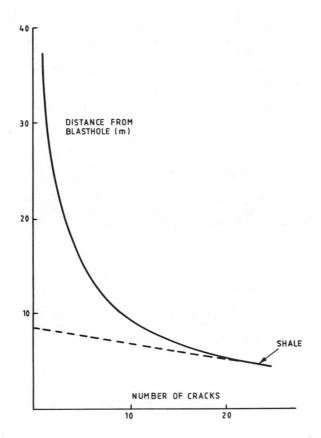

Fig. 3.   Crack Intensity versus Crack Length

Calculation of Particle Velocity

The strain ε and particle velocity V at any point are related by

$$V = \varepsilon\, Vp \qquad\qquad -(3)$$

where Vp is the longitudinal sound velocity in the rock.

The particle velocity at a distance R from a blasthole is therefore

$$V = (VpKb/R)\, \exp(-\alpha R/b) \qquad\qquad -(4)$$

Equation (1) of the paper by Dimock and Clayton (1977) can be written

$$V = C(R/b)^{-z} \qquad\qquad -(5)$$

where $C = K/(\rho\pi)^{-z}$

and ρ is the density of the rock.

Although equation (4) is an exponential equation and equation (5) is a power equation, over the range of interest, equation (4) can be approximated by a straight line on log-log paper (see Fig. 4).

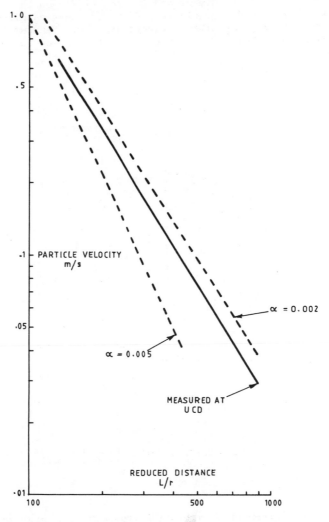

Fig. 4. Comparison of calculated and measured particle velocities.

The curves have been drawn for

K   =   0.0244m/m
Vp =   5000m.s^{-1}
r   =   0.155m

the slope of the straight line is −1.90
when ∝ = 0.005 and −1.45 when ∝ = 0.002.

To put the equation of Dimock and Clayton (1977) on the
same graph R was found from Figure 3 of their paper assuming W
= 272kg and putting b = 115mm. (These are their figures for
230mm diameter blastholes.) Their curve falls between the other
two curves.

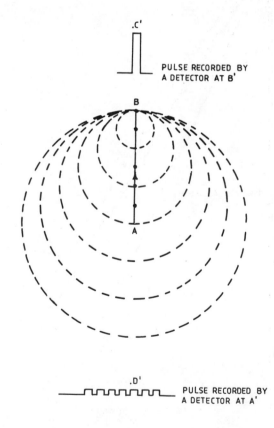

.C'

PULSE RECORDED BY
A DETECTOR AT B'

B

A

.D'

PULSE RECORDED BY
A DETECTOR AT A'

Fig. 5.   Reinforcement of vibration from single line of charges
           when rock and detonating cord velocities are equal.

Firing With Multiple Charges

Consider first the case of a row of charges fired in a rock with a longitudinal sound velocity of $7000m.s^{-1}$. The charges are connected by detonating cord with a velocity of detonation of $7000m.s^{-1}$. It can be seen from Fig. 5 that if the detonating .cord is fired from A to B then the pulse recorded at D will have a much lower amplitude (maximum particle velocity) than the pulse recorded at C.

More generally if the rock velocity is a fraction X of the detonating cord velocity the line of maximum amplitude is that normal to a line at an angle of $sin^{-1}X$ to the line of the charges as shown in Figure 6 where the rock velocity is $5000m.s^{-1}$ (5/7 of the detonating cord velocity).

There will be, in general, two directions in which the amplitude is a maximum. This will be the sum of the individual amplitudes from each blasthole. When firing en echelon at 45° one of these directions will be behind the blast (see Fig. 6).

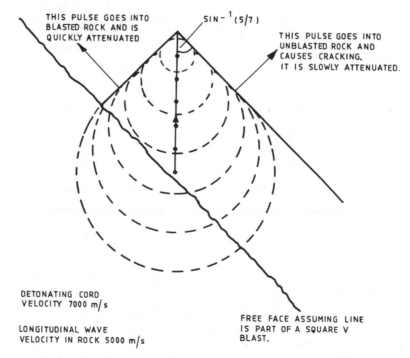

Fig. 6.   Reinforcement of vibration from single line of charges when rock velocity is less than detonating cord velocity.

The amplitude varies with the number of blastholes fired. For a seven row blast the amplitude of the strain wave in these two directions can be seven times that expected from the firing of a single blasthole. From equation (4) the distance to which this reinforced strain-wave can cause cracking is 51m.

As the rock can be cracked to distances of more than 50m by normal blasts the present system of blasting should not be used within 55m of the final wall. (Dimock and Clayton have used a particle velocity of $0.33m.s^{-1}$ and not more than 4 blastholes fired simultaneously.) When blasting nearer the final wall all blasts should consist of single blasthole firings. This practice can be achieved with down-the-hole delay detonators as used by Dimock and Clayton, (1977). Alternatively, detonating relay connectors can be used (see Fig. 7) to give a blast of any size in which no two blastholes will fire simultaneously.

APPLICATION TO BLASTING FINAL LIMITS IN MT. McRAE SHALE

These concepts have been applied to the blasting of final limits in two large Australian iron ore mines. At both these sites the haematite is conformably underlain by a weak shale known locally as the Mt. McRae Shale which is the footwall at both mines.

Typical properties of this shale are:

Young's modulus	11.0GPa
Poisson's ratio	0.35
mean ultimate compressive strength	12MPa
tensile breaking strain	38μ strains
mean angle of friction	26°

Seismic surveys were made at two sites across benches which had been created by known blasts.

At one site where the shale had been blasted using ANFO (density $0.8t.m^{-3}$) in 310mm diameter blastholes and seven blastholes/delay the shale showed a low velocity band up to 23m thick.

At a second site where the shale had been blasted with ANFO in 380mm diameter blastholes with an average of 12 blastholes/delay seismic surveys gave evidence of very low velocity layer about 5m thick and a lower velocity layer to a depth of 40m.

It can be seen from Fig. 3 that there is a sharp change in the number of cracks versus length of the cracks graph at the point where the number of cracks is 11. Assuming that this marked change in slope gives the degree of cracking which characterises the usually marked change in velocity found experimentally it

can then be calculated that 1 crack could extend to 100–150m.
It is also assumed that the extent of cracking to the side of a
blast is the same as that below it.

TABLE 1

Effect of ANFO Density on Cracking from 310mm Blasthole

ANFO Density t.m^{-3}	Blasthole Pressure kbars	Distance to which one crack extends m	Distance to which 11 cracks extend m
0.82	24.9	37.3	8.56
0.80	23.7	37.5	8.66
0.75	21.1	36.5	8.24
0.70	18.6	35.8	7.98
0.65	16.3	34.9	7.65
0.60	14.2	33.9	7.27
0.55	12.2	32.7	6.85
0.50	10.4	31.3	6.37
0.45	8.8	29.8	5.84
0.40	7.3	28.0	5.29
0.35	6.0	25.9	4.67
0.30	4.8	23.6	4.03
0.25	3.7	21.0	3.36
0.20	2.8	18.0	2.68

Using the above rock properties the distances to which 1 and
11 cracks extend have been calculated in Table 1 for a range
of ANFO densities in a 310mm diameter blasthole. For ANFO at a
density of 0.8t.m^{-3} the distance to which 11 cracks extends is
8.66m. This distance is less than measured. These blasts,
however, consisted of firing 7 blastholes simultaneously. Using
the square root scaling law recommended by Nicholls et. al (1971)
the distance over which the vibration and the strain and hence
the cracking would be expected to extend is $(7)^{\frac{1}{2}} \times 8.66 =$
22.9m which is good agreement with experiment.

As all distances are scaled to the blasthole radii for
12 x 380mm diameter blastholes the distance to which cracking
would be expected to extend is $(12)^{\frac{1}{2}} \times 380/310 \times 8.66 = 36.8$m
which is also in good agreement with experiment.

The distance to which 1 crack would extend is 100m when using
7 x 310mm diameter blastholes and 150m when using 12 x 380mm
diameter blastholes. It is known that attempts at pre–splitting
50m behind a blast were unsuccessful. As the rock has been
fractured by previous blasting to at least this distance the
failure of pre–splitting would be expected.

It is obvious that firing one blasthole/delay will have a marked effect on reducing vibrations and cracking. However, even firing one blasthole/delay when using ANFO at a density of $0.8t.m^{-3}$ will still give cracking or open cracks and joints to a distance of 38m. In view of the effects of natural joints it is unlikely that one explosive induced crack will be propagated to this distance but the strain induced by blasting could open up and even extend joints to this distance.

To reduce the extent of cracking further it can be seen from Table 1 that explosives with a lower explosion pressure than ANFO should be used.

Explosives Type

ANFO is too powerful an explosive for the Mt. McRae shale. It can be seen from Table 1 that the bulk power of the explosive has to be drastically reduced before a diminution of strain and cracking is noticeable. To reduce the extent of cracking to somewhere near the dimensions of a reasonable berm the ANFO density has to be decreased to $0.2t.m^{-3}$.

It has been found more effective to reduce the density of the explosive rather than reduce the strength of the ANFO by reducing the fuel oil. Reducing the fuel oil reduces the weight strength of the explosive but not the density. Using ANFO 98.5/1.5 the available energy is $2293J.g^{-1}$ compared to $3780J.g^{-1}$ or ANFO 94/6 and the densities will be almost identical. ANFO 98.5/1.5 will give 1 crack extending to a distance of 33m which from Table 1 is equivalent to normal ANFO 94/6 at a density of $0.55t.m^{-3}$.

The extent of cracking, which scales to the blasthole diameter, can be reduced by using smaller diameter blastholes and normal ANFO 94/6 at a density of $0.8t.m^{-3}$. This practice does not, however, reduce the blasthole pressure and the compression of the final wall. This pressure can force gas into the backward facing cracks and when the burden starts to move and the gas vents cracking by release of load can also occur. As reducing the density of the explosive also reduces the blasthole pressure it reduces the intensity of all the mechanisms which can cause cracking.

Low Density Explosives. There are three ways of reducing the density of an explosive in the blasthole. They are all equivalent theoretically.

(1)   using foamed polystyrene beads. Three volumes of polystyrene and one volume of ANFO will give a density of $0.2t.m^{-3}$. This system is used underground in large diameter holes to protect cemented rock fill in adjacent filled stopes.

(2) using air decking. Short decks of ANFO separated by air decks
     can also give low effective densities. Provided the decks
     are short the explosive will not run up to full VOD (velocity
     of detonation) and although the decks are fully coupled the
     expected explosion pressure will not be attained and the
     pressure will rapidly decay to the pressure which would be
     realised by the effective density which includes air decks.
     In small diameter blastholes <165mm detonating cord downlines
     with a charge weight of $10g.m^{-1}$ can cause side initiation
     of ANFO at low VOD.

(3) decoupling the charge. Putting the explosive in a smaller
     diameter pipe also reduces the effective density. A
     cylindrical charge with a diameter of 150mm in a 310mm
     diameter blasthole has an effective density of $0.2t.m^{-3}$.

The first two alternatives have been used for Mt. McRae Shale.

Blast Design

Drilling Pattern. In the shale normally charged blastholes fired
one hole 310mm diameter/delay with 7m burden, 9m spacing drilled
staggered are used when they are more than 37m from the foot of
the slope. These charges will create 11 cracks to a distance
8-9m behind and to the side of the blasthole. This cracking is
intense enough to allow the shale to be dug by the shovel leaving
30m to be blasted. The pattern used for the 310mm diameter
blastholes charged with ANFO/polystyrene (at a density of 0.2
$t.m^{-3}$) is 3.5m burden, 5m spacing drilled staggered and this
pattern should be drilled between 18m and 28m from the foot of
the slope.

Initiation Sequence. It has been shown above that firing one
blasthole/delay will significantly reduce vibration. A large
number of blastholes can be fired singly using detonating relay
connectors. The sequence shown in Fig. 7 has been successfully
fired using 310mm diameter holes charged with ANFO/polystyrene
(density $0.2t.m^{-3}$) with a burden of 3.5m and a spacing of 5m.

Stemming. To reduce vibration to a minimum the low density
blastholes should only be lightly stemmed, 3m should be sufficient.

Sub-grade. There should be no sub-grade drilling in the shale.
The last two rows should be drilled to 1m above grade.

Double Benching. In order to maximise the slope angle two 15m
benches can be joined together. The shovels usually dig 3-4m
behind the final toes and the final batters are contoured with
bull-dozers. At both mines 30m high benches have now been mined.

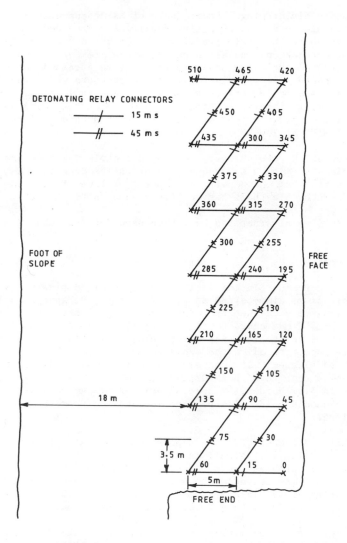

Fig. 7. Initiation sequence — one blasthole per delay — low bulk
strength ANFO.

The overall powder factor used in the final blasts at both
sites whether using ANFO/polystyrene or ANFO/spacers is
0.12kg.t^{-1}. The overall powder factors for all blasting at both
sites is 0.23 - 0.26kg.t^{-1}.

GEOLOGICAL STRUCTURE

In order to achieve the most stable final slope the final design blasts are blasted parallel to the strike. For a bedded deposit exhibiting relatively low angles of friction, the design of necessity lies slightly below the mean dip of the bedding planes. It is therefore essential that the blasting does not undercut these bedding planes. Ideally the final limit blasts should be three rows or less and the toe should be cleaned up. This will ensure that the burden can move freely and prevent the explosive gases pressurising the backward facing cracks and so extending them. If the burden does not move at all bulling of the hole will occur causing undercutting of the structure, bad fragmentation and difficult digging.

CONCLUSIONS

It has been shown that the calculated strain and the related crack length and particle velocity are in reasonable agreement with seismic velocity surveys and particle velocity measurements. From these measurements an estimate of the distance to which blasting can weaken slopes either by creating new cracks or opening existing cracks has been made.

A method has been developed to relate this distance of rock damage to the desired overall slope angle by not allowing blast induced damage behind the toe of the final slope.

It has been shown that firing one blasthole/delay has a marked effect on the degree of blast induced damage. This cracking can be further reduced by using low density explosives, decoupling or air-decking.

Using these concepts successful final limit blasts have been fired in a weak shale at two large iron ore mines in Australia.

ACKNOWLEDGMENT

The author extends his thanks to ICI Australia Operations Proprietary Limited for permission to publish this paper.

REFERENCES

Dimock, R.R., and Clayton, G.D., 1977. Kenecott's Delayed Blasting Techniques Cuts Costs, Improves Pit Stability. Mining Magazine pp.37-40.

Harries, G., 1973. A Mathematical Model of Cratering and Blasting, Aust. Geomech. Soc. Nat. Symp. on Rock Fragmentation, Adelaide. Papers pp.41-54.

Hoek, E., and Bray, J.W., 1977. Rock Slope Engineering. The Institute of Mining and Metallurgy, London.

Nicholls, H.R., Johnson, C.F., and Duvall, W.I., 1971. Blasting Vibrations and Their Effects on Structures. U.S. Bureau of Mines Bulletin 656.

Question

You were striving for very low density ANFO for reasons of back-fracture reduction. Would this not require significant increases in drilling (tighter patterns) and costs to achieve necessary fragment-ation.
If so, how would you justify these higher costs to a tight-fisted mine manager.
Have you noticed any problems on these extensive patterns with inacur-acies in the short period delay times (15 msec. delays).

Answer

2-3 times more holes have to be drilled - explosive costs are reduced by about 1/3.
No increased digging costs as fragmentation is just as good. Tighter muckpile, less clean up, less shovel movement. Cheaper than pre-splitting - no new drills etc.
As long as there are multiple paths to each hole, no trouble with delays.
What is the cost of not doing anything about it?

Question

Did you find the slope crest damage was significantly effected by the type of stemming used, e.g. drill cuttings vs crushed rock aggregate.

If you did not experiment with type of stemming what is your evaluation of its possible significance.

Answer

Experiments with stemming in progress. Normally in production blasts we would use 7m of drill cuttings. In final blasts we only use 2-4m. We will be trying no stemming. We will have to use air decking.

Question

Have you tried using decoupled charges to reduce the peak borehole pressure.
If so, what were the results.

Answer

Not tried - were considered but considered too labour intensive.

Question

When you deck the ANFO in effect shortening the column, do you find that you are approaching simulated spherical charges. (Max L/D 6)

Answer

The individual charges are approaching cratering charges but are distributed up the hole.

Question

As to slope stability would you consider small grid more favourable to large grid.

Answer

Better to have closely spaced grid with lightly charged holes. Aim to get best 3D distribution of explosive.

Question

In your seismic surveys, how did you differentiate damage from blastholes with stress release damage around the pit.

Answer

We don't try to differentiate between damage from blasting and that caused by stress relief. From experience with extensometers, underground rock movement seems to occur for an hour after blasting.

Question

What is the handling quality of the muck after reducing the powder density (muck size). See Answer to first question

Question

Please describe the powder column of the loaded hole - before and after charging design.

Answer

Before Charge	After Charge
7m burden	3.5m burder
9m spacing	5m spacing
17.5m holes	14m holes
7m collar	7m collar
0.8gcm^{-3} explosive density	0.2gcm^{-3} explosive density

Question

Do you think this technique suggested by you can be used in porphyry copper mines production blasting say of 15 meter bench, 254mm to 305mm blasthole diameter.

Answer

Yes.

Question

In your experience does porphyry copper mines rock strength which usually varies between 12000 to 20000 psi have a P-wave velocity of 10,000 ft/sec to 15,000 ft/sec.

Answer

Yes.

Question

What blasting mechanisms (e.g. borehole radial cracking, extension of existing cracks by detonation gases, etc.) would you say are responsible for the extensive weaker zones of rock that you indicated outside of the blast area.

Answer

Definitely not cracks from holes. Extension of cracks greater than 6m long -Griffiths Crack Theory.

Chapter 23

BLASTING PRACTICES FOR IMPROVED COAL STRIP MINE
HIGHWALL SAFETY AND COST

Francis S. Kendorski and Michael F. Dunn

Engineers International, Inc.

Downers Grove, Illinois

ABSTRACT

The fall of rock from strip coal mine highwalls continues to be
the largest single source of fatal accidents, so methods to improve
highwall stability through improved blasting practices were
investigated in the field.

This paper covers the results of a project to improve, using
existing and proven technology, unstable highwall conditions through
better blasting practices in Appalachian strip coal mines. Better
borehole blasting agent distribution and optimization of burden,
spacing, and delay periods allowing satisfactory breakage with
minimal backbreak were the objectives. At a selected field site,
geologic conditions were closely examined with regard to joint
orientation and spacing, and resulting slope stability considerations. Utilizing surface blasting concepts of trying to achieve
flexural failure of the rock mass both horizontally and longitu-
dinally by appropriate relationships between burden, spacing, and
hole length, as well as delay time sequence to achieve optimal
confinement and breakage, successful results were achieved. A test
series of eight blasts was completed and the resultant highwall
monitored by repetitive photography. Actual blockfalls could be
measured and the stability for each blast design thus quantified. The
study of drilling and blasting economics along with equipment
performance led to an economic model of blast parameters and the
identification of blast design cost benefits.

The work demonstrated that small strip mining companies are able
to achieve a more stable highwall excavation utilizing presently
available technology, equipment, and explosive products at improved
operational cost and safety.

515

INTRODUCTION

Background

Contour strip coal mining creates a highwall of unexcavated rock
rising above the work area. Any unstable blocks and rock masses on
the highwall threaten the miners' safety and efficiency. One method
to improve the highwall stability is to minimize the blasting damage
to the rock using a blasting technique that gives an undamaged,
stable highwall and properly fragmented overburden.

A prior study (Barry, 1974) listed falls of highwall as the
largest single cause of fatalities in surface coal mining. The goal
of Engineers International, Inc.'s work was, therefore, to demonstrate blasting techniques to reduce the creation of hazardous
highwall conditions while remaining within acceptable economic
limits.

Accidents

Rock falls, by their very nature, could be expected to cause great
bodily harm, which is borne out by Figures 1 and 2, comparing the
sources of fatal and nonfatal accidents. Falls of highwall that lead
to an accident are much more likely to lead to a fatal accident than
an injury.

It was observed that 35% of the accidents were preceded by foul
weather such as heavy rain, or freezing and thawing. Another 25%
occured during seasonal periods of unfavorable weather.

CONTOUR STRIP MINING OF COAL

Method

In areas with a flat coal seam and mountainous topography, the
coal is mined around the outcrop line and back into the hill until
too much overburden must be removed to realize a profit from coal
sales. Present reclamation regulations require backfilling, grading,
and re-seeding the mined area, so that there are many different
mining methods that can be used. The most common method, haulback
contour strip mining, employs trucks and loaders to handle the
overburden and haul it to the reclaiming area. Mining begins along
the cropline of the coal and continues into the hill, Figure 3, with
the initial overburden spoiled, the coal removed, and the spoil from
the second pit being placed in the first pit. The successive pits
progress into the hillside until the economic limit of the operation
is reached.

Contour strip operations are generally small scale and often
operated on a "father and son" basis, though larger companies may
have a great many separate such mines. Accordingly, equipment used

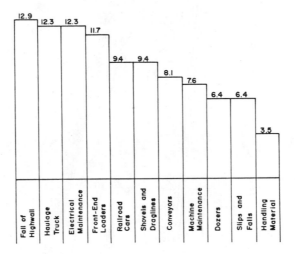

Fig. 1. Percent fatal accidents by type of accidents. From Barry (1974).

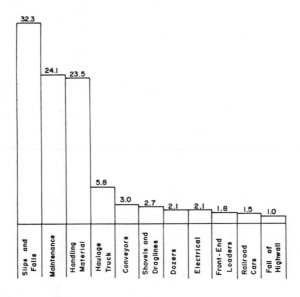

Fig. 2. Percent non-fatal accidents by type of accidents. From Barry (1974).

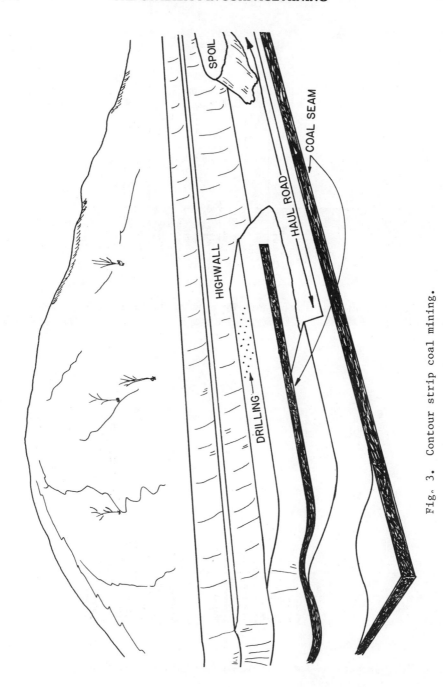

Fig. 3.   Contour strip coal mining.

is variable and often not adequately sized to the job, but was
available at a good price. Thus the situation may arise where the
mine is tailored to the equipment rather than the reverse as more
commonly practiced in larger mines.

Where topographic conditions are suitable, the mountaintop removal
method can be employed. This is accomplished by first making a
conventional contour cut around the hillside, and then working
successive cuts, keeping with a fixed orientation, while spoiling
into the previous cut as in area type coal mining. This sequence is
continued until the entire mountaintop has been removed as shown in
Figure 4.

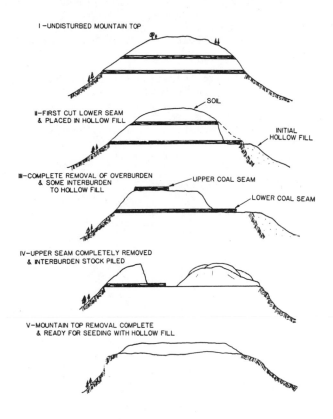

Fig. 4. Mountaintop removal coal mining

In general, contour strip mining and its variations account for
24% of the coal mined by surface methods in Appalachia. The related

mountaintop removal method accounts for another 20% of Appalachian surface coal mined (Cook and Kelly, 1976).

Blasting Practices

A few generalities can be made about contour mining blasting practices used in the Appalachian region. Hole diameters observed range from 130 mm to 230 mm. Vertical holes are drilled until coal appears in the cuttings and the holes are then backfilled a small amount with drill cuttings to protect the coal from the blast. Hole patterns can have burdens and spacings with a square pattern of 3.1 m x 3.1 m or 3.7 m x 4.3 m, or even larger. Most operations use ammonium nitrate and fuel oil (AN-FO) as the primary blasting agent. AN-FO is used in the form of 23 kg bags which are slit and their contents poured into the holes, or in a few cases bulk AN-FO loaded from trucks is used. For wet holes, AN-FO is prepackaged in cylindrical plastic bags.

Most operations use electric blasting caps, some in conjunction with a sequential timing device that will expand delays between holes. Cast primers or sticks of gelatin dynamite are used as primers sometimes attached to a length of detonating cord running the length of the drill- hole. Sometimes extra primers are randomly added to "sweeten the hole" or provide a "kick." Figure 5 illustrates a typical borehole in cross section.

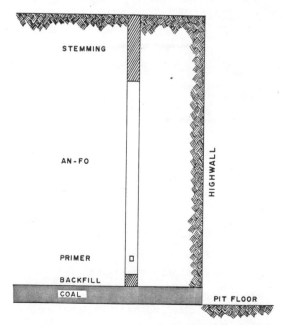

Fig. 5. Typical strip coal mine blast hole.

In most cases side or face benches are heavily buffered with previously shot rock not providing adequate relief for rock movement. The buffer may be purposeful to help maintain loading equipment operations when the drill breaks down.

Highwall Instability

A highwall in contour strip mining of coal is an excavated rock slope and the principles used in rock slope stability analysis can be applied. The stability of rock slopes has been found to be controlled by the orientation and mechanical characteristics of joints, faults, bedding planes, and other discontinuities in the rock mass. Sliding along these planes of weakness or intersections of planes (wedges) is the principal cause of instability in most highwalls.

In general, falls of rock are most prevelant after rainstorms, since the flowing water may wash out fracture fillings or hydrostatic forces may change the stress situation.

Mining equipment can be used to deal with highwall instability. A loader bucket can scrape or hit the highwall or a miner in the bucket can use a scaling bar. Loose rocks near the crest will be hard to notice from the coal level and remedying the situation may be a major under- taking. Repeated scraping of the highwall with a dragline bucket can clean the crest of loose rock, but only within the dragline's immediate reach. Clearly, such scaling operations are very hard on the equipment, dangerous for the equipment operator, and non-productive.

MSHA and state mining regulations deal specifically with highwall conditions owing to the accident records clearly indicating a problem. Persons are not supposed to work near unstable highwalls, and loose rocks are to be removed. At some of the operations visited during this project, the miners constantly watched the highwall, and would not approach closely, thereby decreasing operational efficiency.

Present Practices

Project engineers and geologists visited nine contour strip coal mining operations in Appalachia, and collected details on their mining and blasting practices, as well as their highwall experiences.

Figures 6, 7, and 8 show some of the practices and conditions observed at these mines, where blasting quality is given little attention.

Table 1 summarizes the mines visited and their blast parameters. These figures are based on designs reported by the companies and not upon those actually measured in the field. In Table 1, various ratios among the geometric parameters of the blast designs are given. Acceptable values based upon experience (Dick, 1976) are also given,

and may be used to judge the blasts expected efficiency and
effectiveness.

Fig. 6. Typical highwall conditions in an Appalachian strip
          coal mine.

Using these ratios as a guide, very few design parameters of the
operations visited fit into the acceptable ranges for good results.
Hence, design parameters used should be revised. In general,the
bench blasts observed did not provide adequate relief for rock
movement, and the explosive energy was poorly distributed in the
borehole. The burden and spacings as accurately measured in the
field show very little consistency with the blasting pattern provided
by the company, in (Figure 9) since these holes were laid out by
pacing, "eyeballing," or using the drill trucks as a measuring scale.

Field Site

    The Appalachian mine that was selected for this project (no. 5 in
Table 1) was the only one visited that used a dragline for overburden
removal. Management was very cooperative, and the company was in the
process of installing a computer system for cost accounting. It was
thought that this would prove beneficial for a cost analysis of the
effects of blasting changes in the operation. Factors considered in
the selection process were that the site chosen may to be able to
benefit from the project; geology was not too unfavorable; the
existing blasting practice appeared reasonable; and the labor was
competent and stable.

Fig. 7. Unstable con-
ditions observed in an
Appalachian coal strip
mine highwall.

Fig. 8. Drilling a "vertical" blasthole at an Appalachian strip
coal mine.

TABLE 1. Blasting Parameters

MINE OPERATION	HOLE DIAMETER (mm)	BENCH HEIGHT (m)	BURDEN (m)	SPACING (m)	STEMMING (m)	POWDER FACTOR (kg/m³)	BURDEN / HOLE DIA.	STEMMING / BURDEN	SPACING / BURDEN	HOLE DEPTH / BURDEN	FACE MIS-ALIGNMENT	ACTUAL MILLISECOND DELAY PER METER OF BURDEN	MINIMUM MILLISECOND DELAY PER METER OF BURDEN
1	170	7.0	2.5	3.0	5.5	0.35	16.07	1.94	1.11	2.44	45°*	36.5	3.6
2	170	5.0	3.0	3.0	3.3	0.65	17.86	1.10	1.00	1.70	28°	16.4	3.6
3	170	15.0	3.5	4.3	3.3	0.98	21.43	0.92	1.17	4.08	42°	17.7	4.3
4	170	8.5	3.5	4.0	3.3	0.56	21.43	0.92	1.08	2.33	0°	0.07	3.9
5	220	21.0	3.5	3.5	10.0	0.98	16.46	2.75	1.00	5.67	44°	27.2	4.3
6	230	11.0	3.5	4.5	4.5	0.90	16.00	1.25	1.25	3.08	42°	22.6	3.9
7	170	20.0	3.0	3.5	3.3	0.47	17.86	1.10	1.20	6.50	15°	9.8	4.9
8	130	9.0	3.5	3.5	3.0	0.59	27.43	0.83	1.00	2.50	12°	6.9	3.3
9	170	2.5	3.0	3.5	1.8	0.40	17.86	0.60	1.20	0.90	0°	8.2	3.3
DESIRABLE RANGE:							20 to 40	0.7 to 1.0	1 to 2	1.5 to 5.0	0°		

*No Favorable Alignment with Existing Geologic Structure

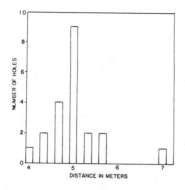 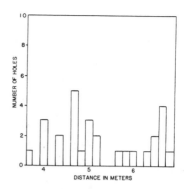

Fig. 9. Actual measured burden distances, left, and spacing
distance, right.

The highwall at the field site contains pronounced joint trends
that form wedges of rock that fall from the face. Figure 7 shows a
rather dramatic example. In order to fully understand the geologic
setting, the rock mass characteristics were examined in detail. It
was learned that the highwall was originally oriented along a major
joint trend, but the strata dipped out of the face at a shallow
angle. Inspecting authorities thus suggested that mine management
re-orient the highwall to avoid this problem, but in so doing cut
across both prominent joint trends at about 45°.

BLASTING MECHANICS

Blast Designs

In consultation with the Project Blasting Consultant, Dr. R. L.
Ash, and the Bureau of Mines TPO, Mr. L. R. Fletcher, EI decided that
the best approach to producing stable highwall conditions using
readily available equipment and techniques would be to first work
with the operator to achieve high quality blasting practices, and
then, utilizing the flexural failure concepts of Ash (1973) to
produce a stable highwall with minimal overbreak.

The flexural failure concept of Ash (1973) is illustrated in
Figure 10 along with descriptions of the basic geometric terms used
in this paper. Referring to Figure 10, when the explosion-produced
gases begin to expand and push the rock out, if the burden and hole
length are such that the block of rock formed in the face area is a
slender beam, rock mass failure is facilitated, since failure is
largely in tension where rock is weakest. If the burden and length
are similar, a rather square block is formed, which must fail in

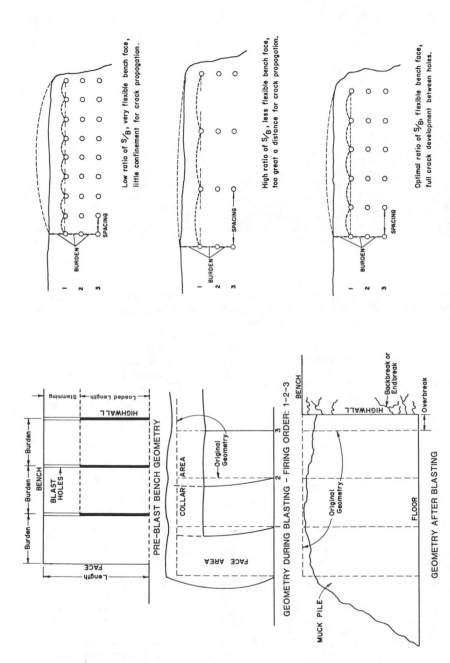

Fig. 10. Flexural failure concepts of bench blasting. Left, cross-section. Right, plan view.

compression and shear under which rock is stronger. The same concept can be applied to the collar area, so that the unloaded upper bench reaches flex during blasthole detonation and the rock readily breaks up. If most of the charge were located at the hole bottoms, the collar area would be inflexible and difficult to break up.

Whereas the ratio of hole depth to burden controls the flexibility of the bench in two dimensions, the ratio of spacing to burden extends the control to the third dimension. Any beam not only has a thickness (burden) but it also has a width (spacing), as well as a length (hole depth or bench height). Also, radial fracturing must develop from each hole, and breaks the rock between holes. A high spacing to burden ratio (Figure 10) helps confine the blast by decreasing flexibility and encouraging cracking, while a low ratio increases flexibility and deemphasizes between-the-hole cracking owing to less confinement.

Blast Design Ratios

Blast design concepts were aimed at satisfying proven and accepted ratios among the geometric blast parameters. Figure 10 should be referred to for sketches of the blast geometry.

The ratio of burden to hole diameter controls the extent and direction of radial fractures for a hole or row of blast holes. A satisfactory ratio is needed also to improve rock movement while still controlling the heave and throw of the muckpile. In general, the lower the ratio, the less control there is over explosives distribution. Also, the chances of violence, lack of confinement, and excessive explosives use increase. A ratio between burden and hole diameter of 20 to 40 has been found satisfactory (Dick, 1976).

The ratio of stemming length to burden contols prevention of cratering at the hole collar. Also, referring to Figure 10, excessive stemming length may result in an inflexible hole collar region. A ratio of stemming length to burden of 0.7 to 1.0 has been found satisfactory.

The hole depth to burden ratio is a measure of bench flexibility. The more flexible the bench (the higher the ratio) the better the rock fragmentation. However, the explosive length in the hole must not be too short (as would happen in a large diameter hole) or flexibility is lost as the almost point source charge acts as a cratering charge. This ratio of hole depth to burden should not be less than 2.0 (Ash, 1973) or more than 5.0 (Dick, 1976) for satisfactory results.

For bench face flexibility and charge confinement along the face length, the spacing to burden ratio must be optimized. For satisfactory results considering both the effects of flexibility and confinement, Dick (1976) has found a range of 1.0 to 2.0 as suitable, while Ash (1973) finds a ratio of 1.4 optimal for holes over 130 mm in diameter.

Face Misalignment

A blasting parameter that is exceedingly important to highwall stability as well as to satisfactory breakage is face misalignment. Ash (1973) recognized its importance, as have most quarry operators. This parameter is the angle in degrees that the face alignment differs from the contolling geologic structures. When only one structural trend such as jointing occurs, the best misalignment is 0°, the maximum is 90°. In Appalachia, most often there are two contolling structures approximately at right angles to one another, so there are two "best" directions and a maximum misalignment of 45°. If the controlling structure parallels the face, rock breakage is improved and a cleaner, tighter highwall face may result since there exists a plane to which the rock may break.

Timing and Relief

For good blasting results, the rock mass blasted by each hole or row of holes needs space in which it can move as well as to give a free face for the rock mass behind to be broken. The delay plan must include this or else the rock mass in back rows of holes will be hemmed in, and the explosive forces will cause backbreak and flyrock. If the front row of holes does not have enough relief because of buffering with previously shot rock or too large a burden for the delays used between rows, then the problem is compounded from the beginning. In general, approximately 3 milliseconds per meter of burden has been found adequate to allow a long enough time for relief by removing the rock in front and a short enough time to make sure the rock in front is still moving and not piling up at the face.

BLASTING TESTS

Implementation of a good blast design is as important to a successful blast as the design itself. Wandering drill steel, haphazard hole placement, or holes not drilled deep enough can create poor highwall conditions by creating improper and irregular explosives distribution that leads to violence and overbreak. Therefore, the first step for achieving a good blasting demonstration would be to use the operation's present design and ensure its correct application.

During the field tests, borehole locations were marked ahead of the drill using Brunton compass and tape and spotted with a painted rock. The verticality of the drill mast was also controlled, as an angled hole would result in improper explosives distribution at the bottom. This control is necessary in order to analyze the results from a given design rather than from driller workmanship.

Hole depth was also noted so as to make adjustments in the powder column and stemming regions; especially if particularly hard or soft strata are encountered, or if the hole caved after drilling.

We observed two typical blasts at the field site that had no control from project personnel. Two more blasts were laid out and controlled by project personnel using the mine's stated design. These two controlled, but not redesigned, blasts wee termed "Quality Assurance" blasts. If sufficient improvement could not be obtained, then the blast design would have to be altered. These Quality Assurance tests were the first small step toward achieving highwall stability.

The two typical mine blasts and the two Quality Assurance blasts are summarized in Figure 11 showing the layout of the four blasts with the numbers referring to the millisecond delay number. Table 2 lists the appropriate blast parameters. Overbreak occurred from each blast and fragmentation appeared to be about the same. From these four blasts it was clear that the mine's design would have to be altered to achieve a more stable highwall.

Constraints on Test Blasts

Before describing the planning and execution of the test blasts that were carried out to achieve improved highwall stability, the constraints imposed on the project by the mine's operations must be understood. The mine's existing practices resulted in a very ragged and uneven highwall that had considerable damage to the highwall rock. Thus the first row of holes often had variable burden and the possibility of venting through open cracks. Also, large blocks of rock remained in front of the face and could not be drilled. These then "went along for the ride" and ended up in the muck pile.

The highwall is over 300 m long and it takes the dragline many months to progress along the full length. Therefore the results on the middle and lower portions of the highwall could not be examined and the next blast revised within the project schedule. Blast results were consequently judged based upon blast observations, muck pile characteristics, and the condition of the upper part of the remaining highwall.

The mine labor did not appreciate the importance of quality assurance and blast hole delay timing techniques to achieve optimum confinement and breakage, and unless our crews were present immediately readopted their usual practices of estimating hole locations, shooting with no relief, closely spacing holes, and multiple priming.

The mining company consistently used a drill capable only of 230 mm diameter holes or larger that resulted in overloaded holes and poor powder distribution. A smaller diameter drill was made available on a temporary basis when not needed elsewhere, and we

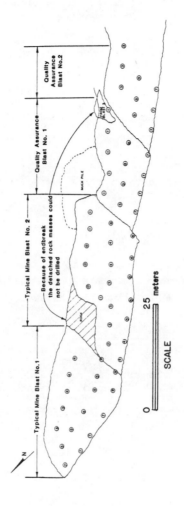

Fig. 11.   Plan of typical mine blasts and Quality Assurance blasts.   Circled numbers are blasthole locations and delay numbers.

TABLE 2. Test Parameter Summary

BLAST NO.	HOLE DIA. (mm)	DEPTH (m)	BURDEN (m)	SPACING (m)	STEMMING (m)	VOLUME m³
1	230	18	4 to 5.5	4 to 5.5	6	7,865
2	230	20	4 to 5	4.5 to 6.5	6	5,560
3	230	21	4.5 to 5	5.5 to 7.5	6	10,450
4	230	21.6	4.5 to 5	5 to 7.5	6	7,785
5	230	22	5	5.5 to 7	6	10,100
6	160 to 230	21	3.5 to 5.5	5.5 to 7.5	4.5 to 6	20,460
7	160 to 230	20	3.5 to 5.5	3.5 to 7.5	3 to 10	10,565
8	160 to 23C	17.5	3.5 to 5.5	4.5 to 7.5	3.5 to 7.5	8.135
INIT 1	230	20 to 22	4.5 to 7	5.5	6	9,775
INIT 2	230	21 to 21.5	4 to 5	4.5 to 6	6	10,940
Q A 1	230	19 to 21	4.3 to 6	5.5	6	6,220
Q A 2	230	18 to 20	5	5.5	6	4,375

BLAST NO	POWDER FACTOR (kg/m³)	BURDEN TO HOLE DIAMETER	STEMMING TO BURDEN	SPACING TO BURDEN	HOLE DEPTH TO BURDEN	m³ OF HOLE PER m³ OF ROCK x 10⁻⁴	ACTUAL MILLISECOND DELAY PER METER OF BURDEN	MINIMUM MILLISECOND DELAY PER METER OF BURDEN
1	1.03	17.3 to 24	1.1 to 1.5	1 to 1.4	3.2 to 4.5	17	10.5	3.9
2	1.02	17.3 to 22.7	1.2 to 1.54	1.2 to 1.5	3.8 to 5	18	13.0	3.4
3	0.825	21.3 to 22.7	1.2 to 1.25	1.13 to 1.41	4.0 to 4.3	12	17.4	3.4
4	0.87	20 to 22.7	1.25 to 1.33	1.13 to 1.6	4.2 to 4.7	13	12.5	3.4
5	0.81	22.7	1.33	1.05 to 1.35	4.2	12	12.8	3.9
6	0.71	23.04 to 24.7	1.25	1.5 to 1.33	3.82 to 5.73	11	15.4	4.3
7	0.72	23.04 to 24.7	0.83 to 1.83	1.25 to 1.33	3.61 to 5.42	13	10.8	3.9
8	0.60	23.04 to 24.7	1.0 to 1.39	1.25 to 1.33	3.2 to 4.8	13	11.2	4.3
INIT 1	0.90					15	14.4	4.3
INIT 2	0.87					15	16.4	4.3
Q A 1	0.91					16	22 0	4.3
Q A 2	0.86					15	22 0	4.3
Desirable Range:		20 to 40	0.7 to 1.0	1 to 2	1.5 to 5.0			

ultimately had to hire a drill contractor and buy a 160 mm bit.

All of these constraints, plus the already mentioned unfavorable geologic conditions, made the achievement of the stable highwall impossible in only a few tests.

Test Series 1

The main objective of the blast trials was to develop a blasting technique that would be compatible with the operation's existing equipment and that would leave a tight and straight highwall face. We wanted to improve blasting efficiency by better controlling the explosive energy and increasing breakage in flexure. The elimination of violence and flyrock, poisonous fumes from burning AN-FO, and noise, and lowering explosives costs were secondary goals, but also provided clues for judging the success of a blast.

To accurately measure overbreak, stakes were placed behind and parallel to the back row of holes, and 11 m behind and parallel to the alignment of the end hole in each row. The difference between these distances and that measured after the blast would determine the amount of overbreak.

Figure 12 details Test Blast 1, which was little different from the mine's normal practice. Holes in rows were each timed individually, and rather close burdens and spacings were utilized. As the photograph in Figure 12 shows, the blast was quite violent, with considerable overbreak. Figure 13 illustrates a later Test Blast, No. 5, with wider burdens and spacings and holes timed the same in each row with a delay element skipped at the rear and ends. This blast was not observed, due to heavy fog, but was examined later.

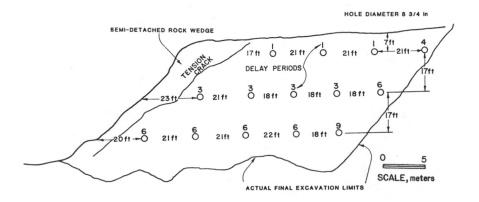

Fig. 13. Test Blast 5.

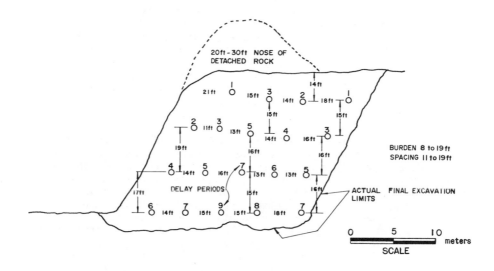

Fig. 12.   Test Blast 1.

From these five tests the blasting behavior of the rock was determined, powder factor was reduced, overbreak was reduced, highwall damage lessened, the highwall was becoming less ragged, poisonous orange-brown fumes were eliminated, and violence was controlled. The exhibited overbreak was still excessive and the damage to the highwall was not acceptable. The concepts of optimizing rock flexure by adjusting the burden and spacing, and generally improving blasting conditions by controlling relief and explosives distribution were working, and would be extended during the next test series.

After these initial tests, a good deal was learned about the mining conditions and some improvements noted. Explosives cost has been reduced slightly while yielding acceptable fragmentation. The overall blast violence and fumes were drastically reduced, all of which indicated a more efficient use of AN-FO which can lead to better breakage and balanced blast forces that will not adversly affect the highwall. Other than the increased delay interval of the back row and the last hole to detonate in each row, the patterns were very similar to what was already being done in the operation. Spacings and burdens were expanded between adjacent holes only of equal delay and delays expanded as well so as to maintain AN-FO confinement and encourage complete detonation. However, after the first five tests it became apparent that only small additional changes could be made to improve highwall stability with the large diameter blastholes releasing more energy than was controllable. Too large a spacing produced a ragged face, while too small a spacing caused severe backbreak and loss of AN-FO confinement and performance due to holes cracking toward each other. No optimum spacing could be found that would reduce overbreak. In any event, judgement of the success of any blast test must be reserved until the dragline has dug through the muckpile, exposing the highwall.

Nevertheless, it was believed that small hole diameters could be used to complete project goals of increased blasting efficiency and improved highwall stability, and arrangements were made for using a smaller diameter drill for the next series.

Test Series 2

This test series began seventeen days after Test Blast 5 with the period in between being a scheduled drilling and blasting idle period.

As before, hole locations were carefully laid out with Brunton compass and measuring tape. Each hole was measured for depth and for water. To maintain proper drill hole orientation, the mast was checked for verticality with a 1 m carpenter's level. Blasting caps with 30 m leg wires were used to assure bottom initiation for the 21 m deep holes. Markers were again used to determine overbreak.

The objective of Test Blast 8, Figure 14, was, using the successes

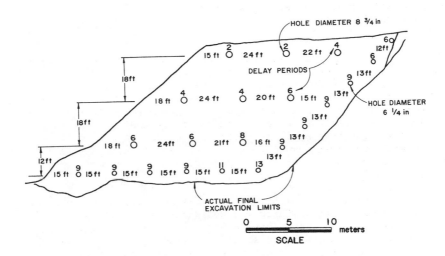

Fig. 14.   Test Blast 8.

of the previous blasts, to create still more delay time between the back row and the main blast and increase relief. The same delays were also employed for the end holes to try to obtain a more even end wall.

Holes of large diameter were laid out as before, as was the back row of holes. The end holes were placed closer together and used two delay periods. The front two having one period and the next four with another longer period. All holes were extremely wet, so deck loading was not used instead 120 mm x 11 kg pre-packaged tubes of AN-FO were used which reduced the amount of explosives, but they occasionally got hung up in the hole making loading difficult. Consequently the test was conducted under less than ideal conditions. Stemming was unchanged for the larger holes, and increased by 0.6 m in the smaller holes.

This was the least violent blast of all the tests. Flyrock was minimal, as were fumes. Throw was well controlled and the muckpile was well heaped. Breakage appeared to be good. If dragline digging were to improve, this pattern should be tried again with deck loading. There was every indication of a well-balanced blast. More delineation of back and side walls may have been possible if the powder column were brought higher with deck loading. Overbreak ranged from 2 m to 3 m at the rear and was 1 m to 3 m at the end making for a distinct improvement.

From these 3 tests using smaller diameter drill holes at the limits of the blasts, overbreak was considerably reduced, fragmentation improved, highwall damage minimized, violence reduced, and fumes eliminated. The powder factor was reduced as well, and a satisfactory blast design concept realized.

Test Blast Discussion

Test parameters have been summarized in Table 2. It can be seen that powder factor alone has been reduced from 1.03 kg/m^3 to 0.60 kg/m^3. This shows that there is more efficient use of explosive energy. Also, the desirable range of blast design ratios has been met. A range of values has been given in Table 2 for each ratio because of the variable burden and spacing for the two hole diameters or the hole locations such as at the end of a row. Table 3 is summary of measured overbreak resulting from the test blasts.

The test blasts ended with Test Blast 8, but if tests could have been continued, they would be conducted at one end of the pit where the hard massive sandstone was located. This area could challenge the effectiveness of a final blast design. Hence it would have been ideal to repeat Test Blast 8 in different rock as a Test Blast 9, and drill a round with only small diameter holes as a Test Blast 10.

TABLE 3. Overbreak Summary

Test	Backbreak (m)	Endbreak (m)	Powder Factor (kg/m³)
1	2.3	1.7	1.03
2	3 to 5	6 to 10	1.02
3	3.5 to 4.25	7 to 7.5	0.825
4	2.75 to 5	3.5 to 4	0.87
5	4 to 4.5	3	0.81
6	2.5 to 3	2.5 to 5	0.71
7	1.25 to 2	1.5 to 3.5	0.72
8	1.5 to 2.75	1.25 to 2.5	0.60

TABLE 4. Cost Summary

BLAST	METERS DRILLED 230 mm	METERS DRILLED 160 mm	HOURS TO DRILL 230 mm	HOURS TO DRILL 160 mm	EXPLOSIVES TUBES (kg)	EXPLOSIVES BULK (kg)	m³/m OF HOLE	EXPLOSIVE $/m³	DRILL $/m³	TOTAL $/m³	% CHANGE
Standard	-	-	-	-	-	-	26.58	0.145	0.110	0.255	-
Test 1	320	-	34	-	680	7,275	24.60	0.166	0.152	0.318	+24.6
Test 2	258	-	24	-	230	6,300	21.50	0.157	0.165	0.322	+26.2
Test 3	334	-	28	-	860	7,570	31.27	0.135	0.110	0.245	- 4.1
Test 4	259	-	15	-	860	6,745	30.06	0.129	0.102	0.231	- 9.2
Test 5	306	-	17.5	-	-	8,185	32.97	0.120	0.093	0.213	-16.4
Test 6	422	269	20	6	4,920	9,500	29.59	0.149	0.085	0.234	- 8.2
Test 7	218	298	8	6	1,235	6,435	20.46	0.129	0.111	0.240	- 5.6
Test 8	155	197	7	6	3,400	1,470	23.12	0.165	0.103	0.268	+ 5.1
Final	158	218	7	6.5	610	5,085	20.54	0.126	0.120	0.246	- 3.6

COST BENEFITS

The cost benefit aspect of our blasting demonstrations is one of the most important factors of this project. We know that if the mining industry is to adopt the project techniques, the incentive must come from an economic advantage.

Unfortunately, many cost savings are difficult to segregate from the overall mining operation without examining the operation for long periods of time in great detail. For example, it is dfficult to determine how long the dragline may spend cleaning a highwall, or how often a drill must take extra time to set up a hole due to an irregular or fractured face. Consequently, care must be taken that cost estimates reflect only what can be quantified with few or no assumptions made.

Drilling and Blasting Costs

A summary of blasting and related costs from the field tests are presented in Table 4. Hours of drilling time are converted to $/m^3 by summing fuel, labor, and bit costs for each size hole. Explosives costs are separated according to the amount of prepackaged tubed AN-FO for wet holes, and bulk AN-FO used for the rest of the blast; then totaled for both 230 mm and 160 mm holes. The drilling and blasting cost is then totaled and expressed as a percentage relative to a typical mine blast observed prior to the testing.

This direct comparison is not entirely valid in that the standard blast contained all dry holes and therefore did not require more expensive pre-packaged tubed AN-FO. Consequently, the test blast using the pre-packaged tubed AN-FO appear to be more expensive in total cost per cubic meter. On the other hand, powder factor is reduced with the use of tubes in that they do not fill the borehole as completely as bulk AN-FO thereby resulting in less powder per meter of borehole. The Final Design shown on Table 4 is based on experience from the test blasts. The amount of pre-packaged tubed AN-FO is based on two tubes per hole, with decking in all the small diameter holes except in the corner hole. The tubes raise the explosive cost but lower the powder factor resulting in an overall cost reduction of 3.6% less than the standard operation blast used as a reference. The elimination of tubed AN-FO would lower the cost per cubic meter to $0.237 or a reduction of 7.2%.

However, there are other cost savings as well. For example, the reduced powder factor means that more holes can be loaded by the bulk AN-FO truck. It is more likely that a blast round can be loaded with one truck load of AN-FO, thus eliminating a trip back to the magazine to get another load. Downtime for this task at the field site was over an hour. This involved wasted time for the blasting crew of three men.

Any time savings is a benefit in that this one blasting crew handles the blasting for five mine operations. Hence, any time lost at one can jeopardize the blasting schedule at another. Time delays can cause caving blastholes which will affect blast performance.

Another hidden cost savings is the decreased probability of leaving hanging wedges or rock left in the front corner of a bench by endbreak. This rock mass cannot be drilled safely, and most likely will be carried in its entirety to the pit by the next bench blast. This can create extra wear on the dragline, and increase cycle time. No large masses of rock were left after a test blast.

Dragline Cycle Times

To help determine the effectiveness of a blast, cycle times of the dragline were recorded. It was thought that measurement of digging times would provide some basis for judging the degree of blast fragmentation. Unfortunately, too may variables existed that affected the data. Rock characteristics varied three times over the length of the pit and this alone can affect fragmentation and thereby digging time. The cycle times also cannot differentiate the digging of an exceptionally difficult area in the muck pile from the digging of the complete muck pile of a particular blast. Digging times may also vary from summer to winter, which were the time periods when the test blasts were dug.

The cycle time statistics for each blast, including the two Quality Assurance blasts indicate an increase in dragline cycle time with each blast, which is the opposite of what was expected. However, F-test comparisons indicated that there was no difference in the data, and that almost all blast cycle times were drawn from the same population. Apparently, there are too many variables that influence the actual length of the dragline time such as:

- The dragline operators' methods of operation
- Location of dragline at the time the cycles were being measured
- Weather conditions
- The operating condition of the dragline itself.

Therefore, the dragline cycle times do <u>not</u> indicate an improvement in overburden handling characteristics with different blast patterns. Two conclusions are therefore possible: firstly, that the different blasting did not alter the handling characteristics, or, secondly, that dragline cycle times are an inadequate measure of overburden handling improvements owing to other uncontrollable factors.

Dragline Fuel Consumption

Dragline fuel usage was examined on a daily basis for all of the test period. The thought was that easy digging would show up in lower fuel consumption. If large rocks are to be handled, or the highwall scraped this lost time may not show up in digging cycle times, but should appear in fuel consumption.

Analysis of these data for fuel daily consumption while digging in each blast revealed no discernable trend, and all data appeared to have been drawn from the same population.

As before, the indications are that the test blasting had no effect on fuel consumption or that fuel consumption is not an effective measure of altered blasting practice.

Cost Benefit Summary

Many mining phases were examined for favorable cost reductions as a result of optimum blast design. Areas studied for cost changes were:

- Explosives
- Drilling
- Equipment maintenance
- Fuel consumption
- Dragline performance.

Of these areas, only the first two could be quantified with a high degree of accuracy. Reduced costs are evidenced in lower powder factor and therefore less explosives; and in an expanded drill pattern resulting in fewer holes thus lowering drilling costs. With less explosives and fewer holes, loading time is reduced, thereby making more efficient use of personnel and allowing them to do other tasks.

Other operations involving equipment maintenance, performance, and fuel consumption contained too many variables in order to quantify their performance in relation to the blast tests. It is hoped that in the long run, highwall stability will add to overall operational efficiency.

SAFETY BENEFITS

Stable highwalls add increased safety to virtually every aspect of contour strip or similar mines. However, some benefits are hard to quantify such as increased productivity due to higher morale among employees who work in the pit.

Highwall Monitoring

To measure the degree of highwall stability, the highwall was monitored over a long period of time and under a variety of weather conditions. Camera stations were set up whereby the highwall could be photographed repetitively from the same vantage point.

Blasting tests began with two camera stations mounted on the dragline spoil pile across the pit. As tests progressed, two more stations were added to make the total coverage of over 240 m. The camera stations consisted of a wooden post concreted into the spoil with a camera mounting fixed on top of the post. These mounts were fully adjustable to view the part of highwall desired, and then fixed in place. A 35 mm camera was then slipped into the mount for photographing the same spot on the highwall. Photographs were taken every other day. Also, a rain guage was placed at one of the stations to help find a correlation between rock falls and rainfall.

To judge if any falling rock occurred, the highwall pictures were blown-up to 250 mm by 360 mm. Pictures of the same area, but taken at different times, can then be compared by overlaying on a light table and examining closely for any differences such as fallen or missing rock.

Close examination of camera station pictures, Figure 15, showed several rock falls. Possible immediate causes are rain and blast vibrations loosening unstable blocks. Table 5 shows the highwall monitoring results of all the blasts which include the standard mining blasts (INIT 1 and INIT 2), the Quality Assurance blasts (QA 1 and QA 2) and the 8 test blasts.

TABLE 5. Highwall Monitoring Results

BLAST AREA	TOTAL EXPOSURE TIME	ROCK FALL APPROX. WT.	RAINFALL (mm)
INIT 1	–	–	–
INIT 2	56 Days	580 kg	76
Q A 1	58 Days	1,740 kg	76
Q A 2	48 Days	7,700 kg	81
TEST 1	–	–	–
TEST 2	14 Days	3 blocks 290 kg each	81
TEST 3	39 Days	None	135
TEST 4	39 Days	None	81
TEST 5	23 Days	None	2
TEST 6	23 Days	None	2
TEST 7	Not Monitored	–	–
TEST 8	Not Monitored	–	–

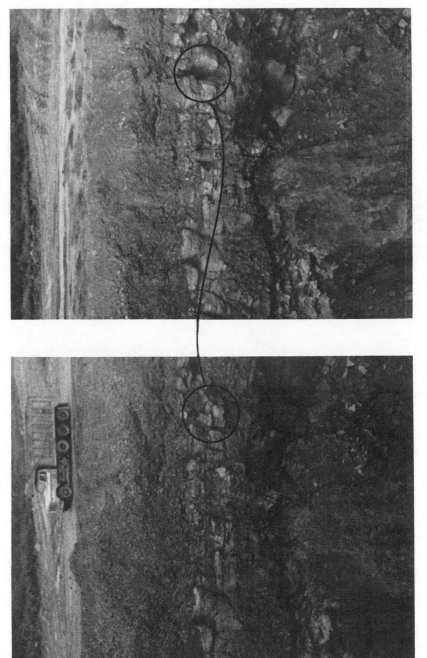

Fig. 15. Results of repetitive highwall photography. Photographs are taken nine days apart. Note rocks seen in earlier photograph are missing in later photograph.

The Test Blasts were only monitored for 39 days at most as compared to up to 58 days for the typical mine blasts and the Quality Assurance blasts. Nevertheless, some conclusions can be drawn. For example, rock falls occurred between 4 and 47 days of exposure; most fell within 15 days. As seen in Table 5, Test Blasts 3 through 6 apparently had few or no rock falls, even though they were monitored for longer than 15 days.

Even though a significant amount of rainfall occurred while monitoring the latter test blasts, and apparently had little effect, it is clear that there was significant highwall stability improvement. Unfortunately the exposed highwall from Test Blast 7 and 8 was not examined for any extended length of time due to project schedule and budget restrictions. Falling rock is difficult to quantify, and it must be pointed out that the rock in Test Blasts 7 and 8 was a flaggy sandstone as compared to the more massive sandstone present at the earlier tests. Obviously, the flaggy rock falls would be harder to detect.

The existing highwall for Test Blasts 7 and 8 was checked by site visit after it had been exposed for some time. From visual observation it is obvious that the highwall had improved over what was first encountered in similar areas of the pit where the rock is of the same type. However, this is based on judgement, and is not definitely quantified. It must be noted that Test Blast 8 was considered the optimum blast design and therefore highwall results would carry much more weight than results of other tests.

Safety Results

The successful results of this project are dramatically shown in Figure 16, where the regularly employed blasting created the loose, ragged highwall shown in the foreground while Test Blasts 6, 7, and 8 created the tight, smoother highwall further along. We feel that the conditions have demonstrably become much safer.

CLOSURE

A survey of contour strip coal mining practices throughout the Appalachian region revealed very variable levels of technology. Some operations – generally those operated by large companies – had highly qualified personnel who carried out safe and efficient blasting. On the other hand, some smaller operations used practices that were wasteful and occasionally dangerous. Almost all mines visited had highwall instability problems aggravated by their blasting practice. Improved contour strip and mountain top coal mine blast designs intended to achieve highwall stability have been developed and presented in this paper.

Jointing of rock, varying weather conditions, and dip of bedding planes can all contribute to highwall instability. However, improper blasting practices can aggravate the existing rock conditions. Close

Fig. 16. Results of improved blasting practice on coal strip mine highwall stability. Foreground is typical mine practice; background is improved practice.

hole spacings, and large hole diameters lead to overloading of a blast round, and if inadequate relief exists, the extra explosive energy is directed toward the highwall and creates overbreak and damage to the highwall rock. A potentially dangerous situation then arises from loosening rock blocks and wedges that may fall after removal of the muckpile.

During the course of this work, eight test blasts were conducted at an Appalachian strip mine with very poor geologic conditions and restricted highwall geometry. Each test incorporated minor changes in blast design parameters such as hole spacing, delay timing, and hole diameter. The concept of trying to achieve flexural failure of the rock mass by proper proportioning of the ratios between burden, spacing, stemming, and hole depth was utilized with success.

The results of the testing showed that highwall stability can be improved with existing, cost effective, blasting technology. Smaller hole diameters result in better explosive distribution and powder factor reduction. An extra delay being given to the back row, end holes, and corner hole also cut down the degree of overbreak considerably by giving more time for burden movement and the development of relief. Overall, the blasts were much less violent and produced much less fly rock and poisonous fumes than before.

Cost savings were realized from reduced powder factor and reduced drilling requirements. Other savings from improved safety, worker efficiency, and dragline performance were examined, but these are more difficult to quantify.

Operating Recommendations

Based upon the field survey of contour strip mines and the test blasting conducted, the following points, it is felt, will substantially improve the cost efficiency and safety of contour strip mining of coal in Appalachia.

When planning the mine, the highwall orientations should be selected to coincide with natural rock fracture trends where possible. Even in contour strip mining with its sinuous highwalls, the almost universal presence of two widely separated fracture trends in the rock can be utilized by making the highwall follow more of a zig-zag along the joints than a wiggle along the topography.

The blasthole drill should be selected <u>after</u> the general mine plan is established so that the optimum burden and spacing for the hole diameter and depth can be established based upon sound blasting principles.

The use of a "bottom line" accounting system hides the true costs. With rapidly escalating explosive costs, accurate accounting of all blasting costs should be adopted.

When a blast pattern is established for a given area, more accurate blasthole layout procedures should be utilized as very often too many holes are drilled with consequently excess powder loaded, or improper fragmentation may result from widely spaced holes.

The delay timing of a bench should be planned out logically so that maximum relief is given to holes. The planner should visualize the bench after each delay element has fired and consider the result.

ACKNOWLEDGEMENTS

This work was funded by the United States Bureau of Mines under Contract No. HO282011, and this financial support is gratefully acknowledged and appreciated. Mr. Larry R. Fletcher of the Twin Cities Research Center was the Technical Project Officer for the Bureau of Mines and Dr. Richard L. Ash was project blasting consultant, and without their contributions, this project could not have succeeded.Dr. Lee Saperstein served as project mining consultant, and his efforts materially aided this study. Sincere appreciation and thanks are due to the management and workers of the cooperating mining company, whose release of sensitive information precludes their identification. The support of the management of Engineers International, Inc., is greatly appreciated, as is the permission granted by the Bureau of Mines and Engineers International, Inc., to publish the work.

REFERENCES

Ash, R. L., 1974, "The Influence of Geological Discontinuities on Rock Blasting," Ph.D. Dissertation, Univ. of Minn., University Micro Films, Ann Arbor, MI, 286 pp.

Ash, R. L., and Smith, N. S., 1976, "Changing Borehole Length to Improve Breakage: A Case History," Proceedings of the Conference on Explosives and Blasting Techniques, Society of Explosives Engineers, pp. 1-12.

Barry, Theodore and Associates, 1974, "Industrial Engineering Study of Hazard Associated with Surface Coal Mines," Final Report U. S. BuMines Contract No. HO230004, 24 June 1974, NTIS No. PB-235927, 265 pp.

Cook, Frank, and Kelly, William, 1976, "Evaluation of Current Surface Coal Mining Overburden Handling Techniques and Reclamation Practices," Final Report on U. S. BuMines Contract No. SO144081, 24 December 1976, NTIS No. PB-264111, 317 pp.

Dick, R. A., 1976, "Practical Design of Blasting Rounds," Presented at Univ. of Minn., Rock Mechanics Short Course, Design Methods in Rock Mechanics.

Dick, R. A., and Olson, J. J., 1972, "Choosing the Proper Borehole Size for Bench Blasting," Mining Engineering, Vol. 24, No. 3, pp. 41-45

MSHA Technical Support, 1978, "Surface Mine Fatalities Related to Unstable Benches and Highwalls," U. S. Dept. of Labor, Health and Safety Analysis Center, Denver, CO, 11 pp.

Chapter 24

PRODUCTION BLASTING AND THE DEVELOPMENT
OF OPEN PIT SLOPES

John P. Ashby

Associate Mining/Geotechnical Engineer
Golder Associates Inc.
Seattle, Washington

ABSTRACT

Mine production blasting is a process of destruction of rock masses in order that ore may be extracted. Many open pit operations are faced with the apparently conflicting requirements of providing large quantities of fragmented rock for the processing plant and of minimizing the damage inflicted upon the surrounding pit slopes. A reasonable compromise between these two conflicting demands may often be found by means of simple engineering of production blasts.

Many operations tend to rely too heavily upon traditional wall control techniques to alleviate a problem already created by the production blasts.

Various production trends including use of larger blastholes and patterns have tended to aggravate the situation. These practices only serve to emphasize the need for engineered improvements.

Central to blasting engineering is the process of systematic trial and evaluation. Changes to the blasting system should be made singly, starting with the simplest.

This paper discusses simple changes involving drill pattern control, effective use of blastholes and explosives, adequate delaying including use of down-the-hole sequential techniques, effective utilization of free faces, and avoidance of "choked" situations. By directing explosive effort where it is needed, such changes generally result in a significant improvement of slope conditions and of production blasting cost and performance.

INTRODUCTION

Blasting is, by its very nature, a destructive process. The open pit blaster is often faced with the apparently conflicting demands of providing large quantities of well fragmented rock and of minimizing the amount of damage inflicted upon the rock slopes that remain.

Lack of attention to blasting adjacent to pit slopes can lead to difficult digging and pit walls that are psychologically uncomfortable and even dangerous to work beneath. There is evidence to suggest that a substantial number of slope failures have been aggravated and some even precipitated by poor blasting practice.

At the very outset of being confronted with the problem of development of pit slopes, many blasters will consider the use of traditional controlled blasting techniques such as presplitting. They may test out the technique, often with marginal success, and then conclude that the technique is too expensive, too demanding of drilling capacity, or just unsatisfactory. In many cases, the reason poor results are obtained is that the damage has already been caused by previous blasting in front of the wall control shot. For this reason, the author considers that wall control blasting (in the strict sense) should be used as the last and optional step in a process requiring careful production blasting. This paper discusses those changes that can sometimes be made to the production drilling and blasting system that result in improvement of the wall conditions. Most of these changes can be made with minor effort and little extra cost and invariably result in generally improved fragmentation. These modifications are often necessary, merely for the development of good blasting technique.

SYSTEMATIC MODIFICATION OF THE DRILLING AND BLASTING SYSTEM

Any changes that are made to the drilling and blasting system may significantly affect other aspects of the operation, to its benefit or detriment. Therefore, the first step in rationalizing a blasting system should be to ensure good record keeping of each blast and the results of that blast.

Unfortunately, precise measurements of the effect of blasting changes upon the operation are difficult and time-consuming. However, honest evaluation can be achieved by observation of the results in the field. Successful blasting can often be identified by observing the conditions of the muckpile after the shot. Figure 1 illustrates some important features to look for.

Once the blast is excavated to pit wall limits, it is necessary to evaluate the wall conditions. Table 1 can be used to classify blasting damage levels commonly observed on bench faces.

TABLE 1   LEVELS OF BLASTING DAMAGE COMMONLY OBSERVED ON PIT WALLS

| | | Observed Conditions of the Wall | |
Arbitrary Damage Level	Joints & Blocks	Dip Angle Appearance and Condition of Face	Digging Condition at Face (Electric Shovel)
1 Slight	Joints closed, infilling still welded.	>75° circular sections of wall control holes seen.	Scars of shovel teeth seen in softer formation, further digging not practical.
2 Moderate	Weak joint infilling is broken, occasional blocks and joints slightly displaced.	>65° Face is smooth, some hole sections seen. Minor cracks.	Some free digging possible, but teeth "chatter."
3 Heavy	Some joints dislocated and displaced.	>65° Minor spalls from face. Radial cracking seen.	Free digging possible for <1.5m (<5 ft) with some effort.
4 Severe	Face shattered, joints dislocated. Some blocks	>55° Face irregular, some spalls, some backbreak cracks.	Free digging possible for <3m (<10 ft).
5 Extreme	Blocks dislocated and disoriented. Blast-induced fines or crushing observed.	55°>37° Face highly irregular, heavy spalling from face. Large backbreak cracks.	Extensive free digging possible for >3m (>10 ft).

Successful drilling and blasting involves the complex interaction of many factors. Therefore, when making modifications, the blaster should endeavor to change only one thing at a time, with the simplest change being made first. Side-by-side evaluation of a change in the same ground is the most affective test of a modification.

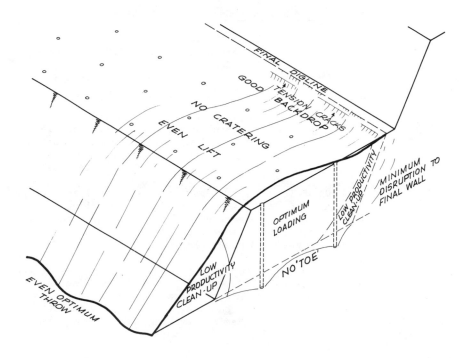

Fig. 1.   Features of a satisfactory production blast.

MODIFICATIONS TO THE DRILLING AND BLASTING SYSTEM

Modifications that may lead to improvement of pit wall conditions
and overall blasting efficiency are described under the following
headings in this paper.

- Blasthole Pattern
- Explosive Types
- Charging Policy
- Front Row Control
- Delaying Practice
- Production Blasting and Excavation Adjacent to Pit Walls
- Controlled Wall Blasting (in the Strict Sense).

Upon study, many of the components of a particular drilling and
blasting system will be found to be in order. However, the following
portion of this paper describes some of the important components of
mining systems that are frequently found to be deficient.

Blasthole Pattern

- Blasthole patterns should be carefully planned, pegged, and drilled. Pacing-out of patterns should be avoided in favor of more accurate methods such as use of tape or survey.

- Use of square or staggered patterns seems to be a matter of personal preference. Use of staggered patterns leads to improved charge distribution. However, square patterns are easier to lay out and are therefore more precise.

- Spacing should be selected to ensure effective hole utilization and adequate stemming. When selecting hole spacing, particular consideration should be given to the problems of charging of the front row holes adjacent to the free bench face (Figure 2A).

- Patterns should be designed with particular attention to the free bench face. Excessive front row burden is a common problem and can be overcome by reduction of pattern spacing. Locally, problems of excessive front row burden can be overcome by the drilling of fill-in or "easer" holes (Figure 2A). It is the opinion of the author that many problems encountered, particularly within or behind a blast, are a function of inadequate attention to front row control.

- Hole depths should be controlled to yield optimum design sub-grade. Hole depths, as drilled, should be measured and, if overdrilled, should be backfilled to optimum depth. In certain situations, subdrilling can be avoided altogether.

Explosive Types

- Selection of explosive type is generally made on the basis of availability, cost, personal preference, and what products have traditionally been used at the particular operation. Although the methods of comparison of the energy output of explosives is a contentious issue, cost comparisons of different explosives should always be made on an energy equivalent basis. Explosives having higher weight strengths may appear excessively expensive until they are compared on an energy-equivalent basis with cheaper but weaker products.

- High-strength explosives such as those containing aluminum, TNT or other promising nitrated fuels such as nitromethane are useful as a bottom load in operations where drilling costs are high and in situations where locally tough ground conditions are met. High-strength explosives are also valuable where the drillhole spacing or burden are too large, such as heavily burdened front row hole, a short hole, or where an error in the blasthole pattern has been made.

A. DRILLING

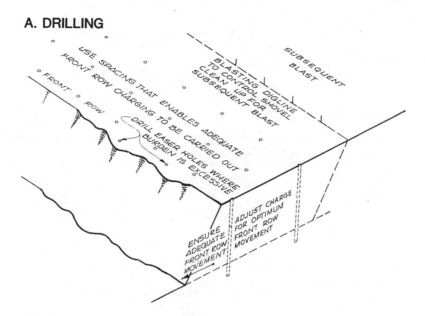

B. FRONT ROW CHARGING

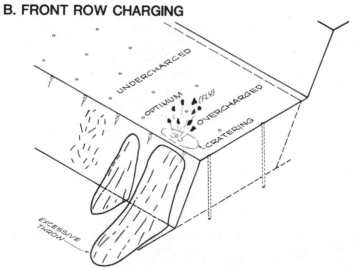

Fig. 2.   Front row control.

- Low-strength explosives are particularly useful where improved charge distribution is required. Therefore, use of low strength explosives should always be considered adjacent to a pit wall where good charge distribution is so critical.

- From experience in slope engineering, the author has found that very few open pits are "dry," especially in proximity to the pit walls. Care should always be taken when using explosives such as ANFO that are susceptible to moisture attack. Such explosives should be fired as soon as possible after loading. Holes in low-permeability ground that appear dry during drilling may become wet with time. Holes should always be checked for standing water prior to loading. Explosives susceptible to moisture attack should always be used with caution, since a reduction of ANFO efficiency by only 20 percent is often all that is needed before it may prove more economical to use a more expensive, water-resistant explosive.

- The advent of continuous, coiled, cap-sensitive explosives such as Iremite has simplified the job of charging large-diameter wall control holes. Where such explosives are unavailable, stick explosives can be "strung" together with detonating cord using continuous "socks" of plastic netting. Such a product, used for vegetable packaging is marketed under the name "Vexar" craft netting.

- Use of explosives that produce predictable results is essential to the engineering of any blast and particularly blasts adjacent to pit walls.

Charging Policy

The reader may ask why production blasting patterns and hole-charging policy should be mentioned in a paper on pit wall control. First, it is the author's belief that good wall control blasting requires careful design, planning and placement of explosive charges. Second, there is a natural tendency to try to solve blasting problems by using more explosive to be on the "safe side." Obviously, quite apart from being wasteful, for a given blasting system, more explosive produces more pit wall damage.

Properly designed charge weights are fundamental to the success of a blast. A method of design of charges using powder factor, preferred by the author, is described in the following section and by Hoek and Bray, 1977.

The Use of Powder Factor as a Simple Design Tool

Many operations treat powder factor (specific charge or energy factor) as an accounting tool, either for cost prediction or for reporting purposes. Fewer operations use powder factor to design patterns and charges, and those that do not, often manage well without. However, for efficient use of explosive and to produce consistent results, the author has found the technique useful for producing a controlled product.

In the main text of this paper, the importance of control over the production blast has been stressed. Efficient use of explosives to produce a consistently good product is the key, not only for good economical fragmentation but also for control of wall damage.

Selection of Design Powder Factor

The term "design powder factor" as used in this paper is the charge weight of a particular explosive divided by the burden volume (i.e., burden X spacing X bench height) (Figure 3).

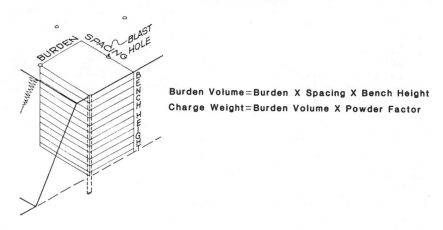

Burden Volume=Burden X Spacing X Bench Height

Charge Weight=Burden Volume X Powder Factor

Fig. 3.   Burden volume and the design of production charge.

Powder factors should always be adjusted to account for the differences in weight strength of the explosives used. Unfortunately, reliable relationships between powder or energy factor, rock type, and style of blasting do not appear to exist and must always be evaluated at the blast site. Trial, evaluation, and recording of the results are the most valuable tools for blasting control.

From the author's work in rock slope design and blasting engineering, an intersting empirical relationship has been developed (Figure 4). The relationship was derived by the recording and optimization of several hundred blasts in a porphyry copper pit that exhibited a wide range of natural fracturing. The relationship has since been evaluated in a wide range of geological conditions with fair success. This relationship indicates that "natural" fragmentation or block size of the rock mass and the surface characteristics of the blocks may significantly influence the amount of explosive effort required in blasting. The relationship is included in this appendix to stimulate thought and criticism but is not intended to be used without site specific evaluation.

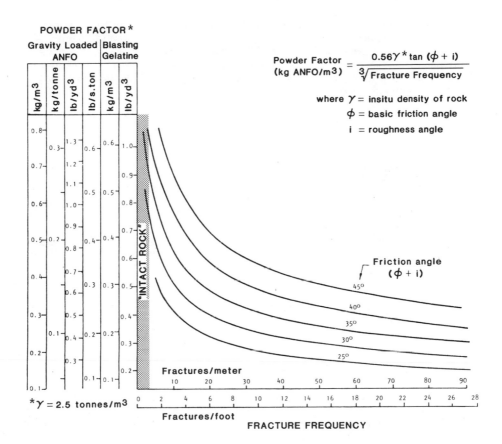

Fig. 4. Empirical relationship between powder factor, fracture frequency and joint shear strength developed at Bougainville Copper Mine.

Determination of Charge Weights

The first step is to select an appropriate powder factor for the rock conditions encountered in drilling the pattern. In an operating mine, this can be done by referring to the records of practice in similar ground, often from the bench level above. The blasthole pattern and the bench height are measured and the charge weight for each hole is calculated or determined by using tables. The process assures consistent charging, especially where pattern irregularities or variation in rock type occurs.

Error in selection of an appropriate front row charge weight may result in a "choked" blast or excessive throw. Some experienced blasters can successfully "eyeball" front row burdens and estimate charge weights. However, measurement at the crest, calculation of front row burden volumes, and the use of the powder factor, usually provide much more consistent results (Figure 5).

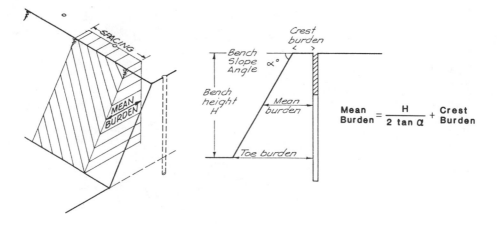

Fig. 5.   Design of front row charges.

A method of estimating buffer charge weight is illustrated in Figure 6. The method requires the calculation of the volume of the wedge of relatively unbroken ground that occurs beneath the backbreak zone and in front of the pit slope.

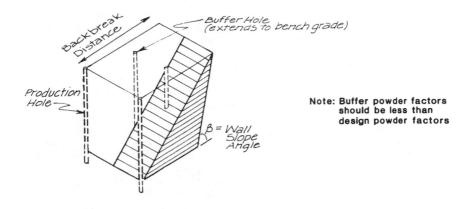

Note: Buffer powder factors
should be less than
design powder factors

$$\text{Average ''Buffer Burden''} \approx \frac{\left(\begin{array}{c}\text{Backbreak} \\ \text{Distance}\end{array} - H \cot \beta\right)}{2}$$

Fig. 6. Design of buffer charges.

<u>Front Row Control</u>

The front row is perhaps the most cirtical portion of the blast (Figure 2). When the front row burden moves out adquately, there is an excellent chance that the remainder of the shot will succeed. Good forward movement of the blast is only possible with an adequately charged front row and a good free face to move to.

Provided that front row holes are not excessively over- or under-burdened, adequate front row charging can be achieved (Figure 6B). Measurement and calculation is the most reliable method of determining front row charge weights. One such method is described in the previous section.

- Because the result is wastage of explosives and eventual wall damage, "choked" situations (i.e., blasting into broken or unbroken muck piles) should be avoided whenever possible.

- Shovel cleanup of a good free face should be encouraged over as great a length of the blast as possible. Blasting diglines placed by visual evaluation of the shot are excellent for controlling shovel development of the subsequent free faces and for preventing over-excavation of subsequent blasthole patterns (Figure 2A).

Delaying Practice

Proper initiation and delaying of a blast is of critical importance in ensuring optimum movement of the muck pile, good fragmentation, optimum explosive utilization, comfort of the neighbor, and maintained integrity of the pit slopes.

- Correct use of delays of a suitable length ensures adequate movement of burden. By this means, the tendency for "choking" of that portion of the shot adjacent to the pit walls can be minimized. One of the most promising developments in this area is the development of practical sequential blasting techniques (illustrated in Figure 7). The technique was perfected with the development of the sequential electric system which has in turn been followed by non-electric methods (Tansey, 1979). The advantages of the method are that only a limited range of down-the-hole delays are needed. Shorter delays are potentially more accurate than the longer delays that must be used when blasting with conventional delaying methods (Winzer et al., 1979). In many sequential blasting situations, the risk of trunkline cutoff tends to be reduced, and hence longer delays can be used than would be possible with conventional delaying.

- Careful delaying can also be used to avoid certain choked situations by breaking ground from awkward patterns toward free bench faces.

- The common use of delays to divide the blast energy into smaller "packets" in order to reduce ground vibration and noise is also of particular importance in controlling damage to pit walls. The empirical relationships between maximum "instantaneous" charge weight per delay interval, distance, and the resulting peak particle velocity are well known. However, what happens to the pit wall in close proximity to a blast? A crude estimate can be made using the method shown in Figure 8. A rather better estimate can be made by use of BLAST, a computer program developed by Golder Associates. The program utilizes an empirical relationship developed by Korman et al., 1971 and can be used to predict peak particle velocity at any point on a planar grid or to print out a simulated seismogram at a specified location. The program was used to estimate peak particle velocities within a pit wall for an actual blast (Figure 9A). For comparison and to illustrate use of an improved delaying technique. Figure 9B was derived by modification of firing sequence and it illustrates the reduction of particle velocities behind the blast.

Although the prediction of vibration and damage to rock masses is in its infancy, it is obviously prudent to be considerate to the pit wall as well as the neighbor by careful use of delayed blasting.

A. ELECTRIC SEQUENTIAL DELAYING

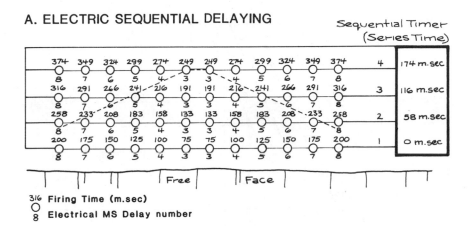

316 Firing Time (m.sec)
O
8 Electrical MS Delay number

B. NON-ELECTRIC SEQUENTIAL DELAYING

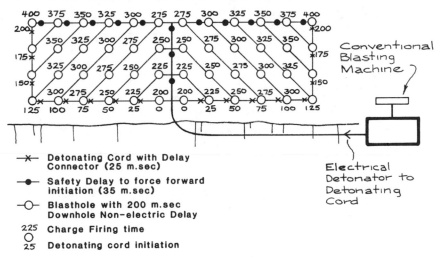

—✗— Detonating Cord with Delay
Connector (25 m.sec)

—●— Safety Delay to force forward
initiation (35 m.sec)

—O— Blasthole with 200 m.sec
Downhole Non-electric Delay

225 Charge Firing time
O
25 Detonating cord initiation

Fig. 7. Sequential delaying.

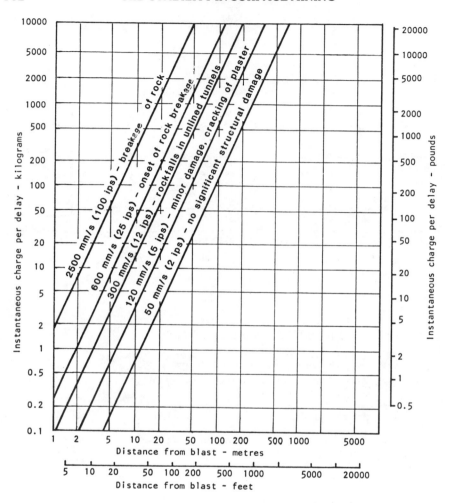

Fig. 8. Relationship between maximum charge weight per delay, distance and peak particle velocity. (After Langefors and Kihlstrom, 1973).

Production Blasting and Excavation Adjacent to Pit Walls

Close interaction between drilling, blasting, and excavation exists in any hard rock mining situation. When approaching pit walls, the need for careful blast engineering must be emphasized. Some of the common errors made when blasting adjacent to pit walls are illustrated in Figure 10A. More difficult digging can, and perhaps should, be tolerated adjacent to critical walls than would be accepted in the main production areas.

One common method of attaining steeper slopes is to develop multiple bench slopes by pushing back the benches to leave a large safety berm on every second or third bench. This method involves mining at normal bench height, but it does not leave the small, often inadequate safety berms typical of single benching methods. The multiple bench configuration must obviously be stable and especially during initial trial and excavation it should be monitored for any sign of slope failure. In addition, the bench faces should be developed and cleaned with special care if troublesome ravel is to be avoided. In certain extreme situations, scaling of slopes by means of a crane is used as an additional safety measure.

The following factors should be considered when blasting and excavating near interim or final pit walls:

- Careful shovel cleaning of pit walls is usually justified, especially when developing steep slopes or multiple benches.

- Considerable success has been reported from Bougainville Copper Mine where excavation or trimming to slope limits has been achieved by means of bulldozers working towards the shovels. In soft or fractured rock masses the method may achieve a smooth, sound bench face without overhangs, enable a reduction in powder factor and inefficient shovel clean up operations, and may in such circumstances eliminate the need for wall control blasting. Trial and economic evaluation would be necessary to determine the feasibility of dozer trimming at other operations.

- Backbreak or overbreak beyond the final wall limit is one of the major causes of wall damage and is obviously a wasteful activity. One of the simplest modifications that can lead to improvement in wall conditions is that involving adequate allowance for backbreak behind the final wall blasts. The backbreak distance behind a blast is a function of the blast and the rock mass conditions. Once an appropriate level of blasting control has been achieved, variation of backbreak will be a function of rock type and structure. Diggable backbreak should be reserved for each set of geological conditions. Diggable backbreak distance can be used as a design parameter in determining the distance that the back of the blasthole pattern should be offset from the final wall. In view of the importance of "back row" design adjacent to the pit walls, design of the blasthole pattern from the back to the front of the shot may be justified, provided that front row control can be maintained.

- In most production blasting and particularly when blasting adjacent to pit walls, multiple-row blasts should be avoided. Multiple-row or deep shots tend to "choke", fire irregularly and cause overbreak. Approximately three rows are optimum and the author feels that more than five rows should be avoided. Figure 11 illustrates how, subject to operating width constraints, a pit expansion can be mined-out in a series of relatively small steps involving drilling, blasting, and excavation back to ultimate pit limits. The sequence does require slightly more shovel mobilization than is necessary for "single" pass mining, but the results generally justify use of the method.

- Sub-drill should be minimized and in many cases can be completely eliminated when drilling holes adjacent to pit walls. When loading holes adjacent to pit walls, maintenance of optimum explosive rise and proper hole depths is at least as important as loading normal production patterns. To maintain adequate charge distribution and powder factor control, use of low-strength, low-density explosives is favored.

- Good charge distribution is also achieved by using smaller-diameter buffer charges than for the main production pattern. To maximize charge distribution and blasthole utilization, smaller-diameter holes are drilled on a closer spacing. At many operations, the drilling of one or more rows of buffer holes placed between the back of the main production pattern and the wall provides a significant improvement in wall condition. Buffer charges are fired in delayed sequence with the main production pattern and have the major advantage over traditional perimeter wall control because the rock is actually blasted rather than trimmed or split.

- A major problem facing many operations that are confronted with the need for wall control work is that of flexibility and capacity of the drilling system. Drilling using large, high productivity rigs such as the Bucyrus Erie 60R and Gardner-Denver 120 and hole sizes ranging from 250 mm (9-7/8 in) to 380 mm (15 in) is commonplace at many open pit mines. Drilling of smaller holes than are used for normal production is often inconvenient, time-consuming, and expensive and for scheduling reasons, almost impossible.

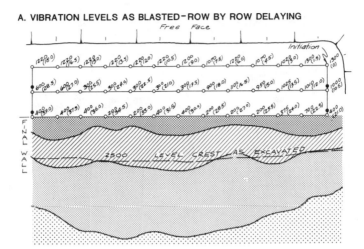

A. VIBRATION LEVELS AS BLASTED-ROW BY ROW DELAYING

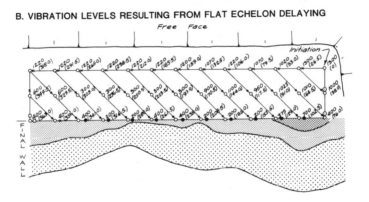

B. VIBRATION LEVELS RESULTING FROM FLAT ECHELON DELAYING

Legend

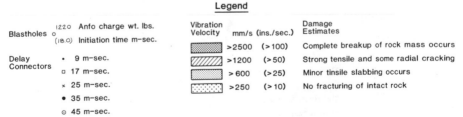

| Blastholes ○ | 1220 Anfo charge wt. lbs. |
| | (18.0) Initiation time m-sec. |

Delay Connectors	• 9 m-sec.
	□ 17 m-sec.
	× 25 m-sec.
	● 35 m-sec.
	⊙ 45 m-sec.

Vibration Velocity	mm/s (ins./sec.)	Damage Estimates
	>2500 (>100)	Complete breakup of rock mass occurs
	>1200 (>50)	Strong tensile and some radial cracking
	>600 (>25)	Minor tinsile slabbing occurs
	>250 (>10)	No fracturing of intact rock

Fig. 9. Comparison of peak particle velocities generated behind a pit wall by different delaying techniques.

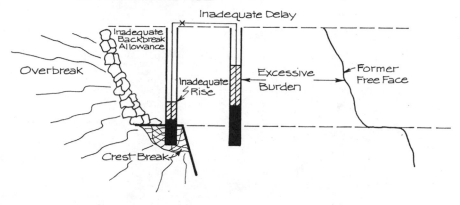

A. POOR PRACTICE

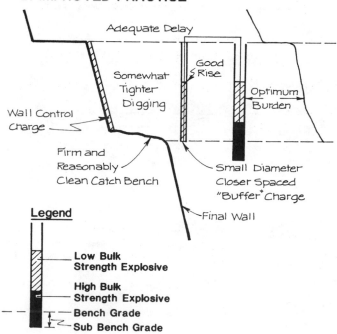

B. IMPROVED PRACTICE

Fig. 10.   Pit slope damage from production blasting and suggested remedies.

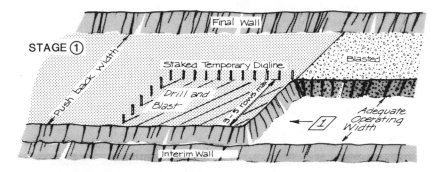

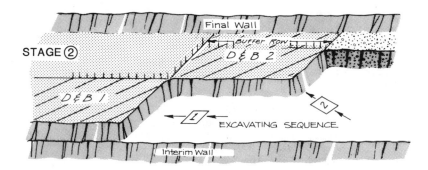

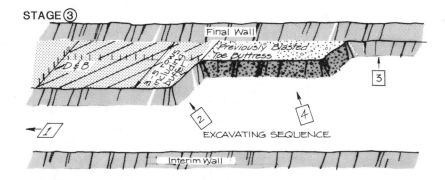

Fig. 11. Preferred drilling/blasting/excavating sequence during mining to final slope limits.

- Small decoupled buffer charges can be placed in the larger holes as a matter of expediency and for trial purposes. The advent of small-diameter, continuously coiled, cap-sensitive explosives such as Iremite has simplified loading of such holes. However, apart from cost and storage considerations, the process has a major drawback--it makes poor use of the blasthole volume!

- Once a functioning operation has demonstrated the benefits of wall control blasting, modification of existing equipment or purchase of new drill rigs should be considered. One promising technique that has been tried at a few operations and is available as an option from at least one major manufacturer. The technique involves minor modification of drill pipe racks and handling tools that enable the driller to easily switch from, say, 310 mm (12-1/4 in) to 250 mm (9-7/8 in) diameter holes adjacent to pit walls.

- New operations and those planning steep wall mining should give serious consideration to purchase of drilling equipment that is flexible enough to cope with the job of wall control as well as production blasting. The advantages of reduced stripping ratios or additional ore exposure can usually justify the expense.

- Slope failure is a time-dependent phenomenon and is frequently precipitated by removal of benches at the toe of a critical slope. Slope monitoring by survey and visual means will usually provide warning of impending failure, provided that the rate of failure is fairly gradual.

  To improve the safety of shovel operations beneath critical faces, progressive excavation back to final configuration is recommended. Stage 3 of an excavation sequence illustrated in Figure 11 includes the provision of a temporary toe buttress. The purpose of such a buttress is as follows:

  - To allow time and partial relief of the toe of a failure to occur so that the failure can be identified by monitoring and remedial action taken

  - To afford maximum protection to the equipment working on the bench

  - To provide maximum space for the excavation equipment to operate

  - To aid in equipment retrieval by assuring adequate operating space around a failure should it occur.

Wall Control Blasting (in the Strict Sense)

In the strict sense, the term "wall control blasting" or "perimeter blasting" is used to describe techniques such as presplitting which was developed for relatively small trial excavations. Ample references exist on the subject of perimeter blasting including those by Canmet 1977, Holmberg and Persson 1978, Kihlstrom 1972, and Langefors and Kihlstrom 1973.

The techniques of perimeter blasting have been applied to large-scale blasting in open pits, often with mixed success. The reasons for this limited success often lie in the scale and nature of the production blasting process (often the damage has been done before the perimeter blast has been fired). Once the open pit blaster is satisfied that the production and buffer blasting system is "engineered" and functioning well, smooth-wall or presplit trials may be evaluated using established methods of design. In many production situations, smooth-wall blasting, in which the perimeter charges are fired in delayed sequence towards free faces created by the main blast, appears to hold more promise than the presplit technique.

ACKNOWLEGEMENTS

The author wishes to thank Golder Associates for its assistance in preparation of this paper, Bougainville Copper Limited for providing the author with the opportunity to develop a basic practical understanding of blating principles, and the many clients of Golder Associates who have provided access to their open pits and an opportunity to evaluate the techniques discussed in this paper.

REFERENCES

Canmet, 1977. "Perimeter Blasting" Chapter 11, Pit Slope Manual, Canadian Department of Energy, Mines and Resources, Ottawa.

Hoek, E., and Bray, J.W., 1977. Rock Slope Engineering, Rev. 2nd ed., Institution of Mining and Metallurgy, London, pp. 271-308.

Holmberg, R. and Persson, P.A., 1978. "The Swedish Approach to Contour Blasting," Proceedings 4th Conference on Explosives and Blasting Technique, Society of Explosives Engineers, Morgantown, W. Va., pp. 113-127.

Kihlstrom, B., 1972. "The Technique of Smooth Blasting and Pre-Splitting with Reference to Completed Projects," Nitro Nobel AB Technical Memorandum, No. B672.

Korman, H.F., Mow, M.C.C., and Dai, P.K., 1971. "An Empirical Ground Motion Prediction Technique for a Buried Planar Array of Explosives in Rock," Proceedings 12th Symp. on Rock Mechanics. AIME, New York.

Langefors, U., and Kihlstrom, B., 1973. The Modern Technique of Rock Blasting, 2nd ed., John Wiley and Sons, New York.

Tansey, D.O., 1979. "The DuPont Sequential Blasting System," The Canadian Mining and Metallurgical Bulletin, Vol. 72, July, pp. 80-85.

Winzer, S.R., Montenyohl, V.I., and Ritter, A., 1979. "The Science of Blasting," Proceedings 5th Conference on Explosives and Blasting Technique, Society of Explosives Engineers, St. Louis, Missouri, pp. 132-142.

Question

Do you have any particular recommendations for blasting practices in operations where virtually all production blasts are of necessity "choked".

Answer

First, I would question the necessity for "choked" blasting. I would attempt to demonstrate the benefits of free face blasting by careful excavation to the next drill pattern by use of blasting diglines, careful front-row control, long drill patterns and scheduling of the drilling and blasting within the production cycle. From my experience, more consistent digging, controlled backbreak and reduced drilling and blasting costs will result from free face blasting.

Second, when faced with poor wall conditions but forced by scheduling, blending, space constraints, or by an unconvinced production department, I would attempt to free face long "trim" blasts adjacent to the pit slopes. Trim blasts would typically consist of three rows of holes including the buffer row adjacent to the slope.

Question

Your statement that blastholes should be fully loaded precludes the use of decoupling and decking in control blasting. What alternate methods do you propose.

Answer

When wall control blasting at many large open pits, many operators resort to decoupling or decking, primarily as a means of improving charge distribution and reducing powder (energy) factor within production size (>250mm diameter) holes. Where drilling equipment and capacity are available, smaller holes can be used to achieve the same ends, without the costs associated with the loading of decoupled or decked charges and the drilling of oversized holes that must be subsequently backfilled.

From my experience in large scale open pit blasting, the decoupling effect is far less important than achieving optimum charge distribution through use of properly spaced, smaller diameter, continuous charges that can be loaded efficiently.

Question

Is there any correlation between the direction of initiation of the explosives column e.g., bottom or top priming, and overbreak within the bench floor.

Answer

I am not aware of any direct correlation. My major concern when designing a priming system is to ensure optimum and consistent charge initiation. Proper charge initiation as well as the many other factors including consistent charging and drillhole pattern, lead to predictable results within given ground conditions. Blasting predictability provides many benefits including enabling a reduction of subdrill and good breakage to bench grade.

Question

Please comment on the fracture frequency versus ANFO powder factor relationship shown in Hoek and Bray's <u>Rock Slope Engineering</u> and attributed to you.

Answer

The relationship, also presented in this paper was determined empirically by me while working as blasting engineer at the Bougainville copper mine, Observation of the wide variation in fracture frequency in the porphyry deposit coupled with a corresponding variation in powder factor led to development of the empirical relationship which suggests that joint block size and shear strength are important in determining the effort required to swell and move already naturally fractured rock masses. Slurry explosives were used and the powder factors were calculated in terms of ANFO equivalent. Subsequent testing of the relationship in a wide range of geological conditions has been fairly successful.

Rather than providing a rigorous tool, the intent in presenting the relationship is to provide a guide for blasting design in naturally fractured rock processes and to attempt to provide an alternative to the "intact" or "continum" concepts of blasting design.

Question

Please could you give some details on the objectives and procedures of your BLAST computer program.

Answer

The objective of developing the program was to predict peak particle velocities around and at fairly close proximity to blast patterns consisting of a large two dimensional array of blastholes with varying charge weights fired at varying times. The program overcomes some of the problems associated with use of scaled distance relationships, in particular the point source assumption and the wave interaction problem. In the various situations tested, program Blast provided better estimates of peak particle velocity at points remote from the blast site than with the use of scaled distance.

The most useful aspect of the program is the generation of graphical output that enables a comparison of vibration level to be made between different delay sequences (Fig. 9).

3
Case Examples

Chapter 25

PRACTICAL ASPECTS OF WALL STABILITY
AT BRENDA MINES LTD., PEACHLAND, B.C.

G.H. Blackwell   and   Peter N. Calder

Chief Mine Engineer,
Brenda Mines Ltd.,
Peachland, British Columbia

Professor and Head, Mining
Engineering Department
Queen's University
Kingston, Ontario.

ABSTRACT

The development of an open pit slope monitoring
system, from equipment selection and justification to
complete computer data storage and analysis, is described. Methods of overcoming the limitations of the
electronic transit/distance measuring device and survey
system selected are discussed.

Results of the monitoring program showed that blasting
was a major cause of instability, and, after much experimentation, suitable blast patterns for mine production and pit wall integrity were designed. The experiences gained with double benching and wall cleanup procedures are discussed.

The monitoring program also indicated that water was a
contributor to instability, and methods of dealing with
water in a rock with a low storage coefficient and transmissibility are described.

Because of the danger of falling rock, photogrammetric
methods are used for joint mapping and analysis, this
work being contracted out. Joint strengths are measured
on-site using a portable shear tester.

Finally, a discussion of a complex failure is given.

The experience gained over the life of the mine contributed to the successful removal of high grade ore despite
the instability of the haul road and wall immediately
above the working area.

INTRODUCTION

    The Brenda Mines open pit is situated 40 km west of
Kelowna in the southern interior of British Columbia as
indicated on Figure I. The mine elevation is at 1500 m
and temperatures vary from -30 to +30C degrees with
400-800 centimetres of snow falling annually. A heavy
spring runoff occurs during April, May and June when the
volume of water pumped from the pit reaches 120 L/sec
from the nominal 3 L/sec over the rest of the year.

    After a two year construction period, production
started in 1970 at 7 million tonnes of ore and 7 million
tonnes of waste per year. Present production is 9 million tonnes of ore and 9 million tonnes of waste per year.
The mining plan consists of a series of three ever expanding shells being sunk into the core of the orebody,
giving higher grades and rates of return in the early
years.

    The low grade copper and molybdenum ore body occurs in
a relatively homogenous quartz diorite rock mass of
Jurassic age known as the Brenda Stock. Mineralization
and later faulting, shearing and rock jointing occurred
in many stages, often overlapping, (Oriel, 1972 and
Soregaroli, 1974). This has resulted in numerous
mineralized clay gouges which can have major effects on
wall stability. The major jointing system dips at about
70 degrees to the south, and strikes approximately east-
west. This has resulted in a relatively simple rock
structure as indicated in Figure 2. The major jointing
systems of the pit are visible in Figure 3, a photograph
of part of the north wall.

    The pit wall slopes were originally designed at 45
degrees before roads. Discontinuous stepped jointing has
reduced the south wall slope to 40 degrees. Although the
working bench height is 15 m, the final walls are double
benched so that there is 30 m between berms. The slopes
of the local berm faces vary from 50 to 70 degrees,

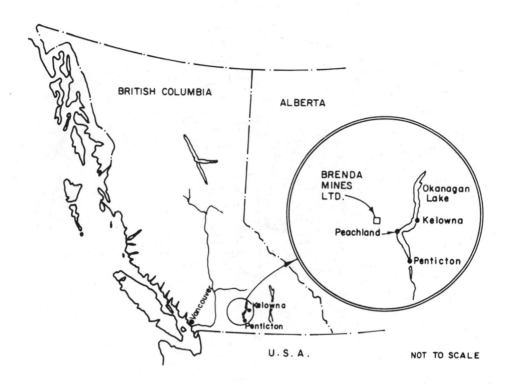

FIGURE 1: Location of Brenda Mines Ltd.

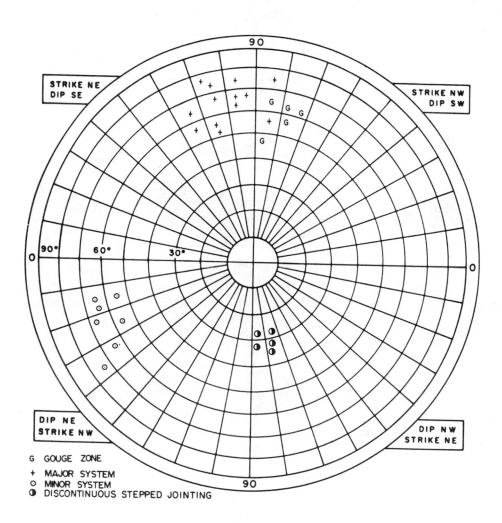

FIGURE 2:   Stereo Net Showing Typical
            Structural Systems

FIGURE 3: Photograph of the Pit North Wall

leaving berms that are usually wide enough for access.

STRUCTURAL CONSIDERATIONS

Mapping of Joints, Faults and Other Features

Two problems exist when mapping structural features by hand. First the notes are hard to follow some years later, and secondly, it can be dangerous. The danger is much greater when mapping the lower section of 30 m wall. For these reasons, photography is recommended as a mapping technique. Ordinary colour photographs complemented by field notes are a great help, but photogrammetric methods can be used to measure the dips and strikes of features on the rock face, and provide a lasting record of the wall for later comparison in quantitative terms. Unfortunately, this work requires equipment and expertise outside the scope of most mines, and is best contracted out on an annual basis.

The photogrammetric method involves the taking of two photographs which form a stereo pair. The process is repeated several times to ensure that all sections of the pit wall are included in at least one pair of photographs. The camera positions are surveyed using a Geodimeter. A set of targets are also used as check co-ordinates in the photographs. These targets consist of 1.2 m square pieces of plywood mounted on frames, and placed against the pit wall, at least two per stereo pair. A large black letter is painted on each plywood target. The lettering must be 50 - 100 mm thick to be clearly visible in the photograph. A Geodimeter reflector is placed on the top left corner of the target and surveyed with the Geodimeter to find its position.

Each stereo pair of photographs is then stored with the contractor, and a duplicate set kept at the mine. When analysis is required, the mine contacts the contractor, informing him as to which photographs and features are to be analysed. Calculation of the coordinates of these features is carried out using a stereo comparator. Figure 3 is a reproduction of a typical photogrammetric plate used for this purpose.

Estimating Angles of Friction

Because of its simplicity and portability, a Robertson Research (Hoek) shear tester (Hoek and Bray, 1972) was

purchased in 1974 to find the residual angle of friction
of the discontinuous joint plane shown in Figure 2.
Great care had to be taken in collecting samples to
ensure a proper fit in the casting moulds while the joint
plane remained undisturbed. The maximum normal stress
used was the equivalent of 150 m of rock. To obtain low
normal and shear pressures, weights were used as normal
loads, and a spring balance for shear loading. The
residual angle of friction was found to be about 26
degrees.

DEVELOPMENT OF A MONITORING SYSTEM

In 1974 it was decided that a slope monitoring system
had to be developed to monitor pit walls planned at final
heights of 400 m or so. The system had to be simple,
effective and of low cost. Its primary purpose would be
to warn the mine management of any impending pit wall
stability problems. Further, it would allow the mine to
operate near failing areas once the causes of failure had
been understood, and allow the safe extraction of ore and
removal of equipment. After consideration of alternative
monitoring methods, it was decided to adopt a system of
surveying known target points on the pit walls on a
regular basis. In order to obtain movement vectors of
the targets from one set of sightings, distance from
survey instrument to target, horizontal angle between the
target and a stable base station, and vertical angle from
survey instrument to target must be known. This meant
that a combination of electronic distance measuring unit
(EDM), and theodolite would be necessary, and special
targets would be required. Should serious stability
problems develop, requiring numerous measurements, a
system utilizing a computer would be added to reduce
manpower requirements and the time taken to analyse
survey readings.

After trials, the instrument chosen was the AGA 710
Geodimeter, which arrived at the mine in late 1974. This
unit measures angles electronically, and uses a red light
beam to measure distances. The visible red light beam of
the EDM allows readings to be taken at night. The readings are shown on a digital electronic display beside the
controls. An optional Geodat punch tape recorder was not
purchased until 1977, when the calculations required
became a major user of time and manpower. Early experiences with the AGA 710 are described (Backwell, Pow and
Keast, 1975). The Geodimeter is capable of measuring
angles to +/-5 seconds and distances to +/-5 millimeters

in a moderately sized open pit.

The majority of targets in use are MRM 25 mm cube corner reflectors, but some similar 60 mm AGA reflectors are used on important stations such as baselines. Attaching reflectors is accomplished in one of the following ways:

1) By bonding a strip of rubber tape between the rock face and the back of the reflector with instant glue. The rock face and sides of the reflector are then coated with quick setting epoxy resin to give longer lasting adherence to the pit wall.

2) By screwing a reflector onto a piece of pipe cemented in an "Air Track" hole.

3) A threaded wedge is driven and glued into a small crack in the rock face and the reflector screwed on.

If possible a numbered identification plate, (300 by 150 mm) is placed near the reflector, if not the number is painted on the rock face for identification. It is most important that the reflector be high enough up the rock face and sheltered by a small overhang or cover to prevent burial in snow or icing over during the winter.

A base triangle was constructed consisting of three concrete monoliths with instrument or target platforms at distances well outside the influence of the pit. Any of these base stations can be used as a reference or backsight when operating the Geodimeter from one of several similar concrete stations around the pit walls. The instruments and survey station are shown in Figure 4.

Readings with the Geodimeter are taken using the reiteration procedure, and as soon as the distances and angles are satisfactory, the data is recorded on the punch tape in the Geodat. The tape is then brought to the computer for processing and comparison with previous readings taken to the same targets.

Originally the Geodat punch tape was read using a back-up Fisher CP212/Interdata computer. This unit was on standby for a similar computer controlling many critical operations in the Brenda Mill. Two major programs were written in BASIC type computer language, including trigonometrical function routines. The first program read the tape and processed the data required for

FIGURE 4: Photograph of the Instruments on a
Main Survey Station

calculation of target coordinates. The second program
compared the new data with the old, drew visual display
terminal plots of readings versus time, and plans and
sections at any scale required. The operator could then
determine if and how fast a target was moving, which
events may have caused movement, and in which direction
movement was taking place. A paper copy was available
from the line printer.

With the expansion of mill computer requirements in
1979, a Texas Instruments DS990/10 computer was purchased
for mine engineering use in slope monitoring, grade
control, and production control, and to provide a general
purpose interdepartment computer for small tasks. Also
purchased was a Tektronix 4663 Digitizer/- Plotter as a
peripheral device.

Figure 5 shows the computer calculated coordinates for
a particular target (114) on the crest of a failing area
involving a major haul road. This failure is discussed
in more detail later. The accuracy of coordinates depends on the product of angle and distance and varies
considerably with the geometry of the measuring station
and target point. On average the accuracy of coordinates
is about +/-12 mm. Referring to Figures 5, failure
occurs during the spring runoff, starting suddenly in
April and slowing down in June. Examination of the
nearby targets in plan in Figure 6 shows the pattern of
movements along a fault plane. At the east end (target
104) movement is on the apparent dip of the fault, and
becomes movement on the true dip between targets 116 and
111. At the west side target 111 shows movement being
forced down another apparent dip by the buttress effect
of the pit wall.

Figure 7 is a section looking West. Here the movement
of the upper targets is seen to be moving on the fault.
The movement of the lower targets (130, 137, 141) indicates failure is on another plane of weakness, pushing up
the pit floor by the downward movement of the wedge
shaped section of rock over which the haul road is situated. This type of drawing is most useful in analysis of
rock slides, and is produced on the plotter for any
number of targets and time period.

The limitations of the data management system are few.
In the event of a Geodimeter breakdown, a simpler AGA 78
EDM is rented and used with a Wild TIA theodolite.
Readings are noted on paper and keyed into the computer.

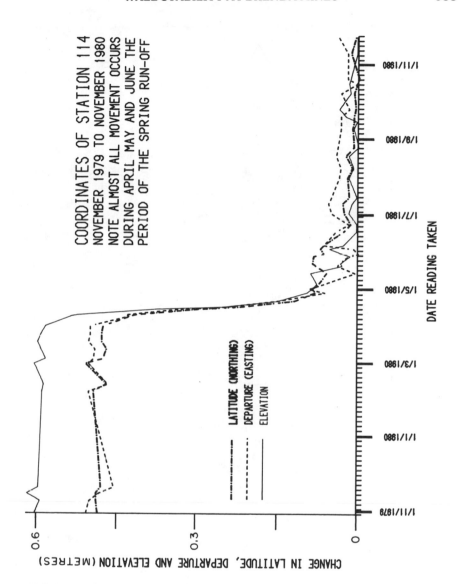

FIGURE 5: Coordinates of Target Reflectors, Late 1979 to Late 1980

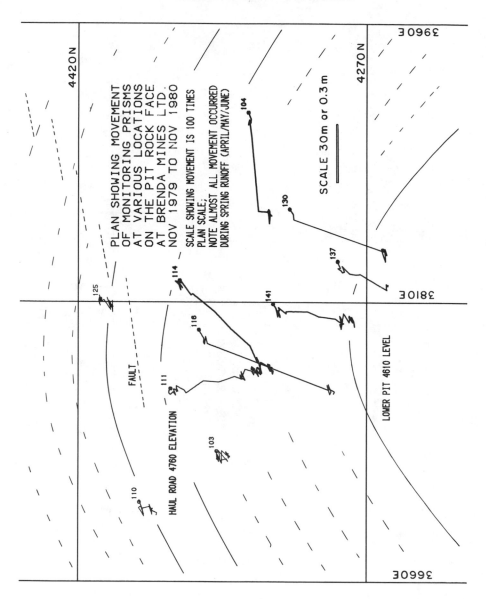

FIGURE 6:   Plan Showing Movement of Targets

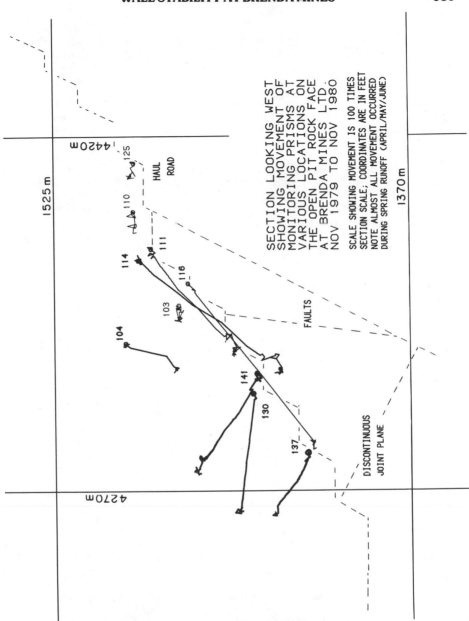

SECTION LOOKING WEST
SHOWING MOVEMENT OF
MONITORING PRISMS AT
VARIOUS LOCATIONS ON
THE OPEN PIT ROCK FACE
AT BRENDA MINES LTD.
NOV 1979 TO NOV 1980.

SCALE SHOWING MOVEMENT IS 100 TIMES
SECTION SCALE; COORDINATES ARE IN FEET
NOTE ALMOST ALL MOVEMENT OCCURRED
DURING SPRING RUNOFF (APRIL/MAY/JUNE)

FIGURE 7:   Section Showing Movement of Targets

In the event of a computer breakdown, calculations can be carried out manually, and compared with the hard copies of previous readings stored in files. Although the system appears to be quite complex, it is managed by a Mine Technician who has no computer background, but is a trained surveyor. However, the system could never be operated by personnel with no surveying training, and is dependent on the skills normally available in a mine engineering office.

There are some important limitations to the survey system. In the fall and spring there are short periods of dense fog at the mine, and heavy snows in winter. These prevent readings from being taken. Problems of ice forming on target faces can be prevented to a large degree by using covers and locating targets in sheltered positions. Also the accuracy might be inadequate when working under or on a failing area. Safety specification might require several readings per shift or require continuous monitoring. For these reasons, specialized monitoring systems are designed to meet the requirements of the particular failure.

A description of a specialized system developed by Brenda is given. One end of a beam was anchored onto the stable side of a fault, and a concrete (Calder and Backwell, 1980) pad anchored below the other end of the beam in the failing section as shown in Figure 8. Circular potentiometers were used to measure vertical movement between pad and beam. By using the potentiometers as pulleys, movement of wires anchored to the pad turned the pulleys giving a movement readout of +/-0.05 mm.

Unlike many extremely accurate measuring devices, the beam could be adjusted as the failure progressed. Thus as the fault dropped many feet, the system could still read to the accuracy required. Another feature of the system was the simple digital readout, which allowed any mine employee to inspect the beam and see how many thousandths of an inch the fault had moved, Figure 9. It is important to note that this system was used to supplement the Geodimeter survey system, not replace it. The two systems have two separate functions, the one to provide data over a large area for analytical purposes, and the other to precisely and continuously monitor movements in a critical situation.

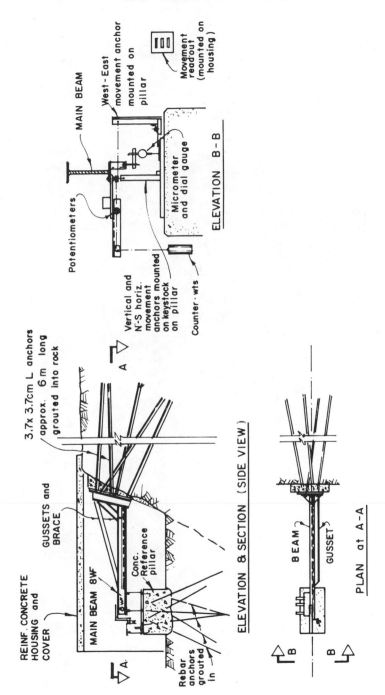

FIGURE 8: Basic Construction of the Beam and Pillar Monitoring System

FIGURE 9:   Photograph of the Beam and Pillar
            Monitoring System

RESULTS OF THE MONITORING SYSTEM

Blasting

The first and possibly most important result of the monitoring program was the discovery of just how blasting affects the stability of the pit walls. Uncontrolled blasting of thirty 300 mm blastholes per delay in a blast pattern was normal practice. The south wall of the pit moved up to 50 mm per day into the pit immediately after some of these blasts, and then stopped within a few days. The use of scaled distance to predict peak particle velocity or acceleration is well known. (Bauer andd Calder, 1970). The scaled distance [(distance from blast centre in feet)/(square root of pounds of explosive per delay)] was calculated and it was found that for scaled distances of less than 6, movement occurred. Thus by increasing the number of delays in a blast, the scale distance could be increased and the likelihood of in- stability decreased.

Figure 10 shows data for a particular station on the haul road in a failing area of the pit, and the in- stability caused by two blasts with scaled distances of 1.8 and 2.2 in late June 1978. By increasing the number of delays per blast and increasing the scaled distance to over 6, a corresponding improvement in stability can be seen for the months of July and August. Early in Septemer, several days of increasing rainfall were re- corded, and the slope, already weakened by blasting, began to fail because of the buildup of water behind the rock face.

However, the readings for the more accurate beam and pillar system showed that all blasting below the failure area caused movement of the failure. This displacement is shown in Figure 11, and scaled distances as high as 7.25 gave vertical movements of 0.317 mm.

One of the problems encountered in increasing the number of delays is the loss of fragmentation, and danger of misfires. Several months of trials were required to find the best system for blasting the rock at Brenda. For a two row blast with 24 holes per row, 75 millisecond surface delays are used between rows, and 15 millisecond delays used every third hole along the row. In most cases this increases the scaled distance by a factor of 3, with no decrease in fragmentation.

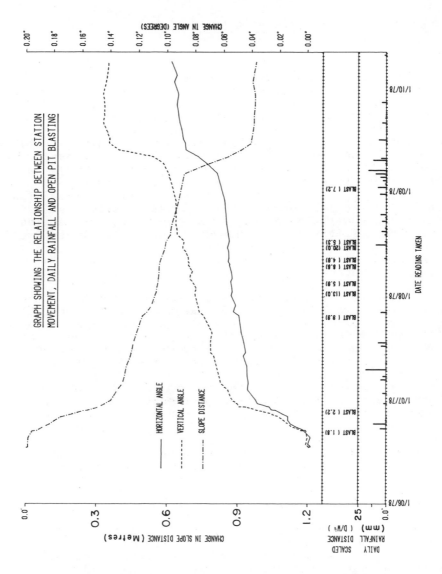

FIGURE 10:   Relationship Between Daily Rainfall,
Blasting Level and Movement

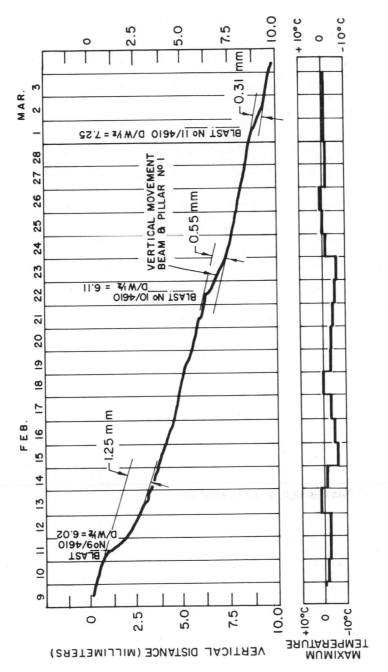

FIGURE 11: Typical Movement Versus Time Chart for the Beam and Pillar System

Wall control blasting is essential when operating with berm heights of 30 m. The system developed at Brenda consisted of placing 110 mm diameter waxed cardboard tubes inside the 300 mm blastholes and loading the tube. A small 70 kg toe load was necessary to enable the shovel to dig back to the planned wall. The spacing and burden on the trim row is 5 m and 6 m respectively. For the north wall with a 70 degree south dipping prominent joint plane, results are excellent. No modifications were required here when the 30 m berm height was used. However on the other three walls, problems were encountered because the berm tended to stand at too steep an angle, and cracking on the upper berm was evident. To prevent this a second trim row was placed behind the first. Spacing was increased to 6 m for the original trim row, and a row of 11 m deep holes were placed between these holes with a burden of 1.5 to 2 m. This shallow trim row had no toe load, and gave excellent control of the berm face angle. Figure 12 shows a typical example of this type of pattern and firing system.

Cleaning of the berm face when 30 m high is also essential if safe access is to be maintained. At Brenda, a scaling chain is dragged accross the face by a Caterpillar D9 bulldozer operating on the berm level above. Figure 13 shows the chain clearing debris and loose rock. Usually 150 m of face can be well cleaned in 8 hours. The operation is repeated when necessary.

Water

The second major cause of instability was found to be water. Examination of the rainfall data given in Figure 10 shows that with the exception of the large rainstorm of early September 1978, rainfall has not caused any major problems before or since. However, the spring thaw starting in early April has a significant effect on wall stability as seen in Figure 5.

The principal method of water control at Brenda is by open ditching around the pit to remove excess surface runoff; and by pumping from the pit bottom. Pumping rates are usually less than 3 L/sec until mid April, when an increase occurs culminating in a 120 L/sec peak in the first week of May. The rate slowly dies down to 60 L/sec by mid May and reaches the low of 3 L/sec by July. From pumping tests, the storage and transmissibility coefficients of the Brenda rock are so low that deep well pumping systems would not prevent the spring surge of

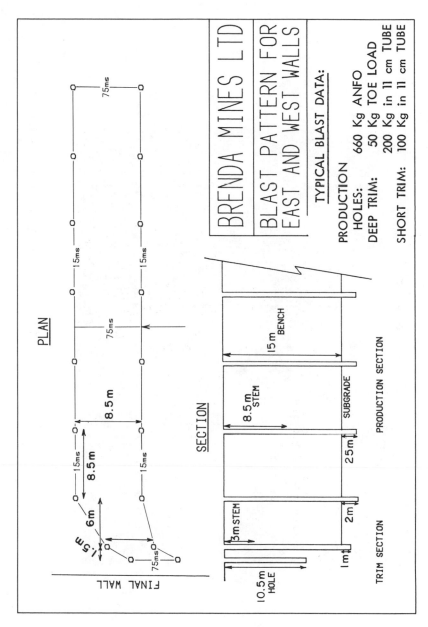

FIGURE 12: Typical Blast Pattern, East and West Walls

FIGURE 13: Photograph of a Bulldozer Scaling the Berm Face

water into the pit. However, it is the uplift force of
the water pressure rather than the quantity of water that
causes instability. It was thought that horizontal drain
holes might relieve water pressure. Two 150 mm diameter
drain holes were drilled up at 5 degrees to the
horizontal under the south wall. Although the rates of
flow were never more than 3 L/sec, the wall has remained
stable.

Some pumping in shallow holes (less than 30 m) has
been used to stabilize areas where water is a problem.
The Bucyrus Erie 60R Drill is used to bore 300 mm vertical holes for a depth of up to 30 m. The hole is cased
with slotted 270 mm pipe, and Webtroll 352S159 100 mm
diameter 1 L/4 HP 9 stage pumps used to dewater a
"curtain" behind a failure zone. The pumps often do not
draw down any water in other holes as close as 15 m away,
and have to be operated with automatic level floats in
each of the pumped holes.

DISCUSSION OF A COMPLEX ROCK FAILURE

The haul road leading to the active pit bottom intersected a major gouge zone on the North wall at the 4860
feet elevation during the Spring of 1978. This gouge
zone strikes east/west and dips 50 degrees south, as
indicated on Figure 14. Large amounts of water were
encountered when the road was sunk through the gouge
zone; however, this was neither abnormal nor unexpected.
Mapping of the gouge zone prior to sinking the haul road
indicated that the zone would pass along the road, but
would not daylight into the pit. The dip and strike were
such that failure would be most unlikely unless some
other structures, as yet unseen, were present.

In June, 1978, cracks were noticed on the centre line
of the haulage road. Figure 15 is a plot of movement, in
feet versus time, including blast and rainfall data, for
the period from June to October 1978. By mid June, the
cracking had become so bad that the inside of the road
was 0.7 m lower than the outside and a single-lane traffic pattern was adopted. Monitoring continued to indicate that production blasting was the cause. Figure 16 is
a photograph of the area at that time. The displaced
rock mass did not break up and, viewed from below, there
was no apparent change in the integrity of the mass which
had shifted. Mining continued below the area until the
end of August. Truck haulage continued behind the displaced zone using the single-lane traffic pattern. In

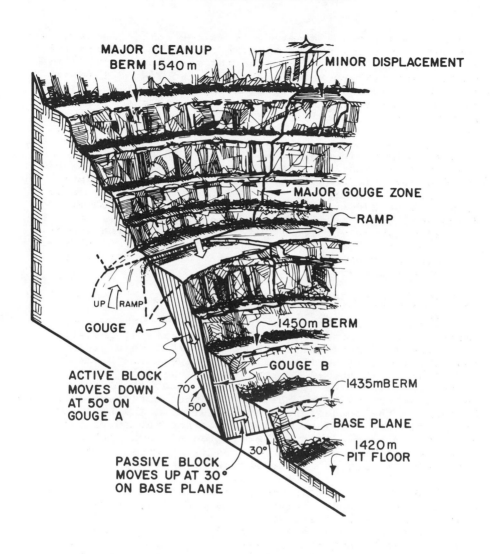

FIGURE 14:   Isometric View of Haulage Road
Displacement, Showing Major
Structural Features

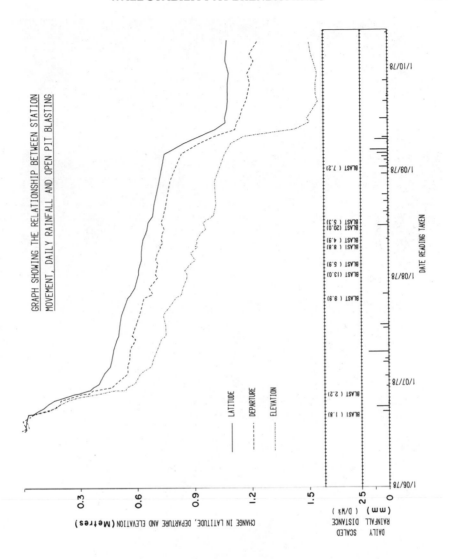

FIGURE 15:   Angles and Distances to Target
Reflectors, Summer 1978

FIGURE 16: View of North Wall Haulage Road Displacement

midSeptember, a continued period of heavy rainfall resulted in a rapid acceleration in the movement and a
vertical drop of 0.7 m in a matter of days. It was
decided that mining would not be resumed in the area
unless and until safe working conditions could be assured.

Referring to Figure 14, two major infilled fault zones
pass through the area approximately parallel to the road
in the displaced zone; these are referred to as Gouge A
and Gouge B. These gouge zones consist of a soft
molybdenum-clay mixture which has the consistency of a
solid lubricant. A bedding plane, dipping at 30 degrees
into the wall, indicated as the base plane, is common in
the area. The definition of the controlling structural
systems was aided by the movement data from various
points on the face. The active block in the center of
the slide is moving down a 50-degree dip to the south.
The passive block is moving up a 30-degree dip to the
south. In addition, the magnitude of movement of the
passive block is controlled by the magnitude of movement
of the active block, the predicted magnitudes for this
joint orientation are confirmed by the movement records.

Past experience during mining operations has confirmed
that gouge zones of this type form impermeable barriers
to ground-water flow. Figure 17 is a crosssection
through the displaced zone, divided into an active and a
passive block, as indicated.

The water-pressure distributions acting on the two
blocks during fully saturated conditions was defined as
indicated in Figure 17. The two gouge zones were assumed
to act as impermeable barriers. The water behind Gouge A
was not transmitted through to the passive block due to
the gouge thickness; the water acting on the base of the
passive block was equivalent to the head measured along
Gouge B to the surface.

Samples of the fully saturated molybdenum-clay gouge
indicated that no reliable degree of cohesion or friction
was present; these parameters were given zero values in
the analysis.

It was clear from all the evidence that the passive
block was being pushed up the 30-degree dip of the bedding plane forming the base. The active block was moving
down the dip of Gouge A. Part of the weight of the active
block was being transferred to the passive block and

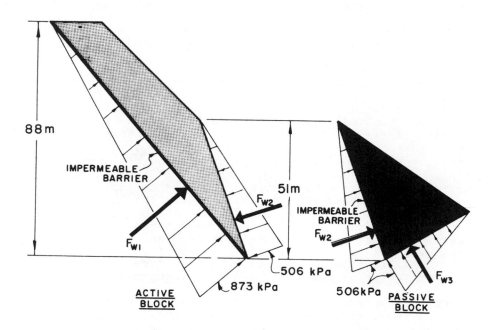

FIGURE 17:   Section Showing Assumed Water Pressures
on a Failing Slope

this, in combination with the ground-water pressure and accelerations due to blasting, was controlling the failure.

A stability analysis of the active and passive blocks was performed using vector diagrams and basic rock slope design methods as discussed. (Calder, 1970)

This indicated an unstable condition for the water pressures indicated and explains the rapid rate of movement experienced during fully saturated conditions. It was obvious that lowering the water table was required to maintain stable conditions. A drill hole placed in the toe area for investigative purposes soon closed off and it was clear that toe drains through the gouge would be impossible to maintain. It was decided to use well holes drilled from the haulage road to lower the water table.

The effect of lowering the water table on stability was analyzed. This analysis indicated that a factor of safety of 1.20 could be attained by maintaining the water table 6.5 m below the road surface.

The next step was to analyze the effect of blasting on stability. During the passage of blast-generated vibrations through the slope, the acceleration field due to gravity has superimposed upon it a second acceleration field. The level of acceleration due to blasting can be measured with conventinal blast monitoring equipment. Based on measurements at Brenda, the following relationship has been developed relating acceleration to scaled distance:

$$A = 295.9 \ (D/W^{1/2})^{-1.022} \quad \dots\dots\dots\dots\dots\dots(1)$$

where A   – peak acceleration level – in/sec*
       D   – distance from the blast – feet
       W   – weight of explosives (AN/FO) in pounds
       $D/W^{1/2}$ – scaled distance

*Note that the more common imperial units for scaled distance are used throughout.

This acceleration acts in a line from the shot point to the center of gravity of the unstable zone, in the horizontal plane.

Figure 18 is a plot of factor of safety versus scaled distance based on this analysis, a scaled distance of 8.3 or greater is indicated to maintain the factor of safety above 1.0 during the passage of blasting vibrations.

A more detailed analysis of this complex failures is available. (Blackwell, Pow and Keast, 1975)

SAFETY

The eventual purpose of any monitoring system is that it should give adequate warning of impending failure. This warning period must be early enough to remove all equipment from affected areas, and allow alternative production plans to be implemented. A plot of the data for both the Geodimeter and Beam and Pillar Monitoring systems is given in Figure 19. The criteria for safe extraction of ore from below the complex failure described was estimated at 120 mm per day or 6 mm per shift. These figures were arrived at on the basis that the maximum rate of movement of the slide had previously been of the order of 150 mm per day, and a safety multiplier of 10 was required.

Referring to Figure 19, on March 7th, it was evident from the Beam and Pillar system that acceleration of the slide was taking place, and plans were made to evacuate the area. By March 10th, the Geodimeter data was confirming the acceleration, which was by now at 6 mm per day. On the 11th of March operations ceased in the lower pit, and all equipment had been removed by the afternoon of March 12th. Although the warning given by the Geodimeter was much less than that given by the Beam and Pillar, both systems had fulfilled their function.

CONCLUSIONS

By installing slope monitoring systems at an early stage of a mine's life, a great deal of information can be obtained on the behaviour of wall slopes subjected to water pressures and blasting shocks. Development and implementation of remedial measures to counteract wall instability are best carried out as soon as possible early in the mine's history, so that when major problems do occur, trained personnel and techniques will be available to deal with them. These must be flexible enough to adapt to the unexpected.

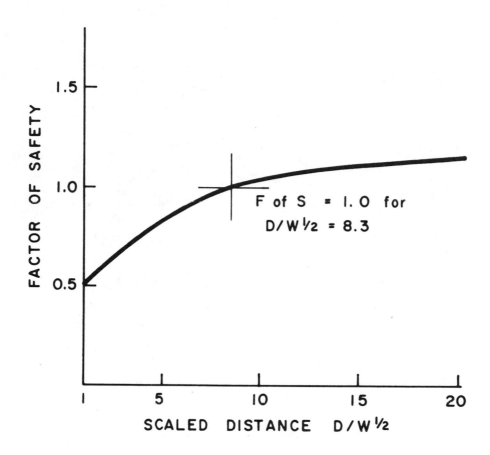

FIGURE 18: Factor of Safety vs Scaled Distance

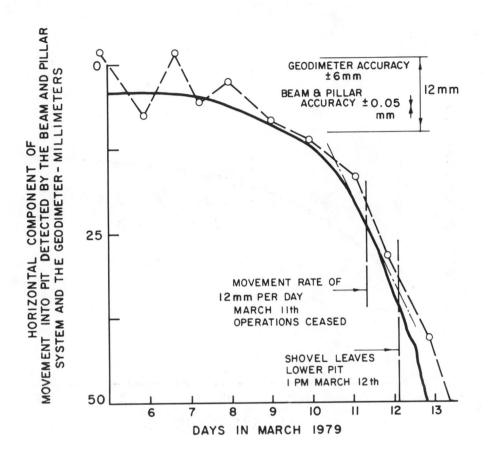

FIGURE 19:   Plot of Geodimeter and Beam and Pillar
Monitoring Data for a Critical Period -
Spring 1979

ACKNOWLEDGEMENTS

The authors wish to thank the management of Brenda Mines Ltd. for permission to publish this paper. The assistance of the mine personnel was appreciated.

REFERENCES

ORIEL, W.M.; Detailed Bedrock Geology of the Brenda Copper Molybdenum Mine, unpublished M.Sc. thesis, University of British Columbia, 1972.

SOREGAROLI, A.E.; Geology of the Brenda Copper-Molybdenum Deposit in British Columbia, CIM Bulletin, Vol. 67, No. 750, October 1974.

HOEK, E., and BRAY, J.W.; Rock Slope Engineering, The Institute of Mining and Metallurgy, London, England, 1972

BLACKWELL, G., POW, D., and KEAST, M.; Slope Monitoring at Brenda Mines. Procedings of the Tenth Canadian Rock Mechanics Symposium, Volume II; Department of Mining Engineering, Queen's University, Kingston, Ontario, September 2-4, 1975.

CALDER, P.N., and BLACKWELL, G.; Investigation of a Complex Rock Slope Displacement at Brenda Mines., CIM Bulletin, Vol 73, No. 820, August 1980.

BAUER, A. and CALDER, P.N.; The Influence and Evaluation of Blasting on Stability. Proceedings of the First International Conference on Stability in Open Pit Mining; Society of Mining Engineers of the American Institute of Mining, Metallurgical, and Petroleum Engineers, Inc.; Vancouver, British Columbia, November 23-25, 1970.

CALDER, P.N., Slope Stability in Jointed Rock, CIM Bulletin, Vol. 63, No. 697, May 1970.

Question

If a lowering of the phreatic line of only 6 m could have stabilized
the pit slide, did you consider the use of closely spaced vacuum well
points.
Could you elaborate on the chosen method of drawdown.

Answer

There was little room for any pump installation. Four wells and
six piezometer holes had taken up all the space available in the
area. The maximum quantity pumped was 0.6 L/sec, and the minimum
0.13 L/sec. Floats were used to control the pumps so that they didn't
run dry. The drawdown characteristics were such that no hole drew
water down in any adjoining hole, the piezometric constants being
extremely low.

The pumps were chosen because of their proven reliability, and
because they will be of use pumping much higher heads from small
diameter holes should the need for them arise again. Vacuum pumps
may not have been adaptable to future dewatering situations.

The water in the pumped holes was kept about 21.3 m below surface.
In the 27.4 m deep holes, 3.0 m was allowed for sludge accumulation
and 3.0 m to allow the control floats to turn the pumps on and off.
The higher water levels between holes would result in an average
drawdown of about 6.1 m. In the event of a sudden inflow of water,
the pumps would have increased the time required for the area to
become saturated because of their excess capacity.

Question

Would you comment on your method of analysis to determine F.S.
in terms of blasting vibrations. As discussed by Messrs Oriad and
Glass, peak accelerations alone give too conservative a F.S. and
therefore frequency or wavelength must be taken into account. This
may be supported by the fact that in some instances your calculated
F.S. were less than 1.0 during blasting.

Answer

This approach was conservative and this was desireable to ensure
the safety of men and equipment.

A velocity recorder was used to obtain the velocity characteristics
for many blasts. The velograms were converted to acceleration, and
scaled distance scatter plots made to find the acceleration for a
particular charge and distance. We do not feel that the frequencies
can be reliably predicted before a blast, so the scatter plot was
used to give a conservative analysis because of the safety
implications.

Prior to the 'Beam and Pillar' system, we used a scaled distance of
6 or more as reasonably safe. This did not cause movements that
could be picked up by the geodimeter survey within a few hours of a

blast under other areas where we have had instability in the past
(see our reference 4). However, the more accurate beam and pillar
records clearly show that the factor of safety was momentarily less
than 1.0 when returning to normal background in less than a day.
The measured agreed with the failure model we developed.

Several problems arise in the definition of factor of safety. 1) The
traditional definition of F.S. is that "failure" will occur at values
less than unity. However the meaning of the term "failure" ranges
from onset of movement to total slope collapse. Perhaps acceleration
of movement should mean a F.S. of less than one, but even this could
lead to major instability being regarded as safe in certain situations.
2) The "actual" F.S. for a complex failure is unknown. What is known
is a "calculated or theoretical" F.S. which is dependent on the
assumptions etc. on which it was made. 3) The F.S. for this
particular slide at Brenda can be transient as the disturbing and
resisting forces (water and blasting) change with time.

Hence in this case, the calculated F.S. was used as an index which
correlated well with measured rates of movement. Naturally this
correlation is specific to this failure only, but a similar treatment
of any slide is possible.

A F.S. of 0.8 to 1.0 would give very low failure rates, say 13 mm per
week, compared to background rates of 13 mm per month at a F.S. of
1.2. A factor of 0.5 would be quite dangerous, and 1.5 very safe.
Our analysis is based on a F.S. of 1.0 being stable because of our
conservative approach to any unknowns. We are opposed to any broad
rules being made because the time element, type and condition is
unique for each failure. For example, a brittle cement holding
large blocky ground would require little movement or time before
failing. For the failure we have described the reverse is true.   In
other words, use engineering judgement based on experience.

Question

Have you thought about using highway reflectors in place of costly
prisms as pitwall targets. The cost of 76mm highway reflectors is in
the range $3 to $6 in comparison to prism costs of $150 and up.
Any comments.

Answer

The 89 mm highway reflector cost 30 - 40 cents (1977). They have a
range of 152 m dirty and 306 m clean. We use them for survey pick
up work in the pit, but they are of too short a range for the
distances across our pit. Our experience would indicate that
expensive reflectors of this type are not better and much smaller in
diameter.

The 25 mm glass cube corner reflectors cost $70 and give excellent
results. We would not use anything of lesser quality, especially if
it was necessary to rappel over the rock face on a rope to install
the reflector. We wasted much time and effort in trying to utilize
the cheap highway reflectors. Until something better comes along,

the 30 or so glass units used each year are not considered
excessively expensive.

Question

Could you expand on your application of photogrammetry. Does it yield
reliable results, and how much does it cost.

Answer

At least every two years a contractor photographs our pit walls
using a phototheodolite to take a series of stereo pairs. The mine
engineering department surveys the camera and wall target positions
with the geodimeter as the photographer moves around. The plates are
stored with the contractor, but a stiff paper copy is kept at the
mine. For a wall length of say 3050 m this costs about $4000.

When a particular joint set, fault, berm slope or geological map is
required, the contractor is informed on which photos the features
can be seen. The contractor uses a stereo-comparator to find the
co-ordinates of any point in the stereo photo pair, and consequently
dips, strikes, etc. can be calculated. It is possible to plot stereo
nets and bench or berm plans directly when the stereo comparator is
interfaced with a computer and plotter. This equipment is generally
too specialized and expensive for the average mine to afford.

Provided well placed and well surveyed targets are visible in all
stereo-photo pairs, the results for each point found in the stereo-
comparator will be good. However, if a feature has little depth in
the photo pair, then results for dip and strike would suffer, but
no more so than readiness taken with Brunton compass.

When ravelled rock covers the structural features, and it is too
dangerous to take measurements of strike and dip, a pair of stereo
photographs of an area taken a year before are worth a great deal.

Question

How does the Beam and Pillar System work.

Answer

One end of a steel beam is anchored on or into the stable side of a
failure. The other end of the beam projects out over the failing
side of a joint plane etc. Below the free end of the beam a concrete
pillar is constructed on the failing mass. The distance between
beam and pillar is measured in three dimensions to define the movement rate and direction. Although not applicable to every situation,
the system is simple and inexpensive, and was well suited to the
failure described in the paper.

Chapter 26

SLOPE INSTABILITY AT INSPIRATION'S MINES

BY James P. Savely and Victor L. Kastner

Chief Geological Engineer
Inspiration Consolidated Copper Company

Geological Engineer
Inspiration Consolidated Copper Company

ABSTRACT

Inspiration Consolidated Copper Company is currently mining in four pit areas; Live Oak, Red Hill, Thornton and Joe Bush Extension, near Globe, Arizona. Small satellite orebodies lying outside the main pits have been mined out in a series of small open pits. Major slope instability has affected production in every pit.

Some consequences of the instability have been;

(1)   Serious interruption of mining plans, which resulted in deferred ore production, lost ore, reduced mining rates and increased stripping requirements.

(2)   Forced replacement of crushing facilities.

(3)   Serious effects on other production facilities.

The introduction of a slope stability monitoring and control function as an integral part of mine planning has been successful in reducing the impact of these failures and will eventually result in economic pit design based on probabilistic slope analysis.

This case history presents a brief review of the problems associated with slope instability at Inspiration's mines followed by a more detailed report on one specific failure area in the Thornton pit where the use of horizontal drain holes for slope stabilization appears to be successful. Relationships between slope movement rate, rainfall, and mining activity are shown.

609

INTRODUCTION

In mining operations a distinction must be made between failure based on analytical solution and slope failure that increases costs of mining ore.   Large-scale movement can be tolerated if the mine has flexibility and contingency planning.   In other situations bench ravelling can become costly.

At Inspiration Consolidated Copper Company near Globe, Arizona (Fig. 1), slope failures have been occurring since the early 1950s when underground block caving operations were stopped and open pit operations began.   Many real failure costs cannot be reconstructed from the records, but from past mining plans and memoranda it is apparent that many mining plans were significantly altered because of large-scale slope movement.   Step-outs, deferred mining, and backfilling were common responses.   Fortunately, the ore body was large and the grade was high and copper prices were high; profits could support costs incurred from slope failures.   Slope failures are still occurring and now current economic trends and lower grades require that these resulting costs be minimized.   Technology is reaching the state where good estimates of optimum pit design can be made based on probabilistic slope analysis and mine economics.

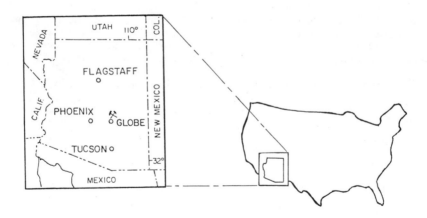

Fig. 1:   Location map for Inspiration Consolidated Copper Company
          near Globe, Arizona.

In October 1979, Inspiration began an intensive slope stability program to provide not only remedial solutions to the current stability problems but also to include rational pit slope design in mine planning. Since the start of the program the challenge has been to develop long-range solutions and at the same time complete quick analyses with recommendations that will lessen the impact of current failures. Immediate problems often seem to interfere with the long-range goals, and progress on long-range solutions moves at a slow pace. However, if the current problems are not solved, the long-range solutions can become needless paperwork.

When data are limited or inadequate and there is little time to collect more data, engineering judgment must be used. It is apparent from experience at Inspiration that there is nothing more useful to a mine geotechnical engineer than complete and accessible base data, and there must be a system developed to keep the base data current.

This paper is a progress report on the slope stability program at Inspiration and the impact slope instabilities have had on mining. The northwest wall of the Thornton pit is described as an example where attempts at stabilization appear to be successful. Probabilistic stability analysis is presented as the basis for future cost-benefit analyses and optimization of the pit design.

GEOLOGIC SETTING

Precambrian Pinal Schist is the major rock unit in the mine (Fig. 2). Moderate to strong deformation of the schist is present which resulted from multiple episodes of intrusive activity, tectonism, and regional folding. The schistosity strikes about N 50 E and dips $40°$ to $45°$ SE, but local variations in orientation have been mapped. There are several facies to the schist unit, but two facies most commonly seen in the mine are quartz-sericite schist and quartz-rich schist.

Two types of Tertiary intrusive rocks are also found in the mine. First is quartz monzonite porphyry, and second, coarser grained granite porphyry and biotite granite. Contacts between the granitic rocks are often gradational. Dacite and two conglomerates also occur in minor amounts.

There is a prominent and economically important north-northeast striking system of normal faults which include the Barney, No. 5, Keystone, Bulldog, and Miami faults. Dips on these faults are $25°$ to $55°$ SE. Vertical displacements of 30 m (100 ft.) to over 300 m (1000 ft.) have been interpreted.

A system of reverse faults has been mapped in the southwest area of the Live Oak pit and this same orientation is probably present throughout the area. These faults have a general west-northwest strike with dips of $40°$ to $45°$ SW.

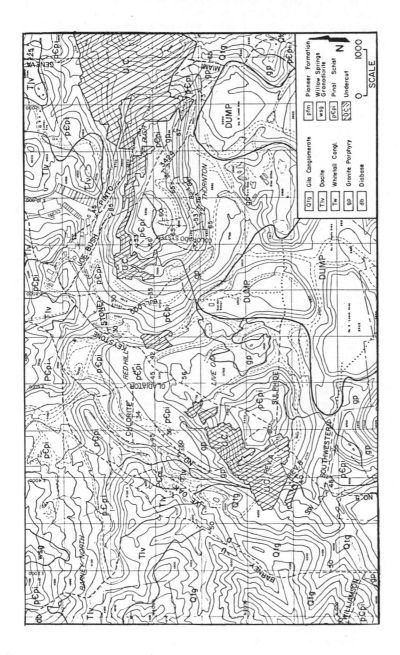

Fig. 2: General Geology of the Inspiration Mine Area.

Another significant major structure trend is represented by the
nearly vertical northwest striking Joe Bush fault. The Pinto fault
is also a northwest striking fault, but has a flatter dip of $35°$ to
$55°$ NE.

DEVELOPMENT OF THE GEOTECHNICAL DATA BASE

Current slope stability work at Inspiration is about a year
and a half old. Slope stability had been studied in previous years
by various mine personnel and consultants, but these studies were
not carried through extended periods nor were the results of the
studies incorporated into the mine planning program. This lack of
continuity has resulted in pieces of information on selected areas.
Priorities for the current studies are compiling basic information
and developing an analytical system for both quick and detailed
slope design. Included in these priorities are to:

1. Develop and upgrade geologic information and establish the
interpretive three-dimensional view of the geology on a scale
compatible with mine planning.

2. Develop routine pit mapping procedures, which include
observations on the geomechanical characteristics of the rock.

3. Compile and process basic structural information.

4. Determine rock and fracture strength properties.

5. Develop and install routine slope monitoring which consists
primarily of a survey network, prisms, and slidewire extensometers,
and most important, presentation of the information.

6. Document failure areas and develop slope failure histories.

7. Revise core logging procedures to include geotechnical
observations and routine sampling of full core for rock strength
testing.

8. Install and monitor water-level observation wells to
develop the piezometry of the pit area.

9. Develop quick analysis methods to aid in remedial solutions
to slope problems.

10. Provide slope-angle guidelines for mine planning based on
judgment and available information until additional data can be
obtained and detailed slope analysis can be done.

11. Install horizontal drains to aid stability in active
mining areas.

12. Develop routine reporting to management and planning personnel.

Slope stability work is maintained on a yearly budget of $US 170,000, which includes salaries of two full-time geotechnical engineers and one technician. Horizontal drain hole work which was not originally budgeted, will have a budget of about $US 250,000 through 1981. More expenditure is usually required in the first years of a slope stability program when procedures are being developed than in later years when the work tends to reach a maintenance level.

At the start of the slope stability work, an assessment of available geology maps showed a need for updating and interpreting. The first task was to begin a conscientious program to obtain current geologic information and to upgrade the maps. Although major ore-control geology consisting of faults which acted as the major control for migrating solutions and displacement of the ore body had been mapped in previous years, the structural geology information was incomplete. Numerous flat-dipping faults could be seen in the pit that were rarely seen on the pit maps. Now with emphasis on applied pit geology and mapping, the geology is beginning to appear more complete, and insights into the mechanics of the slope problems and knowledge of the ore body are being gained.

Pit mapping is routinely done by bench face mapping (Peters, 1978). This information is later transferred to a pit composite map and level maps. Not only is the rock type, mineralogy, and alteration included on the mapping sheets, but also rock hardness is designated based on a classification proposed by Robertson (1970) and by Piteau (1970). An estimate of the fracture intensity is also noted. Geologists carry a structural mapping form in addition to the standard pit mapping sheet (Fig. 3). The structure form is used to record the orientations and characteristics of significant joints and major structures. Recording this information does not seem to add appreciably to the geologist's field mapping time. Structure data are entered into a Hewlett-Packard 9845B desktop computer and filed as part of the routine posting procedure. In this way a current structural data base is available and retrievable without further compilation work.

To develop the capability for quick, accurate slope analysis the geotechnical data must be organized and retrievable. Basic to all geotechnical studies is geologic knowledge. At Inspiration the first task was to improve the geology maps by updating the interpretation. Then a structural data base was developed by compiling all structure information from old underground maps and pit mapping sheets and placing this information into the HP9845B system. Many of these old mapping sheets dated back to the early 1950s. Preliminary structural domain boundaries were defined based on the position of major known faults and lithologic contacts. Compiled structure data were sorted,

and lower hemisphere Schmidt plots were produced for each domain
(Fig. 4). Structure sets have been defined for most of the domains
and statistical distributions of fracture characteristics are being
defined with supplemental structure mapping (Call, Savely, and
Nicholas, 1976). Some of the domain boundaries will probably be
eliminated after further structural analysis.

Some strength testing was done in 1974 by Moore (1977), and
these data provided a base for additional testing. As major structures are exposed in the pits, samples are collected from the
surfaces and tested by large scale direct shear testing. This will
continue to be an on-going process. Results from the testing are
reduced to strength parameters considering both linear and power
failure laws. Statistical parameters associated with these strength
criteria are also determined. Again a basic data file of strength
properties is built on the HP9845B system.

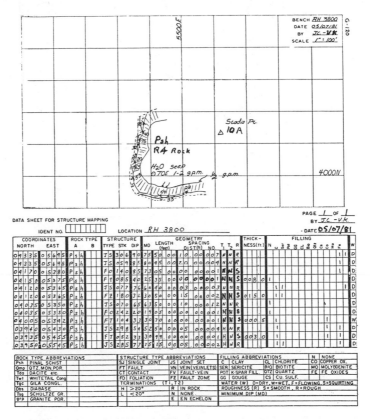

Fig. 3: Example of geology field sheet and structure data form.

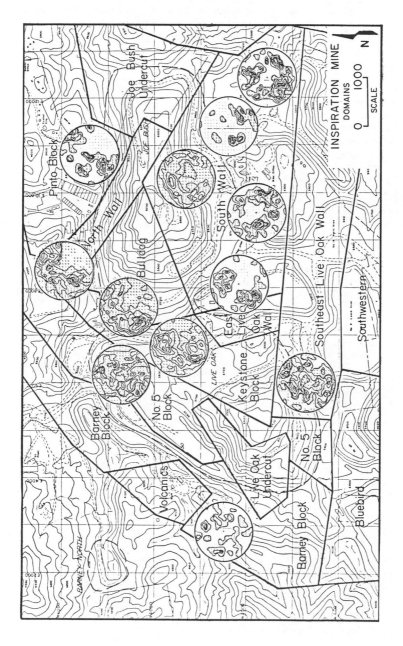

Fig. 4: Preliminary Structural Domains and Lower Hemisphere
Schmidt Plots of Compiled Structure Data.

Slope monitoring is essential in an operating mine because it is one of the best means for early failure detection and warning, and it provides a basis for understanding the failure mechanics. Survey networks give the best information because a three-dimensional vector direction is determined. These vectors can be compared with postulated failure geometries to determine the most likely geometry of the failure. Currently, Inspiration has prisms placed on moving slopes. These prisms are surveyed a minimum of once a week from permanent stations, located opposite the pit wall. Distances from the permanent stations to the prisms range from 200 to 1,000 m (700 to 3200 ft) for a single prism setup. There are several outlying points that tie into primary survey stations. Primary stations are read on a frequency of 6 to 12 months. Most of these stations are 60 m (200 ft) or more beyond the pit perimeter and will be read more frequently if they show movement or when each stage of mining is advanced. Slidewire extensometers are used to monitor tension cracks in unstable areas and to serve as warning devices when mining in the area. The warning devices consist of a switch set to trip and activate a strobe light if slope movement exceeds a specified tolerance. Generally, the tolerance is set for 5 to 15 cm (2 to 6 in.) depending on rock type and failure history. If the light is activated, mining is discontinued until a geological engineer can inspect the slope and assess the stability.

Review of past slope behavior gives an understanding of the failure mechanics and also stability parameters at the time of failure. All memoranda and reports were reviewed and aerial photographs examined to develop a failure history (Wiley, 1981). A map of all known slope movement areas was constructed (Fig. 5). The most serious slope problems have been present almost since the first open-pit mining began. Now, when new failures occur they are documented on a standard form, which describes the geometry and conditions before and after failure. These histories are filed by pit sector and back analysis is done when there is sufficient information. Tension crack mapping is kept up to date to document the extent of failure and to relate slope movement with known geology. Table 1 summarizes the failure history in the mine area.

Core-logging procedures have been revised to include a routine Rock Quality Designation measurement, determination of the number of rock fragments > 2.5 cm (1 in.), the length of the longest piece of core in the run, and a collection of a 30 cm (1 ft) stick of full core in every 15 m (50 ft) for rock strength testing. Rock hardness is also noted using the same classification used in pit mapping.

Very little water-level information and data on the hydrologic characteristics of the rock masses have been available. Botz (1971) reported on pump tests and seeps in the pit and concluded that the schist had a very poor water-bearing capability and transmissivities would be low. He concluded that the porosity of the schist would

Table 1: Slope Failures History of Inspiration Mines 1956 through June 1981.

Map Reference	Slope Design Sector	Failure Date(s)	Failure Mode(s)	Contributing Factors	Impact Of Instability	Comments
1	Joe Bush South Wall	Mar 1980	3 Bench wedge failure along minor faults & major joints.	Mining at toe, possibly high water pressures.	None	Similar structure persists along entire south wall but slope appears stable in most areas.
2	Thornton-Joe Bush	Jul 1956	Probably major wedge geometry along faults & major joints.	Flooding of old underground workings may have caused excessive water pressures in slope.	Lost & deferred ore.	Toe heave, movement southeast.
3	Thornton-Joe Bush	1956 Jun 1959	Rotation-like movement through highly fractured rock & along joint surfaces.	Weak rock mass, adverse joint orientations, high water pressures.	Subsidence in haul roads & building foundations, cracks present under Tertiary Crusher foundation.	Early warning for #7 failure mode when slope oriented north-south. Slope movement in lower benches evidence for deep seated failure.
4	Thornton Northwest	Dec 1977 Jan 1978	Plane shear on Bulldog faults.	Probably high water pressures, mining at toe.	Plan disrupted, ore deferred.	
5	Thornton Southwest	Mar 1953 Nov 1960 Dec 1960	Major wedge from intersection of Bulldog faults & step geometry on persistent jointing (step wedge).	High water pressures, mining at toe.	Step out required, deferred ore, reduced mining rate, disruption of access to upper levels of Thornton northwest & high grade ore. Possible progression of tension cracks to major haul road & maintenance facility.	Mining currently below toe of slide with movement on 28° slope angle. Casting required to gain access to northwest side of Thornton Pit.
6	Thornton-Joe Bush North Wall	Mar 1960 Dec 1969 Feb 1980	Plane shear on major joints, major wedge from two faults.	High water pressures, rainfall, creek 30 m (100') behind slope gave constant recharge to slope.	Haul road, guest house, admin. building affected, closed access road, step outs required. Increased stripping, deferred ore.	Slope very sensitive to rainfall. Toe heave on benches 60 m (200') below crest, slope flattened to 18°.

	Location	Dates	Failure Mechanism	Conditions	Consequences	Remarks
7	Thornton Northwest	Mar-1957 Oct 1958 Dec 1961 Mar 1963 Apr 1967 May 1969 Oct 1969 Jan 1972 Dec 1976	Plane shear on Bulldog fault. Three dimensional wedge on major joints Fault zone 60 to 90 m (200 to 300') wide. Faults dip 20° to 40°.	High water pressures, possible blasting effects, rainfall increased movement rates, mining at toe.	Deferred ore, interruptions in mining plans increased stripping, reduced production from Primary Crusher, replacement of crusher. Power poles affected.	Slope flattened to 22° to control movement. Movement in southeast direction.
8	Black Copper Pit	June 1972	Probable major wedge between 2 faults.	High water pressures, mining at toe.	High cost of removing slide mass, caused lost ore.	Failure occurred while stripping overburden and pit abandoned before ore extraction.
9	Red Hill North End	1980	Plane shear on fault dipping 25° to 35°.	High water pressures, chloritic fault gouge has low shear strength.	Caused re-establishing powerline. Lost ore. Tension cracks within 20 ft. of major haul road.	Slope flattened to 25°. Toe heave 30 m (100') below crest.
10	Red Hill North End	1980	Plane shear on 25° dipping fault, side release from steep dipping fault.	High water pressures, low shear strength, mining at toe.	Required re-establishing & rerouting power line, tension cracks within 15 m (50') of major haul road.	Slope flattened to 25°. Toe heave 30 m (100') below crest.
11	No. 5 Sector	1978	Major wedge on faults.	High water pressures.	Minor repairs to haul road.	Some toe heave in haul road 30 m (100') below crest.
12	No. 5 Sector 550 Sector	Nov 1960 Jan 1961	Major wedge on faults.	High water pressures, mining at toe, continuous major fault.	Haul road disruption, increased stripping.	Hanging wall of No. 5 fault has moved over 30 m (100') vertically, as much as .45 m (1.5') toe heave 197 m (650') below crest.
13	550 Sector	1976 to 1978	Major wedge on faults.	High water pressures, mining at toe.	Haul road disruption, increased stripping.	Probably progression of #12 failure. Slope is currently moving in response to mining.
14	Bluebird Schist Sector	Nov 1974 Mar 1975	Probably plane shear on fault.	Side release from major structure.	Interfered with short & long range mining plans.	Probably related to #5 fault structure.
15	Barney North Pit	Nov 1979	Plane shear on fault & foliation.	Low shear strength, 20°-40° dipping fault planes.	Lost ore, interfered with water management plans.	Slope failed shortly after stripping overburden.
16	Barney North Pit	Dec 1979	Major wedge along dacite-schist contact.	Low shear strength tuff unit between competent vitrophyre and schist, water pressures.	Loss of water storage reservoir, lost ore.	Slope failed on approximately 8° failure surface, toe heave on every bench.

only be about 3% and that water from one year's storms would be
equivalent in volume to about 8% of the effective porosity of the
entire rock mass. Thus, water pressures would tend to build along
faults and in fractures and reduce stability.

One measurement for water level that has been consistently
recorded over the years is the water levels in the blastholes. This
gives quite detailed information on where water levels were in the
pit but says very little about current conditions. But by plotting
blasthole water levels on the pit topography and determining where
water was occurring some assumptions on the water level configuration can be made. Observation wells were drilled in selected areas
and water levels were measured. A preliminary water-level contour
map was made using data from all sources. Eventually, this map will
be improved by installing and monitoring more wells and piezometers,
and water-level regimes will be defined to delineate areas of
similar hydrologic characteristics.

Because emphasis in mining is production it is often necessary
to make quick stability assessments to keep mining on schedule.
When the data is computerized, and stability programs are operational,
a good analysis can be done in a short time. During a crisis, there
is little time for additional data collecting and analysis must be
done using available information. These "quick-dirty" analyses
should be recognized as needing detailed analysis later with recommendations made on design changes that will eliminate similar problems from future mine plans.

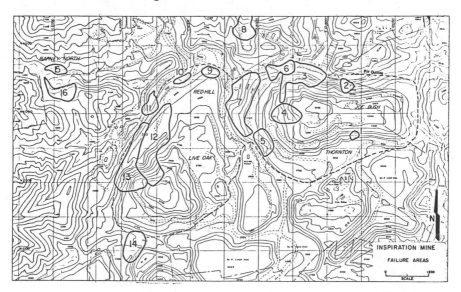

Fig. 5: Unstable areas as of May 1981.

When slope stability work was initiated in late 1979, a need for slope-design criteria had been present for several years because the failures had been affecting the mine plans, often at critical moments. An immediate requirement was slope-angle guidelines for the planning. At that time there was neither the capability nor the time to do detailed slope design in-house. However, the structural geology data compilation work initiated by consultants was nearly completed. The ultimate pit plan was divided into design sectors (Fig. 6). Slope geometry, known geology, and structural domain information were used to make evaluations on slope design for each pit sector based on kinematics. These slope-angle guidelines were provided for planning while more detailed study is being done.

High water levels in the slope, immediate slope movement after rainfall, wet slope faces, wet and sometimes flowing blastholes and development holes were all indicators of saturated slope conditions. A drainage program for the most critical slope in the Thornton pit was initiated immediately and is reported on in a later section. Planning and managing this work have taken considerable time in this first year of slope design work.

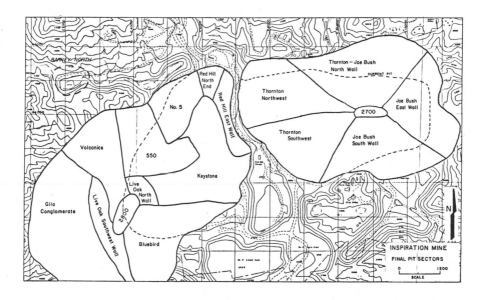

Fig. 6: Slope design sectors on an ultimate pit plan.

Unless important facts and recommendations are communicated much of the slope stability work goes unnoticed or has little impetus. A slope stability report which details areas of slope movement and implications of this movement on mining is provided to management and planning personnel each month. Recommendations on measures to alleviate production delays are made through active participation in weekly production meetings. Much of the effectiveness of the slope work must be credited to supportive management and cooperation from pit operations.

THORNTON SLOPE HISTORY

The present pit was sectored not only for use in design, but also as a means for referencing activities such as mapping, monitoring, and back analysis (Fig. 7).

The geotechnical data base and slope stability technology were applied recently in the Thornton Northwest Sector where faults intersect with faults and fracture sets to cause large-scale three-dimensional wedge failure and wedge backbreak on benches. In addition, water levels measured in observation wells indicated that water pressures were building behind the fault surfaces and aggravating an unstable condition. An immediate recommendation was made to control water recharge sources and implement slope drainage using horizontal drain holes.

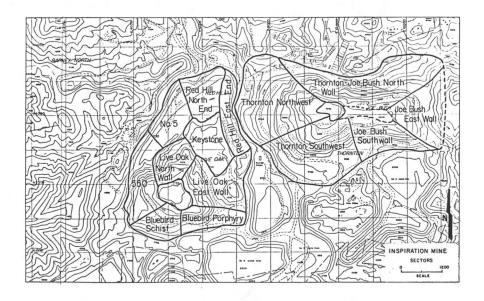

Fig. 7: Slope design sectors on the present pit.

Quick action was necessary because plans called for intensive mining with few alternative mining locations for the next two years.

Rock substance in the sector is two schist facies   a low-strength quartz-sericite schist with a uniaxial compressive strength of 41 to 48 MPa (6,000-7,000 psi) and a quartz-rich schist with a uniaxial compressive strength of 63 to 102 MPa (10,000-15,000 psi).

Moore (1977), reported difficulty in coring intact schist specimens because the specimens would tend to break along the foliation. During the testing he found that there was considerable strength anisotropy with uniaxial compressive strength of 47 MPa (6820 psi) parallel to foliation and over 124 MPa (18,000 psi) normal to foliation. It is believed these were samples of the quartz-rich facies.

The schist rock mass can be described as blocky, with fracture spacings on the order of 1 to 3 m (3 to 10 ft.), although there are areas where progressive failure on the benches make the rock appear quite broken and blocks 15 cm (6 in.) and less in diameter can be seen. Development of the foliation is variable and is often obliterated by alteration or is incomplete or absent. Gneissic character is sometimes seen.

Three times in the last two years mining attempted in the sector resulted in slope movement above the working level and excessive backbreak on the benches. All mining attempts resulted in delays and extra stripping. In some areas the bench could not be mined to plan because progressive bench backbreak involved several benches in an increasing failure volume. Some ore was lost, and the plan for mining the next lower bench affected. Slidewire extensometers were established around the head of the slide area and the movement monitored (Fig. 8). The slope responded quite expectedly to mining at the toe, but rainfall also caused immediate and significant increase in movement rate. A mining attempt in early 1980 was delayed two months because rainfall in combination with the mining caused a movement rate exceeding 6.5 cm/day (.15 ft/day), which was considered unacceptable at that time. This criteria was based on work reported by Zavodni and Broadbent (1978). Now after developing some movement records it is believed this value for progressive stage failure is conservative for the schist, but may be valid for porphyry and quartz monzonite in the area.

The Joe Bush fault and foliation intersect with the Bulldog fault and produce major wedge geometry (Fig. 9). Movement caused by this geometry has been occurring since the mid-1950s, and tension cracks at that time developed within 30 m (100 ft) of the old primary crusher, which was located near the present crest of the Thornton pit west wall. By 1969 the tension cracks had progressed to within 3 m (10 ft) of the primary crusher and within 9 m (30 ft.) of the old tertiary crusher which was located on the

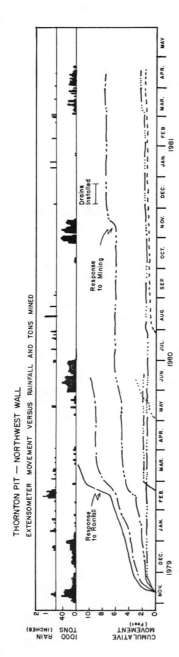

Fig. 8: Relationship Between Slope Movement Determined From Slidewire
Extensometers, Tons Mines, and Rainfall in the Thornton Northwest Sector.

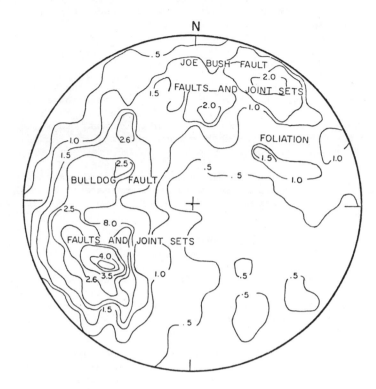

1477 Observations

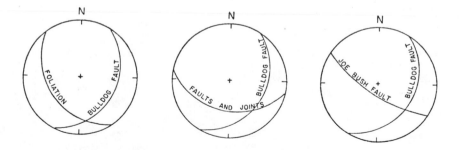

Fig. 9: Lower hemisphere Schmidt plots illustrating the
structure orientations in the northwest wall.

north slope in the Thornton-Joe Bush North Wall Sector. Plans were started to replace the crushing facilities and the slope was flatten-ed from 37 degrees to 22 degrees along the west wall and to 18 degrees along the north wall. An approximately 30 m (100 ft) step out, the "Buffer Zone", was left at the 3800 bench which was about 45 m (150 ft.) below the crest. Sloping the ground behind the crest and covering tension cracks was done to prevent water flowing on the surface from entering the slope. All of the measures taken apparently slowed the movement rate but did not reduce the failure's potential effects on future mining. All crushing facilities had been replaced by 1976 and the tension cracks had progressed under-neath the old primary crusher and ruptured the foundation. The crusher was demolished in 1980.

Attempts to mine the slope in 1979 and 1980 were delayed by movement and in October 1980 a decision was made to install horizon-tal drain holes to stabilize the slope and improve operating conditions.

STABILITY ANALYSIS OF THE THORNTON NORTHWEST WALL

Stability analysis showed that if drainage could be achieved and maintained 60 m (200 ft) behind the face the 364 m (1200 ft) high final slope could be mined at $25°$. A typical cross section consider-ed to show the relationships of Bulldog fault planes to the pit wall was constructed (Fig. 10).

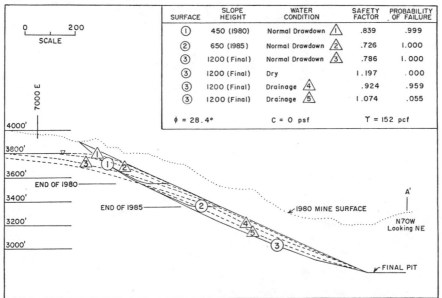

SURFACE	SLOPE HEIGHT	WATER CONDITION		SAFETY FACTOR	PROBABILITY OF FAILURE
①	450 (1980)	Normal Drawdown	/1\	.839	.999
②	650 (1985)	Normal Drawdown	/2\	.726	1.000
③	1200 (Final)	Normal Drawdown	/3\	.786	1.000
③	1200 (Final)	Dry		1.197	.000
③	1200 (Final)	Drainage	/4\	.924	.959
③	1200 (Final)	Drainage	/5\	1.074	.055

$\phi = 28.4°$ C = 0 psf Υ = 152 pcf

Fig. 10: Cross section of Thornton northwest wall.

Side release is caused by foliation, Joe Bush fault orientations, and fracture sets. The location of the last tension crack was also used as a portion of the failure surface. Results from tests by Moore (1977) on remolded fault gouge were used for the analysis. These parameters are: $Tan \emptyset=.540$; $\emptyset=28$; $\underline{SD} Tan \emptyset=.024$; $\underline{SD} \emptyset=1.4$; $c=0$; rock = 2.44 g/cm^3 (150 pcf).

The friction angle value ($\emptyset$) seemed high and cohesion (c) low considering that the gouge contains abundant clay as well as rock fragments. This points out the need for additional testing to determine the variability in shear strength.

For evaluating future mining plans various slope heights corresponding to sequences in the mining plan were considered. Probability of failure versus slope height and slope angle was estimated using Monte Carlo sampling of the distribution of strength parameters, which in this case was the friction.

Probabilities of instability for an undrained slope in the Thornton Northwest Sector are calculated to be:

Slope Height		Slope Angle					
m	ft	20°	23°	25°	35°	45°	65°
15	50	.005	.005	.050	.360	.887	.995
30	100	.005	.005	.050	.360	.900	.995
61	200	.005	.005	.381	.693	.983	.995
137	450	.005	.005	.995	.995	.995	.995
183	600	.005	.285	.995	.995	.995	.995
213	700	.005	.533	.995	.995	.995	.995
364	1200	.016	.724	.995	.995	.995	.995

(.005 and .995 are minimum and maximum cut off values).

The conclusions from the analysis were that a 20° undrained slope would be moderately stable over the 1,200-foot final slope height, however, stripping this amount of material would not be economic.

The probability of failure is calculated to be .995 for an undrained final pit slope 364 m (1200 ft) high and a 25° overall slope angle. If drainage is achieved to produce water-level configuration 5, shown in Fig. 10, the probability reduces to .055 and the mining plan could be met. The question remained whether the rock was permeable enough to drain to the required water level configuration.

HORIZONTAL DRAIN HOLE PROGRAM

On October 21, 1980 horizontal drain-hole trials were begun. Observation wells were drilled first above areas planned for the trials. Ten drain holes with depths of 150 m (500 ft) were planned. Initial flows of 0 to 0.019 m^3/min. (0 to 5 gpm) were produced. Before the trials could be fully evaluated the Thornton slope showed signs of accelerated movement due to mining and tension cracks appeared within 30 m (100 ft) of the present conveyor transfer. A complete program to depressurize the slope by installing drains from the 3700 bench was recommended and approved. By the end of December 1980, 42 drain holes totalling 5,694 m (18,790 ft) of drilling had been installed. An estimated 7500 m^3 (2 million gallons) of water had been drained from approximately 3.4 x 10^6 m^3 (9.6 million tons) of rock above the 3700 bench. Water levels in the observation wells had been lowered in excess of 9 m (30 ft) and were equilibrating (Fig. 11). The flattening of the drainage curves indicated that additional drainage would probably not be achieved with the present system and mining was continued with close monitoring of slope movement. The 3700 bench was successfully completed in March 1981, about three months behind the original plan. The slope showed some creep, but the large accelerated movement that had occurred in previous mining attempts did not occur: the tension cracks opened only about 5 cm (2 in.) by visual observation. Some minor adjustments occurred within the sliding mass.

A second row of horizontal drains was established in April through mid-May 1981 from the 3550 bench to predrain the slope in advance of mining, which would lower the failure probability and increase the chances for completing the mining plan without further delays. The maximum depth of these drains is about 180 m (600 ft) with an average depth of 127 m (420 ft). Depths were not sufficient to penetrate the Bulldog fault zone as interpreted from geologic information, but the slope immediately behind the face would drain and improve conditions for future mining. When mining is completed on the 3550 bench, drains will again be installed to increase the stability of the slope above the deep mining.

Twenty one holes have been drilled from the 3550 bench and initial flows of 0.120m^3/min. (30 gpm) have developed near failure area 5 (Fig. 5). Observation wells above the drains are again showing a drop in water level which is evidence that drainage and depressurization are feasible.

Downhole air hammer and rotary tricone drilling was required to advance the holes. The air hammer was most effective in the quartz-rich schist, whereas the rotary tricone was better in the softer quartz-sericite facies. Slotted 1½-inch diameter, Schedule 80 PVC pipe with 0.020 in. slot width and 3 rows of slots around the diameter of the pipe was installed in each drain hole. An unslotted collar pipe was left to attach to a collector system of 4-inch-diameter Driscoll pipe, which was laid along the toe of the bench.

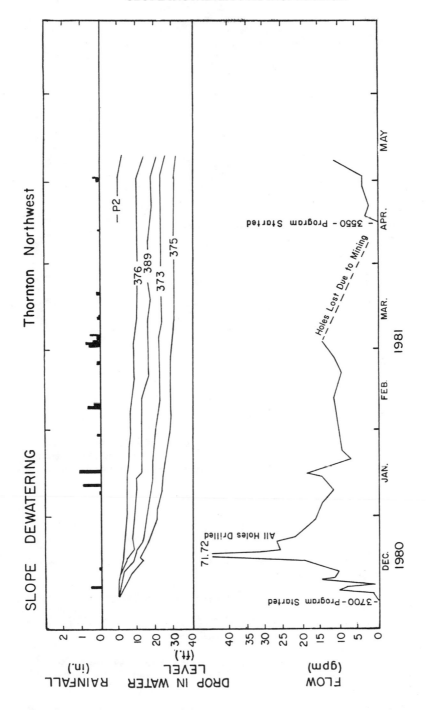

Fig. 11: Water level and total flow from the horizontal drains in the Thornton Pit.

This system was used to carry the water off of the bench to lower levels and underground workings beneath the pit where the water would eventually be pumped up the old Inspiration shaft and removed from the area.

The following costs were incurred for the drainage work.

Item	Number Of Holes	Length m	Length ft.	Costs $ US	Comment
Observation Wells	11	279	920	7,800	
Drilling:					
Trials	11	1,170	3,860	68,800	
3700	31	4,524	14,930	235,500	
3550	21	2,333	7,700	107,800	
Standby				1,400	14 hours
Mobilization				7,125	3 times
Demobilization				7,125	3 times
Site Preparation				3,500	
Collector System				1,000	
TOTALS	63	8,027	26,490	440,050	

FUTURE WORK

Engineering judgment has been used because information on critical design parameters was not available. Some of the future work planned to improve knowledge and increase mining efficiency are:

1. Continue geologic development work.

2. Investigate the hydrologic characteristics of the rock mass and install additional observation wells and piezometers.

3. Study alternatives such as drainage galleries developed from old shafts and workings.

4. Plan and implement water management programs.

5. Test fracture and fault strength properties with more large scale laboratory direct shear testing.

6. Study RQD, fracture frequency, and rock hardness data to determine if a rock classification system can be developed for blasting design and prediction or rock mass behavior.

7. Drill oriented core to further define fault zones and fracturing at depth.

8. Expand the slope monitoring system with additional survey points.

9. Analyze slope movement to determine if there are predictive characteristics.

10. Analyze each design sector in detail and determine probability of failure.

11. Simulate mining using current mine plans and the calculated failure probabilities to determine chances of obtaining ore.

12. Optimize pit design based on cost-benefit analysis similar to the procedure proposed in the CANMET Pit Slope Manual (1977).

13. Study of blasting effects on the pit walls.

14. Apply geostatics to study spatial dependence of fracture characteristics and rock mass properties.

Obviously, the geologic characteristics of the Inspiration ore bodies are such that slope instability will be a feature of operations for the life of the mine. During periods of heavy rain, such as occurred in 1978 and 1979, the problems will be particularly severe. Successful mining of the ore body demands that this be recognized. Slope stability, rather than being a peripheral function, is being incorporated as an integral part of the mine planning process.

ACKNOWLEDGEMENTS

Supportive management is the primary reason for success in the geotechnical endeavors at Inspiration. In particular the efforts of J. B. Winter, J. J. Ellis, and T. R. Couzens in reviewing and approving publication of the manuscript are appreciated. Work of this nature is never the result of one or two individuals and credit for data collection and presentation belongs to the members of the Mining and Geology department at Inspiration.

REFERENCES

1. Botz, M.K., 1971, "Groundwater and Slope Stability in the Thornton Pit", report to ICCCo., 7 pp.

2. Call, R.D.; Savely, J.P., and Nicholas, D.E., 1976, "Estimation of Joint Set Characteristics From Surface Mapping Data" Proc. 17th U.S. Symp. on Rock Mech., Snowbird Utah, Aug. 25-27, 1976, preprint volume pp. 2B2-1 to 2B2-9.

3.   CANMET, 1977, Pit Slope Manual, Ch. 5, Supplement 5-3, CANMET, Ottawa, Ontario, Canada.

4.   Moore, J. A., 1977, <u>Analysis and Design of Open-Pit Slope Angles at Inspiration, Arizona</u>, M. Sc. Thesis, Dept. Mining and Geol. Eng. Univ. of Ariz., Tucson, Az., 146 pp.

5.   Peters, W.C., 1978, <u>Exploration and Mining Geology</u>, John Wiley, New York.

6.   Piteau, D.R., 1970, "Geological Factors Significant to the Stability of Slope Cut in Rock", in <u>Symposium: The Theoretical Background and Planning of Open Pit Mines</u>, Johannesburg, S. Africa Aug. 31 - Sept. 5, 1970, A.A. Balkema, pp. 33-53.

7.   Robertson, A. Mac G., 1970, "The Interpretation of Geological Factors for use in Slope Theory", in <u>Symposium: The Theoretical Background and Planning of Open Pit Mines</u>, Johannesburg, <u>S. Africa</u>, Aug. 31 - Sept. 5, 1970, A.A. Balkema, pp. 55-71.

8.   West, R. J., 1979, letter report to ICCCo. 14 pp.

9.   Wiley, K.L., 1981, internal report, ICCCo. 14 pp + map (1" = 200' scale).

10. Zavodni, Z.M. and Broadbent, C.D., 1978, "Slope Failure Kinematics", paper presented at the 19th U.S. Symposium on Rock Mechanics, Lake Tahoe, Nevada, Univ. of Nev. - Reno, pp. 86-94.

Question

It appears that blasting is not an essential element required at the Inspiration Pit. Seriously though, is the failure problem at the pit more of a problem due to too steep a design, or due to lack of ground drainage.

Answer

Blasting effects on both bench and full slopes are quite noticeable. We are currently studying these effects and hope to incorporate better blasting design and practice soon. The majority of the slope failures that can be seen in the pits are caused by the combination of steep slope design, adverse failure geometry, and active water pressures. We will really know if these designs are too steep when we complete economic optimization. It is likely that the optimum design will, in fact, have to accept considerable slope failure. This means we must increase flexibility and contingency planning.

Question

How did you convince management and the unions to mine in an area that was moving at a rate of 1 foot per day.

Answer

This failure occurred in a small satellite pit which had a final slope height of under 300 feet. The movement vector for the sliding mass was within 20 degrees of being parallel to the strike of the slope and the vector had a plunge of 8 to 11 degrees. The hanging wall of the failure surface was blocky dacite that readily rubbelized upon movement. Consequently, the dacite ravelled off the side of the failure and into the pit. Some toe heave was noticed in an isolated area near one corner of the pit where the failure surface daylighted.

By routing haulage around the immediate rockfall areas, establishing a warning system on the extensometers, and by using mobile equipment, that is rubber tired front end loaders and 50 ton haulage trucks, mining could be done safely.

Communication of the problem and solutions was the major factor in convincing all parties that continued mining was possible.

Question

Does your monitoring system only consist of extensometers. Are you planning on expanding the system to include other monitoring means such as EDM-Prism monitoring, inclinometer monitoring and piezometers.

Answer

Currently, we have established a survey net consisting of permanent survey points, permanent observation stations, and prisms on active slopes. Extensometers are established in tension crack areas to add redundancy to the system. A continuous recording device is connected to one extensometer. In addition, level lines and steel pins are set in some moving areas. Water levels are recorded periodically in observation wells around the pits. Warning strobe lights are often attached to extensometers in moving areas where mining is progressing.

Future plans will focus on upgrading the survey network. More sophisticated or electronic monitoring devices would only be used under highly specialized circumstances, such as definition of a failure surface when there is no apparent structural cause for the failure. The main emphasis has been to keep the monitoring system simple and reliable.

Question

Who did the horizontal drain hole drilling for you.

Answer

Soil Sampling Service, Inc. of Puyallup, Washington has done the majority of our drilling. Ten of the 63 holes were drilled by Jensen Drilling Company of Eugene, Oregon. Both rotary tri-cone and air hammer drilling techniques were used.

Question

It seems that a more extensive slope dewatering program might help solve some of your problems, if so have you considered other methods of dewatering besides horizontal drains(e.g., perimeter pump wells, etc.).

Answer

Two other alternatives to horizontal drains have been considered at Inspiration: 1) perimeter pump wells, as you have suggested, and 2) an underground drainage gallery. Both of the alternatives would take more time to install and would likely delay the slope drainage in the short term. Pump wells would have to be maintained and would add to the cost. Because of mining priorities, we felt we did not have time to install and maintain pump wells, much less establish a drainage gallery. However, pump wells and a drainage gallery are being studied for possible utilization in a long term dewatering program. Because Inspiration was originally an underground mine, the possibility of re-establishing old workings makes a drainage gallery worthy of consideration. Cost studies will be done by next year on these alternatives.

Question

Could you give an approximate figure of the average flow obtained in the horizontal drains at Inspiration Mine.

Answer

Initial flows and sustained flows from our drain holes were quite variable. The chart below represents "average" flows for the two schist faces.

	Initial (l/sec)			Sustained (l/sec) (after 30 days)		
	Max	Avg	Min	Max	Avg	Min
quartz-sericite schist	0.75	0.1	dry	0.15	0.06	dry
quartz-rich schist	7.5	1.1	0.4	1.1	0.5	0.2

Most of the drains stopped flowing after an extended period of time.

Chapter 27

OPEN PIT SLOPE STABILITY INVESTIGATION
OF THE HASANCELEBI IRON ORE DEPOSIT, TURKEY

Caner Zanbak*, Kemal A. Erguvanli**, Erdoğan Yüzer**, Mahir Vardar***

*Assistant Professor, Department of Geology
Kent State University, Kent, Ohio

**Prof. Dr., ***Assistant Professor
Engineering Geology and Rock Mechanics Department
Mining Faculty of Istanbul Technical University, Istanbul, Turkey

ABSTRACT

The Hasancelebi iron deposit in Eastern Turkey is a very low grade (average 20% oxide) magnetite body consisting of 720 million longtons of proven reserve at a 10% cut-off limit. The ore-bearing scapolitic rocks cover an area of approximately 4300m x 300m, and the thickness of the outcropping deposit exceeds 300m.

Prefeasibility project studies revealed that a one degree decrease in a $45°$ final slope angle of the proposed pit would require excavation of approximately 10 million cubic meters of excess waste material, for a projected pit bottom 300 meters below the valley floor. Therefore, determination of the final slope angles of the open pit was considered to be the most important task in the feasibility analysis of this project.

In this investigation, the main emphasis was given to collection, compilation and evaluation of the engineering geological, hydrogeological and rock mechanics data of the highly fractured and weathered rocks. Due to the presence of intense fracturing, the rock wedge analysis method was used for projected 30m-high benches and the circular failure analysis method was used for the stability of the projected final pit slopes. The proposed bench slope angles ranged between $60°$ and $80°$, whereas the projected final slope angles varied between $30°$ and $48°$, depending on the slope heights and groundwater levels.

INTRODUCTION

The Hasancelebi iron ore deposit is a very low grade (average 20% oxide) magnetite occurence consisting of 720 million long tons of proven reserve at a 10% cut-off limit. The deposit is located 80 kilometers north of the provincial city of Malatya, in eastern Turkey (Fig. 1). The iron ore has no technical problems in terms of benefication and agglomeration other than minimizing a few impurities in the concentrates. Due to the low grade and large reserves, development of an open pit mine requires a very large amount of investment which will affect the feasibility of a poten-tial mine. However, the deposit is considered to be a significant future source of iron as consumption of this metal continues to increase. It is planned to upgrade the ore to 64% Fe pellets after the impurity problems are solved (Anon, 1977).

Fig. 1 - Location of the Hasancelebi Iron Ore Deposit

The deposit is located in a mountainous region of Turkey where the richest part of the ore crops out along the floor of a valley 1200 m above sea level. The altitudes of peaks in close proximity to the deposit are as high as 1650 m above sea level, and the average natural slope angle of the hills is 23°. The ore bearing rocks under-lie an area approximately 4300m x 300m. According to drill data, the thickness of the outcropping deposit exceeds 300m at the valley bottom in the eastern part of the ore body.

In addition to the low grade of the ore, there are a number of geographical disadvantages which considerably affect the feasibility of developing a mine, namely the presence of a main road, a railroad, a pipeline and intersection of two rivers over the ore body. The main road, railroad and the pipeline are to be rerouted from the mine site as mining operations progress. Under study are investigations concerning river diversions, and construction of a dam and a diversion tunnel for surface water control and water supply for the ore preparation plant.

Geophysical, geological and exploration studies of the Hasancelebi iron ore deposit were initiated in 1970 by the Mineral Exploration and Research Institute of Turkey (MTA). An area of approximately 75 square kilometers has been investigated, 148 diamond drill holes, (totaling 35000m) drilled, two adits were driven and 150 surface trenches were cut for ore grade and reserve determinations. In addition, 11 boreholes (totaling 2400m) were drilled for hydrogeological purposes. Upon completion of the surveys and prefeasibility studies of the ore deposit by MTA in 1975, exploitation studies were assigned to a management group of the Turkish Iron and Steel Works (TDCI). Three more adits (totalling 1000m) were driven, to assure the representativeness of the pilot plant ore samples and 23 additional boreholes (total length of 7345m) were drilled by the Hasancelebi Management Group in 1978-79. In addition to the 11 hydrogeological pump test boreholes drilled by MTA, 6 more boreholes were drilled by the State Water Works of Turkey for further hydrogeological investigation of the mine site.

The prefeasibility project study prepared by the Hasancelebi Management Group revealed that a one degree decrease in a 45° final slope angle would require excavation of approximately 10 million cubic meters of excess waste material, for a projected pit bottom 300m below the valley floor in the area investigated. Therefore, determination of the final slope angles of the open pit was considered to be the most important task in the feasibility analysis of this project.

The engineering geology, hydrogeology and open pit slope stability investigation for the assigned pit bottom at 900m elevation was executed by the Engineering Geology and Rock Mechanics Department of the Mining Faculty of Istanbul Technical University, during 1979-1980 period. Five progress reports and a final report were submitted to the Hasancelebi Management Group of TDCI (Anon, 1981). The final feasibility project of the open pit is under study.

GEOLOGY

The area of the Hasancelebi iron deposit has been investigated since 1930 by numerous geologists (Kurt and Akkoca, 1974). According to these investigations, an ophiolite complex (melange), emplaced prior to early Upper Cretaceous, constitute the basement rocks with which the ore deposit is associated. Upper Cretaceous pyroclastic sediments rest upon this basement, and these are unconformably overlain by an Eocene sequence of conglomerate, shale and limestone. At

the end of the Eocene, the ophiolite sequence was overthrust north-
ward on the Upper Cretaceous and Eocene sediments. The pyroclastic
rocks beneath the allochthonous ophiolite complex were changed to
scapolite-rich rocks due to heat and gases associated with intrusive
igneous activity and high pressures related to tectonic stresses.
The scapolite-rich rocks crop out in an area of approximately 15
kilometers x 3.5 kilometers. All of the aforementioned rocks were
ultimately buried by Oligocene and Miocene sediments and Neogene age
extrusive rocks. The regional geological map of the deposit is given
in Fig. 2.

The outcrop dimensions of the ore deposit which developed in the
scapolite-rich rocks are approximately 4300m x 300m and the deposit
has a trend parallel to the long axis of the scapolite-rich rock
zone. These rocks, range compositionally from marialite to meionite,
have various magnetite contents (5-46 % Fe_3O_4) and also contain diop-
side, garnet, tourmaline, biotite, pyrite, titanite, ilmenite, etc.

The geologic structure of the region formed by the Alpine movements during the Tertiary Period. Thrusts trending E-W and tear-
faults (strike-slip) trending N-S changed and displaced lithological
boundaries, and ore bearing zones within a highly fractured area.
The formation of the ore deposit is related to the Pyrenean phase of
the Alpine Orogeny.

HYDROGEOLOGY

Hydrogeological studies of the Hasancelebi deposit were aimed at
assessing (a) groundwater levels; (b) quantity and seasonal variation
of groundwater discharges; (c) quality of groundwater; and (d) drainage characteristics of the rock masses.

The hydrology and hydrogeology of the deposit has been investigated previously by four different institutions. During these
investigations, a total of 4 pumping and 13 observation wells were
drilled in several locations of the deposit to obtain information for
the hydrogeological parameters of the scapolite-rich rocks.

The two rivers (Ulu and Koy) intersecting over the ore deposit,
are the main sources of surface water in the area. The Ulu River,
which enters the study area from the northeastern part of the proposed
open pit, has a 374 square kilometer drainage area whereas the Koy
River has a 27 square kilometer drainage area. Prior to 1976, hydrological data were obtained from five meteorological stations around
the drainage areas of the rivers. Recent cumulative flow data of the
Ulu River and Koy River were obtained from a gauging station built on
the Ulu River in 1976. According to statistical evaluations of the
flow data, the maximum flow rate of the incoming surface water
entering the proposed pit area varies between 14 cubic meters per
second (m^3/sec) to 25 cubic meters per second (m^3/sec) for 5 and 100

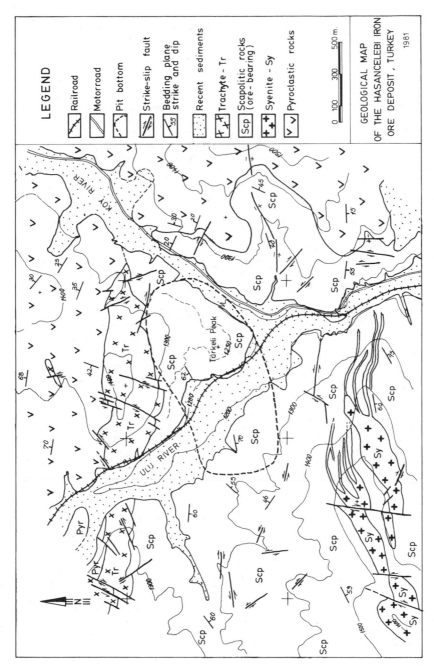

Fig. 2 – Geological map of the Hasancelebi iron ore deposit

year periods, respectively. The average cumulative water flow of the Ulu and Koy rivers is 45 million cubic meters per year (approximately 1.5 cubic meters per sec.

Rain water observations at the Hekimhan station (18 kilometers south of Hasancelebi) for the last 24 years reveal that the maximum rainfalls occur in April-May (maximum of 151 millimeters) and no rainfall activity takes place in July-August. Annual arithmetical averages of the rainfall values are maximum 538 millimeters and minimum 197 millimeters. Because of great variations in monthly rainfall values, hydrological studies of the area were based on daily rainfall records. The statistical maximum values of daily rainfalls show a variation between 39 millimeters and 63 millimeters for 5 and 100 year periods, respectively. Floodwater studies will be completed after compilation of the hourly based rainfall data.

In the prefeasibility study of the Hasancelebi Project, a dam construction was planned on the Ulu River approximately 1.5 kilometers north of the proposed pit area. Due to its close proximity to the pit and potential siltation problems of the dam during the lifetime of the open pit, alternative solutions of river drainage and planning of smaller dams are suggested by the authors.

All of the springs and the surface waters in the area investigated were observed in terms of their seasonal variations in discharge rate and water quality. Total discharge values showed a variation between 5.3 liters per second and 3.0 liters per second, for June and September of 1979. Approximately 90 percent of the surface water sources were found appropriate for irrigation and industrial uses.

The 1/2000 scale geological map was reevaluated for the hydrogeological purposes. Aquifers in the area were classified into three groups according to the porosity, permeability and transmissibility of the rock types, namely:

a)   Unconsolidated material (alluvium, terraces, slide-debris) - Very Permeable.

b)   Jointed rock with moderate porosity, permeability and transmissibility (scapolitic rocks, volcano-sedimentary rocks) - Permeable

c)   Jointed rock with low porosity, permeability and transmissibility (trachytes and syenites) - Impermeable.

The very permeable alluvium material will be excavated and drained after the start of the excavation process to minimize its effect on the open pit operations. The impermeable rocks are not considered as a water bearing material, however, they may cause confinement boundaries for the permeable aquifers, which have to be drained during the pit operations.

During the groundwater investigation of the potential open pit area, the major importance was given to the hydrogeological parameters of the permeable aquifers. Such aquifers consist of jointed scapolitic rocks and volcano-sedimentary rocks and were also reclassified as "Permeable" (down to 100m depth from the surface) and "Semipermeable" (More than 100m depth from the surface) according to the results of Lugeon Tests.

Calculated average values for the hydrogeological parameters of the permeable scapolite-rich rocks are:

Permeability, K: 1.3×10^{-5} meters per second

Transmissibility, T: 1.3×10^{-3} meters squared per second

Coefficient of Storage, S: 0.015

The detailed hydrological and hydrogeological investigation of the area revealed that groundwater levels in the Hasancelebi open pit region show a continuous decrease due to the excess discharge of groundwater from the area (total 0.09 million cubic meters per year). According to evaluations of the groundwater inventory of the area for 1974-1980, the recharge of the open pit area is 0.64 million cubic meters per year and the discharge is 0.73 million cubic meters per year.

During operation of the proposed open pit, the total amount of groundwater entering the pit bottom was determined to be less than 100 liters per second. A drainage investigation of the open pit is presently under study.

ENGINEERING GEOLOGY

Geological studies in the area investigated were based on 1/2000 scale topographical maps. Because of abrupt changes in the direction and magnitude of the magnetic field due to the magnetite bearing scapolitic rocks in the area, a geological compass could not be utilized. In addition, a sun compass was not considered because of its dependence on sunny days. The orientation of the discontinuities was measured by means of a simple device used in coordination with a transit theodolite. The use of a theodolite during joint data collection eventually provided better accuracy for plotting the location of discontinuities and lithologic boundaries on the map.

The previous 148 borehole cores obtained by MTA were not available for this investigation. For this study, 7345m of cores from 23 boreholes (located in the eastern portion of the proposed pit area) were evaluated in terms of their core recovery, joint frequency, RQD,

joint dips, extent of weathering and the physicomechanical properties
of the rock materials. Color photographs of these cores were taken
for archival purposes. The results of the engineering geological
studies are summarized below.

Rock Types

     The rock units in the Hasancelebi iron ore deposit area are
classified into five genetic groups.

     a - Pyroclastic rocks

     b - Syenite porphyry and microsyenites

     c - Trachyte, diabase and lamprophyre

     d - Scapolite-rich rocks

     e - Unconsolidated materials

     The geologic map (Fig. 2) shows the locations of these rock
types in the region. Many of the details of the rock types, vein
rocks, joints and normal faults plotted on the 1/2000 scale geologic
map   had to be omitted from Fig. 2 due to the change in scale.

     The pyroclastic rocks in the area include marine and continental
facies. Marine pyroclastic rocks are located north of the ore body
whereas continental pyroclastic rocks cover the eastern region of the
deposit.

     The syenite and microsyenites are the intrusive rock units mainly
located in the south and southeastern part of the ore deposit. Micro-
syenite dikes in various thicknesses (0.5cm to 1m) are abundant
throughout the scapolitic rock unit.

     The trachytes occur as stocks, sills and dikes and crop out at
the northern boundary of the scapolite-rich rocks. The trachytes
have a highly porous texture and the pores are filled with oligist.
Also in this group are mainly diabase and lamprophyre dykes.

     The ore bearing scapolite-rich rocks are the differentiation
products of pyroclastic volcano-sedimentary rocks under thermal
pneumatolic processes and tectonic events. The scapolite-rich rocks
are highly fractured, intersected by the syenite and diabase dikes
and highly weatherable as observed in recent road cuts.

     The unconsolidated materials are mainly Quaternary and Holocene
sediments (terraces, alluvium) and slope debris with thicknesses
less than 25 meters.

Discontinuities

Discontinuities in the ore deposit area were surveyed intensively.
Due to the magnetic character of the ore-bearing rocks, orientation
of the discontinuities was measured by means of a compass-like device
used along with a transit theodolite. Approximately 8000 discontinuity observations were made. The types of discontinuities observed
in the area are:

a. Bedding planes

b. Joints

c. Faults

d. Shear Zones

e. Dikes

f. Mineral veins

Locations of the discontinuity surfaces were determined by
means of surveying instruments. The recorded characteristics of the
discontinuities were:

a. Coordinates (location)

b. Orientation (strike, dip)

c. Type of the discontinuity

d. Rock type containing the discontinuity

The information for each observed discontinuity was coded and a
computer data bank was prepared to be used in determination of predom-
nant discontinuity sets in selected areas.

Borehole Drilling

Boreholes drilled in the area prior to this investigation had
been planned for grade and reserve determination of the deposit, thus
they were mainly located on the ore body. The borehole cores obtained
by MTA during 1970-1975 period were not available for this study.
The NX size cores obtained by Hasancelebi Management Group were intact
and available for the slope stability study. However, as mentioned
above, those diamond drill cores had limited use for the representation of the final pit slope materials because of their locations.
During this study 18 boreholes (total length of 2240m) located on the
projected slopes around the open pit, were requested by the authors
to obtain data on the slope materials (Fig. 3).

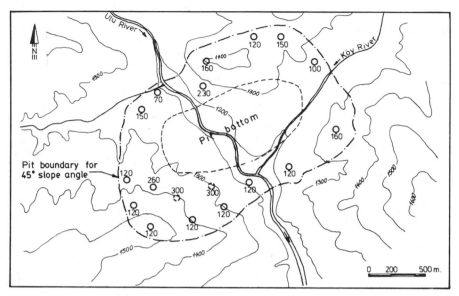

Fig. 3 – Location of the boreholes drilled for the slope stability investigation

Borehole Logging

    Logging of the diamond drill cores (average 90 percent core
recovery) provided invaluable information on jointing and the
material properties of the rock mass. Variations in core recovery,
RQD index and joint frequency value were recorded for each borehole.
Dip values of joints in the cores were also recorded as rosette diagrams for 20m depth intervals, to find a correlation between surface
and underground discontinuities. However, due to the intense jointing
and the weathered nature of the ore-bearing scapolite-rich rocks, no
such correlation could be made. The RQD values showed a variation
between 10% to 60% whereas the general joint frequency values were
greater than 15 joints per meter. According to the borehole surveys
(along with their mechanical properties) the rocks in the proposed
open pit area were considered to be poor – fair quality (Deere, 1977)
or as very weak – weak rock (Muller, 1963).

    Color photographs of the cores were extremely helpful in evalua-
ting the borehole logs prepared in the field. Taking color photographs of the cores is highly recommended in rock engineering projects
for their future invaluable usage.

Natural Slope Angles

    The natural slopes and present mass movement activity in a pro-
ject area can provide some information for the design of artificial
slopes. Natural slopes as high as 450-500m in the area investigated

do not show any sign of considerable mass movement. Therefore, the natural slope angles were considered as the lowest boundary for the selection of the final slope angles of the proposed open pit, for very long term stability. These data can provide guidelines for the design studies, however their use in slope stability analysis will cause an unacceptable engineering overconservatism.

The natural slope angles in the northern section of the two intersecting rivers in the study area were treated statistically to obtain relative densities of the slope angles in three different predominant rock types. The average natural slope angles in scapolite-rich and volcanic-sedimentary rocks were similar and varied between 21°-24°, whereas trachytes showed higher values (25°-37°) as seen in Fig. 4. These results reflect the similarities in the behavior of scapolitic and volcano-sedimentary rocks during the geological slope formation processes. Higher natural slope angles in trachytes are related to their petrographic features which provide slightly greater durability against weathering and erosion as compared to the scapolitic rocks.

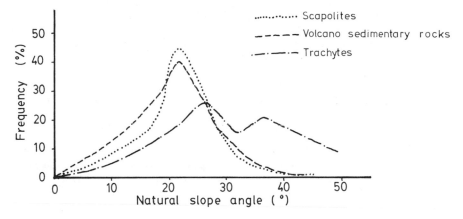

Fig. 4 - Frequency Distribution of the Natural Slopes

Seismicity

The Hasancelebi iron deposit is located in the third degree earthquake zone of the Anatolian Peninsula. There are five earthquake epicenters in the region, with magnitudes between 4.3-4.7 recorded in the last 75 years within a 100 km radius. A seismic risk analysis study of the area (Ada and Ata, 1979) reveals that the maximum horizontal ground accelerations are:

a - 0.11g with a 1/900 year frequency (for an open pit life time of 50 years)

b - 0.125 with a 1/1950 year frequency (for an open pit life time of 100 years)

MECHANICAL PROPERTIES OF THE ROCK MASS

The assessment of the shear strength parameters of the jointed rock mass in still one of the most questionable tasks in the rock engineering practice. Especially, for the design of high open pit slopes, where no previous information on the mechanical behavior of the rock mass is available, the determination of shear strength parameters becomes more difficult. The potential Hasancelebi iron ore mine is an example of this.

The projected open pit bottom is overlain by a minimum of 300 meters of overburden and the maximum final slope heights reach 600 meters in some sections of the potential open pit. The main focus of determination of the rock mass properties were based on the engineering geological studies and mechanical properties of intact rock samples obtained from the borehole drill cores.

The borehole cores were sampled at approximately 10 meter intervals for systematic mechanical testing of the intact rock and discontinuity surfaces. A summary of the physico-mechanical properties of the intact scapolite-rich rocks is given in Table 1.

Unit Weight kN/m^3	Porosity %	Slake Durability %	Uniaxial Comp. Strength MPa	Cohesion MPa	Internal Friction Angle (^{o})	E Modulus 10^{3}MPa	
						Static	Dynamic
27 - 31	1 - 5	80 - 98	78 - 100	18 - 29	60 - 42	7 - 20	30 - 60

Table 1 - Generalized Physico-mechanical Properties
of the Intact Scapolitic Rocks

Friction angle values of the joints tested in the scapolite-rich rock at low normal pressures (0-300 kPa) are found as $34^o \pm 3^o$.

Estimation of the shear strength parameters of the jointed rock mass using the available data was based on the general decreasing trend of cohesion and the internal friction angles with increasing jointing. The results provided by several researchers were used as guidelines for the assessment of the shear strength parameters of the jointed scapolitic rocks (Hoek and Bray, 1977; Stimpson and Ross-Brown, 1979; Manev and Avramova-Tacheva, 1970).

Engineering approaches to the determination of the rock mass cohesion by Stimpson and Ross-Brown (1979) and Manev and Avramova-Tacheva (1970) present the relationship between the ratio of rock mass cohesion-cohesion of the intact rock and the intensity of jointing (Fig. 5). These diagrams were found appropriate to use at the feasibility stage of the Hasancelebi open pit slope design. The general intensity of jointing obtained from the borehole cores is greater than 15 joints per meter for more than 80 percent of the logged boreholes. The rock mass cohesion ratio corresponding to 15 joints per meter intensity is 0.02, as can be seen in Fig. 5. Using the average cohesion value of the intact scapolitic rocks (25 kPa), the apparent cohesion value of the jointed scapolitic rock was selected as 490 kPa.

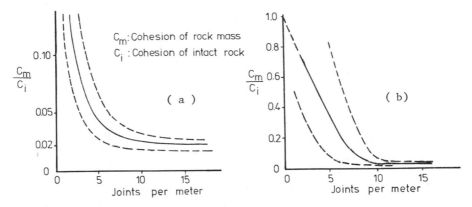

Fig. 5 - Relationships Between the Rock Mass Cohesion and Intensity of Jointing. a) Manev and Avramova Tacheva,1970, b) Stimpson and Ross-Brown,1979.

The selection of the apparent friction angle of the scapolitic rock mass was based on the present review and discussions on the shear strength parameters of the jointed rock masses (Hoek and Bray, 1977). Evaluations of the several criteria and back analysis results for the shear strength of the metamorphosed rock masses and the judgement of the overall engineering geological study of the area, led the authors to the use of 35° for the apparent friction angle value of the scapolite-rich rock mass in the Hasancelebi iron deposit area.

SLOPE STABILITY

The determination of the potential failure modes is one of the most important criteria for the design of rock slopes. The general mode of slope failures in rock masses are classified into four groups.

a - Planar failure

b - Wedge failure

c - Toppling

d - Circular failure

Kinematic conditions of the rock blocks surrounded by Systematic discontinuity sets, control the planar, wedge and toppling failures in rock masses. Circular failures mostly occur in soils and highly fractured and weathered rock masses where individual blocks are very small compared to the size of the slope.

The potential mode of failures and the design criteria used for the potential Hasancelebi iron mine are discussed in the following paragraphs.

Kinematic Study of the Open Pit Slopes

Approximately 8000 discontinuity orientations were collected from the area investigated to obtain structural information for the rock mass. The open pit area was divided into 14 subregions to determine the regional predominant discontinuity orientations which control the stability of the rock slopes (Fig. 6, Table 2). The kinematic studies were based on the assumption of underground continuity of the surface plane of weaknesses. Even though this may not be the case, the assumption does provide an approach to a first estimate of potential slope stability problems in the preliminary design of pit slopes. The verification of continuity of the plane of weaknesses is projected as the excavation operations proceed in the projected mine.

Kinematic studies of each subregion reveals that the potential slope failure mechanisms in the open pit area are:

a - Planar failure

b - Wedge failure

c - Circular failure

Due to the highly fractured nature of the rocks, which do not show columnar structure in the ore deposit area, the toppling mode of failure was not considered as a predominant failure mechanism for

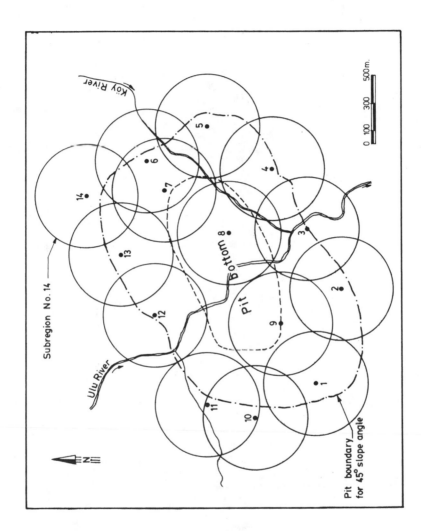

Fig. 6 — Locations of the selected subregions in the potential Hasancelebi open pit. (Radius of the subregions is 375 meters.)

Subregion No.	Predominant Discontinuity Orientations
1	(26,81) ; (54,35) ; (107,65) ; (187,63) ; (195,36) ; (218,76) ; (283,48)
2	(9,88) ; (52,84) ; (108,80) ; (239,48) ; (354,59) ; (358,32)
3	(26, 3) ; (46,36) ; (106,84) ; (268,57) ; (281,85)
4	(37,75) ; (111,81) ; (217,20) ; (260,82) ; (341,69)
5	(12,55) ; (25,85) ; (196,36) ; (205,67) ; (291,29) ; (332,85)
6	(14,57) ; (27,84) ; (111,58) ; (172,83) ; (175,28) ; (192,57) ; (292,87) ; (322,47) ; (331,86)
7	(91,57) ; (159,26) ; (173,83) ; (191,58) ; (264,46) ; (281,86) ; (321,47) ; (329,87) ; (344,29)
8	(59,84) ; (107,68) ; (131,84) ; (180,83) ; (304,83)
9	(54,31) ; (191,82) ; (327,88) ; (338,85)
10	(191,47) ; (201,83)
11	(9,82) ; (35,67) ; (74,87) ; (89,59) ; (190,57) ; (209,82)
12	(9,89) ; (19,45) ; (108,75) ; (111,56) ; (129,26) ; (292,61) ; (305,72)
13	(71,87) ; (282,84) ; (294,41) ; (302,66)
14	(13,88) ; (27,28) ; (66,86) ; (89,67) ; (160,25) ; (165,62) ; (221,75) ; (306,86)

Table 2 - Orientations of the predominant discontinuities in the subregions.

the open pit slopes.

Orientations of the predominant discontinuity surfaces in each of the selected subregions of the open pit area are given in Table 2.

Potential Planar and Wedge Failures

Stability investigations of the daylighting rock wedges and rock blocks were based on the assumptions mentioned below.

a. Assumptions made for three-dimensional rock wedge or planar failure analysis are valid.

b. Friction angles on both discontinuity surfaces are equal and discontinuity surfaces have no cohesion.

c. Such failures more likely occur on the open pit benches.

d. No groundwater on the benches.

Because of the limited mechanical data for the discontinuity planes in the feasibility stage of this project, the absolute values of the safety factor of the daylighting rock blocks for a selected average 45° final slope angle would not be meaningful. Therefore, instead of calculating the factor of safety values of the rock blocks and wedges, the required friction angles of the discontinuities for the limiting equilibrium condition were calculated for each kinematically unstable wedge. Under the probable effects of the assumptions made above, the wedges requiring friction angles of 25° or more were considered to have stability problems.

A summary of the stability evaluations of the daylighting rock wedges and blocks is given in Table 3. The determination of the bench slope angles (Table 4) were based on the orientation of the predominant discontinuities (Table 2) and the stability evaluations given in Table 3. According to the data in Table 4, the generalized slope angles for the open pit bench slopes were selected to be:

Northern section	Eastern section	Southern section	Western section
75°	70°	60°	75°

Subregion No.	1	2	3	4	5	6	7	8	9	10	11	12	13	14
Slope angle (^{o})	75	70	65	70	70	60	60	45	70	80	60	75	75	60

Table 4 - Design angles for the bench slopes

Subregion No.	Slope Surface	Probable Wedge Failure	Probable Planar Failure
1	(360,45) , (80,45)	--	(54,35)
2	(330,45)	--	--
3	(330,45)	(106,84)-(46,36);(281,85)-(46,36)	
4	(330,45) , (340,45)	--	--
5	(330,45) , (270,45)	--	--
6	(240,45)	(331,86)-(192,57)	--
7	(240,45)	(329,87)-(264,46)	--
9	(360,45) , (90,45)	--	(54,31)
10	(100,45)	--	--
11	(90,45) , (150,45)	(74,87)-(209,82);(9,82)-(89,59) (89,59)-(190,57)	--
12	(150,45) , (170,45)	--	--
13	(150,45)	--	--
14	(160,45) , (230,45)	--	(160,25)
8	Subregion No. 8 is located in the center of the proposed open pit. In order to prevent the occurence of daylighting wedges, an excavation slope with N30W strike is selected. Due to the presence of discontinuities with dip angles greater than 60°, planar failures are expected on the benches.		

Table 3 - Potential unstable rock wedges and blocks in the selected subregions of the potential Hasancelebi open pit. The numbers in parantheses denote the dip direction and dip of the discontinuity planes.

Design Concepts for the Final Pit Slopes

The stability of soil slopes and slopes in highly jointed rock masses has been studied extensively and are based on the circular failure analysis. A modified version of Janbu's method was used in the design criteria of the highly jointed rock slopes in the Hasancelebi project (Hoek and Bray, 1977).

The determinations of the effective shear strength parameters of the jointed rock mass were based on test results obtained from the borehole cores and joint surveys. In the application of conventional circular arc analysis, it is assumed that the highly jointed scapolitic rock has a plastic behavior and that the stresses occurring along the potential failure surface will be redistributed.

In most mining engineering projects, the stability of an existing slope is of prime concern. In such problems the geometry of the slope is generally known and a factor of safety value of the slope is generally calculated to obtain relative information for the safety. However, for the Hasancelebi open pit project, the problem is the reverse. The aim was to determine the slope angles for a selected factor of safety. Because of the mountainous morphology of the area, for a predetermined pit bottom boundary, the heights of the final slopes changed locally as a function of the slope angle (Fig. 7). Therefore, as an input to the circular stability analysis, variations of slope heights versus slope angles were graphically determined for 26 selected radial cross-sections of the open pit (Fig. 8).

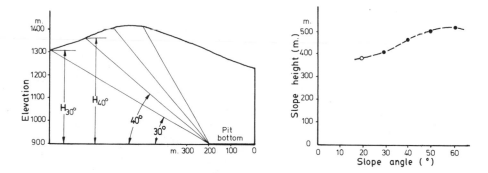

Fig. 7 - Dependence of the slope heights on the slope angle

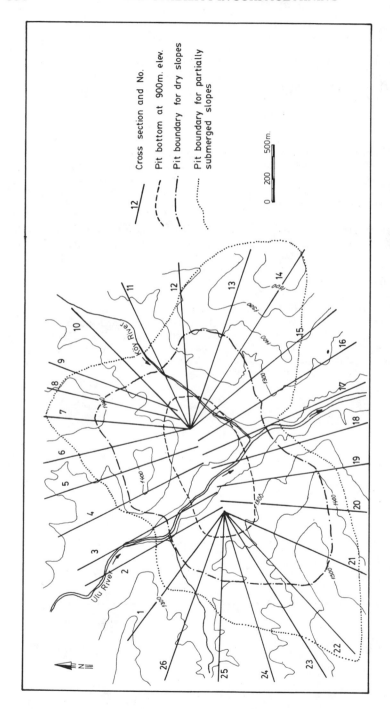

Fig. 8 – Selected cross-sections for the pit slope design and the potential pit boundaries

Despite its controversial nature in the slope stability investigations, the factor of safety is still a valid concept. In the Hasancelebi Project, 1.3 was selected as the factor of safety for the final slopes. The charts prepared by Hoek and Bray were used to determine the relationships between the slope height-slope angles. The design chart is given in Fig. 9.

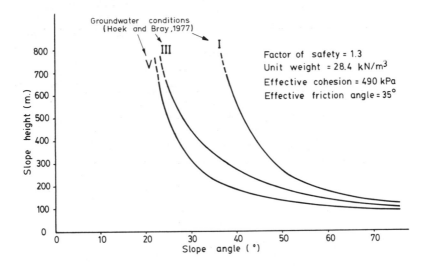

Fig. 9 - Slope design chart for the Hasancelebi project.

The geometric inputs for the design chart (Fig. 7) were provided by the slope angle-slope height relationships prepared for the selected axial cross-sections of the pit area (Fig. 8). The design slope angles for various groundwater conditions were then determined from the combined use of the diagrams given in Fig. 7 and Fig. 9 (Fig. 10).

The resultant design slope angles for the selected cross-sections of the overall potential open pit are summarized in Table 5. Considering the equal aerial weight of each cross section, the general final slope angles of the open pit were determined as $43.5^{o}\pm3^{o}$ for the completely dry slopes (Case I), $30.5^{o}\pm3.5^{o}$ for the partially submerged slopes (Case III), and $26^{o}\pm2^{o}$ for the completely submerged slopes (Case V). Pit boundaries for the completely dry slopes and partially submerged slopes are shown in Fig. 8.

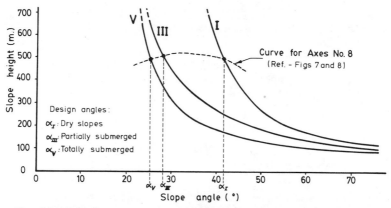

Fig. 10 – Use of the design chart to determine the slope angles.

Slope angle (°) Cross section No.	Groundwater Conditions (Hoek and Bray, 1977)		
	I	III	V
1	47	33	26
2	48	37	29
3	49	38	29
4	42	32	27
5	43	33	28
6	42	30	26
7	41	28	25
8	41	28	25
9	42	29	26
10	46	33	27
11	47	34	28
12	41	28	25
13	41	20	20
14	40	24	22
15	41	28	25
16	46	31	26
17	48	37	30
18	47	34	28
19	42	31	27
20	41	29	26
21	39	28	25
22	39	24	22
23	42	28	24
24	42	29	26
25	44	31	27
26	47	32	26

Table 5 – Design angles for the final pit slopes

CONCLUSION

The design of final slopes and the stability of excavated slopes are two of the major concerns of open pit mining projects. Rock, as an engineering material, has a highly complex nature. The behavior of the rock mass is controlled mainly by the induced stresses as a result of slope excavation and the mechanical properties of the slope material. However, the induced stresses in a slope are also controlled by the structure and the mechanical properties of the rock material. Failure of the slopes is caused by excess shear stresses in the slope along predetermined surfaces (discontinuities) or curved surfaces through jointed material, where the rock mass cannot provide enough shear strength for compensation. Decreasing the slope angle is one of the methods used to decrease shear stresses in slopes. However, this procedure has adverse effects on mine economics. Therefore, the design of slopes becomes an engineering decision problem, namely, to create maximum stresses in the slope by increasing the slope angle to a level that the rock mass can stand without failure.

In order to make an engineering approach to the design of rock slopes, continuum and discontinuum stability analysis methods have been developed. The most important input data for these methods are the engineering geological and mechanical properties of the rock mass. Precision of the stability analyses is highly dependent on interpretation of the field and laboratory studies, and the judgement and past experience of the investigators.

The main purpose of the project outlined was to investigate the geology, hydrogeology and engineering geology of the Hasancelebi iron ore deposit and to make engineering approaches to the determination of the final slope angles for the feasibility study of the potential open pit mine.

Geotechnical data are usually limited at the feasibility stage of the open pit mine projects. For large projects, like the potential Hasancelebi open pit mine, the purpose of slope stability investigations is to provide minimum slope angle data for the final design of the overall exploitation project. Detailed geotechnical studies and observations after the start of the excavation will provide more precise information on the mechanical properties of the rock mass, which will also enable the design engineers to reevaluate the slope angles outlined in this article.

ACKNOWLEDGEMENTS

The authors greatly appreciate the cooperation and support provided by director Mr. Necati Eray on behalf of the Hasancelebi Management Group of the Turkish Iron and Steel Works. Our appreciation also goes to Dr. Ihsan Seymen and Dr. Yüksel Aydin for their cooperation in the preparation of the geological maps and to the personnel of the

Engineering Geology and Rock Mechanics Department who were essential during the investigation.

REFERENCES

Ada, E. and Ata, M., 1979, "Earthquake Risk Analysis Report of Hekimhan - Hasancelebi Iron Deposit Region", Turkish State Water Works, Ankara, Turkey (in Turkish).

Anon, 1981, "Engineering Geology, Hydrogeology and Open Pit Slope Stability Investigation of TDCI Hasancelebi Iron Ore Deposit," ITU Mining Faculty, Department of Engineering Geology and Rock Mechanics, Final Report, Project No. P1/79, (in Turkish). Also, a Preliminary and four Progress Reports (1979-1980).

Anon, 1977, "General Description of Hasancelebi Iron Ore Deposit, " Turkish Iron and Steel Works, Hasancelebi Iron Ore Development Management Group, Ankara, Turkey.

Deere, D.U., 1977, "Geological Considerations", Rock Mechanics in Engineering Practice, K.G. Stagg and O.C. Zienkiewicz, ed., Chap. 1, John Wiley, New York.

Hoek, E. and Bray, J., 1977, Rock Slope Engineering, Inst. Min. and Metall., London.

Kurt, M. and Akkoca, A., 1974, "Geological Report of the Malatya-Hekimhan-Hasancelebi Iron Deposit", Mineral Exploration and Research Institute of Turkey, Ankara, Turkey (in Turkish).

Manev, G. and Avramova-Tacheva, E., "On the Valuation of Strength and Resistance Condition of the Rocks in Natural Rock Massif," Proceedings, 2nd International Congress of Rock Mechanics, Belgrade, Vol. 1, Paper No. I-10.

Muller, L., 1963, Der Felsbau, Ferdinand Enke-Verlag, Stuttgart.

Stimpson, B. and Ross-Brown, D.M., 1979, "Estimating the Cohesive Strength of Randomly Jointed Rock Masses," Mining Engineering, SME, February, pp. 182-188.

Question

You mentioned a sensitivity analysis for the design slope angles with respect to the selected shear strength parameters of the jointed rock. Could you please give some more information on this analysis?

Answer

The design slope angles of the Hasancelebi open pit can be obtained from the slope angle-slope height diagrams derived from the design charts which were developed by Hoek and Bray. The selected input data for the potential Hasancelebi open pit mine were based on the results of detailed engineering geology and laboratory testing studies. A factor of safety value of 1.3 was considered to be appropriate for the overall pit slopes.

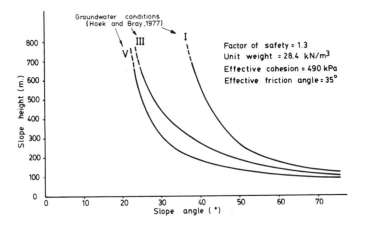

Fig. A1 -- Slope design diagram for the selected rock mass properties and groundwater conditions.

The design diagram given in the paper shows the relationships between the slope angle-slope height for various groundwater conditions. This diagram (Fig. A1) is valid only for the assigned cohesion and friction angle values of the jointed rock mass and has limited general use and does not present the sensitivity of the design slope angles to the changes in the input shear strength parameters. Therefore, a new set of charts were prepared to reevaluate the diagrams in Fig. A1 to study the sensitivity of the slope angles to cohesion and friction angle of the rock mass. The procedure for the preparation of the sensitivity charts is as follows:

a- Select a circular failure chart for a groundwater condition (example: Chart Number 3 - Hoek and Bray).
b- Select a factor of safety, unit weight and a range of friction angle values (example: FS=1.3, Unit weight=2.9 t/m^3 (28.4kN/m^3) $\emptyset$:25°, 30°, 35°, 40°).

c– Calculate each Tan∅/FS for the selected ∅ values (example: Tan25/1.3=0.36, Tan30/1.3=0.44, etc.).

d– Using the Tan∅/FS value on the chart, read the corresponding c/ɤHTan∅ value for the selected slope angles (example; for Tan25/1.3=0.36 and slope angle of 30°, c/ɤHTan∅=0.182).

e– Calculate ɤTan∅ for the selected ɤ and ∅ values (example: ɤ=2.9 t/m³, ∅=25° then ɤTan∅=1.352).

f– Calculate H/c value by substituting ɤTan∅ in c/ɤHTan∅ (example: c/1.352H where H/c=4.1 m³/t).

Graphical representation of an example solution of the above mentioned procedure for ɤ=2.9 t/m³ (28.4 kN/m³), FS=1.3, ∅ angles of 25°, 30°, 35°, 40° is given in Fig. A2.

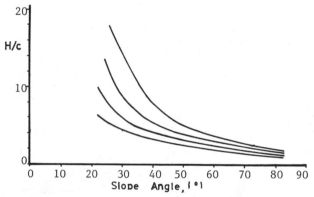

Fig. A2 –– Graphical representation of slope angle – H/C relationship for the example given above.

Slope angle–H/c diagrams for different groundwater conditions can be prepared following the similar procedures mentioned above. Cohesion of the slope material enters into these diagrams as a variable in the H/c ratio. Addition of a H–H/c diagram, using the cohesion values as a parameter, to Fig. A2 provides a simple evaluation of the slope angle variations as functions of the shear strength characteristics of the slope material (Fig. A3).

Variations in slope angles–slope height as functions of cohesion and friction angle of the slope material for three different groundwater conditions are shown in Fig. A3.

Sensitivity of design slope angles to a tolerance limit on selected cohesion and friction angles can be obtained from the diagrams in Fig. A3. For example, the design slope angles of a 300 m. high slope for c=50 t/m² (490 kPa) and ∅=35° with 10 percent tolerance limits are obtained as:

	Groundwater Conditions		
	I	III	V
Design slope angles, (°)	48.5±5.5	37.5±7.5	30±4

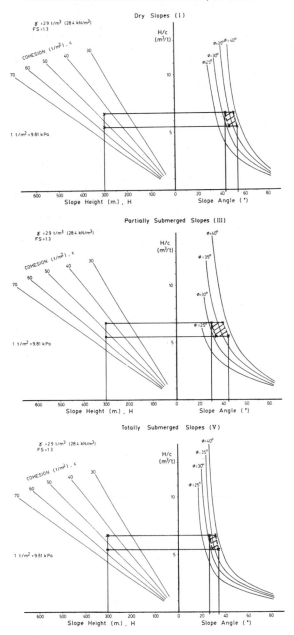

Fig. A3 -- Diagrams for design slope angle sensitivity on the shear strength parameters of the rock mass.

Question

The locations and dimensions of design subregions appear arbitrary. Were various subregions ultimately combined? If so, on what basis? Was the discontinuity cohesion measured? How did the measured cohesion value compare with the cohesion value which you used for the rock mass utilizing a reduction factor on the intact rock cohesion?

Answer

The location of the centers and radius of the circular subregions were selected on a trial and error basis to determine the orientations of the local predominant discontinuity sets over the potential open pit mine. In selecting the locations and dimensions of the subregions, the criteria was to cover the overall potential open pit and to obtain statistically reliable discontinuity clusters. For the case of overlapping discontinuity clusters, (which were not statistically acceptable), the centers of the subregions were shifted.

The measured cohesion values of the discontinuities were equal to zero. However, the potential mobilized shear strength through the jointed rock mass along a circular failure surface is not purely controlled by the shear strength parameters of the discontinuities. The present state of the art in determination of the rock mass shear strength parameters has not yet reached an universally accepted approach, as discussed throughout this conference. Our approach to assign a cohesion value to the rock mass was based on the assumption of decreasing nature of the rock mass cohesion as a function of joint intensity, as suggested in the previous studies. The resultant design slope angles, which were based on the selected shear strength parameters, were found appropriate for the prefeasibility stage of the potential Hasancelebi iron ore mine.

Chapter 28

REDESIGN OF THE WEST WALL
KANMANTOO MINE, SOUTH AUSTRALIA

Barry K. McMahon

Principal, McMahon Burgess and Yeates.
Chatswood, Australia

ABSTRACT

Following minor slope failures at the commencement of mining the
west wall of the Kanmantoo Mine was redesigned from an overall slope
between haul roads of 55° to 46°. This decision was based on a
probabilistic analysis which indicated that 46° was the optimum
economic slope.

INTRODUCTION

The Kanmantoo Mine is located in South Australia approximately
100 km east of Adelaide. The mine recovered lenses of chalcopyrite-
pyrrhotite ore averaging 1% copper within a host rock of garnet
chlorite schist. Surrounding the host rock and forming most of the
walls of the pit is an unusual metamorphic rock described as garnet-
andalusite-biotite-schist. (Figures 1, 2, and 3).

The mine is situated in undulating grazing country with a rainfall of around 500 mm per year.

The mine was commenced in August 1970 and is roughly circular
in shape with a diameter of about 500 m and a maximum planned depth
to 226 m.

Initial studies by Jaegar (1970) led to a recommendation that
the pit be excavated at slope angles between haul road segments of
55° on all walls, subject to review after the fresh rock was exposed.
In 1972 after the pit had been excavated to a maximum depth of 48 m,
minor slope failures on the western wall caused sufficient concern
that the writer was retained to carry out a further investigation
which essentially confirmed Jaegar's recommendations on the north,
south and east walls but resulted in a revised recommendation of 46°
for the west wall. (Figure 2).

Following re-excavation of the west wall to 46° during 1973 the
mine floor was lowered by 1976 to a depth of 110 m. At this time
following a period of low copper prices the mine was placed on care

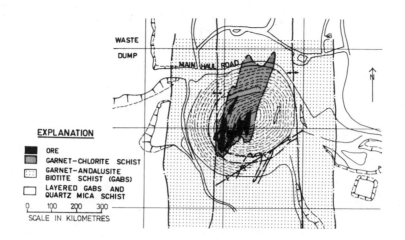

Fig. 1.   Generalised geological plan of Kanmantoo Mine.

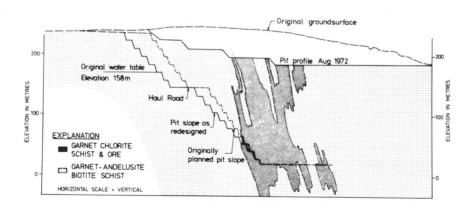

Fig. 2.   Generalised Cross-Section through the west wall showing
originally planned pit slope at 55° and redesigned pit
slope at 46°.

Fig. 3. Photograph of the west wall taken August 1972.

and maintenance and has not yet been re-opened.

This paper describes the procedures used in the review investigation which essentially followed the methods described by McMahon
(1971, 1974, 1975).

METHODS OF INVESTIGATION

The review investigation followed the following steps:

(1)   Detailed geological mapping within the pit for the dual
      purposes of reserves evaluation and slope stability. This
      mapping was carried out on a daily basis by the pit
      geologist and recorded the locations of rock and ore type
      boundaries, mineralised veins, major mappable fractures such
      as shears, crushed and altered zones and faults and major
      mappable joints.

(2)   Statistical sampling within the pit of the orientations,
      exposed lengths, terminations, spacing, roughness and
      coatings of fractures too numerous and repetitive to be
      mapped as individuals. These consisted of foliation and
      joints. The conclusions based on pit sampling were spot
      checked at depth using oriented diamond drill cores.

(3)   Close study and back-calculation of minor slides in the
      pit. These provided estimates of friction angles at low
      stresses and insight into the probable mode of failure of

the slope.

(4)   Samples of matched joint surfaces were tested in direct
      shear. In addition, results of triaxial tests on jointed
      rock specimens were available from the original investiga-
      tion. Some unconfined compression tests were also carried
      out on selected cores.

(5)   Intersections of the groundwater table in reserves
      boreholes were recorded and all springs in the pit walls
      were mapped.

(6)   A probabilistic analysis of the above data was carried out
      to estimate the relationship between slope height, slope
      angle and probability of failure.

(7)   An economic analysis was carried out to estimate the
      optimum economic slope.

GEOLOGICAL CONDITIONS AFFECTING THE WEST WALL

Rock Types

    The rock types within the west wall consist of garnet-andalusite-
biotite-schist and garnet-chlorite schist (Figures 1 and 2).

    The garnet-andalusite-biotite-schist is a grey-green foliated
rock and consists of large (5 to 10 mm) porphyroblastic andalusite
crystals making up an average of 8% of the rock set in a coarse
grained schistose matrix of biotite (52%), quartz (27%), almandine
garnet (55) and chlorite (5%).

    The garnet-chlorite-schists is a poorly foliated fine grained
rock which is gradational with the garnet-andalusite-biotite-schist
and is thought to be a metamorphic alteration of it due to
chloritizition of the biotite and garnet, replacement of andalusite
by staurolite and increase in garnet content.

    The unconfined compressive strengths of the two rock types is
dependent on the angle between the direction of loading and the
foliation as shown in Figure 4.

Mappable Fractures

    Mappable fractures fall into two groups:

(1)   Faults striking NE with near vertical dips parallel to
      joint set 4 shown in Figure 7. These were mostly concent-
      rated on the south wall and contained between 10 and 300 mm
      of clay gouge.

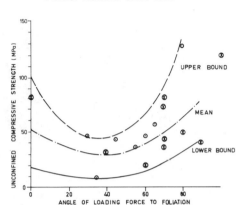

○ GARNET CHLORITE SCHIST
⊗ GARNET ANDALUSITE BIOTITE SCHIST

Fig. 4. Unconfined compressive strength related to foliation
direction.

(2) Major inclined fractures of which 95% strike roughly north-
south (N30W to N30E) and dip between 26° and 57° (mean 42°)
towards the east.

These fractures had unstable orientations with respect to the
west wall as shown in Figures 5 and 6. They were oriented in the
direction of fracture set 2 as shown in Figure 7.

These fractures had rough clean surfaces with waviness (departures from mean dip measured over lengths greater than 1 m) varying
from 0° to 20°.

During the period August 1972 to August 1973 the benches shown
in Figure 2 were cut back towards the west close to final limits.
During this time it was found possible to map the major inclined
joints in the bench faces and the bench floors over a distance of
335 m. In many cases this was made easier by marked heaving of the
rock above the inclined fractures as a results of blasting. This
mapping enabled the strike-lengths of the fractures to be measured
as discussed later.

Statistical Fracture Studies

The orientation, spacings lengths, roughness and waviness of
joints and foliation surfaces were sampled statistically using both
line sampling and point sampling methods. Results of the two methods
were judged to be reasonably consistent and they were merged to give

Fig. 5.   Inclined mappable joint of set 2 exposed in working face

Fig. 6.   Small scale failure due to sliding on joint of set 2
          accompanied by tensile separation on foliation surface.

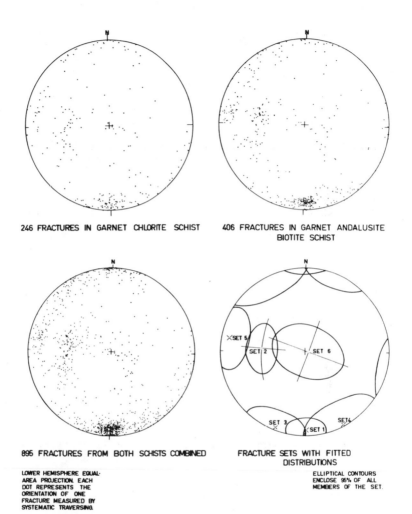

246 FRACTURES IN GARNET CHLORITE SCHIST

406 FRACTURES IN GARNET ANDALUSITE BIOTITE SCHIST

895 FRACTURES FROM BOTH SCHISTS COMBINED

FRACTURE SETS WITH FITTED DISTRIBUTIONS

LOWER HEMISPHERE EQUAL-AREA PROJECTION. EACH DOT REPRESENTS THE ORIENTATION OF ONE FRACTURE MEASURED BY SYSTEMATIC TRAVERSING.

ELLIPTICAL CONTOURS ENCLOSE 95% OF ALL MEMBERS OF THE SET.

Fig. 7. Orientations of fractures.

the results reproduced in this paper.

Joint orientations throughout the pit in both rock types fell into one domain as shown in Figure 7. The sets were contoured as described by McMahon (1971) and the 95% iso-probability contours are shown. These fall into six sets of which fracture sets 2, 5 and 6 contain members that are potentially unstable on the west wall. Sets 1, 3, and 4 formed side boundaries to minor slides in the working benches on the west wall but were too steep to cause problems on the north, east and south walls.

The foliation was parallel to joint set 5 and formed the backs of minor slides in the working benches as shown in Figure 6.

The information collected by statistical sampling on fracture spacing lengths, roughness and waviness was ultimately not used in the revised design procedure in this particular pit. However, as discussed below the information on major inclined fracture lengths obtained from the mapping was of critical importance.

Study of Minor Slides

Minor slides in the working faces as noted above and shown in Figures 5 and 6, were the events which caused sufficient concern to the mine management that they implemented the design review. The slides were either simple plane shear failures due to sliding on the set (2) joints until these terminated against a foliation joint (set 5) which formed the back of the slide as a surface of tensile separation. (Figure 8).

Prominent inclined fractures undercut by the working benches were subdivided into 3 groups:

(1) Stable fractures along which the overlying rock showed no signs of movement except in some cases minor disturbance due to blasting.

(2) Metastable fractures along which very slow movement was initiated by blasting and continued afterward.

(3) Unstable fractures along which rapid movement took place following blasting.

The dips of these three groups are shown in Figure 9. The flattest unstable dip recorded was 37°.

Shear Strength Tests on Fractures

A large number of shear tests by both direct shear by the writer and by triaxial testing by Jaegar were carried out on joint surfaces selected as typical. These gave the results shown in Figure 10.

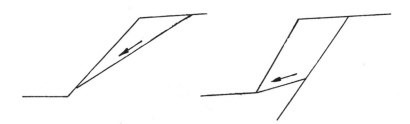

Fig. 8. Modes of failure on the west wall.

Groundwater

Measurements of standing water levels in drill holes indicated that the original groundwater table behind the west wall was as shown in Figure 2.

Some springs encountered in the south wall behind the NE trending fault indicated that the water table in this area was about 30 m higher.

Design Analysis

Given that the presence of inclined joints had cast the original design into doubt the problem was to find a reasonable basis for selecting an alternative slope. The following possibilities were considered:

(1) To assume that an extensive inclined joint in the worst likely orientation would intersect the toe of the slope and excavate the slope parallel to this fracture. On the basis of the information in Figures 9 and 10, it would be reasonable to assume a value of $35°$ as a design friction angle. This would lead to an estimate of critical dip (McMahon 1974) of between $27°$ and $33°$ depending on the assumptions made with regard to groundwater draw down. As an estimated 2.5% of the joints of set (2) dip flatter than $26°$ the critical dip values of $27°$ to $33°$ would control.

On these assumptions a pit slope of between $27°$ to $33°$ would have a safety factor of 1.

(2) To apply a safety factor of 1.3 to the above criteria as recommended by many authorites. This would produce a design slope between $21°$ and $26°$.

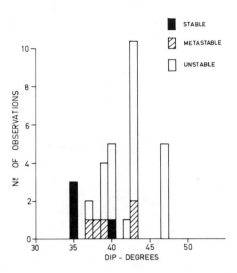

Fig. 9. Relationship between dip of inclined joints undercut by
         working benches and the stability of the overlying rock.

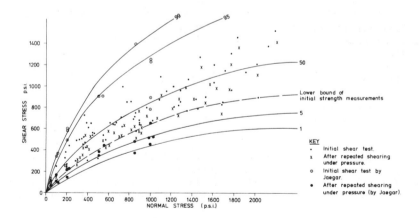

Fig. 10. Results of shear tests on rock fractures. Contour values
         are labelled in percentiles. (note 1MPa = 145 psi).

(3)   To excavate parallel to the mean dip of the fractures. This would produce a slope of 42° between haul road segments but it would be noted that such a slope would be undercut by 50% of the joints.

(4)   To assume that a combination of joints of sets 2, 5 and 6, could approximate a slip-circle. Friction angle could be estimated as the friction of the joints at about 35°. Cohesion values would have to be guessed and a safety factor added to satisfy tradition. Depending on the values so chosen any slope between 21° and 55° could be justified.

(5)   To analyse the slope on a probabilistic basis assuming that:

    (a)   The modes of failure of the slope would be as shown in Figure 8.

    (b)   The probability distribution of orientations would be as shown in Figure 7.

    (c)   The friction angle of the joints is 35°. This value was included deterministically as it was considered that criteria for correlation between laboratory tests and field friction angles were not available (i.e. should peak or residual friction angle be used or something in-between) but that the bulk of the field and laboratory data indicated that 35° was a reasonable lower bound value. Thus estimated probabilities of failure would be on the high side.

    (d)   Groundwater pressures under anticipated drawdown reduce the critical dip to 33° at the toe of the maximum slope. This was also input on a deterministic basis.

    (e)   Fracture lengths in the strike direction could be predicted on a probabilistic basis as described by McMahon 1974, p. 806. These gave the results shown in Figure 11.

During the study it was noted that the inclined joints were rectangular rather than square in shape and that a common ratio of dip length to strike length was 0.6. As it is the dip lengths which affect the slope angle these were estimated from the strike lengths using this factor.

This procedure reflected a practical judgement that the stability of the slope would be most sensitive to variations in fracture orientations and lengths and much less sensitive to variations in water table and joint friction at this particular mine.

The probability of failure of slopes of varying heights and angles

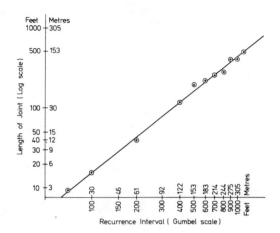

Fig. 11. Plot of maximum strike length of inclined (set 2) joints
vs. recurrance interval on a Gumbel extreme value probability scale (reprinted from McMahon 1974)

was calculated using the method described in McMahon (1974) (it
should be noted that equation 6 McMahon (1974) and equation 10
McMahon (1971) are incorrect and the correct equations are given by
McMahon (1975). However, the earlier equations provide reasonable
approximations at failure probabilities less than 25% although
they are seriously in error at higher values).

The results are shown in Figure 12 which illustrates the well
recognised effects that as slopes become steeper and higher they
have a greater risk of failure due to a greater risk of undercutting
unstable combinations of fractures. The same results are presented
in a different form in Figure 13.

A cost optimization was carried out using the method described
in McMahon (1974). In this case the costs of initial excavation
were estimated as 25 cents per ton and the cost of slope repair
was estimated at 46 cents per ton.   This would be the cost incurred
if the full resources of the mine were diverted to shifting the
landslide.

No allowance was made in the analysis for the time value of
money. In effect this implied that other avenues of investment
open to the mine would not provide a rate of return greater than
inflation.

The results of the cost optimization are shown in Figure 14.
This shows an optimum economic slope of 46° and this value was used
in the redesign for the slope angle between haul road segments as
shown in Figure 2. The total cost curve also shows that there is

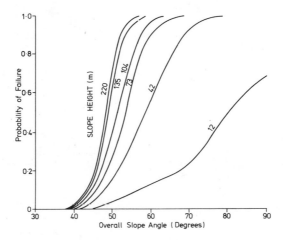

Fig. 12. Plot of relationship between slope angle slope height and probability of failure for the west wall. (From McMahon 1974 with correction).

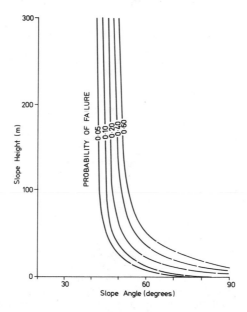

Fig. 13. An alternative representation of the information in Fig. 12.

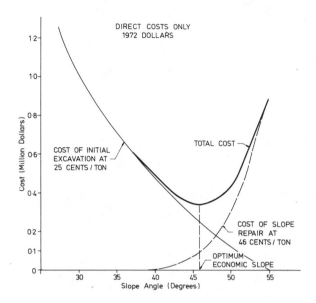

Fig. 14. Cost Optimization Curves for the west wall.

very little economic difference between slopes in the range 43° to 49° but a substantial economic penalty for slopes flatter or steeper than this range.

REVIEW OF INITIAL DESIGN

The initial design by Jaegar was based on the judgement that the joint pattern was stable and that the slopes would be controlled by the ratio of the compressive strength of the rock substance to the maximum stress concentrated at the toe up the slopes. This was judged to be greater than 2 when the slopes were 55°. This analysis was confirmed during the review on the north, east and south walls and the slopes have remained stable at this angle during mining.

The fracture orientation information available at the time of the inital design is shown in Figure 15. Allowing for the different method of contouring these do show the presence of the inclined joint sets that caused concern on the west wall. However, it was not possible until the mine was excavated to tell how long these joints were and if they had been very short the design angle of 55° would have been considered appropriate during the review.

CONCLUSION

At this time the slope has been excavated to a depth below the

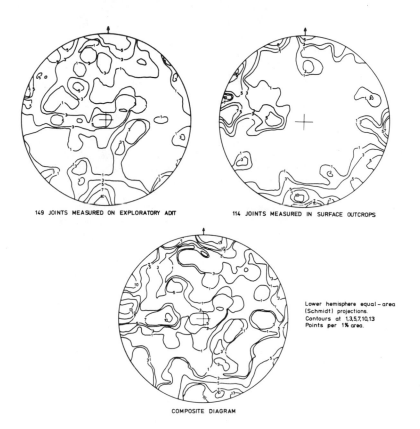

149 JOINTS MEASURED ON EXPLORATORY ADIT

114 JOINTS MEASURED IN SURFACE OUTCROPS

Lower hemisphere equal-area
(Schmidt) projections.
Contours at 1,3,5,7,10,13
Points per 1% area.

COMPOSITE DIAGRAM

Fig. 15. Fracture information available at the time of the initial
design.

haul road at 46° and has not failed.   It is not possible to say
whether or not it would have failed if it had been excavated at 55°
but the review analysis indicated that the risk of attempting
that angle was uneconomically large.   The key factor in the review
analysis was evaluation of the lengths of the unfavourably
oriented joints.

ACKNOWLEDGEMENTS

The assistance of Mr C. Johnstone, Mr M.C. Bridges, Mr R.
Stephenson and Mr J. Treloar in carrying out this study is gratefully
acknowledged.   The study was supported by Kanmantoo Mines Ltd.

REFERENCES

Jaegar, J.C., 1970, "Interim Report on rock properties, stresses
and factor of safety in Kanmantoo open pit", Unpublished report
to Broken Hill South Ltd.

McMahon,B.K., 1971,"Statistical Methods for the design of rock
slopes", Proc. First Australia New Zealand Conference on
Geomechanics, Vol 1, pp 314 - 321.

McMahon,B.K., 1974, "Design of rock slopes against sliding on
pre-existing fractures", Prox. Third Congress Inter. Soc.
Rock Mechanics, Denver, Vol 2, Part B, pp 803 - 808.

McMahon,B.K., 1975, "Probability of Failure and Expected Volume
of Failure in High Rock slopes" Proc. Second Australia New
Zealand Conference on Geomechanics, pp 308 - 313.

Chapter 29

DESIGN EXAMPLES OF OPEN PIT SLOPES SUSCEPTIBLE TO TOPPLING

Douglas R. Piteau, Alan F. Stewart and Dennis C. Martin

D. R. Piteau & Associates Limited
West Vancouver, B.C.

ABSTRACT

Three examples of open pits where toppling failure controls the stability and design of the slopes are described. Two examples involve the design of overall slopes in base metal mines. The third example concerns the analysis and design of remedial measures to prevent toppling failure from occurring in a critical slope in a coal mine.

The analytical techniques for assessing toppling failure on a rational basis are discussed for the three examples. This includes an assessment of the important parameters which govern the stability for the toppling failure mode.

The geological setting and input data for each example are described, as are the important external or limiting factors pertinent to the sites in question. A statistical cumulative technique was used to assess one of the most important input parameters in the two examples involving overall slope design. Both deep-seated toppling failures, involving the entire slope, and shallow bench failures are considered. Results of the stability analyses and their related engineering significance with regard to slope stability and design are discussed. A sensitivity analysis of various parameters is effectively utilized. Design charts are developed and evaluated accordingly. In the example where toppling failure was unavoidable, the rationale for design of stabilization techniques to prevent toppling is discussed.

679

INTRODUCTION

Discussed in the following are case history examples of the design of three open pit slopes which are susceptible to flexural toppling. Two case histories involve the design of bench and overall slopes in iron ore mines. The third case history involves the analysis and design of remedial measures to prevent toppling of a critical slope in a coal mine. In all three cases, thin tabular columns or slabs of rock, which dip steeply into the slope, topple by bending or breaking in flexure as a consequence of loss of support due to oversteepening. Oversteepening results in the centre of gravity of individual blocks acting outside the toe of the blocks, thus causing the overturning moments to exceed the resisting moments.

The toppling analysis technique in all three cases is based on the method initially described by Goodman and Bray (1976) and Hittinger (1978). Both deep-seated toppling failure of the entire slope and shallow bench failures of a local nature are considered. Results of the stability analyses, and the related engineering significance with regard to slope stability and design, are evaluated. For purposes of rational analysis of the potential toppling failures, sensitivity analyses of various parameters were effectively utilized and design charts were developed accordingly.

DESCRIPTION AND BASIC MECHANICS OF TOPPLING FAILURE

Toppling can develop in natural and excavated rock slopes wherever relatively closely spaced, steeply dipping discontinuities occur approximately parallel to the slope. In its simplest form toppling of a single block occurs when the centre of gravity of a rock block overhangs a pivot point at the lowest corner of the block. As shown in Fig. 1, toppling is closely related to sliding in that both failure modes rely on the dip of the basal surface. Sliding can occur whenever the dip of the basal surface exceeds the friction angle of this surface (i.e. $\psi > \phi$). Toppling is related to block geometry and only occurs when the width to height ratio is less than the tangent of the dip of the basal surface (i.e. $b/h < \tan \psi$). Fig. 1 graphically illustrates the conditions for simple planar sliding and single block toppling on an inclined basal surface having an angle of friction of $30°$. Combination of the two conditions, (i.e. $\psi = \phi$ and $b/h = \tan \psi$), results in four areas on the diagram in Fig. 1 within which a single block may be stable or may fail by sliding, toppling or a combination of sliding and toppling.

Although the basic mechanics of toppling appear to be relatively straightforward, toppling of a natural assemblage of blocks generally is much more complex than indicated above. Quantification of rational stability analyses is more involved than the somewhat uncomplicated failure modes such as plane and wedge failures. Where analysis of both plane and wedge failures essentially do not consider interaction of adjacent blocks during failure, this is not the case for toppling

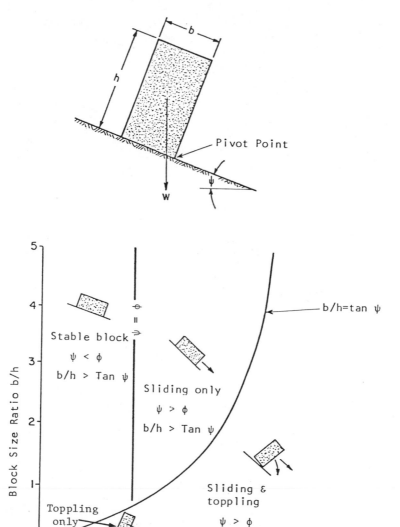

Fig. 1   Conditions for sliding and toppling
          of a block on an inclined plane(after
          Hoek and Bray, 1977).

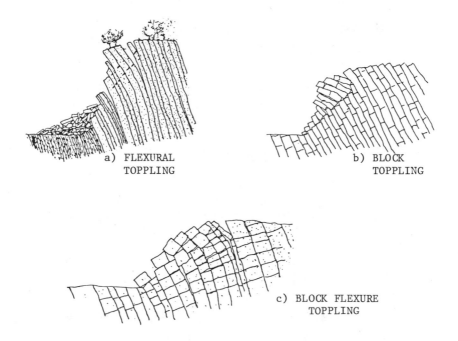

a) FLEXURAL
   TOPPLING

b) BLOCK
   TOPPLING

c) BLOCK FLEXURE
   TOPPLING

Fig. 2. Typical types of toppling failure (after Goodman and Bray 1976)

failure. For toppling failure, a complex transfer of loads from one block to another is considered as well as variability of block size in section.

Recent investigations by Goodman and Bray (1976), de Freitas and Watters (1973), Piteau and Martin (1981) and others, have described and classified various types of toppling failures. The three most common forms of toppling are flexural toppling, block toppling and block-flexure toppling (see Fig. 2), although several variations of these failure mechanisms can develop and have been described. In any event, a feature common to most toppling failures is that they contain three distinct regions of behaviour. As discussed by Ashby (1971), these three regions are as follows:

1. A region of sliding near the toe.
2. A region of toppling or toppling and sliding in the centre of the failure mass.
3. A stable region at the top of the slope immediately behind the toppling region.

MECHANICAL STABILITY ANALYSIS METHOD FOR FLEXURAL TOPPLING

Basic Approach

Analysis of flexural toppling was carried out using a limit equilibrium analysis method in which the slope was subdivided into a series of columnar blocks as shown in Fig. 3. The various forces due to gravity and water pressure acting on each block, and the shear resistance between blocks, are resolved and the net force acting on each block is computed to determine whether each particular block is stable, or if it could fail by toppling or sliding.

A computer program developed by Hittinger (1978) was modified and used for the analysis described below. The toppling analysis assumes that a given slope contains continuous joints which have a uniform dip $(90-\alpha)$ and spacing (Δx) and form the boundaries of discrete columnar blocks as shown in Fig. 3(a). As toppling commences, a stepped basal failure surface at an angle β is developed, below which toppling of the individual blocks does not occur. Examination of toppled slopes at two iron mines indicates that the basal failure surface is generally formed by induced tensile fracturing of each block or developed along pre-existing joint surfaces (see Figs. 4 and 5). The average dip of the basal failure surface appears to be reasonably consistent for a particular site.

As shown in Fig. 3(a) groundwater conditions are incorporated into the analysis by constructing a water table with a horizontal phreatic surface. The phreatic surface is assumed to fall uniformly in the slope from h_w at a point vertically below the slope crest to 0 at the toe of the slope.

The analysis is performed on the highest block in the slope first, and all resultant forces are transferred to subsequent blocks progressively down the slope. For each block, the various parameters, such as forces due to gravity, water pressure, and frictional properties along the basal failure surface and between the blocks, are calculated and resolved. Each block is analyzed for possible failure modes involving both planar sliding on the basal failure surface and toppling about the lowest corner of the block and the resultant force, P_n (if any), for each mode of failure is determined as shown in Fig. 3(b). The failure mode with the largest resultant force, P_n, is assumed to define the failure mode of the block, and that resultant force is assumed to be transferred to the lower block immediately adjacent.

The failure mode and resultant force determined for each block is recorded, as shown on a typical computer printout of the output in Fig. 3(c). The computer output shows the largest resultant force, P_n, and the mode of failure for each block, as well as various other forces calculated to be acting on each block.

Slope instability is assumed to occur when there is a positive resultant force, P_n, for the lowest block at the toe of the slope. Where slopes are unstable the possible support required to insure

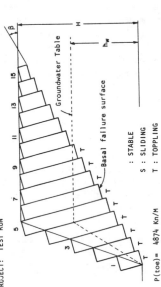

GEOTOP-LIMIT EQUILIBRIUM ANALYSIS OF TOPPLING FAILURE IN JOINTED ROCK

PROJECT: TEST RUN

P(toe)= 4874 Kn/M

: STABLE

S : SLIDING

T : TOPPLING

a) Model for limit equilibrium analysis of toppling.

b) Conditions for toppling and sliding of the nth block (after Goodman and Bray 1976).

Sliding $P_{n-1_S} = P_n - \dfrac{W_n(\cos\alpha\,\tan\phi - \sin\alpha)}{1 - \tan^2\phi}$

Toppling $P_{n-1_T} = \dfrac{P_n(M_n - \Delta x\,\tan\phi) + (W_n/2)(Y_n\,\sin\alpha - \Delta x\,\cos\alpha)}{L_n}$

GEOTOP - LIMIT EQUILIBRIUM ANALYSIS FOR TOPPLING FAILURE IN JOINTED ROCK

PROJECT: TEST RUN

MATERIAL DENSITY= 2700 Kg/M3   INTERNAL FRICTION ANGLE= 30 DEG
FAILURE SURFACE ANGLE= 30 DEG   FRICTION ANGLE= 35 DEG
SLOPE HEIGHT= 50 M   SLOPE ANGLE= 70 DEG
BEDDING PLANE SPACING= 7 M   DIP ANGLE= 75 DEG
WATER TABLE DEPTH= 25 M
ADJUSTED BEDDING PLANE SPACING= 6.1

N	PN,T	PN,S	PN	RN	SN	SN/RN	MODE
15	-386	-491	0	676	181	.268	
14	-317	-861	0	1187	318	.268	
13	-249	-1232	0	1698	455	.268	
12	-181	-1602	0	2208	592	.268	
11	-112	-1973	0	2719	729	.268	
10	-44	-2343	0	3230	865	.268	
9	25	-2714	25	3726	977	.262	
8	124	-2969	124	4068	1006	.247	T
7	287	-3103	287	4353	1029	.236	T
6	500	-3173	500	4646	1065	.229	T
5	986	-3194	986	4809	877	.182	T
4	1698	-703	1698	3563	1065	.299	T
3	2585	433	2585	2535	493	.194	T
2	4436	1640	4436	1064	-920	-.865	T
1	4874	3810	4874	965	42	.044	T

c) Typical output for toppling analysis.

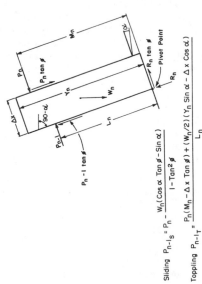

Fig. 3. Limit equilibrium analysis method for flexural toppling

Fig. 4. View of toppling failure. Note the well developed basal failure surface which dips about 27° in this case.

Fig. 5. View of toppling failure. Note the well developed basal failure surface and rock beneath the failed zone which has not undergone any movement.

stability can be determined based on the magnitude of the resultant force, P_n, which has been calculated. The block by block analysis also may be used to assess the effect of removing or excavating one or more blocks at the toe of the slope. In this stability analysis approach, no factor of safety is calculated. However, a resultant force of $P_n = 0$ on the bottom block would be equivalent to a Factor of Safety of 1.0. A factor of safety could be calculated by assessing the required friction angle (ϕ_{req}) for equilibrium, i.e.

$$F = \frac{\tan \phi_{available}}{\tan \phi_{req}}$$

The analysis method has been developed so that a large number of iterations may be carried out. The results are recorded in terms of the various parameters assumed and the calculated resultant force, P_n, on the toe block. Analyses are normally carried out for different discontinuity dips, spacings, slope heights, slope angles and groundwater conditions in the slope in order that comparative plots can be prepared for assessing the sensitivity of the slope to each of the parameters considered.

Limitations of the Analysis

The analysis method described above for flexural toppling is considered to be somewhat conservative. The analysis assumes the existence of the basal failure surface which often appears to form as a result of tensile failure and cracking of the rock mass columns. In using limit equilibrium analysis in the manner described above, consideration has not been given to the possible variation in behaviour of the tabular blocks due to varying stresses and strains at different locations in the slope. That is, the results imply that the interaction between blocks is the same along the length of the block regardless of location in the slope. There is no means of assessing the behaviour of the rock mass under load, effects of lateral constraint of adjacent rocks, or possible changes of shear strength with depth from the surface of the slope. In this regard, the application of this analysis method for design of high slopes may require more sophisticated analyses. Because of the inadequacies of present knowledge of toppling behaviour, evaluation of the degree of reliability of analyses results requires an assessment of either trial slopes or documentation and back analysis of toppling failures in existing slopes in similar rocks in which the design is being considered.

TOPPLING ANALYSIS AT AN IRON MINE IN AUSTRALIA

This case history of an open pit slope susceptible to toppling failure pertains to both bench design and overall slope design of a high wall at an iron mine in Australia. In this case, toppling occurs along well developed foliation joints along schistosity. It could be inferred from geologic sections developed across the pit that the potential for toppling generally increases with depth.

Geological Conditions

The general geology in the area consists of a north-south trending, steeply dipping sequence of basic volcanics, serpentinites and magnetite-sulphide-silicate ores flanked on the east and west by schistose metasediments. Significant toppling had occurred on large portions of the west wall in the western schist. The schist and associated rocks consist mainly of quartz, feldspar and actinolite with varying amounts of chlorite, sericite and carbonate minerals. With the exception of extremely weathered schist in the upper 20 to 40m of the pit walls, the unconfined compressive strength of the schist is estimated to range between 7 and 55 MPa.

Schistosity is a laminated or foliate structure developed parallel to bedding in the metasediments, along which are formed well developed foliation joints with low shear strength properties (i.e. $\phi = 25^{\circ}$, $c = 0$). Schistosity strikes approximately north-south, approximately parallel to the pit wall in most areas of the pit. Foliation joints form the long thin steeply dipping tabular columns of rock which topple. Schistosity dips vary from 40° toward the pit to about 70° away from the pit. Variations in orientation of schistosity are related to the folding and faulting in the metasediments.

The apparent steepening, and in some cases overturning of schistosity dip on the west wall of the pit indicated from the surface mapping, is verified by statistical analysis of schistosity dips observed in diamond drill cores from the west wall. Statistical analyses of schistosity dip, as measured at close intervals on cores from diamond drillholes which intersect the schist, were carried out using the cumulative sums technique (Piteau & Russell, 1971).

The cumulative sums (cusums) technique provides a rapid and precise method to:

1. determine a reliable estimate of the current mean dip of the schistosity above and below the mean dip of the schistosity data.
2. detect general changes in current mean schistosity dip above and below the mean dip of the schistosity data.
3. predict the average dip of schistosity in parts of the rock mass where drillhole structural data is not available.
4. construct graphical geological sections from which changes in schistosity dip can be determined and thus reconstruct the geometry of the schistosity in the area of the proposed cut slope. Variations in dip of steeply inclined schistosity can significantly effect the slope stability analysis as will be shown below.

Results of the combined analyses of bench mapping and core logging of schistosity clearly indicate that the schistosity, which dips moderately to the east on the upper benches of the west wall, gradually steepens and dips steeply west at depth within the proposed pit limits. At the north end of the pit the moderately dipping schistosity is not apparent.

In addition to foliation joints, four other well developed joint sets and occasional random joint sets occur within the pit. With respect to the toppling failures, however, these joint sets are relatively unimportant and will not be discussed further.

Mechanical Stability Analysis and Related Results

The slope stability analysis was carried out using the method described above. For this case, analysis results are presented as plots of resultant force on the toe block versus schistosity dip for various slope angles or bench face angles, groundwater conditions and spacing of foliation joints. Given in Figs. 6 and 7 are typical plots showing the results for both the overall slope and benches, respectively. These plots show that for a given slope height and slope angle, the slope stability is extremely sensitive to dip and spacing of foliation joints and height of the water table. Also, in terms of a given slope angle, spacing of foliation joints and height of the water table, the resultant force, P_n, varies considerably with change in dip of schistosity. The highest values of P_n occur for schistosity dips of about 70° to 80° into the slope. In general, the values of P_n are lower when the dip of schistosity is less than 70° or greater than about 80° to 85°, as shown in Figs. 6 and 7. This results in the characteristic peaked shape of the curves shown in these figures.

The spacing of the foliation joints, which defines the size of the tabular blocks, is extremely important to stability. As the spacing increases, the columnar blocks tend to become "squatter" (i.e. b/h increases) and the moments tending to cause overturning decrease. The spacing of throughgoing foliation joints at the mine varies from 30cm to 1.5m. It is difficult to determine which of the foliation joints are naturally occurring pre-existing throughgoing features and which of the joints have been opened up by effects of blasting or toppling failures. Based on field observations, for analysis purposes it was assumed in terms of bench evaluations that throughgoing foliation joints have a spacing of 1.5m. On the other hand, it was considered not likely that there are many throughgoing foliation joints of the scale of the slope which could effect the stability of the overall slope. Also, it is very likely that the shear strength along the foliation joints increases considerably deeper in the slope. Based on these interpretations and assessments, it was assumed that throughgoing joints, which could be involved in toppling of the overall slope, have a spacing of about 10m.

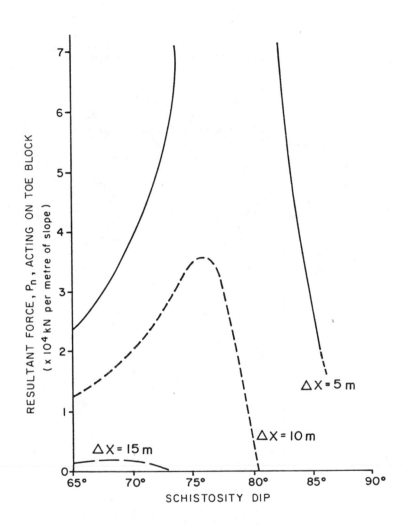

OVERALL SLOPE ANGLE = 45°

FIG. 6. Typical results of analysis of flexural toppling for 200 m high slope.

SYMBOLS

SPACING OF JOINTS SUBPARALLEL TO SCHISTOSITY $\triangle x$	GROUNDWATER CONDITIONS IN BENCHES		
	$h_w = H$	$h_w = .5H$	Water Below Failure Surface
.5m	——H——	——.5H——	—— O ——
1.0m	---H---	---.5H---	---- O ----
1.5m	—— H ——	—— .5H——	—— O ——
2.0m	——·H——·	——·.5H——·	——· O ——·
3.0m	—·· H —··	—·· .5H—··	—··— O —··—

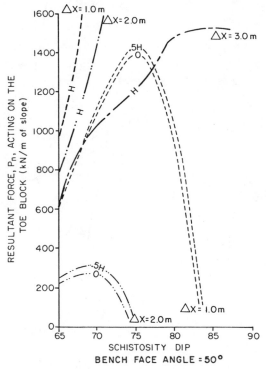

FIG. 7. Typical results of analysis of flexural toppling for 21.35 m high benches.

The effect of water pressure on toppling depends on the height and angle of the slope and the water level, h_w, in the slope. If the water level is below the basal failure surface there is no effect of water on toppling (see Fig. 3). If the water level occurs above the basal failure surface there can be a significant effect on the slope stability. In general, if the water level is less than about one-half the slope height, (i.e. $h_w < .5H$), the analyses indicate that there is little effect due to groundwater (see Fig. 7).

Analysis Results for Overall Slope. The analysis results for an overall slope 200m high are summarized in Table I. These analyses were carried out for a water level $h_w = .5H$ as this was considered to be the groundwater level in most of the west wall. Thus, as discussed above, groundwater was considered to have little or no effect on the overall slope in these analyses.

The analyses indicate that for a spacing of foliation joints of 10m, an overall slope angle of about 45° is appropriate for design for schistosity dips (i.e. foliation joints) of 80° to 90°. That is, provided the slope is reasonably drained. As mentioned earlier, these results may be somewhat conservative in that the behaviour of the blocks is assumed to be the same regardless of their location in the slope.

Analysis Results for Single and Double Benches. Toppling analyses were carried out for bench heights of 10.67m and 21.35m to assess the suitability of single and double benching. Based on the available geological information, dip intervals of 10° were considered reasonable for these analyses. Therefore, the analyses results are presented for dip intervals of dips of schistosity or foliation joints of 70° to 80° and 80° to 90°, in that these particular intervals were most pronounced. The analysis results for the benches are summarized in Table II. These results clearly indicate the necessity to excavate individual bench faces at an appropriate angle to provide buttresses in the slope to prevent toppling.

TABLE I

SUMMARY OF TOPPLING ANALYSIS RESULTS
FOR 200m HIGH OVERALL SLOPE

DIP OF FOLIATION JOINTS (SCHISTOSITY)	SLOPE ANGLE REQUIRED FOR A 200m HIGH SLOPE WITH $h_w = 0.5H$			
	Spacing $\Delta x = $ 5m	Spacing $\Delta x = $ 10m	Spacing $\Delta x = $ 15m	Spacing $\Delta x = $ 20m
80°–90°	$<45^\circ$	45°	51°	$>51^\circ$
70°–80°	$<<45^\circ$	$<45^\circ$	$<45^\circ$	47°

TABLE II

SUMMARY OF TOPPLING ANALYSIS RESULTS FOR BENCHES

BENCH HEIGHT (m)	DIP OF FOLIATION JOINTS (SCHISTOSITY)	BENCH FACE ANGLE REQUIRED								
		Joint Spacing $\Delta x = 0.5m$			Joint Spacing $\Delta x = 0.5m$			Joint Spacing $\Delta x = 0.5m$		
		$h_w = H$	$h_w = 0.5H$	$h_w = 0$	$h_w = H$	$h_w = 0.5H$	$h_w = 0$	$h_w = H$	$h_w = 0.5H$	$h_w = 0$
10.67	$80°-90°$	$\ll 50°$	$\ll 50°$	$\ll 50°$	$\ll 50°$	$60°$	$65°$	$<50°$	$70°$	$75°$
	$70°-80°$	$\ll 50°$	$\ll 50°$	$\ll 50°$	$\ll 50°$	$50°$	$50°$	$<50°$	$55°$	$55°$
21.35	$80°-90°$	$\ll 40°$	$40°$	$40°$	$<40°$	$55°$	$60°$	$40°$	$65°$	$65°$
	$70°-80°$	$\ll 40°$	$40°$	$40°$	$<40°$	$45°$	$45°$	$40°$	$50°$	$50°$

Table II indicates that bench face angles as low as $40°$ may be required if throughgoing foliation joints are closely spaced and high water levels exist. The analyses also indicate that steeper bench faces, as is to be expected, could be maintained with single benches than with double benches. Clearly, the results in Table II indicate the importance of knowing reasonably accurately the dip and spacing of foliation joints and the water levels in the slope.

Slope Design

In the areas of the pit where flexural toppling is possible, single 10.67m high benches are suggested to minimize the overturning forces acting in the slope. Also, single benches minimize unfavourable effects of water pressure in the benches. Therefore, for portions of the slope where schistosity dips between $80°$ and $90°$ into the slope and the slope is considered to be drained, a $75°$ bench face angle is suggested. Similarly, where the schistosity dips between $70°$ and $80°$ into the slope, a bench face angle of not more than $55°$ appears reasonable. As important as the bench face angle itself, is the need to ensure that the benches remains stable after being excavated. This could best be achieved by preshearing the benches to minimize the destabilizing effects of blasting.

With regard to berm widths it was determined that to meet safety requirements and operating needs of the mine, 7.4m wide berms would be required. Overall slope angles between haulroads for portions of the slope where schistosity dips between $80°$ and $90°$ into the slope and between $70°$ and $80°$ into the slope could then be geometrically determined.

TOPPLING ANALYSIS AT AN IRON MINE IN ONTARIO

The second design example pertains to the design of a 300m high open pit slope in steeply dipping metasedimentary rock at an iron mine in Ontario. Toppling occurs along well developed foliation joints in the metasediments. These failures are of particular concern on the west side of the open pit where most foliation dips steeply to the west. This discussion is concerned with the design of the west wall.

Geological Conditions

The metasedimentary rocks consist of Schistose Magnetite Ore, Main Massive Magnetite Ore, Outer Massive Magnetite Ore and Wall Rock which form a distinct interbedded sequence that is repeated throughout the mine area. All of these rocks are hard and tough, generally displaying consistent mechanical and physical properties and well developed foliation joints. Each separate rock type appears to have a characteristic spacing of foliation joints, which is of fundamental importance in evaluating toppling failure and the related slope design.

The metasedimentary rocks are tightly folded into two synclines and a central anticline. Folding in the mine is very tight and more than one stage of deformation is evident. Many minor folds from a scale of a few centimetres to tens of metres have been mapped, particularly in the ore bearing sequence in the lower portion of the west wall.

Foliation trends approximately parallel to the proposed final pit slope, striking north-northeast and dipping steeply southwest in many areas of the west wall. Within individual areas of the pit there are marked variations in foliation dip due to the presence of minor folds. As a result it is extremely difficult, at least on a local basis, to predict variations in foliation dip or to delineate areas of consistent dip within the slope. In order to "smooth out" local variations in foliation dip due to minor folds, statistical methods, including the cumulative sums technique which was mentioned earlier, were used. The cumulative sums technique was used to develop a three-dimentional geologic structural representation of foliation dip on plan as well as on composite geological sections. A typical composite geological section developed from geological mapping and core logging information is shown in Fig. 8. In general, there appears to be a steepening of foliation towards the centre of the pit. The metasedimentary Basal Wall Rock appears to have an average foliation dip of less than 80°W. Foliation in Main Massive Ore and Schistose Ore rocks in the central part of the pit is generally steeper dipping.

As can be seen in Fig. 8, $2\frac{1}{2}^{\circ}$ increments of dip from 70° to 90° were selected in delineating areas of approximately equal dip on

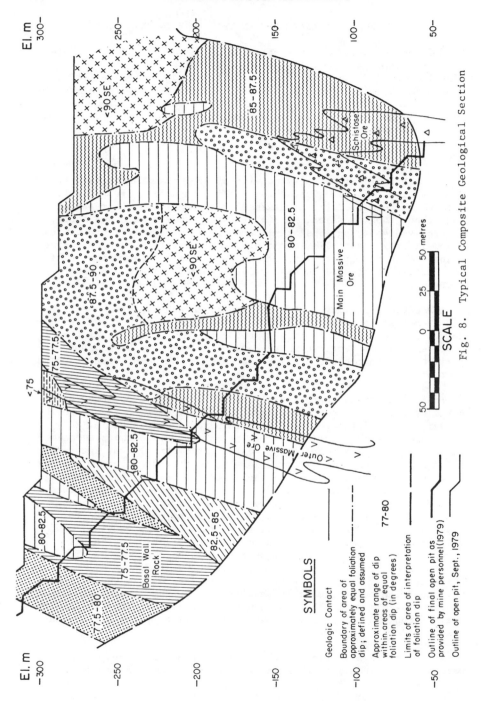

Fig. 8. Typical Composite Geological Section

individual sections through the west wall. Because of the limited amount of information in many areas, these increments were considered to be reasonable.

Average values of spacing of foliation joints were estimated from the range of average joint spacings. In general, Basal Wall Rock, Inner Wall Rock and Schistose Ore are considered to have a relatively close spacing of natural foliation joints. Basal Wall Rock is assumed to have an average spacing of foliation joints of about 0.9m. Because of the interbedded nature of the Inner Wall Rock and Schistose Ore, these rocks have been considered together as a single unit, having an average spacing of foliation joints of about 1.1m. Foliation joints in Outer Massive Ore and Main Massive Ore appear to have an average spacing of about 1.2m and 2.4m, respectively.

Effective length of foliation joints is generally limited by the presence of pervasive minor folds, which provide tight asperities analogous to crenulations or contortions which would disrupt otherwise continuous foliation joint planes. For this reason, it was assumed to be unlikely that continuous foliation joints which could lead to toppling could develop for greater than one or two benches at the most (i.e. 10.67 to 21.35m).

In addition to foliation joints, three pronounced joint sets and several miscellaneous joint sets were mapped in the pit. None of these joint sets were considered relevant to the flexural toppling analysis.

Documentation of Failed and Unfailed Slopes

In order to obtain a better appreciation of the rock mass conditions where toppling failure develops, both stable slopes and unstable slopes susceptible to toppling were carefully documented and evaluated. Evaluating slopes in this manner proved extremely useful in assessing the basic mechanisms of toppling failure. Also, these results provided a basis for comparative analysis of the behaviour of the different slopes with respect to variations in the relevant parameters discussed below. It was assumed that all slopes had been excavated vertically using similar blasting practices. Relevant characteristics, such as lithology, structural geology parameters (i.e. mainly foliation dip and foliation joint spacing) and slope geometry, were recorded for over thirty locations on the west wall. Typical toppling failures on benches on the west wall are shown in Figs. 4 and 5.

For all cases examined in the field, the basal failure surface dips at about 30°. Most failures have well developed cracks on bench crests.

It should be noted that this analysis applies to benches that were established prior to the mine improving their terminal blasting techniques. Improved terminal blasting has minimized the damage to the ultimate wall, and indications are that the opening of joint planes is not as serious as initially observed. Continued improvement in term-

inal blast techniques, should minimize the opening of foliation
joints and a more stable condition may result.

The relative distribution of the various documented slopes in
terms of plotting dip versus spacing of foliation joints is given in
Fig. 9. Only 10.67m high single benches were available for documentation. Shaded symbols represent toppled slopes, whereas unshaded
symbols represent stable slopes. The importance of the interrelationship of dip of foliation joints and spacing of foliation
joints in terms of stability of the bench faces is clearly obvious
in Fig. 9. It can be seen that the boundary conditions for stable
and unstable (or toppled) benches with regard to dip and spacing of
foliation joints appear to be demarcated by a well defined line.
Both an average and a lower bound line have been interpreted for
stable compared to toppled slopes in Fig. 9.

The relationship and related data indicated in Fig. 9 is remarkably consistent with the observed behaviour of toppling. Where
toppling is possible, as foliation dip steepens and/or spacing of
foliation joints increases, the tendency to topple decreases. It is
difficult to determine if the various rock types have an effect on
the observed relationship, in that the range of dips and spacings of
foliation joints for each rock type is often limited (see Fig. 9).
In general, however, it appears that the spacing of foliation joints
of each rock type is reasonably consistent. Also, only a limited
number of slopes were available for documentation in Schistose Ore
and Outer Massive Ore. Other parameters, such as friction angle of
foliation joints and unit weight, have little effect on the toppling
analysis. Most slopes documented had little evidence of water in the
immediate vicinity of the benches. Hence, it is assumed that these
slopes are drained.

Mechanical Stability Analysis and Related Results

Based on the analysis method described earlier, the following
toppling analyses were carried out.

Analysis of Overall Slope. Based on the high compressive strength
of the rock, the presence of minor folding and general lack of major
throughgoing discontinuities with unfavourable orientations, it
appears that deep-seated toppling failures involving large sections
of the whole slope are unlikely for the range of slope angles considered. Hence, analyses for the portion of the wall subject to
possible toppling failures have been carried out with regard to
determining the optimum overall slope angle based on the appropriate
design of benches.

Analysis of Single Benches. Toppling analyses were carried out for
different bench face angles, dips and spacings of foliation joints
for 10.67m high single benches. For a particular foliation dip and
bench face angle, the spacing of foliation joints corresponding to

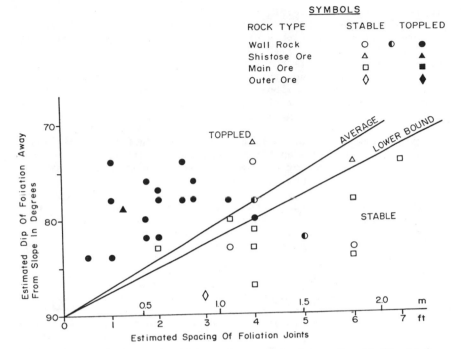

Fig. 9. Distribution of stable and toppled slopes for 10.67m high benches.

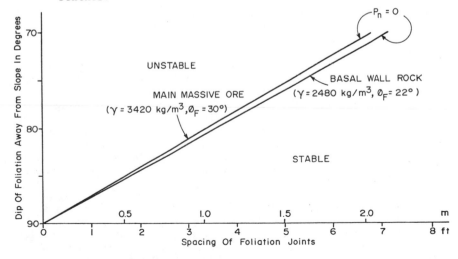

Fig. 10. Toppling analysis results for a 10.67m high drained bench at an angle of 70°.

zero resultant force (i.e. $P_n = 0$) at the toe was determined from the
stability analysis. The results for Basal Wall Rock and Main Massive
Ore were plotted on a graph of foliation dip versus spacing of foliation joints, as shown in Fig. 10, for a 10.67m high bench in a dry
slope with a 70° bench face angle. The resulting lines represent
$P_n = 0$. Below these lines the foliation dip and spacing of foliation
joints increases and the slope is stable with respect to toppling.
Above these lines the foliation dip and spacing of foliation joints
decreases so that the slope is unstable (see Fig. 10).

Fig. 10 suggests there is little difference in the analysis results
for the Main Massive Ore and Basal Wall Rock, indicating unit weight
and friction angle are not as important as the geometrical characteristics of foliation joints.

Bench face angles were varied and accordingly analyzed to develop
a set of design curves as shown in Fig. 11(a). These curves indicate
the bench face angle required to prevent toppling in terms of various
combinations of foliation dip and spacing of foliation joints.

Comparative Analysis: Stability Analysis Results vs Analysis Results
from Documented Slopes for Single Benches. Documented stable slopes
and unstable toppled slopes indicating the actual behaviour of the
slopes in question (see Fig. 9) can be compared to the theoretical
toppling analysis results obtained from the mechanical stability
analyses (see Fig. 11(a)). It can be seen that the estimated lower
bound of documented toppled slopes shown in Fig. 9 is in close agreement with the analysis results for an 89° bench face angle shown in
Fig. 11(a). This remarkably close correlation between the theoretical
model and empirical data from documented slopes indicates the validity of the theoretical analyses results as a reasonably reliable
method for design.

Analysis of Double Benches. Similar analysis methods were used to
develop design charts for 21.35m high double benches as summarized
in Fig. 11(b). An increase in slope height from 10.67m to 21.35m
high benches increases the potential for toppling by reducing the
relative spacing to length ratio of individual toppling blocks. As
a result, the design lines for $P_n = 0$ for the various bench face
angles are shifted downwards. Thus, slopes with a particular combination of foliation dip and foliation joint spacing which are stable
for a single bench could possibly topple for a double bench unless
excavated at a shallower bench face angle.

Effects of Groundwater on the Toppling Analysis. The effects of a
groundwater level equivalent to half the height of the bench ($h_w = H/2$)
for 10.67m and 21.35m high benches are shown in Figs. 12(a) and 12(b).
Groundwater increases the overturning moments which reduces stability.
Hence, the respective design curves for $P_n = 0$ move downwards compared to curves in Figs. 11(a) and 11(b).

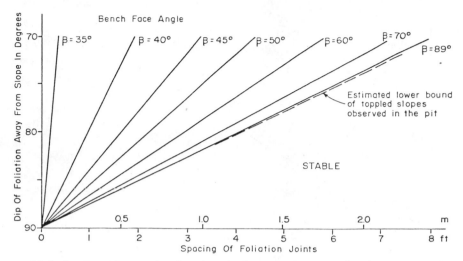

FIG. 11a). Design chart showing bench face angles required to prevent toppling for 10.67 m high single benches for a dry or drained slope.

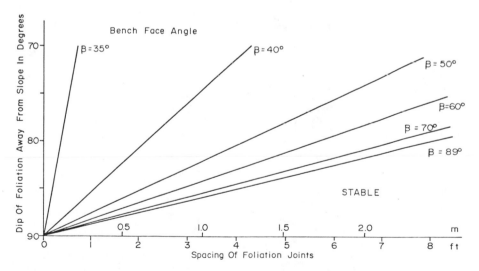

FIG. 11b). Design chart showing bench face angles required to prevent toppling for 21.35m high double benches for a dry or drained slope.

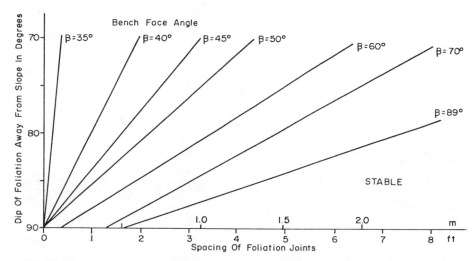

FIG. 12a). Design chart showing bench face angles required to prevent toppling for 10.67m high single benches for a water table 5.34m above the toe of the slope.

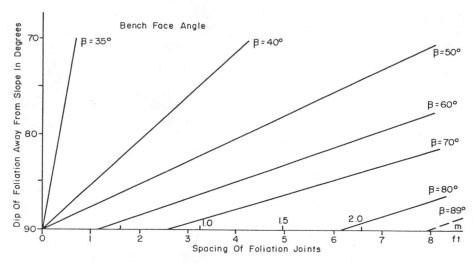

FIG. 12b) Design chart showing bench face angles required to prevent toppling for 21.35m high double benches for a water table 10.67m above the toe of the slope.

Summary of Toppling Analysis Results. The toppling analysis results
agree with the observed behaviour of 10.67m high benched slopes which
have been fully documented. The design charts developed, therefore,
can be used with reasonable confidence for design of bench face angles
in areas where toppling of individual benches is kinematically possible. The design charts have been summarized in Table III for
typical dips and spacings of foliation joints, 10.67 and 21.35m
bench heights and possible groundwater conditions which could be
encountered.

Slope Design

Because the possibility of deep-seated toppling failure appears
to be low, the overall slope design is a function of the bench design
only.

Use of Design Charts to Select Bench Face Angle. The design charts
shown in Figs. 11(a) and 11(b) have been used to determine the bench
face angles required to prevent toppling for 10.67m and 21.35m high
benches. For example, where 77° dipping foliation joints occur
10.67m high benches with 50° bench face angles would appear feasible.
Because of possible local variations in spacing and/or dip of foliation joints due to minor folding or blasting etc., it would appear
that bench face angles should be no steeper than 80° for 10.67m
benches and 70° for 21.35m benches.

The bench face angles determined for this design example are
based on a drained slope condition. If significant groundwater is
encountered on the benches (i.e. the height of water above the toe
of the bench is greater than about twenty-five percent of the height),
shallower bench face angles or, alternately, drainage control measures
would be a consideration. As for the previous design example, preshearing of bench faces is suggested as a means of minimizing the
destabilizing effects of blasting. It is suggested that in critical
areas of the wall where shallower bench face angles are extremely
impractical, the use of dowels across foliation could be used to tie
the rock mass together, thus increasing the effective spacing of
foliation joints. Dowels would have to be installed before the
slope is excavated in order to prevent toppling.

Selection of Berm Width. For 10.67m high benches the berm width must
be sufficiently wide that when two single benches are considered, the
resulting slope angle is not greater than that required to prevent
toppling of a double bench. The minimum berm width (ℓ_{min}) for a
single bench required to prevent toppling of the double bench is
calculated as shown in Fig. 13. This comparison is only carried
out for 10.67m and 21.35m high benches, as it is unlikely that deep-
seated toppling could occur over more than two benches for reasons
explained earlier.

TABLE III

SUMMARY OF BENCH FACE ANGLES REQUIRED TO PREVENT
TOPPLING FOR VARIOUS DIP AND SPACING OF FOLIATION
JOINTS, BENCH HEIGHTS AND GROUNDWATER CONDITIONS

Spacing of Foliation Joints (m)	Dip of Foliation	Bench Face Angle Required to Prevent Toppling			
		Bench Height = 10.67m		Bench Height = 21.35m	
		Dry Slope	hw = 5.33m	Dry Slope	hw = 10.67m
0.9 (Basal Wall Rock)	75°	47°	47°	39°	39°
	77	52	52	41	41
	80	61	57	46	46
	82.5	90	62	49	49
	85	90	70	62	56
	87.5	90	89	90	63
	90	90	90	90	72
1.1 (Inner Wall Rock and Schistose Ore)	75	52	52	42	42
	77	57	55	44	44
	80	68	60	48	48
	82.5	90	68	54	52
	85	90	78	69	58
	87.5	90	90	90	67
	90	90	90	90	72
1.2 (Outer Massive Ore)	75	57	54	44	44
	77	62	58	46	46
	80	90	63	49	49
	82.5	90	71	58	54
	85	90	85	89	61
	87.5	90	90	90	69
	90	90	90	90	74
2.4 (Main Massive Ore)	75	90	80	58	56
	77	90	85	64	59
	80	90	90	89	66
	82.5	90	90	90	71
	85	90	90	90	77
	87.5	90	90	90	82
	90	90	90	90	90

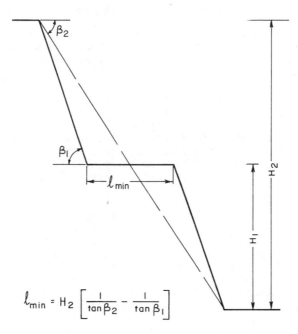

$$\ell_{min} = H_2 \left[\frac{1}{\tan \beta_2} - \frac{1}{\tan \beta_1} \right]$$

Fig. 13.   Calculation of minimum berm width (ℓ min) required for
a single bench in order that the bench face angle for a
double bench is not exceeded.

   Benches that are excavated by preshearing at a specific angle to
prevent toppling will probably have considerably less rockfalls than
those excavated by more normal methods. Consequently, in terms of
controlling rockfalls, berms probably can be significantly narrower
than would normally be envisaged.

DESIGN OF REMEDIAL MEASURES TO PREVENT TOPPLING

   The possibility of toppling failure was recognized in an open pit
coal mine where bedding had steepened to the point of being over-
turned. The bedding in the area of concern appears to be folded,
resulting in bedding joints which dip steeply into the slope.
Bedding dips were not accurately known; however geologic sections
indicate dips could vary from about 80° into the slope to vertical.
Thus, if all the coal is removed at the toe of the slope, an overhanging condition could develop. The bulk of the rock below the
coal appears to be hard sandstone. Spacing of bedding joints is
estimated to be about 0.6m. The height of wall to be exposed along
steeply dipping beds is about 24m, below which the bedding appears
to flatten considerably and dip towards the pit.

Mechanical Stability Analysis and Related Results

The basic analysis of toppling was carried out as described
earlier. However, because the objective of the study was to provide
remedial measures to prevent toppling, the problem had to be approached in a different manner.

Analysis of Bedding Joint Spacing Required to Prevent Toppling. A
plot illustrating the resultant force, P_n, acting on the lowest block
in the slope, for various bedding dips and spacings of bedding joints
is shown in Fig. 14. Where P_n = 0 the factor of safety would be 1.0
and toppling would not occur. Interpolation of the spacing of
bedding joints equivalent to P_n = 0 (see Fig. 14) yields the effective spacing of bedding joints required to ensure a stable vertical
slope.

The results in Fig. 14 indicate that bedding joint spacings of
2m, 4m and 6m would be required to prevent toppling for bedding dips
into the slope of 85°, 80° and 75°, respectively. However, the
observed bedding joint spacing in the potentially unstable slope in
question is only 0.6m, which is significantly less than the bedding
joint spacings indicated above to prevent toppling. Groundwater in
the slope would increase the forces contributing to toppling, thus
requiring even greater joint spacings than indicated above.

Toppling may be prevented in two ways:

1. Flatten the slope angle to reduce overturning moments
    and increase natural buttressing effects in the toe of the
    slope.

2. Install artificial support to counteract overturning moments
    by, in effect, providing a buttress column at the front
    of the slope.

Assessment of Artificial Support Requirements to Prevent Toppling

Slope flattening was not feasible and, therefore, artificial support
measures appeared to provide the only viable alternative to the
problem. The objective of artificial support is to bind the tabular
columns of rock together, and thus prevent interlayer slip which
would take place along bedding joints between individual rock columns
at the front of the slope.

Analysis Method. Analyses were carried out assuming that the slope
is vertical and 30m high. Spacing of bedding joints, or alternately
the width of columns, was estimated to be about 0.6m. Detailed
analyses for bedding dip angles of 80° and 85° were conducted. In
these analyses the net force on each rock column in the slope is
calculated. In both cases the forces were found to be extremely high
on columns near the front of the slope. As expected, the forces

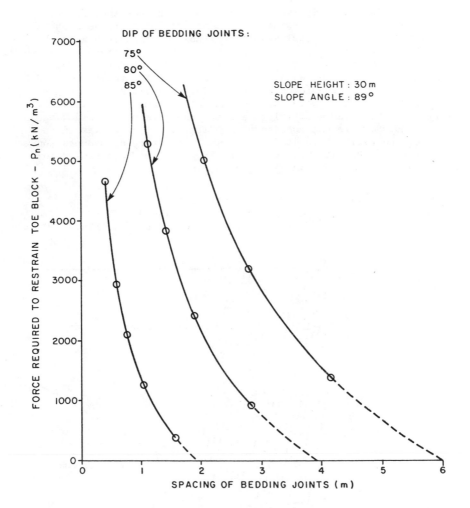

FIG. 14. Toppling analysis results for 30 m high
subvertical slope showing resultant force P_n
on the lowest block in the slope for various
dip and spacing of bedding joints.

diminished behind the slope crest.

The toppling analysis results were used to assess the effects of
tying the individual 0.6m wide columns at the front of the slope
together. By assuming that several columns immediately behind the
slope face were tied together, toppling forces and factors of safety
were recalculated manually to determine a factor of safety against
toppling. That is, the weight of the theoretically reinforced block
was determined and the force on this reinforced block generated by
the columns behind it were obtained directly from the detailed
analysis. The effects of friction along the back of the block and
water pressure were not considered. A factor of safety was calculated as the ratio of resisting moments to overturning moments about
a pivot point, P_v, at the toe of the slope. Slope models used for
this analysis are shown in Figs. 15 and 16.

SLOPE HEIGHT:	30m
SLOPE ANGLE:	89°
DIP OF BEDDING JOINTS:	80°
SPACING OF BEDDING JOINTS:	0.6m

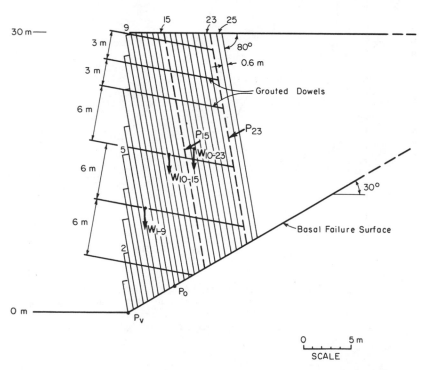

Fig. 15.   Stability analysis model and recommended artificial support
for slope with bedding joints dipping at 80° into the slope.

SLOPE HEIGHT: 30m
SLOPE ANGLE: 89°
DIP OF BEDDING JOINTS: 85°
SPACING OF BEDDING JOINTS: 0.6m

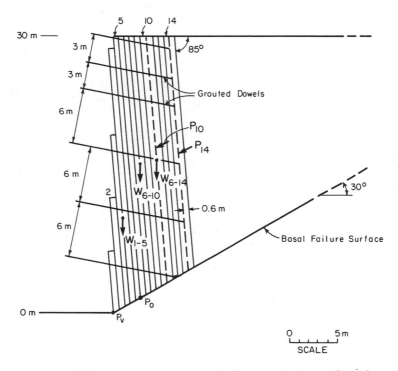

Fig. 16. Stability analysis model and recommended artificial support
          for slope with bedding joints dipping at 85° into the slope.

Evaluation of Size of Reinforced Buttress Block Required. Analysis
results are summarized in Table IV. For design of artificial support
for this stability problem a factor of safety of 2.0 was used. As
would be expected, as successive tabular blocks or columns are tied
together the factor of safety against toppling increases. Results
in Table IV indicate that about 15 columns would have to be tied
together to achieve an acceptable factor of safety if bedding dips
at about 80°. On the other hand, for a more favourable situation
where bedding dips at 85° for example, only about 10 columns would
have to be tied together.

It was considered that where bedding dips into the slope removal
of the coal would undercut the vertical slope. Even without complete
removal of the coal, it was felt that relatively little support
would be provided by the coal. Therefore, the overhanging slope

that would develop would be even more susceptible to toppling.
Hence, analyses also were carried out for the case of the overhanging slope. Assuming that the slope is a free standing overhang,
the pivot point for toppling would occur at P_O, as shown in Figs. 15
and 16. The results of these analyses are also summarized in Table
IV. It is noteworthy that the length of reinforcing dowels required
to tie the front buttress block is relatively uniform for the overhanging slope condition.

TABLE IV

SUMMARY OF STABILITY ANALYSIS RESULTS
FOR SLOPE WITH BEDDING DIPPING AT 80° AND 85°
INTO THE SLOPE

BEDDING DIP INTO SLOPE ($^\circ$)	SLOPE ANGLE ($^\circ$)	NUMBERS OF COLUMNS ASSUMED TO BE JOINED	FACTOR OF SAFETY	LENGTH OF DOWELS REQUIRED TO TIE BLOCK FOR F=2.0 (m)	RECOMMENDED LENGTH OF DOWELS (m)
80	Vertical (90)	1 to 9 1 to 12 1 to 13 1 to 15 1 to 17 1 to 18	0.3 0.92 1.18 1.84 2.30 3.68	4 to 8	9
	Overhang (−80)	10 to 18 10 to 20 10 to 22 10 to 23	0.20 0.76 1.64 2.22	8.5 to 9	
85	Vertical (90)	1 to 5 1 to 7 1 to 8 1 to 9 1 to 10	0.18 0.70 1.10 1.58 2.28	3.5 to 6	6
	Overhang (−85)	6 to 10 6 to 13 6 to 14 6 to 15	0.24 1.39 2.16 2.78	6	

Design of Artificial Support and Drainage. Based on the above analysis it would appear that the effective spacing of the bedding joints near the face should be increased by use of dowels. However, the design and related specifications of the dowelling system for a problem of this nature is difficult to rationalize without field trials and/or an accurate knowledge of the geologic structural conditions. For preliminary estimating purposes, however, it was suggested that dowels approximately 9m long be installed in the rock face at successive levels as the coal is being removed. A suggested typical arrangement of dowels is shown in Fig. 15 and 16. Some of the design specifications are as follows:

1. Dowels to consist of clean steel bars, reinforcing bars, cables or similar material (minimum 25mm diameter) placed in drillholes which are fully grouted.
2. All dowels should be inclined at about -10° below horizontal and placed perpendicular to the slope direction (i.e. at right angles to bedding strike).
3. Dowels to be placed in rows with a 3m spacing of dowels along rows.
4. Top row of dowels to be placed as near the slope crest as is feasible.
5. Second and third rows of dowels to be placed at a vertical distance of 3m and 6m below the upper row of dowels, respectively.
6. Chain link mesh to be placed as required to prevent rockfalls in areas where the slope overhangs, etc.

Sub-horizontal upward inclined drainholes would be beneficial at the toe of appropriate lifts to ensure that hydrostatic pressures do not build up in the slope.

ACKNOWLEDGEMENTS

The authors wish to acknowledge with thanks the cooperation and assistance of the mine staff at the mines where the foregoing slope stability analyses and/or remedial measures were carried out.

REFERENCES

Ashby, J., 1971, "Sliding and Toppling Modes of Failure in Models and Jointed Rock Slopes," M.Sc. dissertation, University of London, Imperial College.

de Freitas, M.H. and Watters, R.J., 1973, "Some Field Examples of Toppling Failure," Geotechnique 23, No. 4, pp. 495-514.

Goodman, R.E. and Bray, J.W., 1976, "Toppling of Rock Slopes," Rock Engineering for Foundations and Slopes, Vol. II, Proceedings of ASCE Specialty Conference, Boulder, Colorado, pp. 201-234.

Hittinger, M., 1978, "Numerical Analysis of Toppling Failure in Jointed Rock," Ph.D. Thesis, Univ. of California, Berkeley.

Piteau, D.R. and Martin, D.C., 1981, "Mechanics of Rock Slope Failure," Third International Conference on Slope Stability in Surface Mining, Vancouver, B.C., June 1-3, 1981.

Piteau, D.R. and Russell, L., 1971, "Cumulative Sums Technique: A New Approach to Analyzing Joints in Rock," Thirteenth Symposium on Rock Mechanics, ASCE, Urbana, Illinois.

Question

Is it possible to have toppling failures in rock that is overlain by 100 feet of glacial till, the till itself being stable until the rock underneath is exposed.

Answer

Yes. The glacial till could act as a surcharge load on the rock after the rock is exposed.

Question

Do you know of cases of massive slope failure related to toppling having serious operational consequences for the mine, as opposed to untidy surface ravelling shown in most of your slides.

Answer

In general, overall slopes in open pit mines are too shallow to be susceptible to massive toppling failures. However, in Idaho, the side of a mountain that forms one wall of a valley appears to be failing by a form of toppling. It should be noted, however, that if all the benches fail, the slope cannot be mined and essentially is analogous to an overall slope failure in that the pit becomes inoperational from an efficiency and safety point of view.

Question

From my own work at the one mine you discussed in Australia, I concluded that the primary mode of failure was a rotating wedge mechanism and that the toppling was a secondary phenomenon. Would you please comment.

Answer

At the time of your work the portion of the wall that is now susceptible to toppling failure was not exposed. We concur that a rotating wedge mechanism is the primary failure mode in the area of the pit wall with which you are familiar.

Question

Are you suggesting that on a practical benefit - cost analysis it is

practical to alter the overall slope angle for these one bench
toppling failures.

Answer

No, not necessarily. However, if all of the benches are lost due to
toppling failures, it is difficult to protect the base of the pit.
In many cases, it is sufficient that the design be altered such that
the integrity of the haul roads is ensured. In this regard, the use
of artificial support (such as dowels, etc.) may be adequate to
practically solve the problem.

Question

Would you advise monitoring toppling failures, and if so what kind
of program would be adequate. Do you thing rock anchor bolts as
suggested earlier by Ben Seegmiller will be helpful in retarding
movement.

Answer

With regard to the first question, it is usually adequate to monitor
toppling failures by simple monitoring of survey hubs. As for the
second question, rock bolts can assist in preventing toppling
failures by tying or "knitting" individual rock columns together to
give a greater width to height ratio to the critical blocks in the
lower portion of the slope. It should be noted that rock bolts may
be necessary if careful control blasting (such as preshearing) and/or
drainage of the slope is achieved. Unlike for other types of
failure modes, preshearing acts in an entirely different way in
controlling stability in toppling failures. Preshearing essentially
causes the front part of the bench which is most susceptible to
toppling to be largely detached from the slope and not cause toppling
of the remaining intact rock. In this case rock anchors are essentially required.

Question

Apparently the dowels were not tensioned in the last case example.
Would not tensioning of the dowels have taken advantage of the
friction between columns, thereby reducing deformation of the block
and consequently further reducing movement of the uphill column and
also reducing the number of bolts.

Answer

It was considered that the dowels would prevent interlayer slip, thus
obviating the need for tensioning. However, the dowelling program
was monitored, with the idea in mind that tensioned rock bolts could
be required if the dowels proved inadequate. It should be noted that
sub-zero temperatures prevailed during the dowelling program. Thus,
it was desirable to keep the remedial measures as simple as possible
to minimize any temperature related problems.

Question

In the last case example, did you check the possibility of shearing of the dowels due to inter-block movement or shear. How would you suggest checking this possibility.

Answer

The possibility of inter-block movement shearing the dowels at any location can be checked by evaluating the forces acting between the two blocks at that location. The limit equilibrium analysis method for flexural toppling calculates the force acting between blocks per unit width of block. Depending on the size, type and spacing of dowels, the shear stress in the dowels can then be calculated and compared with the dowels shear strength.

Question

Static analysis of toppling failure is generally based on columns resting on a stepped base. Is there any mechanical approach to study toppling of columns resting on a flat base.

Answer

Other than studying natural field examples of the various types of toppling, probably the best mechanical approach is to use base friction modelling. In order to generate the necessary overturning moments in the modelled slope, one can apply external loads in the slope or alter the column dimensions in such a way that toppling failure is generated.

Question

The implication is that toppling failure is a rather small scale phenomenon. Is the transition to sliding or collapse failure a function of absolute size, slenderness, or some other parameter. It seems that natural failures of cliffs favour perhaps the absolute size parameter.

Answer

It would seem that a number of parameters are important in this matter. In addition to the parameters mentioned above is the effect of water and the largely unknown effects of the degree of confinement on columns at depth. If fractures close up or are not developed at depth, columns become wider and more stable. Other parameters which can effect the toppling phenomenon but which are not fully understood are: i) the effects of lateral constraint; ii) the change of shear strength with depth; and iii) the behaviour of the rock mass under some external loads.

Chapter 30

SUCCESSFUL IMPLEMENTATION OF STEEPER SLOPE ANGLES
IN LABRADOR, CANADA

Om P. Garg

Chief, Long Range Mine Planning
Iron Ore Company of Canada
Labrador City, Nfld.

ABSTRACT

During the period between 1970 and 1974 the Iron Ore Company of
Canada had embarked on a detailed program to evaluate all aspects of
mine design work for its open pit mines around Schefferville, located
both in Quebec and Labrador, Newfoundland, Canada.

One of the main conclusions was that the majority of the slope
angles for the ultimate pit walls could be steeped. During the
period 1975 and 1976, five (5) open pit iron ore mines were redesigned
using the steeper slope angles up to 55 degrees. Initially these
pits were designed at slope angles varying between 37 degrees and 45
degrees due to the generally fractured and leached nature of rocks
in the area.

During the active mining phase between 1976 and 1979, the stability of the steeper ultimate pit walls was monitored. The mining of
ore was successfully completed in 1978 and 1979 in three pits. This
of course, resulted in a significant savings due to a reduction in
the quantity of waste that had to be removed.

The successful pit slope redesign project was a co-operative
effort between the Federal Government, Universities and the Iron Ore
Company of Canada.

INTRODUCTION

The Iron Ore Company of Canada (IOC) concurrently oper-
ates six (6) open pit iron ore mines in the Schefferville
(54º 49' N 66º 50' W) area, Figure 1. These mines are
located both in Quebec and Labrador, Newfoundland. The
open pit mining is done utilizing 50R rotary drills, 7.7 cu
m (10 cubic yard) bucket capacity shovels and 120 tonne
diesel electric haulage trucks. The average crude ore
production between 1973 and 1980 was 10.16 million tonnes
(10 million tons) per year. In addition, an average of
approximately 5.4 million cubic metres (7 million cubic
yards) of overburden and waste rock material per year
were also removed. The overall stripping ratio during
the 1954 and 1976 period is 1.0 bank cubic yards (BCY) of
waste per ton of ore, for a total ore production of 195
million tons.

The rocks in the Schefferville area are sediments of
Proterozoic age consisting of Iron Formation, slates,
quartzites, cherts and dolomites. The strata have been
folded and faulted and subsequently leached to a varying
degree by ground water. The ore bodies ranging in size
from 5 to 30 million tonnes occur in northwest - southeast trending structural troughs, formed either by synclines or by the intersection of synclinally folded strata
and a major strike fault. The general strike of the formations is northwest - southeast with dips varying between
50º - 70º due northeast.

EARLY HISTORY OF SLOPE DESIGN PROCEDURES

The problems associated with open pit mining in geologically disturbed and leached and altered sedimentary
strata in the Schefferville area were recognized during
the early stages of mining in 1955-56. Based on field
experience and on the geological structure of deposits,
design charts were prepared and overall slope angles of
between 37º and 45º were used in the ultimate pit designs
(Rana and Bullock, 1969). Although most of the pit slopes
designed using these overall slope angles have been stable
during the life of the mines, a few slides did occur in
the area of Ruth Lake and French Mines (Coates, 1965).
In fact the Ruth Lake Mine was prematurely closed in 1966
due to a slope failure which destroyed the main haul road
(Coates et al, 1968). Increased water pressure on the
footwall (west wall) was the main cause of the slope failure. Subsequent to this slide, a comprehensive dewatering
program consisting of peripheral and inpit wells was instituted in 1967 for all new mines (Bullock,1972). With

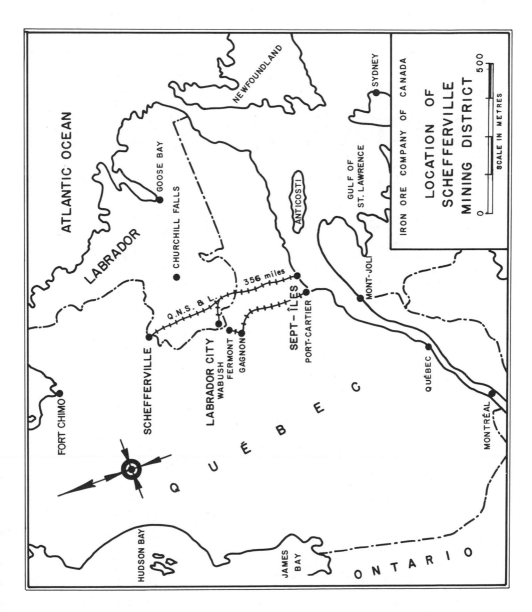

Figure 1
Location of the Schefferville Mining District.

the implementation of the dewatering program, pit walls
were more stable and the mining and stripping operations
have been successfully completed in several mines.

PIT SLOPE REDESIGN PROJECT

During the late 1960's and early 1970's, it was realized that the simple to use design charts mentioned earlier
had limited applications because these did not take into
consideration the variations in structural and strength
properties of rocks along the perimeter of ultimate walls.
Also it was realized that substantial reductions in stripping ratio and quantities could be achieved if the ultimate pit slopes could be steepened even by 1 degree or 2
degrees (Hoek and Bray, 1974).   For example, steepening of final wall slope by 1° from 45° to 46° could result
in a saving of approximately 252,000 bank cubic metres
(330,000 bank cubic yards) of stripping for one mine wall
only 1220 m (4,000 feet) long and 110 m (360 feet) deep.
Therefore, during the period between 1970 and 1974 the
Iron Ore Company of Canada had embarked on a detailed geotechnical program to evaluate all aspects of mine design
work (Nichols, 1968, Garg and Kalia, 1975).   During the
planning phase of the geotechnical program, it was realized that each pit wall must be considered and analysed
separately in order to obtain optimum (steepest & safe)
slope angles.   This is particularly true for the Schefferville area where properties of rocks vary considerably
over a relatively short distance.

Data Collection

The scope of the geotechnical work at the pre-produc-
tion stage of mine development includes:

a)   Detailed geological mapping in the trenches and of
     outcrops in the area and geotechnical logging of
     samples from drill holes.

b)   Collection and testing of disturbed and undisturbed
     (block) samples for shear strength parameters using
     a 10-ton capacity shear box.   Samples of varying
     sizes up to 12" x 12" x 6.3" are tested in the
     shear box.

c)   In situ density tests.

d)   Collection and analysis of surface and groundwater
     data, including the determination of permeability
     values from pressure packer tests and measurements

of water table elevations in the area.

e) Back analysis of past failures to obtain strength properties of rocks.

f) Analysis of structural data using equal area stereographic net. Amongst other items, the Factor of Safety of two intersecting joint sets likely to cause a wedge type of failure is calculated.

Since the details on the above-mentioned geotechnical program have been documented in an earlier report, these are not covered in this publication   (Garg and Kalia, 1975, Garg, 1976).

Analysis of Data

MORGP is the name of a computer program developed by Morgenstern and Price in Fortran language based on the Limit Equilibrium Method of Analysis (Morgenstern and Price, 1967). The details on the application of MORGP computer program at Schefferville are included in an earlier publication (Garg and Devon, 1978). Basically, this program calculates the Factor of Safety for the desired pit slope along each of the pre-determined failure surfaces. A number of possible failure surfaces are tried for a given slope till the lowest Factor of Safety along any of the surfaces is obtained. If the lowest possible value is greater than the acceptable Factor of Safety (of 1.10 for Schefferville), the slope angle is increased in increment of one degree and the entire process is repeated. The final result obtained from the MORGP computer program is the determination of the overall final wall slope angle corresponding to the Factor of Safety of 1.10. The residual shear strength values were used in the analysis.

The recommended final wall slope angles for the various parts of a mine are then used by the mine planning engineer for the ultimate mine design.

Redesign of Mines

Based on these improved methods of data collection and analysis of geotechnical data, steeper slope angles were recommended for Knox, Fleming No. 3, Timmins No. 1, Timmins No. 2 and Timmins No. 6 mines between 1974 and 1979, Figure 2. The recommended slope angles were up to 55 degrees or increased by as much as 10 degrees from the previously used angles in the ultimate pit designs. As an integral part of the implementation of the improved design, it was recommended that:

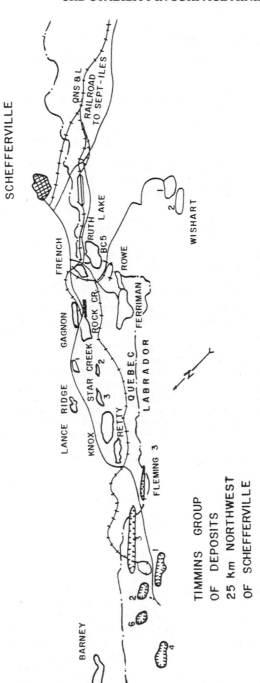

IRON ORE COMPANY OF CANADA

FIGURE - 2
LOCATION OF MINES

SCALE IN FEET

0 4000 8000 12000 16000 20000

a) Blasting in the vicinity of ultimate slopes be controlled to minimize damage to rocks along the final
   pit walls.

b) Slope monitoring be carried out on a routine basis.

c) Water table elevations be kept below the mining
   level so that the wall stability would not be adversely affected.

d) The pit redesigns based on steeper slope angles be
   reviewed at least once a year as more updated geological and geotechnical data become available.

IMPLEMENTATION OF STEEPER SLOPE ANGLES

The recommended steeper slope angles were implemented
for all the five (5) mines mentioned earlier. Figure 3
lists the slope angle data and pit dimensions for Timmins
1 and Timmins 2 mines. One particularly favourable factor
for steepening the final pit walls at these two mines was
the presence of steeply (50° - 65° due northeast) dipping
competent rocks specially along the footwall of the mines.
During the implementation stage, several helpful discussions took place with the CANMET Staff under the direction
of Dr. D.F. Coates. The decision to implement the steeper
slope angles in the Schefferville area was also facilitated due to the work which was done in connection with the
Pit Slope Project sponsored by CANMET during the 1971-76
period. During the data collection and analysis phase of
the work, Dr. N.R. Morgenstern of the University of Alberta and Dr. M.S. King of the University of Saskatchewan
also contributed significantly to the study. The final
decision on steepening the pit slopes in the Schefferville
area was taken by IOC Management.

SAVINGS REALIZED DUE TO REDUCTION IN STRIPPING

The mining was successfully completed in Fleming No. 3
mine in 1978 and in Timmins No. 1 and No. 2 mines in 1979.
Figure 4 shows a substantial reduction of 4.32 million
bank cubic yards in stripping quantities due to the redesign of Timmins 1 and 2 mines, using the steeper slope
angles.

FIGURE-3

TIMMINS 1 & 2 MINES

OVERALL SLOPE ANGLES USED

Wall	Initial	Final
East	45°	50°
West	42-1/2°	50°

PIT DIMENSIONS - FEET

	Length	Width	Depth
Timmins 1	4000	800	540
Timmins 2	2000	800	360

FIGURE-4

REDUCTION IN STRIPPING

DUE TO STEEPER SLOPES

1976-79

Mine	Ore Million Tons	Stripping Ratio (BCY/LT) Initial	Final	% Reduction	Stripping Reduction BCY/LT	Total Reduction Million BCY
Timmins 1	8.0	0.84	0.55	35	0.29	2.32
Timmins 2	4.0	1.90	1.40	26	0.50	2.00

Total Reduction = 4.32
Million BCY

BEHAVIOR OF SLOPES SINCE COMPLETION OF MINING

After the mining was completed in Timmins 1 Mine in September 1979, all the dewatering pumps were removed and as a consequence the water table elevation increased within and around the Timmins 1 pit area. Finally, the hanging wall of Timmins 1 Mine collapsed due to two slides which occurred in October/November 1979. The footwall of Timmins 1 Mine and the hanging wall and footwall of Timmins 2 Mine although basically still intact, have started to show instability as indicated by the appearance of several tension cracks along the crest of the pits.

CONCLUSIONS

As a result of the steepening of final pit wall slopes in five (5) mines during the 1974 - 1979 period, significant financial savings have been realized due to reduction in stripping quantities that had to be removed. This was achieved mainly because of the excellent work done by the team of IOC engineers and an unusually high level of support from the top management of IOC.

ACKNOWLEDGEMENT

The author is thankful to the management of the Iron Ore Company of Canada for providing the excellent opportunity to work on this challenging project, and for the permission to publish this paper. The assistance of geotechnical and planning engineers, and the operating personnel is very much appreciated.

Special thanks are due to Mr. R. Geren, Executive Vice-President of the Iron Ore Company of Canada for providing the encouragement to the team of engineers. Without his support, the recommendation to steepen the pit slopes could not have been implemented.

REFERENCES

Bullock, W.D., 1972, "Development of the Burnt Creek - Rowe Mine
    Complex on the Knob Lake Iron Range", Canadian Mining Journal,
    November 1972.

Coates, D.F., Gyenge, M. and Stubbins, J.B., 1965, "Slope Stability
    Studies at Knob Lake", Proceedings of the Rock Mechanics Symposium
    at the University of Toronto, Queen's Printer, Ottawa.

Coates, D.F.,McRorie, K. and Stubbins, J.B., 1968, "Analysis of Pit
    Slides in some Incompetent Rocks", Transactions AIME, Volume 226
    1963.

Garg, O.P., and Kalia, T., 1975, "Slope Stability Studies in the
    Schefferville Area", Proceedings of the Tenth Canadian Rock
    Mechanics Symposium held in Kingston, Ontario, 1975.

Garg, O.P., 1976, "Application of Geotechnical Studies in Open Pit
    Mine Planning", a Paper Presented at the 11th. Canadian Rock
    Mechanics Symposium held in Vancouver in October, 1976.

Garg, O.P., and Devon, J.W., 1978, "Practical Applications of
    Recently Improved Pit Slope Design Procedures at Schefferville",
    CIM Bulletin, September 1978.

Hoek, E., and Bray, W., 1974, "Rock Slope Engineering", IMM Publica-
    tions, London, England.

Morgenstern, N.R., and Price,V.E., 1967,"A numerical Method for
    Solving Slip Surfaces", Computer Journal, Volume 9, No.4, February
    1967.

Nichols, L., 1968, "Field Techniques for the Economic and Geotechnical
    Evaluation of Mining Property for Open Cast Mine Design, Knob Lake
    Quebec", The Quarterly Journal of Engineering Geology, Volume 1,
    #3, December 1968.

Rana, M.H., and Bullock, W.D., 1969, "The Design of Open Pit Mine
    Slopes", Canadian Mining Journal, August 1969.

Question

Do you think that ice can decrease stability decisively.

Answer

It depends on the type of rock material and the quantity of ice present within the rock mass. If the rocks are competent and ice is present only along the cracks and joints as was the case in Timmins 1 and 2 mines, the stability of slopes is not affected.

Question

Please outline your dewatering program.

Answer

In general, both inpit and peripheral wells are used for mine de-watering. A 30 cm size casing is installed in 37.5 cm diameter holes. Amount of water pumped out varies from well to well and from area to area. In the Timmins area up to 160 meter deep wells were drilled (well below the permafrost bottom limit). The spacing and number of wells are determined by the geological structure, regional water table, rate of mining, etc.

Question

What was the nature of your monitoring program.

Answer

Because of the limited depth and relatively short mining life of our pits, we did not embark upon a sophisticated slope minitoring program. Although EDM devices manufactured by HP (Total Station) and K and E were used during the summer months, we mainly relied on visual observations related to tension cracks and other related indications of slope movement.

Question

Please comment on your success using the Cralius Method of Core Orientation (CCO). Do you feel the CCO method is as good as the "Clay Pot" Method discussed in an earlier paper.

Answer

Oriented cores were obtained successfully using the CCO. It took us about 20 minutes (in terms of drilling delay time) for each measure-ment. We have not tried the "Clay Pot" Method described by Mr. Call, earlier.

Question

In view of the fact that you had no or very little instability dur-ing operation at the mine, do you consider that the use of residual strength or method of analysis were too conservative. In hindsight what strength and slope angles might you have used.

Answer

This is a very good question and the one about which I have spent several nights and days thinking about. Yes, as it turned out the recommended slope angles were too conservative and obviously the slopes could have been designed at steeper angles, perhaps 2° to 3° steeper in our case. However, it must be realized that the approval of the Management and the Government regulatory agency to implement the steeper slope angles was as critical as (if not more critical than) doing the detailed engineering study on the project. In the ultimate analysis I feel that by documenting more successful case histories such as ours, the working life of many geotechnical engineers will become a little easier. In other words the acceptance of technical recommendations may be expedited.

Chapter 31

THE NORTHEAST TRIPP SLIDE - A 11.7 MILLION CUBIC METER WEDGE
FAILURE AT KENNECOTT'S NEVADA MINE DIVISION

Victor J. Miller

Senior Mining Engineer

Dravo Engineers and Constructors
Denver, Colorado

ABSTRACT

The Northeast Tripp Slide is one of the larger slope failures that
can be attributed to open pit mining. It is a 11.7 million cubic
meter (15.3×10^6 yd^3) wedge failure created by two thick gouge-filled fault zones whose intersection plunges into the Tripp pit at 13°.

One unique aspect of this failure is that the slide is now stable
despite having moved an estimated 8.5m (28 ft.) towards the pit and
having reached a maximum overall velocity of 21mm/day (0.07 ft/day).
The slide started to fail in 1968, accelerated until mining in the
slide's toe area was halted in April 1970, then began to decelerate.
With the cessation of mining in the toe area, rubblized rock sloughed
into the pit bottom which added weight to the failures toe constraints
and began the stabilization process.

Measurements from a network of survey stations revealed a correlation between the slide's velocity and heavy precipitation. Runoff
water accumulating in the tension cracks is believed responsible for
this correlation. Moreover, the survey net measurements showed that
during the failure's last stage the vertical component of the slide's
movement greatly increased. It is proposed that this settling was
caused by plastic-like flowage of the clay gouge underlying the failure.

The fault zone's effective shear strength was estimated by a backanalysis on the wedge's geometry and by performing direct shear tests.
The tests, performed on cores taken from samples that were collected on the failure, yielded a residual shear strength of: cohesion 0.61
kg/cm^2 (8.7 lb/in^2), and a friction angle of 9.6°. This residual
shear strength was nearly equal to the back-calculated strength indicating that the controlling fault's effective strength was essentially the same as the residual shear strength of the gouge filling

them. This result supports the argument that residual shear strengths should be used when high clay material is involved in a slope stability study.

INTRODUCTION

The Northeast Tripp Slide is located on the northeast side of the Tripp Pit, which is 9.7 km (6 mi.) from the town of Ely, in east central Nevada. Kennecott Corporation owns the property and operated the Tripp Pit from 1950 to 1971.

Due to the failure's size and impact on mining in the Tripp Pit, considerable work was done on the slide prior to this study. Much of this work, up to 1970, was summarized by R. R. Dimock (1970) in a paper presented at the Open Pit Slope Stability Workshop, April 30, 1970, at the University of Nevada, Reno. Also, two reports dealing with stability in the Tripp Pit were written by K. KO (1970) and C. Broadbent (1975). Both of these reports dealt with the Northeast Tripp Slide and were prepared internally within Kennecott. The U. S. Bureau of Mines initiated one of its first microseismic investigations (Merrill and Stateham, 1970) during the failure's most active period (the first half of 1970). As predicted, the seismic study showed that the highest noise rate corresponded to the failure's highest velocity.

Most of the field work supporting this study was done during the summer of 1977 while the author was employed by Kennecott Copper Corp., Nevada Mine Division. The study formed a part of the author's master thesis completed at the University of Arizona.

MORPHOLOGY

The aerial photograph (Figure 1), looking southwest at the Tripp Pit and the slide area, shows the abandoned shop area located behind the pit's rim. Likewise, the shop area is plotted along with the major scarps and cracks resulting from the failure on the topographic map in Figure 2. As can be seen, the failure involved most of the shop area along with 460m (1500 ft.) of the pit's rim.

The pit face portion of the failing wedge has 168m (500 ft.) of vertical relief and is still basically intact even though four survey stations on the face have moved between 7.9m and 9.5m (26 ft. and 31 ft.) towards the pit.

Northwest of the failure the pit wall has been extensively rubblized, and now stands at between 27° and 35°. It is called the Transition Zone. On the other side of the Northeast Tripp Slide, a related wedge failure called the Morris Slide developed (see Figure 2) and subsequently pushed out the pit face 6m to 9m (20 to 30 ft.).

Figure 1   Aerial Photograph of Tripp Pit and Slide Area.

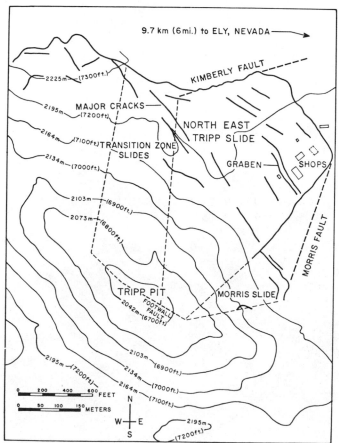

Figure 2 — Topography of Tripp Pit Area and slide location.

The failure is interesting because its scarp has exposed the Morris Fault Zone which also controls one side of the Northeast Tripp Slide.

Many tension cracks can be observed across the top of the failure. Generally, the pit side of these cracks was dropped from 0.05m (.17 ft.) to 0.76m (2.5 ft.) much like a normal fault. The one major exception to this is a 7.6m wide by 2.5 m deep (25 ft. x 8 ft.) graben zone that is situated 168m (500 ft.) behind the pit's rim just in front of the shops (see Figures 1, 2, and the photograph of the zone in Figure 3).

Figure 3   Photograph of the Graben Zone looking North.

The failing wedge's movement towards the pit created side scarps much like a strike-slip fault with a minor dip-slip component. Figure 4 shows the southwest side scarp, where the slide's movement has offset an old railroad grade. On the wedge's other side, the northwest side scarp is less distinct because the area is covered by a 36.6m (120 ft.) high mine dump that has masked the movement.

Figure 4   Photograph of the Southeast Scarp looking into Shop Area.

HISTORY

The slide's movement was first noted in July, 1968 when some cracks developed in the concrete foundation of one of the Kimberly shop buildings. (KO, 1970). At that time, the cracks were attributed to settling. However, significant movement began a year and a half later when "in January, 1970, large cracks in the shop floors began to en-large and it became obvious that the buildings were being pulled apart" (Dimock, 1970, p.6). By this time, mining had deepened the pit to its maximum depth at elevation 2020m (6630 ft.) and the shops were becoming rapidly unuseable.

On January 26, 1970, a 27-station survey net was established in order to monitor the slide's movement. This net was surveyed weekly, and later bi-weekly until February 14, 1971. For this study, it was resurveyed in July, 1977.

By April, 1970, the Transition Zone was rubblizing and the last material was mined from the slide's toe area. Finally, in October, 1970, all mining in the pit ceased even though some potential ore still remained below the slide's toe area.

SLIDE MECHANISM

Geology

When the extent of the failure was recognized in 1970, four diamond drill holes were drilled to obtain more information about the faults controlling the slide. This information combined with data from 22 older churn drill holes, geologic mapping of the Morris underground workings (which lie at the toe of the failure), and some recent surface geologic mapping have made it possible to make a fairly accurate de-termination of the geologic structure controlling the failure.

The wedge geometry controlling the Northeast Tripp slide consists of two major faults, the Kimberly and the Morris, that have an inter-section plunging S48°W at 13 degrees. The Kimberly fault, which lies near the center of the pits north side (see Figure 2), strikes N69°E and dips 33° SE. The fault zone consists of 6.1 to 12.2m (20 to 40 ft.) of gouge.

The Morris fault is the largest of several parallel faults that make up the Morris fault system (see Figure 2). This system, in the northeast section of the pit, strikes N24°E and dips 31 degrees NW. Like the Kimberly, it has a relatively thick fault zone. Several smaller parallel faults similar to the Morris were mapped and they indicate that a system of parallel faults exist.

Cutting across the Tripp Pit's bottom and the slide's toe is the Footwall fault (see Figure 2). It dips steeply to the southwest and is the major ore-waste contact for the pit. Both the Kimberly and Morris faults are cut by the Footwall fault which probably also dragged the two faults to the northwest. Because of this, their exact location at the failure's toe is uncertain.

There are many smaller discontinuities in the slide zone and the Tripp pit in general. For example, previous detailed mapping of the old Morris underground workings 6710 ft. level (at elevation 2045m), which was at the failure's toe area, showed 344 faults. The faults were found to closely parallel the major faults described above. Three detailed line fracture surveys show equally complex jointing, as did the joint-set mapping done for this study.

Above the two controlling faults the wedge consists of interbedded skarns, hornfels, shales, and marbles which strike N37-64°W and dip southwest at 40 degrees to the vertical. During the tectonic his-tory of the area there was extensive thrusting which resulted in the

creation of bedding-plane faults. Probably as a result of the thrusting an out of sequence marble lens was found in the middle of the
Northeast Tripp Slide. Below the two controlling faults all the
diamond and churn drill holes intersected a relatively unaltered
monzonite porphyry.

Slide Velocity

   The velocity of the slide during its early stages is unknown, but
it is believed to have increased gradually in irregular steps. By
February 1970, when the first measurement was taken, the slide had a
velocity of 9mm/day (0.029 ft./day) and an estimated 1.5 m (5 ft.)
of displacement had occurred.

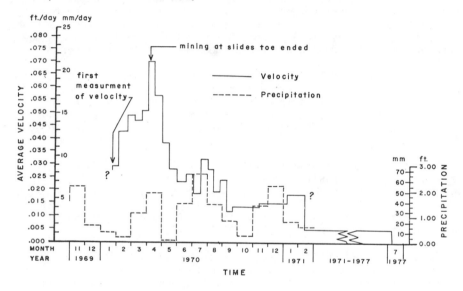

Figure 5   Average Velocity of Failing Wedge and the Monthly
             Precipitation.

   The velocity of the failure is plotted in Figure 5. This velocity
was calculated by averaging the velocities from 13 centrally located
survey stations; four on the pit face and the rest between the pit's
rim and the graben zone.

   Figure 5 also shows the monthly precipitation and the associated
correlation between high precipitation and increases in the slide's
velocity. The first correlation was when the maximum velocity was
reached in April 1970. After that two additional periods of acceleration occurred following high precipitation. Since routine monitoring
ended in the middle of the last velocity increase and no data could

be found showing when the snow melted, the last precipitation relationship is not as significant as the first two.

Mining in the slide's toe area appears to be the key to the failure's acceleration and deceleration. Between February, 1970 and April
1970 the failure steadily accelerated to its maximum velocity of 21mm/
day (0.07 ft./day). When mining at the slide's toe ceased, the overall deceleration trend began and in two months the velocity had dropped to 7mm/day (0.024 ft./day).

Slide Movement

In addition to velocities, the survey results supplied data regarding the direction of movement. Based on this vector type information,
the slide's movement was divided into three stages:

Stage 1 - July, 1968 to April, 1970, Slide accelerating, movement
          parallel to controlling faults' intersection.

Stage 2 - April, 1970 to February, 1971, Slide decelerating, move-
          ment parallel to controlling faults' intersection.

Stage 3 - After February, 1971, Slide decelerating, direction
          (azimuth) still parallel to controlling faults' inter-
          section but vertical component of vector greatly
          increased.

During the first two stages the vector direction paralled the proposed intersection of the two controlling faults. This fact adds
support to the geologic structure used for this study since such a
relationship is characteristic of wedge failures.

Within the failing wedge, the geology influenced the movement.
Figure 6 is a cross section which was made along the controlling
faults' intersection. As can be seen, it is proposed that differential
movement within the wedge occurred along bedding and bedding plane
faults and that this movement was aided by the plastic nature of the
gouge-filled controlling fault zones.

Discussion Of The Failure Mechanics

The Northeast Tripp Slide resulted from several factors that combined to create a factor of safety less than one. First, the geometrical relationship between the Tripp Pit and the geological
structure resulted in a situation where a wedge failure could occur.

Second, the two controlling fault zones were filled with 6m to 12m
(20 to 40 ft.) of mostly clay gouge, which is believed to have a very
weak shear strength. This situation will be discussed later in this
paper.

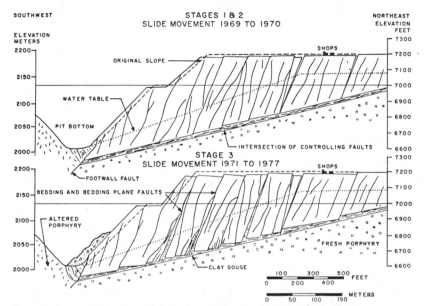

Figure 6 – Cross sections along controlling fault intersection. Illustrating failure mechanism.

Third, based on measurements taken in 1977 the water table within the failing wedge was high. The geology of the pit also supports the presence of a high water table since all drainage into the pit had to cross the bedding in the wedge. In addition, the montmorillonite clay gouge in the two controlling faults has a very low permeability which would essentially contain all water above them. The high water table created hydrostatic pressure along the controlling structures which then reduced the effective shear strength of these structures.

Based on the above conditions, the failure had a factor of safety of less than one and it appears that the toe constraints were respon- sible for bringing the factor of safety to greater than one. Unfor- tunately the exact nature of the controlling faults in the toe area is unknown and is complicated by the Footwall fault (see Figure 6). Prior to April 1970, mining had relieved enough of the toe con- straints that the failure was accelerating (factor of safety less than one). But, when mining in the toe area ceased, the toe's weight was then increased by material which sloughed into the bottom from the Transition Zone. This increased weight apparently brought the factor of safety to greater than one and the slide began to decel- erate.

During the last stage of the failure, the movement was determined by two surveys of the monitoring stations conducted 6 1/2 years apart. Over this period, it is proposed that the vertical settling was caused by plastic flow of the clay gouge and that tension within the wedge probably created zones that had little vertical pressure. The gouge then plastically flowed into the zones due to the weight of the overburden on either side of the tension zones.

FAULT ZONE STRENGTH

The Clay Fault Gouge

The fault zones under the Northeast Tripp Slide are composed of gouge which determined the zones overall strength. X-ray analysis by the Kennecott Research Center in Salt Lake City, revealed that this gouge consisted of 50-70% montmorillonite clay, 10-20% kaolinite clay, 10-15% sericite mica, and 5-10% very fine grained quartz.

Because the zones are relatively thick, undulations of the rock surface on either side of the zone would not add to the zone's overall strength (Barton, 1974). One reason for this is that above and below the major fault zone there are many smaller parallel faults which presumably also exist in the undulations. Before the failing mass could ride up and over such an undulation, it would fail along these smaller faults. In addition, the gouge is a plastic material. For example, gouge tested for this study plastically deformed in the period of a half-hour laboratory shear test at pressures greater than 14.1km/cm^2 (200 psi). Even if an undulation were strong enough to support riding up, the gouge is not; therefore, over a period of time (measured in days and weeks) it would conceivably flow around such an undulation. This flowing would require energy from the sliding block and might tend to dampen the slide, but it would not add to the strength of the fault zone.

On a smaller scale the gouge contains many rock clasts in a matrix of black to greyish-black clay that locally can be a tan-brown color due to limonites. The rock clasts have a "floating" texture where they are usually not in contact with each other. The stringers of different colored clay and the rock clasts are generally oriented within 30° to the orientation of the overall fault zone.

Direct Shear Testing

Obtaining large samples from the controlling faults was not possible, because only a very limited amount of core, which had deteriorated badly after being stored for seven years in core boxes, was available. Instead, a 5-foot thick fault zone within the Northeast Tripp Slide was sampled. This fault parallels the Morris fault and is part of the Morris fault system. Moreover, the surrounding rocks and the physical appearance of the fault gouge were identical to

those in the core taken from the controlling faults. Based on this,
the sampled gouge is believed to have the same strength properties
as the gouge in the two controlling faults.

At the site, fresh gouge was exposed and samples roughly 20cm (8 in)
on a side were collected. These were then protected against drying
and transported to the strength laboratory at the University of
Arizona for shear testing. At the laboratory the tested samples
had a moisture content of 21 to 31% (drying at 65°c) and had an
average plasticity index of 9.6.

Using a circular saw bit the samples were cut perpendicular to the
lineations resulting in a core 25mm to 51mm (1 to 2 in.) long and
33mm (1.3 in.) in diameter. The core was then placed in a Soil Test
direct shear machine. Each core was then tested so that a peak and
residual shear strength was determined for a particular normal load.
When the residual strength was reached quickly (less than 20% displacement) the normal load was increased and additional residual
strengths determined. Figure 7 shows a representative section of the
direct shear test results. From the figure it can be seen that,
generally, the peak strength was twice the residual and that for several tests the actual residual strength was difficult to establish.

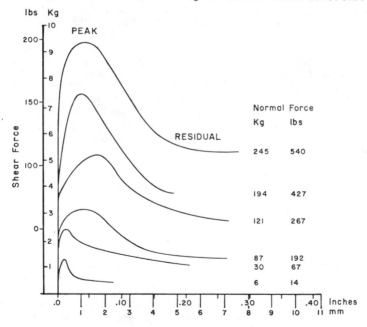

Figure 7 — Representative results of direct shear tests
on montmorillonite fault gouge.
Samples had an area of 8.56cm² (1.33in²).

The results are presented in Figure 8, which has the normal stress plotted against the shear strain. A linear fit to the data that satisfies the coulomb relationship (τ = C + N Tan ϕ) yielded the results shown in Table 1.

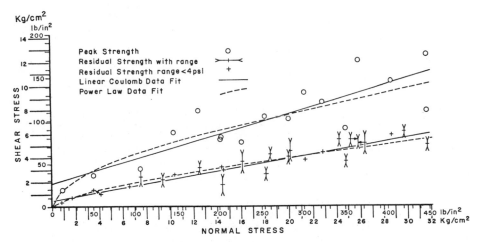

Figure 8 — Direct shear test results on montmorillonite fault gouge.

TABLE 1

LINEAR REPRESENTATION OF DIRECT SHEAR DATA

	Cohesion C Kg/cm^2 (lbs/in^2)	Friction Angle Degrees ϕ	Correlation Coefficient r	Number of Data Points
Peak	1.96 (27.9)	15.8	0.86	17
Residual	0.61 (8.7)	9.8	0.95	27

Coulomb failure criteria may not be the best way to represent the shear strength of the gouge. Jaeger (1971) noted that the power law tends to better fit data that turns down at the lower normal stresses. As can be seen in Table 2, the power equation did fit the data slightly better than a linear equation.

TABLE 2

POWER LAW FIT OF THE GOUGE STRENGTH DATA

$$(\tau = F N^f)$$

	F	f	Correlation Coefficient r	Number of Data
Peak	4.70	0.556	0.93	17
Residual	1.13	0.694	0.97	27

Under low normal pressures and for the peak strength, the power law definitely represented the test data better, but for the residual strength above 3.5 Kg/cm^2 (50 psi) there was no significant difference between the equations. Since the pressures within the Northeast Tripp Slide are expected to be above 3.5 Kg/cm^2 (50 psi) the simpler coulomb equation was used in the slide's failure analysis.

The fault gouge's nature created several problems in performing the shear tests. Two properties of the gouge that could not be adequately controlled in the testing were the gouge's moisture content and its plastic behavior. Both of these factors could have affected the test results, especially those tests performed under higher normal pressures. Generally below 14.06 Kg/cm^2 (200 psi) the tests were run at essentially less than full saturation. At higher pressures, however, the samples sweated water, indicating they behaved as a saturated clay. All the tests were run at a strain rate of 2mm/hour (0.8 in./hr.) to allow the contained water to drain and to avoid an undrained loading effect (Lambe and Whitman, 1969). The gouge's plastic behavior caused samples under high normal pressure to swell laterally during the test. The effect on the resulting shear strength was difficult to evaluate.

Adding to the problems, there was also a great deal of variability between the gouge samples themselves. For example, an unusually long core was sheared, then slid up 8mm (0.3 in.), then sheared again. The first test, with a normal stress of 24.6 Kg/cm^2 (350 psi), yielded a peak strength of 8.4 Kg/cm^2 (120 psi), and a residual strength of 4.9 Kg/cm^2 (70 psi). The second test was performed at a lower normal stress of 19.9 Kg/cm^2 (283 psi, but yielded a higher peak strength of 9.5 Kg/cm^2 (135 psi) and a higher residual strength of 5.6 Kg/cm^2 (80 psi). Despite the above problems the results presented in Figure 8 appear good and are believed to be an accurate representation of the gouge's shear strength.

Back-Calculated Strength

Using Hoek's Wedge Failure Analysis (1977, p. 333-347), the controlling faults' shear strength was calculated based on the following assumptions and input data:

1. Since both faults had thick gouge filled zones, it was assumed that they had an equal shear strength. This also reduced the resulting strength to a function of two variables instead of four.

2. No toe constraints were used in the back analysis.

3. The water table measured in July, 1977 was used (see Figure 8).

4. The tension cracks were assumed to be dry down to the water table.

5. A factor of safety of 1.00 was used.

6. Plane A (Kimberly Fault): Strike, N69°E; dip, 33° SE.

7. Plane B (Morris Fault): Strike, N24°E; dip 31° NW.

8. Pit Face: Strike, N50°W; dip, 37° SW.

9. Top of Wedge: Flat at 2194.6m (7200 ft.) elevation.

10. Toe of Wedge: at 2011.7m (6600 ft.) elevation.

11. Back Tension Crack: 305m (1000 ft.) behind rim; Strike N50°W; dip, 70°SW.

12. Rock density: 2565 Kg/m^3 (160 lb./ft.^{3}).

13. Wedge Volume (from planimetry): 11.7 x 10^6 m^3 (413 x 10^6 ft^3)

Based on the above criteria the back analysis yielded the following linear equation which has the shear strength as a function of the cohesion and the friction angle:

$$1 = C \times 0.000167 + 4.401 \text{ Tan } \phi \quad (C = \text{cohesion } \phi = \text{friction angle})$$

Figure 9 shows the possible solutions to this equation.

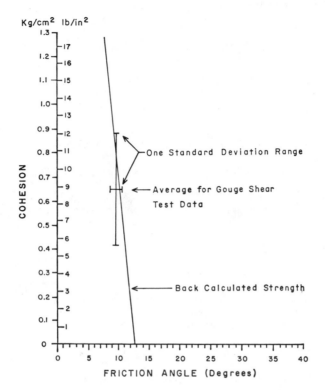

Figure 9 — Comparison of back-calculated
and tested shear strength.

Back-calculated Versus Tested Strength

In addition to the back-calculated strength, Figure 9 shows the residual and peak shear strength from the direct shear testing. As can
be seen, the residual strength falls just below the back-calculated
strength. Since no toe constraints were used in the back-calculation,
the estimated strength resulting from the back-calculation should be
higher than the actual effective shear strength of the fault zones.

The question arises, "Why the tested residual shear strength of a
thick fault zone should be more representative of that zone's effective strength than the gouge's peak strength?" One likely explanation
lies in the nature of the clay gouge. The zone was created by shearing, which left the gouge's flakey clay particles oriented in the
direction of shearing. This orientation is believed to be the major
difference between the peak and residual shear strength of clays
(Skempton, 1964).

Skempton (1964) discusses another possible explanation based on fracture propagation. He proposed that oriented domains within the clays are common and that they can be formed by relatively small strains. Along the oriented domains the shear strength is at the residual value which then transfers additional stress within the clay to some other point causing it to become oriented.

Such a process can be used to explain the transition between the regressive and progressive stages of a failure (Zavodni and Broadbent, 1977). During the regressive stage, the peak strength areas are being loaded and progressively converted to their residual shear strength, thus explaining the irregular movement during this stage. When the last peak area is overcome, the "onset of failure point" is reached and the sliding mass steadily accelerates unless acted upon by an outside force. In the case of the Northeast Tripp Slide, the outside force was the reinforcement of the toe constraints.

CONCLUSION

The Northeast Tripp Slide is a large 11.7×10^6 m^3 (15.3×10^6 yd^3) wedge failure that developed due to open pit mining. The failure was controlled by two fault zones 6 to 12m (20 to 40 ft.) thick which were composed of high montmorillonite clay gouge and whose intersection plunged into the pit at 13°.

Water affected the slide in two ways. First, poor drainage within the wedge created a high water table. Second, after heavy precipitation, the tension cracks behind the failure may have filled with water briefly adding to the net driving force and accelerating the failure.

Based on data from a 27-station survey net, the slide's movement was divided into three stages as follows:

Stage	Period	Velocity	Movement Relative to Controlling Fault Intersection
1	7/68-4/70	Accelerating	Parallel
2	4/70-2/71	Decelerating	Parallel
3	2/71-7/77	Decelerating	Azimuth parallel but vertical component of vector increasing

The change from Stage 1 to 2 (i.e. accelerating to decelerating) was probably due to the cessation of mining at the slide toe and the subsequent reinforcement of the toe by sloughed material. During the third stage, the probable explanation of the increased vertical movement is plastic flow of the clay gouge into tension areas created by the failure, thus allowing the failing wedge to settle.

Direct shear testing of the fault gouge yielded the following strengths:

Peak: Cohesion 1.96 Kg/cm^2 (27.9 lb./in.^{2}),Friction angle 15.8°

Residual: Cohesion 0.61 Kg/cm^2 (8.7 lb./in.^{2}), Friction angle 9.6°

The residual shear strength was found to be very close to the fault zone strength predicted from a back-calculation. This result supports the argument that residual shear strengths should be used when high clay material is involved in a slope stability study.

REFERENCES

Barton, M. R., 1974, A review of the shear strength of filled discontinuities in rock: Norwegian Geotechnical Institute, Publication No. 105, 58 p.

Broadbent, 1975, Slope stability recommendations for the Tripp Pit failure zone: unpublished interoffice memorandum between Kennecott Copper Corporation, Nevada Mines Division, and Kennecott Copper Corporation, Metal Mines Division, Engineering Center, Ely, Nevada, 8 p.

Dimock, R. R., 1970, Slope failure - a continuing problem: unpublished paper presented at the Open Pit Slope Stability Seminar Workshop, April 30, 1971, University of Nevada, Reno.

Hoek, E., and Bray, J. W., 1977, Rock slope engineering, rev. 2nd ed.: London, The Institution of Mining and Metallurgy.

Jaeger, J. C., 1971, Friction of rocks and stability of rock slopes: Geotechnique, v. 21, no. 2, p. 97-134.

Ko, K. C., 1970, Preliminary investigation on the Tripp Pit slope failure: unpublished interoffice report, Kennecott Copper Corporation, Metal Mining Division, Engineering Department, Job. No. 173, 31 p.

Lambe, T. W., and Whiteman, R. V., 1969, Soil mechanics: New York, John Wiley & Sons, Inc.

Merrill, R. H., and Stateham, R. M., 1970, Microseismic investigation of incipient slope movement near the Tripp-Veteran Pit: Denver, U. S. Bureau of Mines, Denver Mining Research Center, Progress Report 55-105.

Skempton, A., 1964, Long term stability of clay slopes. Fourth Rankin Lecture: Geotechnique, v. 14, no. 2, p. 75-102

Zavodni, Z. M., Broadbent, C. D., 1977, Slope Failure Kinematics, <u>in</u> 19th U. S. Symposium on rock mechanics: Reno, Nevada, Conferences and Institutes Extended Programs and Continuing Education, University of Nevada-Reno.

Chapter 32

BACK ANALYSIS OF SLOPE FAILURES IN THE
CERCADO URANIUM MINE (BRAZIL)

Dr. C. Dinis da Gama

Research Coordinator
DMGA - I.P.T.

São Paulo, Brazil

ABSTRACT

Because of the growing importance of back analysis
for the interpretation of rock slope failures, particularly in open pit mines, it is suggested a methodologic
sequence of activities for dealing with those problems.

The principal mathematical tools for the interpretation of the most common mechanisms of slope failures
are reviewed, including methods of stress and   strain
analysis, which seem to be more powerful for
parametrization studies, and for developing   more
realistic explanations of slope behaviour, such as the
measurement of displacements.

An application of those methods is described,
using slope failures detected in a Brazilian uranium
mine (Cercado), thus providing reliable shearing
strength parameters for the rock mass discontinuites,
which were used in the final pit design.

The concept of using backanalysis results to supply
input data for pit design was implemented by means of
the excavation of an experimental trench within mine
limits, where specific instruments were installed, and
several slope failures were monitored and backanalysed,
in order to yield information about sliding mechanisms
in the future open pit.

745

I. - INTRODUCTION

The stability of slopes in open pit mines is a complex problem because it involves geologic, engineering and economic aspects which in general are not simple to characterize and not easy to quantify.

Usually, the compromise between safety in terms of slope stability, and economy in reducing excavation volumes (and in maintaining slope geometry) is a difficult goal to be attained in every sector of an open pit mine and in all phases of mine life. However, the general purpose of economic optimization must be constantly present in the designer's mind by means of which predicted economic penalties due to eventual slope failure are balanced against the savings due to limited stripping and no lost production or landslide repair.

The fact that the majority of open pit slopes are excavated in rock masses demands a series of detailed studies on the mechanics of rock slope stability, which closely depends on the presence of structural discontinuities acting as low strength surfaces along which most sliding occurs. Therefore, the designer is faced with a complicated situation where the effect of discontinuities has to be considered for the establishment of pit geometry, before the excavation reaches and reveals those discontinuities.

In general the mechanical properties of a rock mass discontinuity (be it a joint, a fault, a fracture or any other type) have to be evaluated in order to determine the safe geometry of an excavation, in a moment prior to the begining of the excavation.

The influence of underground water is also essential for the adequate design of rock slopes, and its behaviour in future stages of pit geometry has to be predicted within the desired accuracy.

Those three factors (sliding mechanisms, mechanical strength parameters and water pressures) control the design of open pit mine slopes, and they form the essential input data required for the analysis leading to a reliable solution, accepted under both safety and economic viewpoints.

It is common to characterize slope stability by means of a safety factor established in accordance with classic Soil Mechanics concepts.

However, the validity of safety factors for rock

slopes is subjected to more criticism than for soil
slopes, due to the large amount of different mechanisms
that may produce rock failures, and also because of the
greater number of variables influencing their slidings.

While a great deal of soil slope case studies prove
the validity of the methods used for the materials
characterization and for the analytical models applied
to explain soil failures, the truth is that very few
rock slope slidings have been adequately explained
without using simplifying hypothesis or imaginary situations. It is not common to observe rock slope stability analysis taking in due account the fact that the
shearing strength changes with the displacements suff-
ered by the rock mass.

The effect of those displacements is generally to
reduce drastically the shear parameters of the rock
volume under analysis, therefore diminishing its
safety factor. Seldom is it known how much a safety
factor is reduced after a displacement of a certain
magnitude occurs along a potential shearing surface.
A conservative way to avoid this problem is the utilization of residual values for cohesion and friction
angles, based on laboratory shearing tests performed
on small samples taken from the rock mass, which
usually don't represent the real parameters of the actual
rock slope. Another distinction between soil and rock
slopes is the type of stress analysis that may be done:
due to the high permeability of rock masses it is rare
the occurrence of undrained loadings, thus leading the
analysis to be accomplished in terms of effective stress,
Also water levels and pressures are more difficult to
determine in a rock mass, due to its characteristics of
heterogeneity and discontinuity, and frequent variability of properties.

Therefore, the stability analysis of rock slopes is
not as reliable as for a soil slope study, thus
resulting in the utilization of larger safety factors
for rock slopes and consequently in the acceptance
of greater risks for rock slope design.

This circumstance is critical in open pit mining ,
for most slopes are excavated in rock masses and the
economic constraints require the application of low
safety factors. Those are the reasons why surface
mines require extensive rock mechanics studies, in order
to determine the real conditions of potential failure
mechanisms which control the ultimate geometry of the
open pits, and they should be based on accurate and

realistic rock mass strength parameters.

II. - <u>METHODOLOGY FOR BACK ANALYSIS OF ROCK SLOPE FAILURES</u>

The existing difficulties in determining actual properties of strength and deformability of a rock   mass
require imagination to obtain reliable values for input
data in a mine slope design.

As it is known, a rock mass is composed of two   en-
tities: the rock intact blocks and the surfaces   of
discontinuities, which may be   filled with weathered materials. Because of the much lower strength of
these infilling materials, it is their attitude   and
strength properties that controls potential failures,
and so they must be defined in detail for slope stability studies.

In practice, there are three ways of determining the
shearing parameters of a rock mass discontinuity:

a. - Through laboratory tests accomplished on representative samples taken from the rock mass.

b. - By means of "in situ" shearing tests usually
located over critical joints of the rock mass.

c. - Through backanalysis of previous rock slope failures.

In most cases backanalysis is the most reliable and
representative way of obtaining shear strength
parameters, especially if the slope failure is accom-
panied by comprehensive instrumentation installed to
record the progressive mechanisms of the slide, as well
as several important parameters revealed by this phenomenon.

The concept of monitoring a slope in order to determine the geometric and mechanical aspects of an impending failure in not new in terms of safety, but for the
purpose of achieving backanalysis studies to   supply
realistic data for further design on the same rock mass
is a promising procedure, particularly in surface mining.
The importance of determining the biggest safe inclination of slope faces in a mine, justifies an investment
of that type, which in some case (such as the Cercado
uranium mine) may go up to the point of excavating an
experimental trench with very steep slopes in order to
induce failures that subsequently are backanalysed with
the help of data collected by appropriate instrumentation.

The main types of instruments to be installed in a

mine slope for that purpose, can be:

a. - Apparatus to determine vertical and lateral absolute displacements of slope points, permitting obtain velocities and accelerations of those movements.

b. - Instruments to measure the position of water level, and water pressures within the slope.

c. - Cells for the measurement of "in situ" states of stress and strain.

d. - Apparatus to monitor micronoises and microseisms in a rock mass, with associated alarm systems.

e. - Instruments to record external effects on the slope, such as rains, vibrations from nearby blasting, etc.

In general, the costs involved in installing and reading a set of instruments located in a mine slope, are compensated by future savings in reduced excavation and less slope maintenance costs.

Having this information available, it is possible to develop a complete analysis of an existing slope failure, aimed to obtain the shearing strength parameters of the geological discontinuities that have been responsible for the sliding.

A series of steps are suggested to perform a back-analysis of that type (see Fig. 1).

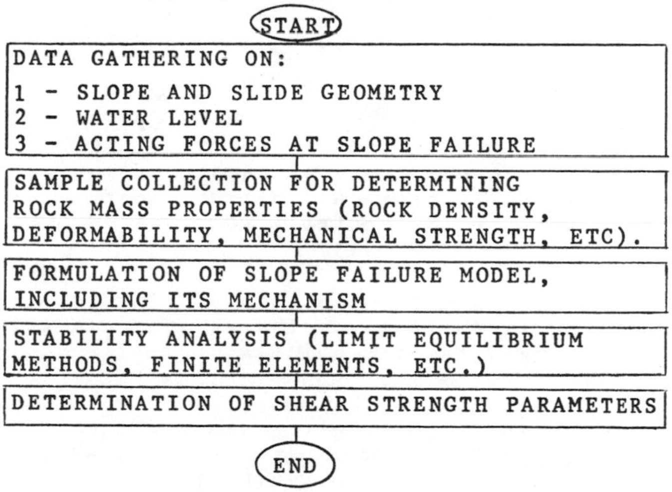

Fig. 1.   Flow chart of the main phases of backanalysis for rock slope failures

Observation of progressive slope failure is also another important aspect to be considered in the process of developing reliable backanalysis studies. It should supply information on the real mechanism that are involved in the sliding, and this makes possible selecting the adequate mathematical formulations to model the failure, in order to furnish the desired shearing strength parameters.

As far as mathematical models of slope failure are concerned, we may distinguish a few groups:

a. - Limit equilibrium methods (ref Hendron et al., Kovari, etc.)

b. - Stress and strain analysis methods - finite differences and finite elements (ref. Desai & Abel, Otter et al., etc.)

c. - Progressive multiple-block failure simulation (ref. Cundall)

A summary of the most common situations of rock slope failure is depicted in Fig. 2 together with the corresponding types of mathematical analysis.

The mathematical expressions to be used for the appropriate calculations are well known, and can be found in several text-books (Attewell & Farmer, Hoek & Bray, etc.). Fig. 3 represents four of the most frequent cases of rock slope failure, indicating the expressions for the dynamic safety factors, obtained by the limit - equilibrium method.

In those formulae effects of water pressure are considered, as well as dynamic forces like blasting vibrations, which are represented by horizontal forces of magnitude KW, where K is a seismic coefficient and W is the weight of the unstable rock volume (the so called quasi-static method).

One of main disadvantages of the limit-equilibrium method is the establishment of stability conditions by means of a safety factor, which gives no credit to the deformability of the rock mass, and is an abstract coefficient that can not be measured or evaluated through instrumentation. Furthemore the safety factor definition in arbitrary and ambiguous, as various authors have shown (see for example Habib, 1979).

Another type of anlytical treatment of slope stability, is the utilization of stress and strain analysis methods, especially the finite element method.

MECHANISMS OF RUPTURE

a) PLANE FAILURE

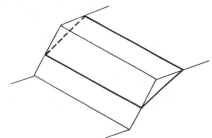

TYPE OF ANALYSIS

Bidimensional analysis.
Limit equilibrium methods.
Finite elements may be
used.

b) MULTI-PLANE FAILURE

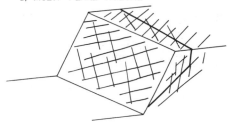

Bidimesional analysis.
Limit equilibrium methods.
Verification of block
instability and rotational
falls.
Multi-block analysis.

c) ROCK WEDGE FAILURE

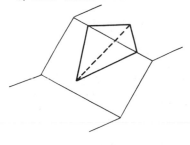

Tridimensional analysis.
Limit equilibrium using
stereographic projections
Finite elements.

d) CIRCULAR FAILURE

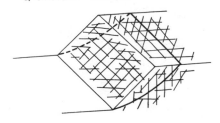

Soil Mechanics methods.
Limit equilibrium for
circular slidings.
Check dominant discon-
tinuities for rock-wedge
failure analysis.

Fig. 2.   Typical rock slope failure situations   and
           corresponding analytical treatment.

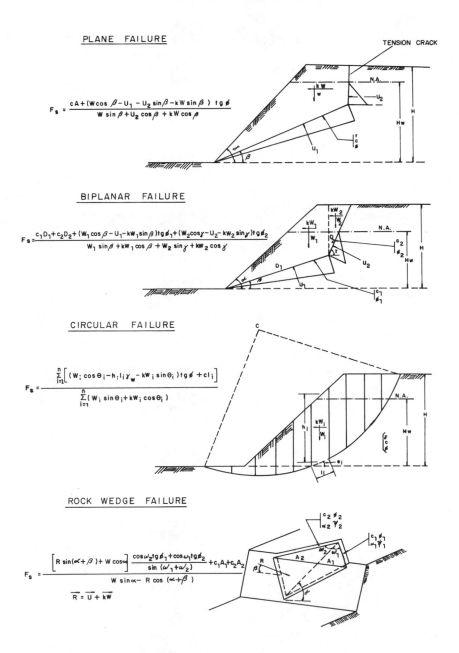

Fig. 3.   Usual rock slope failures and mathematical
expressions of safety factors.

In theory, this method can supply all the necessary information regarding the internal conditions of a three- dimensional rock slope, by means of determining its stress and displacement distributions, incorporating reliable rock failure criteria, so that zones of stability are separated from unstable volumes.

However, the method requires the knowledge of the rock mass stress-strain relation, which is generally nonelastic, nonlinear, anisotropic and heterogeneous. These characteristics are difficult to describe in a three-dimensional rock mass, because of the enormous work required to establish a constitutive law for all elementary volume within the slope. Also, the determination of the "in situ" stress field before excavation is a complicated task.

Modelling joints and other discontinuities is also required for finite element method inputs including shear movements along them, as well as their opening and dilatancy during slope instability. Because of joints' greater deformability and lower strength, emphasis in accurately describing them is recommended by authors (Goodman, St. John, etc.).

The concept of joint material has been very useful to model rockmass behaviour, and not so much detail is now required to characterize intact rock volumes separated by deformable joints.

Once a representative finite element model is developed, stresses and displacements can be predicted with accuracy, and checked by "in situ" measured values. Failure criteria in terms of strain, stress and displacement may then be implemented, providing additional control on the stability of a rock slope by means of installing instrumentation able to recor those parameters.

Recent examples on the application of the finite element method to mine slopes show promising results (Brock and Nilsen, 1979).

Other computer methods, such as the finite difference and the multi-block technique reveal great potential to model realistically the behaviour of rock slopes, indicating a future trend in this direction.

III. - THE CERCADO URANIUM OPEN PIT MINE

III.1. - Early slope stability studies

Located near the city of Poços de Caldas, state of

Minas Gerais, the Cercado open pit is being prepared to
be the first producing uranium mine in Brazil.

The open pit will be approximately 300 m (1000 ft.)
deep and will have about 1 000 m (3200 ft) diameter at
ground level.
Uranium mineralization occurs disseminated within
alkaline rocks covered by a thick overburden layer of
saprolitic soil.

Engineering geology investigations have started early
in the history of the mine, and their first phase was
described elsewhere (Dinis da Gama and Ricardo F. Silva,
1978).

That early work included:
- Geological investigation of mine site with surface
lithologic mapping
- Study of borehole samples (about 50 000 m, 150,000 ft)
to determine vertical distribution of rock types.
- Definition of the principal tectonic lines of the rock
mass
- Geophysical investigation to map underground formations
in terms of seismic velocities
- Mapping of discontinuities intersected by several
exploration galleries opened in to the orebody, and
obtention of stereonets to define dominant families and
their attitudes
- Hydrogeologic characterization of the massif, through
the execution of several deep boreholes where piezometers
had been installed and where pumping tests, as well as
injection tests have been run.

Also the measurement of water flows in the exploration galleries allowed the creation of a model that
simulated in large scale the modification in water level
caused by the opening of the mine.
- Geomechanical characterization of soil and rock types
existing in the orebody, with special emphasis in the
measurement of shear strength in joints and infilling
materials by means of laboratory tests.

Initial slope stability studies provided general
information on ultimate pit geometry, but the preliminary proposition of slope angles was based on limited
geomechanical data (from laboratory shear tests) which
had to be improved. By that time, considerable work had
been done on exploration galleries to map more than 1500
joints which were classified in there groups:

a. - Fractures without infilling materials (rock-rock
joints).

b. - Fractures with infilling materials and unweathered walls.

c. - Fractures with infilling materials and altered walls.

A fourth group of long discontinuities was discovered latter on by its outcrops: a basic intrusion with infilling clays.

It was found that the attitudes (dip and strike) of the dominant families of these joints were the same wherever they had been mapped: in the exploration galleries, in the open pit and in the experimental trench.

Detailed streographic analysis of those joint types was accomplished, revealing their principal families and attitudes, which after the determination of their shear parameters formed the main input data for slope stability analysis.

A summary of the information obtained after the structural characterization of the rock mass is presented in the next table of values, which includes the corresponding shear stress parameters obtained after backanalysis studies.

TABLE 1. Joint properties

JOINT TYPES	NUMBER OF MAPPED JOINTS	ATTITUDES OF DOMINANT FAMILIES	COHESION (MPa)	FRICTION ANGLE
Rock-rock fractures (no infilling materials)	2110	1)N24ºE-46ºNW 2)N14ºE-56ºSE 3)N38ºW-83ºNE 4)N53ºE-89ºNW	0.003	41º
Unweathered walls and infilling materials	265	1)N22ºE-48ºNW 2)N78ºW-23ºSW 3)N80ºE-67ºSE	0.005	31º
Weathered walls and infilling materials	90	1)N23ºE-50ºNW 2)N32ºW-87ºSW 3)N64ºE-88ºNW	0.020	27º
Basic intrusion	9	N28ºE-59ºNW	0.015	36º

Because excavation was not yet reaching the rock mass due to the very thick overburden, it was decided to expose the rock mass, by means of opening an experimental trench.

This excavation was located inside the pit, with an orientation that would intersect the dominant disconti-

nuities of the rock mass with a high probability of
causing failures. Because the purpose was to induce
slidings on this trench, it was excavated with steep
angles of inclination and small berms. Various types of
instruments were installed on the benches of the trench
in order to monitor parameters related with eventual
instability phenomena.

Fig. 4 is a plan view of the Cercado Mine, showing
several instrumented slopes and the description of the
apparatus installed in the benches.

Fig. 5 is a photo of the experimental trench during
the process of its excavation. Final height of that slope
was designed for 60 m (180 ft) with inclinations of 55°
to 65° and a total length of 250 m (750 ft).

III.2. - Instrumentation in the mine and in the experimental trench

After excavation of the first benches of the open pit
was accomplished, it was decided to locate a series of
controlling apparatus to attain the following purposes:

a. - To obtain information on the rock mass behaviour
during the process of deepening the surface mine.

b. - The evalute stability conditions of the various
slopes, and to monitor events related with impending
failures in order to detect and to announce ruptures .

c. - To gather data for elaborating predictions on the
stability of the future pit.

d. - To check hypothesis and previously utilized methods
of computing slope safety factors and related stability
conditions.

In order to attain these goals, a system of different
instruments was installed both in the first open pit
benches (which were mostly excavated in soil) and in the
experimental trench.

Readings and recordings of those instruments were
processed in function of time, giving emphasis to the
following aspects:

a. - Apparent slope stability conditions and examination of
decompression effects in the rock mass as a result of
excavation.

b. - Evolution of large fractures or cracks, mostly on
slope berms, indicating future failures.

c. - Effects of external causes (such and rain and

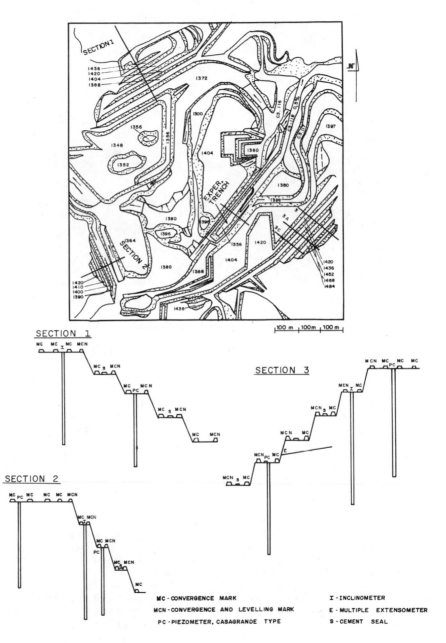

SECTION 1

SECTION 2

SECTION 3

MC - CONVERGENCE MARK

MCN - CONVERGENCE AND LEVELLING MARK

PC - PIEZOMETER, CASAGRANDE TYPE

I - INCLINOMETER

E - MULTIPLE EXTENSOMETER

S - CEMENT SEAL

Fig. 4. Plan view of the Cercado Mine, indicating instrumented sections and apparatus

Fig. 5.   General view of the experimental trench

blasting vibrations) on slope stability.

     Therefore, the main types of instruments which have
been installed were:
- Accurate topographic surveys of bench marks, by means
of levelling,then to measure vertical displacements.
- Convergence measurements to determine mark movements
and their rates, along several sections of the slope.
- Multiple extensometers and inclinometers to measure
internal rock mass displacements, both in the vertical
and horizontal directions, as well as to detect anomalous displacements along discontinuities.
- Piezometers to determine water levels and pressures
inside the rock mass. Together with the observation of
water flows, this provide a continuous control of the
presence of water inside the slopes.
- Pluviometers to record rains in the pit area, which
were correlated with water levels into the rock mass.
- Accoustic transmission and emission monitoring, in
order to measure seismic velocities, and to record
noises from rock slidings.
- Other observation techniques to detect slope movements,
such as the intallation of measuring pins on both sides
of a major crack, and the application of cement seals or
belts for the indication of progressive failures.
- An engineering seismograph was used to measure vibrations levels from rock blasting, which allowed the
formulation of a wave propagation law for the rock mass.

The procedure utilized for the intrepretation of the various parameters recorded by that system of instruments was to plot their individual values on parallel graphs using time as the horizontal axis.

To help the understanding of the meaning of values recorded on the various apparatus, a simple method was implemented in the mine site: three levels of danger were established (green or safety, yellow or precaution, and red or danger) in order to provide an idea of the importance of each parameter on slope stability. Those three levels of safety were suggested as follows:

TABLE 2. Slope safety levels

PARAMETER	LEVEL 1 SAFETY (GREEN)	LEVEL 2 PRECAUTION (YELLOW)	LEVEL 3 DANGER (RED)
Safety factor	> 1.3	< 1.3 ; > 1.1	< 1.1
Probability of failure	< 10%	> 10% ; < 30%	> 30%
Total slope displacement (m)	< 0.1	> 0.1 ; < 2.5	> 2.5
Rate of displacement (m/day)	< 0.005	>0.005; <0.075	>0.075

One example of applying this criterion to a particular case is presented in Fig. 6, where a graph of safety factor vs. slope height is developed, showing the influence of water on the stability of section 03 of the south slope of the experimental trench. This graph was prepared before excavation had started and it proved to be realistic in further stages of bench deepening.

III.3. - Slope failures predicted by means of instrumentation

Several cases of slidings in the experimental bench were detected in early stages of instability as a result of combining systematic observation of the slopes with the interpretation of recordings from various apparatus.

Three case are worth describing:
1) Slope failure in overburden soil, occurred between convergence bases 308 and 309 in Section 3. The significant events before rupture were:
- Beginning of the heavy rain period in Oct. 11, 1978.
- Initiation of a tension crack in the top of the bench, between convergence bases 308 and 309, in Nov. 9, 1978.
- Progressive increase of the displacements between the above mentioned bases.

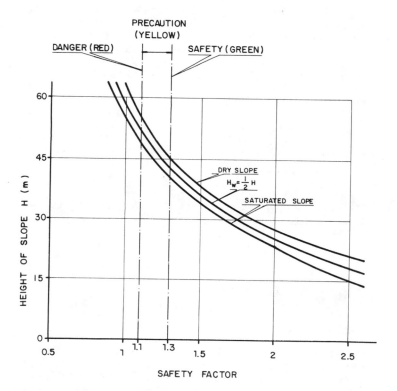

Fig. 6.   Slope stability curves for the south side of
          the experimental trench

- Displacement rates reach about 1 mm/day.
- Continuation of rainy season with continuous raising
  of piezometric levels.
- Failure in Jan. 9, 1979.

     Fig. 7 represents the variations of the principal
parameters recorded with respect to that particular
slope failure.

2) Soil slope sliding between convergence bases A26 and
A27 in the experimental trench. Main events preceding
failure were:
- Start of heavy rains in Dec. 1979.
- Development of a tension crack on top of slope,
  between bases A26 and A27, begining Jan. 4, 1980.
- Slope deformation increases, particularly after
  blasting rounds detonated 16 m (50 ft) below.
- Continuation of rains in the area, and raising water
  levels in the slope.

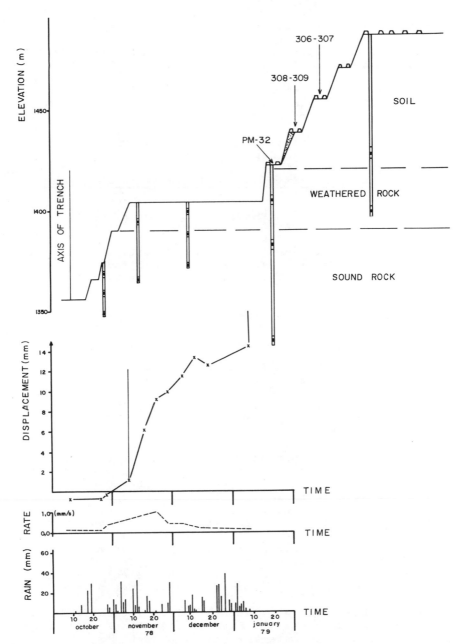

Fig. 7.   Evalution of slope displacements, rates of
displacement, and rains for section 03A.

- Failure in April 11, 1980.

3) Rock wedge failure in the experimental trench between
 convergence bases B33 and B34, preceded by the following
 events:
- Initiation of rainy period in Dec. 1979.
- Verification of unfavourable orientation of sliding
  planes in the slope, with its intersection dipping into
  the trench.
- Influence of vibrations caused by nearby rock blasting.
- Failure of wedge in Jan. 30, 1980.

III.4. - Backanalysis of slope failures in the exper-
         imental trench

   As it is mentioned above, the role of the experimental trench excavated in the Cercado mine was to
induce slope failures that could indicate typical mechanisms of slidings in the future open pit, and, to allow
the determination (by backanalysing such failures) of
shear parameters for the various joint types involved in
those slidings.

   Several ruptures occured in the trench, involving all
joint types detected by the structural study, as well as
their combinations.

   One example of such failures is presented in Fig. 8,
where two benches of have been affected.

   In each detected rupture case, geometric, geomechanic and hydrogeologic conditions were obtained and
processed in accordance with the flowchart of Fig. 1.

   Both limit equilibrium and finite element analysis
were performed with the purpose of determining cohesion
and friction angles for the various joint types. In some
cases a single joint was responsible for the sliding,
but in general a combination of joints caused slope
failure. Typical mechanisms were classified in four
groups, as follous:
1) Plane failures: along one joint alone or in association with a subvertical tension crack.
2) Biplanar failures: resulting from the combination of
two joints of different types, or sometimes of the same
type.
3) Rock wedge failure: when two joints intersected with
an attitude favourable to slide a rock block with wedge
shape towards the interior of the trench.
4) Circular type failures: in zones of high fracture
density, originating a complex combination of joint types
and attitudes which lead the mechanism to be analogous

Fig. 8.   Photos of the experimental trench showing rock
           slides   in two benches

to soil slope ruptures.

    All these four types of failures were present in the
experimental trench and they provided a means to determine shear parameters for the various joints (rock-rock
joints, sound walls and infilling materials, weathered
walls and infilling materials, and basic intrusion). Sometimes only one joint type was involved in the mech-
nism of rupture, but in most cases an association of two
or more types could be detected.

    In simpler cases only one joint type was the cause of
slidings in two or more slopes and the corresponding
backanalysis supplied accurate predictions for cohesion
and friction angles of that joint.

    This is illustrated by means of an example, involving
type 2 joint (sound walls and with infilling materials).
Before and after the sliding data were gathered on its

characteristics, and as a result of readings in nearby instruments.

Two slope failures were caused by the sliding of rock wedges separated from the massif by that joint type, and their principal characteristics are described below.

TABLE 3.

FAILURE NO.	WEDGE WEIGHT (TONS)	FIRST JOINT ATTITUDE	SECOND JOINT ATTITUDE	BENCH STRIKE	DEPTH OF WATER (m)
1	42.3	N34ºE,60ºNW	N80ºW,54ºNE	N26ºE	4.0
2	54.3	N25ºE,58ºNW	N36ºW,46ºNE	N26ºE	4.1

Applying the limit equilibrium method and the mathematical expressions presented in Fig. 3, two relations were obtained:

$$\text{1st failure: } 2.32 = \text{tg } \emptyset + 3.38 \text{ c}$$
$$\text{2nd failure: } 1.51 = \text{tg } \emptyset + 1.82 \text{ c}$$

Solving these equations provided the shear parameters of joint 2 type:

$$\text{Cohesion c} = 0.52 \text{ t/m}^2 = 0.0052 \text{ MPa}$$
$$\text{Friction angle } \emptyset = 29.5º$$

Finite element models were also applied in the backanalysis of several of the failures in the experimental trench.

Fig. 9 shows one of those models regarding a rock wedge failure along two inclined joints.

Basic information contained in this particular model included joint thickness, water pressure, elastic properties of the rock and density. After the model was established, a parametrization study was accomplished by means of supplying different values to c and $\emptyset$ for the joint, and determining rock wedge displacements which were checked by values obtained in the convergence measurements. The criterion of using a high ratio of normal stiffness to shear stiffness for the joint (kn/kt of the order of 100) which corresponds to the adoption of the rigid wedge hypothesis, provided adequate solutions for this problem.

Therefore, upon the application of a rupture criterion for the joint material a parametric study was done in order to find adequate values of c and $\emptyset$ which could be representative of the joint behaviour.

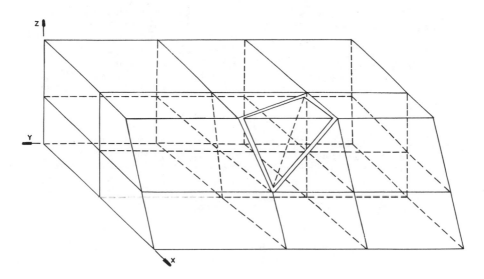

Fig. 9.   Three dimensional finite element mesh of a rock
          wedge failure.

Sometimes finite element solutions have been compared
with limit equilibrium results for the same backanalysis,
thus providing better insight on the problem.

After performing various backanalysis of slope failures in the experimental trench, values of the shearing
parameters of the four joint types were obtained, as
indicated in Table 1, which contains their average
values.

It was observed that c and ∅ values from backanalysis
studies were systematically smaller than the corresponding values as determined in laboratory shear tests.

III.5. - Contributions to ultimate pit design

The main purpose of the geomechanical work developed
in the Cercado mine was to supply reliable data aimed to
suggest slope angles for the open pit.

It was clear from the beginning of this program of
studies that the mine would have different slope angles
for different sectors because of the great influence
exerted by the existing joints and discontinuities.

Another aspect to be considered was the fact that
because of the patterns of uranium occurrence, the pit

will have not a uniform depth but at least three different depths for three different orebodies.

These circumstances, associated with the knowledge of underground water behaviour and the prediction of blasting vibrations influencing slope stability, lead to the proposition of ultimate pit slope angles in a peculiar way.

According to the attitudes of the dominant joint families (see Table 1) the areas of the pit were they can produce sliding have been located, and the types of possible slope failures were indicated. Furthermore the combination of two or more joint types was considered as a cause of other types of ruptures, such as the biplanar and the rock wedge slidings.

In each possible combination, the relative frequency of occurrence of each joint type and or its combinations with other joints, was determined (only frequencies greater than 1% were considered).

Using this criterion the following failure cases were defined, as well as their azimuthal location in the pit and their probability of existence:

Plane failure: 9 cases
Biplanar failure: 21 cases
Rock Wedge failure: 23 cases

In addition to these, a circular failure mode was considered as the "background" mechanism for the all pit, so that it can be chosen for design when in a certain sector there are no other mechanisms of rupture, or the existing ones had a low probability of occurrence, thus leading the designer either to decide on the acceptance of this risk or to install appropriate stabilization measures.

Therefore, the pit (which has an approximate circular shape) was divided in a large number of sectors each of them subjected to a slope stability analysis (see Fig. 10).

In each failure case several situations were analysed:

a. - Six different slope heights between 50 and 300 m (150 and 900 ft)
b. - Two water level positions (1/3 and 2/3 of slope height).
c. - Nine different inclination angles of slope (30,35, 40,45,50,55,60,65 and 70 degrees).
d. - Other slope parameters, such as length of the tension cracks in the plane failure, or the length of

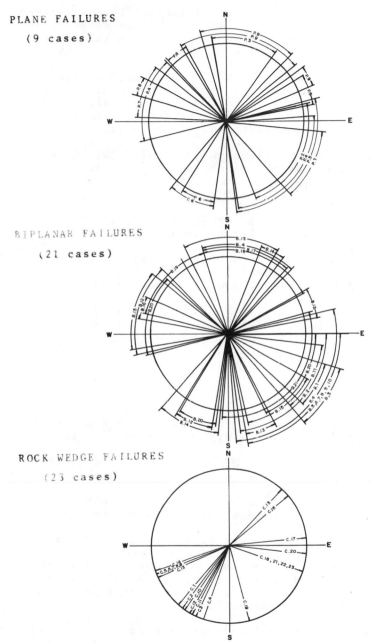

PLANE FAILURES
(9 cases)

BIPLANAR FAILURES
(21 cases)

ROCK WEDGE FAILURES
(23 cases)

Fig. 10. Azimuthal location of possible slope failures
in the future open pit

lower sliding plane in the biplanar failure have been
submited to variations in function of the slope height.
However, for the purpose of analysing results, they have
both been considered equal to 1/4 of slope height.

With these data, a total of about 300 000 safety
factors were computed, and they have been represented
in graphs showing the variation of slope height with
safety factors, for different curves of slope inclinations
and for two distinct positions of the water level.

This information allows the selection, by the designer
of the open pit, of the slope inclinations in all the
sectors, according to his own criteria of safety,
economy and risk.

Fig. 11 ilustrates the graphic variations for the case
of circular failure in the pit, which constitutes the
safest situation among all, to be selected only when in
certain sectors other more critical failure cases have
either a very low probability of occurrence, or might
be subjected to specific stabilization operations.

Thus, with a larger number of stability graphs the
designer is able to select for each pit sector, according to its particular depth and structure, what type of
situation will represent best the future conditions in
the open pit.

IV. - CONCLUSIONS

A series of programs are required to propose a
certain geometrical shape for an open pit mine, a shape
that must satisfy both safety and economic criteria.

When the rock mass is heterogeneous and full of
discontinuities that will control slope stability (such
as in the Cercado orebody) detailed studies are necessary to predict future behaviour of the ultimate pit
slope, as well as intermediate phases of the excavation.

In order to get reliable data to develop accurate
stability predictions it was implemented in the Cercado
mine a concept of excavating an experimental trench,
within the pit and before its deepening. That trench was
designed to fail in several positions thus supplying
valuable data for final pit design.

Backanalysis of failures occurred in the instrumented
slopes of the trench have given reliable information
regarding future behaviour of the open pit, thus providing excellent tools for its final design.

WATER : $H_W = 1/3$ H

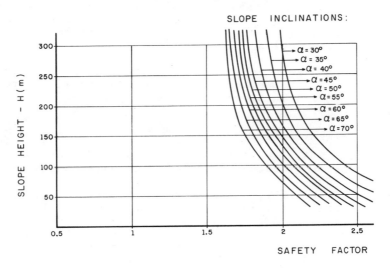

WATER : $H_W = 2/3$ H

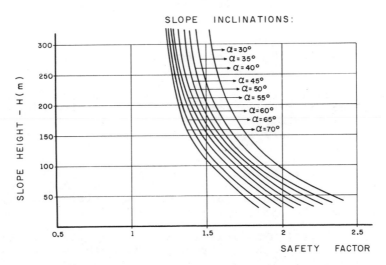

Fig. 11.   Stability curves for the circular type of slope
           failure in the open pit.

V. - ACKNOWLEDGEMENTS

The author expresses gratitude to Nuclebras S.A. for the authorization given to publish this report on the work that I.P.T. has developed in the Cercado mine for the last five years.

Also he acknowledges the helpful discussions with his colleagues Ricardo F. Silva and A.A. Tognon which provided excellent contributions to the work that has been described.

VI. - REFERENCES

Attewell, P.B., and Farmer, I.W., 1976, "Principles of Engineering Geology," Chapman and Hall, London.

Barton, N.R., 1971, "Estimation of In Situ Shear Strength from Back Analysis of Failed Rock Slopes," I.S.R.M. Symposium on Rock Fracture, paper II-27, Nancy.

Broch, E., and Nilsen, B., 1979, "Comparison of Calculated, Measured and Observed Stresses at the Ortfjell Open Pit (Norway)," Proceedings of the 4th I.S.R.M. Congress, Vol. 2, pp. 49-56. Montreux.

Cundall, P., 1976, "Explicit Finite - Difference Methods in Geomechanics," Proceedings of the II International Conference on Numerical Methods in Geomechanics. Blacksburg.

Dinis da Gama., 1978, "Estabilização de Escavações a Céu Aberto," Construção Pesada, Junho 1978. pp. 124-132. São Paulo.

Dinis da Gama., e Silva, R.F., 1978, "Engineering Geological Studies for the Cercado Uranium Mine," Proceedings of the III International Congress of I.A.E.G., paper Sec. III, Vol. 2, pp. 239-250. Madrid.

Goodman, R.E.; Taylor, R.L. and Brekke, T., 1968, "A Model for the Mechanics of Jointed Rock," Journal of the Soil Mechanics and Foundation Division, ASCE, Vol. 94, No. SM3, May 1968. New York.

Habib, P., 1979, "Le coefficient de securité dans les ouvrages au rocher," Proceedings of the 4th I.S.R.M. Congress, Vol. 3, pp. 18-22. Montreux.

Hendron, A.J.; Cording, E.J. and Aiyer, A.K., 1971, "Analytical and Graphical Methods for the Analysis of Slopes in Rock Masses," U.S. Army Engineer Nuclear Cratering Group, Technical Report 36. Livermore.

Hoek, E., and Bray, J.W., 1974, "Rock Slope Engineering," Institution of Mining and Metallurgy. London.

Kovari, K., and Dritz, P., 1975, "Stability Analysis of Rock Slope for Plane and Wedge Failure sith the Aid of a Programmable Pocket Calculator," 16th Symposium on Rock Mechanics. Minneapolis.

Otter, J.R.H.; Cassell, A.C.; Hobbs, R.E., 1966, "Dynamic Relaxation," Proceedings of the Institution of Civil Engineers, Vol. 35, Dec. 1966, pp.633-656. London.

St. John, C.M., 1971, "Three Dimensional Analysis of Jointed Rock Slopes," I.S.R.M. Symposium on Rock Fracture. paper II-9. Nancy.

Chapter 33

CASE EXAMPLES OF BLASTING DAMAGE
AND ITS INFLUENCE ON SLOPE STABILITY

Roger Holmberg   and Kenneth Mäki

Swedish Detonic Research Foundation (SveDeFo)
Stockholm, Sweden

ABSTRACT

This paper describes open pit studies where blasting damage introduced in the remaining rock has been investigated. Results from two
open pit mines are described and discussed. Parallel performed laboratory experiments, where simulated blasting damage has been introduced into samples before shear strength testing took place, are
also reviewed. These studies are tied together in order to throw some
light upon how the blasting damage influences the slope stability.

A mathematical model is also described which will make it possible
to optimize the blast design for a specified damage zone in the rock
mass.

INTRODUCTION

A slope stability analysis based on a detailed line survey, investigations of joint set characteristics, rock mechanics investigations, ground water flow measurements, interpretations of data and
statistical analysis will express a certain probability for slope
failure connected to the slope angle.

The most significant visually observed damage caused by a stock
blast is the resulting backbreak. By practising simple arithmetics
anyone easily can translate the average noticed backbreak to a
theoretical maximum slope angle assuming a certain working width of
berms for access. Fig. 1 expresses the theoretical maximum slope
angle as a function of established bench slope angles and berm widths
for two bench heights adjacent to the pit slope.

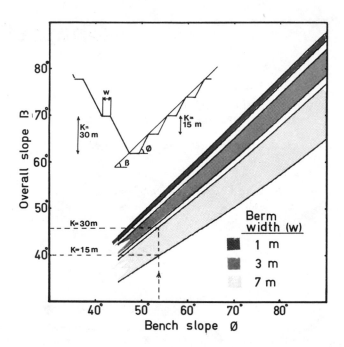

Fig. 1.   Maximum theoretical overall slope angle as a function of the
established bench slope.

If the achieved slope angle is steeper than the slope angle determined by the performed slope stability analysis usually not too many
will bother about the blast design. If the opposite occurs questions
concerning the blast design will be raised.

A direct question is illustrated by Fig. 2 which shows an observed
distribution of backbreak. "What backbreak can be tolerated? Is it
sufficient to consider the average backbreak or must one pay attention to the maximum one?"

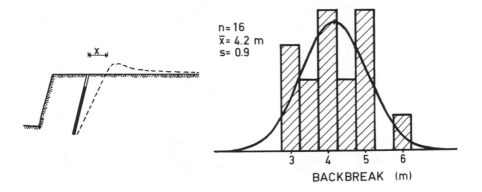

Fig. 2. Backbreak distribution observed from a stock blast.

The answer to this question can sometimes be found in the probability analysis for the actual pit slope. However, often this question raised will demand a need for extensive information about
blasting damage and structural behaviour of the rock mass.

This paper describes field and laboratory experiments carried out
to throw some light upon what damage a blast introduces in the remaining rock mass and how this damage can be taken into account when
the pit slope design approach is carried out. A mathematical model
is described which can be utilized for blast design or for prediction
of the damage zone.

APPLICATION OF THE ROCK DAMAGE
MODEL AT THE BOLIDEN MINERAL AB AITIK MINE

When the Swedish Boliden Mineral AB Aitik copper mine had reached
the stage where attention had to be paid to the final pit slopes,
investigations were initiated to quantitatively examine the damage.

A slope stability analysis was performed by Call et al., 1977.
Utilizing the ordinary stock production blast design, there was
no way to manage to keep the suggested economical pit slopes.

A project immediately started with the aim to determine the damage
zone when production blasts with hole diameters of 250 mm and a high
density aluminized TNT-slurry were used.

Methods selected to survey the damage were core drilling before
and after the blast together with direct measurements of the peak
particle velocities close to the round. Peak particle measurements
had previously been established by Langefors and Kihlström, 1963,

for predicting damage of buildings close to blasting operations.
As damage is caused by the strain in the material and as the particle
velocity is affecting the strain, it was considered interesting to
examine the particle velocities.

Fig. 3 shows the positions of the diamond drilled holes and the
sites where peak particle velocity measurements took place.

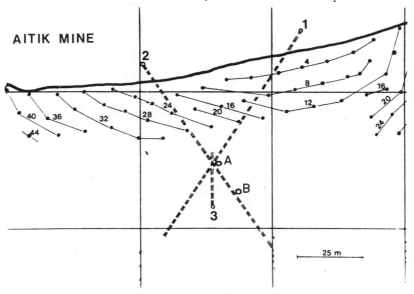

Fig. 3.   Experimental set up for core drilling 1,2,3 and peak
          particle velocity measurement points A,B.

Accelerometers were used together with charge amplifiers and a
high speed FM-tape recorder to monitor the ground vibration signals
from the blast. Numerical integrations of signals made it possible
to evaluate the peak particle velocities. Gauge positions are described in Fig. 4.

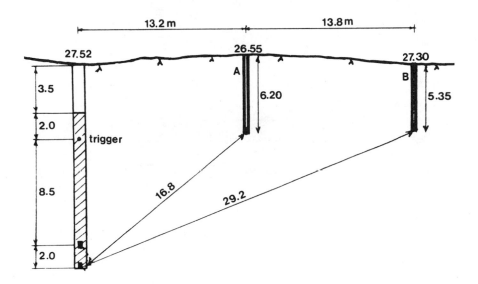

Fig. 4.   Blasting geometry and accelerometer positions.

Cores were drilled parallel to each other before and after the blast. Fracture analysis were performed by Holmberg and Krauland, 1977, and compared. Statistical analysis indicated that damage occured more than 20 meters into the virgin rock mass.

Based upon these observations a mathematical model for blast damage was established (Persson, Holmberg and Persson, 1977). The model was described in detail at the Tunnelling ´79 – meeting in London by Holmberg and Persson, 1979.

The core investigations showed that, according to the model, damage is introduced at a vibration level of 700 – 1000 mm/s. Fig. 5 gives the zone of damage for specified charge concentrations per meter borehole when a charge height of 15 meters is used.

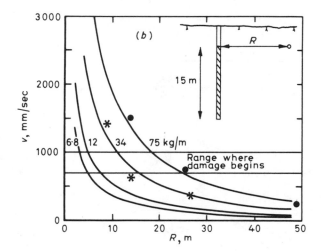

Fig. 5.   Model for estimation of rock damage and for blast design. Experimental data for 34 and 75 kg of explosives per meter are plotted.

The model has later on been verified for hole diameters in the range of 45 - 250 mm.

Today the Aitik mine is utilizing the model to optimize the cautious blasting towards the final pit slopes. An accepted zone of damage has been chosen to about 13 meters. The Aitik mine uses a Nitro Nobel aluminized TNT-slurry for the stock blast production but has decreased the charge concentration per meter borehole in the two rows adjacent to the final pit slope. The first row is blasted with a nonaluminized, low content TNT slurry loaded in 250 mm holes and the final row is loaded with the same explosive in 150 mm diameter PVC-tubes (decoupled charges).

Aitik Geology (Zweifel, 1972)

The rock units of the area are of Precambrian age and consist of metamorphosed sediments occurring in a zone, 40 km long parallel to the general N 20° W strike and with an average width of about 5 km. This zone of metasediments is surrounded by the younger Lina granite and gabbro. One can distinguish between an east and a west part of the Aitik-Liikavaara zone, both according to the geological structure and the type of country rock.

The rocks in the west part have been labelled as Aitik group, divided in the older Aitik formation and an overlaying formation biotite to biotite-amphibole gneisses. The main rock types of the Aitik formation are skarnbanded gneisses, fine grained biotite gneisses sometimes going over into micaschists or quartzites, gneisses with sharnschlieren, amphibolites and coarse grained biotite gneisses.

LKAB LEVEÄNIEMI IRON ORE OPEN PIT MINE

Pincock, Allen and Holt, 1979, performed a pit slope stability analysis for the LKAB Leveäniemi iron ore mine. A joint venture was initiated where LKAB, Pincock, Allen and Holt and SveDeFo contributed to a project where extensive examination of blast damage in the mine was performed.

SveDeFo together with Leveäniemi started a program to investigate in detail by field and laboratory studies, how the blasting disturbes the geology and the stability in sector 4 B in the mine (see Fig. 6).

The laboratory experiments were carried out to investigate how the joint shear strength is affected by blast damage. The field measurements were performed to characterize the disturbance of the rock mass continuity when stock production rounds were blasted.

Geology of the Leveäniemi Open Pit Mine

As a result of the surface mapping by Pincock, Allen and Holt and Leveäniemi geologists, the rock mass forming the slopes was devided into parts with a similar geologic structure. These structural domains,individually considered when the stability analysis was performed,are illustrated in Fig. 6.

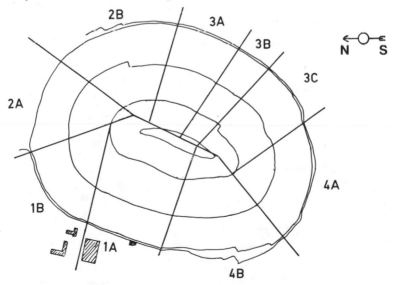

Fig. 6.   Structural domains defined by line mapping.

The bedrock of the area described by Frietsch, 1966, consists of volcanics and sediments with the following stratigraphy. Red leptite (oldest), conglomerate (leptite as ball material), scapolite, mica schist (mostly biotite schist). The ore brecciated the schist and has thus been deposited intrusively no doubt in conjunction with the tectonics which have given the area its synclinal form. The fold axis of the syncline dips on the surface at an angle of 50 degrees to the NNE. The dip becomes less steep with increasing depth. "The shackles" are pressed together which gives the ore body the shape of a sail boat hull.

In sector 4B the foliation of a biotite schist dips about 55° towards the pit center with a strike direction parallel to the pit limit.

SveDeFo´s studies were restricted to the sectors 1A and 4B.

Blasting Geometry

Stock production blasts in the Leveäniemi mine are performed with holes inclined 70° and with a hole diameter of 191 mm, drilled with DM-4 and Hausherr drill rigs. Bench heights are 15 m, burden is 5 m and the spacing is 6 m. A Kimit AB TNT-slurry with a density of 1300 kg/m^3 is used. Row by row initiation with 50 or 100 ms delay is used. Fig. 7 shows a typical round.

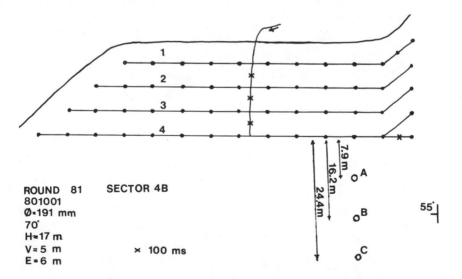

ROUND 81   SECTOR 4B
801001
∅=191 mm
70°
H≈17 m
V=5 m   × 100 ms
E=6 m

Fig. 7. Stock blast in the Leveäniemi mine. A,B and C indicate where particle velocity measurements have taken place.

Field Studies

  Some of the results from field measurements of round 81 (see
Fig. 7) are reviewed. These are from the extensometer measurements,
the seismic measurements and the core logging. Results are in detail
given by Holmberg, 1981.

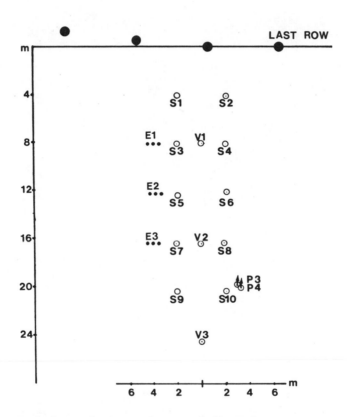

Fig. 8. Experimental set up for round 81. E denotes extensometers,
         S denotes holes for seismic cross hole measurements,
         P denotes core drilled holes and V represents holes for
         peak particle velocity measurements.

Extensometer Measurements

At the distances of 8,12 and 16 meters behind the last row of
blastholes rebars with a diameter of 22 mm were anchored at depths
of 6 and 16 meters from surface. Another short rebar was also grouted
at the surface.

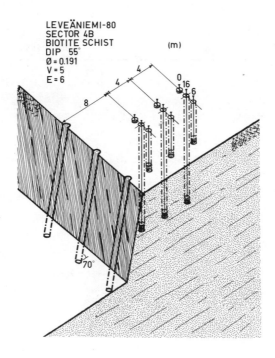

Fig. 9. Positions of rebars for extensometer tests behind
         the round.

Upper end positions of the anchored rebars were monitored before
and after the blast. The vertical positions were determined relative
each other and relative to a fix point situated 60 meters behind
the round.

Fig. 10 describes the measured relative displacement for the
three various distances from the round.

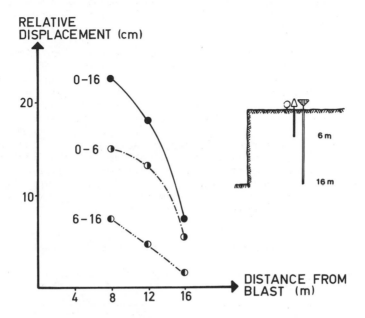

Fig. 10.   Relative displacements after the blast.

According to the figure above, the relative displacement between the surface rebar and the rebar anchored at a depth of 16 meters was as much as 23 cm, eight meters behind the round. This displacement corresponds to an equivalent homogeneous strain of 1.44%. Sixteen meters away from the round this equivalent strain is found to be 0.47%.

The displacements are largest in the vicinity of the free surface. This is an expected result since the vertical stresses are low in this part of the rock mass. With increasing depths the displacements decrease. However, remarkable displacements in the depth interval 6-16 meters still takes place sixteen meters away from the blast. Here, the relative displacement is 18 mm corresponding to a equivalent strain of 0.18%.

In sector 4B where the extensometer measurements took place, fracture mapping, performed by Paul Visca of Pincock, Allen and Holt, indicates an average joint spacing of 0.72 meters measured perpendicular to the dip of the foliation. The mean dip of the foliation in this sector is 55°. A horizontal cross joint set is reported to have an average joint spacing of about one meter.

During the blasting, the quasistatic gas pressure pushes the new face backwards when the burden breaks and the throw occurs. Gaseous

detonation products penetrate into previous existing and newly formed
fractures, separates them, and heaves the rock mass and heavily disturbs the rock mass continuity.

Considering only the horizontal <u>visually</u> observed joints there
will be about 16 fractures from the surface to a depth of 16 meters.
Assuming that only these observed fractures are affected and separated during the blasting, quite large widths of open spaces between
joint surfaces (<u>apertures</u>) would be the result if the measured
relative displacement is divided with the number of fractures.

The core logging indicates an average of about eight fractures per
meter which is considerably more than the mapping shows and will of
course result in smaller apertures. Due to the moment applied during
the core drilling, fractures probably are introduced. However, the
true number of fractures ought to be somewhere in the interval between the above mentioned upper and lower limits of fracture per meter.

In Table 1 the average apertures have been tabled for the two limits
of fractures per meter assuming no apertures existed before blasting.

Table 1. Estimated average apertures of joints in the rock mass
          after the blast

Distance from blast (m)	Depth (m)	Relative displacement (mm)	Aperture (1 fracture/m) (mm)	Aperture (8 fractures/m) (mm)
8	0-6	151	25.2	3.2
	6-16	74	7.4	0.9
12	0-6	131	21.8	2.7
	6-16	49	4.9	0.6
16	0-6	57	9.5	1.2
	6-16	18	1.8	0.2

Seismic Measurements Behind the Round

Seismic cross hole measurements were performed before and after
the blast. At every fourth meter behind the round, up to twenty
meters away, five meters deep twin holes separated four meters
were drilled. Gelled water was poured into the transmitter holes
S2,S3,S6,S7 and S10 (see Fig. 8). This was done to increase the
acoustic impedance when a blasting cap was detonated in the gelled
water. The emitted pulse was monitored with accelerometers positioned in a neighbour receiver hole and the signal was monitored
with an FM-tape recorder. At the same time the time of detonation
was also recorded.

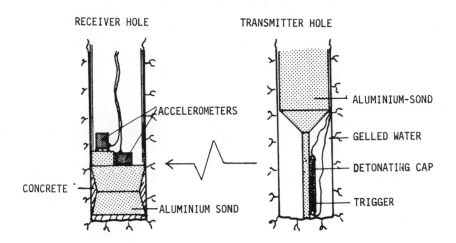

Fig. 11. Method for seismic cross hole measurements.

The reason for this type of measurement was to control how the amplitudes and the transmission times changed after the blast, indicating if any disturbance had occurred in the rock mass. Unfortunately only the transmitter holes S7 and S10 were still intact after the blast.

The results achieved from these holes showed that both the P-wave velocity and the maximal amplitude were reduced after the blast. Frequency analysis of the recorded signals showed a frequency decrease after the blast.

From the number of cross hole measurements it was not possible to determine if the changes were larger or not across the foliation. However, the measurements clearly showed that still at a distance of 21 meters behind the round and at a depth of five meters below the surface, the rock mass continuity was disturbed.

Core Drilling

Before and after the blast core drilling took place behind the round at the P-marked spots in Fig. 8. The holes were drilled from the surface about 20 meters behind the round with an inclination of 45° towards the last row. The length of the cores was about 20 m which means that the end of the cores were situated about 10 m from the last row and at a depth of about 14 m.

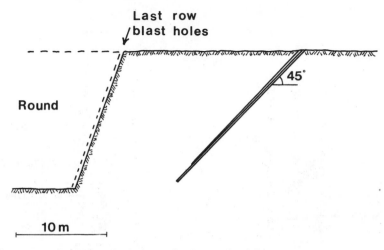

Fig. 12. Core drilling before and after the blast.

From the first five core drilled meters bad core recovery was
achieved and the core logging indicated crushed zones close to the
surface. This was expected as the surface is the bottom part of a
previously blasted round.

Fracture frequencies were logged and the result is shown in Fig.13.

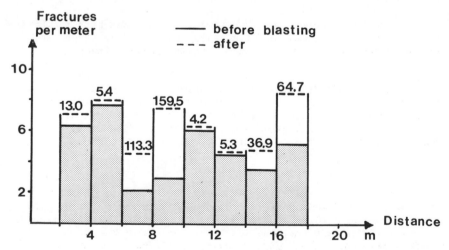

Fig. 13. Fracture frequencies before and after the blast. Percent
of fracture increase is also given.

After the blasting the fracture frequencies are larger than before the blasting for the total core length, although the differance is small in the interval 10-14 m. Two intervals have a very high increase of the fracture frequencies (65-160%). These are the intervals 6-10 m and 16-18 m. The interval 6-10 m is close to the surface (depth 4.7 - 7.1 m) where large residual displacements have been noticed. The interval 16-10 m is located at a depth of 11.3 - 12.7 m quite close to the blasthole bottom.

Point Load Strength Index Test

After core logging was performed the two cores were point load index tested. The point load strength index can be used as a measure of the uniaxial tensile strength. In this test the core pieces were loaded with a load oriented parallel to the foliation.

The core drilled before blasting indicated a steadily increasing point load index along the core, i.e. at further distance from the free surface. There are two possible explanations for this behaviour; First, it is possible that the strength of the biotite schist simply increases with depth in this bench. Such a behaviour could be due to accelerating weathering close to the free surface after the bench above has been excavated. Second, and most likely, a reduction of the strength has been introduced due to the earlier round blasted in the previous bench above.

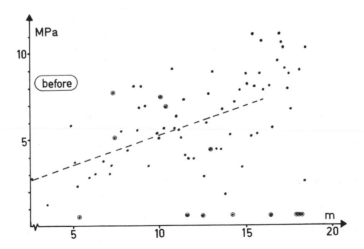

Fig. 14. Point load strength index along the core before blasting. The larger dots indicate unsatisfactory tests that were neglected when the least square fit was performed.

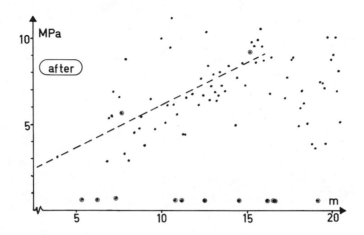

Fig. 15. Point load strength index along the core after blasting.

Examination of the core drilled after the blasting shows a similar increase of the point load strength index up to about 16 m along the core. After 16 m there is a drop of the strength. This part of the core was close to the last row blasthole bottom.

A regression line was fitted to the discrete values. The regression lines indicate that the strength is higher for cores tested after blasting (in the interval up to 16 m). This is a phenomena earlier reported by Mäki and Holmberg, 1980, where cores before and after blasting were tested with uniaxial tensile tests, Brasilian tests and by point load strength index tests.

The explanation for such a behaviour is probably that the biotite schist contains a number of potential weakness planes, each of them having a unique strength. Blasting introduces dynamic loads and several low strength weakness planes will be fractured. The core examined after the blast will therefore contain more fractures. Core pieces selected for the point load index test will obviously not contain the low strength weakness planes and consequently a higher average of the strength will be achieved.

Peak Particle Velocity Measurements

The results from the measurements are somewhat higher than what the damage model predicts. As the model is based upon blasts with in-the-hole delays, the result indicates that row-by-row initiation with detonating cord is unfavourable if rock damage should be depressed. Measurements indicate a damage zone of about 20 meters.

Laboratory Experiments

In order to investigate the shear strength of joints after movements in the joints have occurred, direct shear tests in laboratory
scale were performed. This was made with a specially designed shear
apparatus which basically consists of two steel plates positioned
parallel to each other and at an initial distance of 1 cm from each
other.

The plates have throughgoing holes where three rock core samples
with a diameter of 30 mm can be mounted. Intact rock samples containing geologic structures such as schistocity can be mounted into
the apparatus in such a way that the structures will have an orientation parallel to the plates and to the shearing direction. Small
movements of the plates created by wedging cause single planes of
weakness to be fractured in the samples. Further movements increase
the aperture along the fractured weakness planes. After a specific
desired temporary aperture has been reached the surfaces are repo-
sitioned into contact without creation of further damage and the
apparatus is ready for shearing.

It is also possible to study the effect of movements in pre-
existing joints. Samples containing such joints are kept together
and mounted into the apparatus as in the previous case. The apparatus has facilities for increasing the initial distance between
the plates which may be used to give the joints in the samples a
remaining aperture at the start of shearing.

Results of experiments carried out on rock material from the
Leveäniemi mine are reported by Mäki, Nord and Persson, 1979 and
Mäki, 1981.

Fig. 16 shows a typical plot of the peak shear strength as a
function of the temporary aperture. The large points are results
from successful shear tests. The smaller points give results of tests
associated with problems mostly due to unfortunate fracture orientations.

The results show that the peak shear strength of joints that were
given a temporary aperture of 0.2 mm is almost twice the strength of
joints that were given a temporary aperture of 0.7 mm. Observations
during the experiments showed that although the temporary aperture
was as large as 0.7 mm the planes of weakness did not always consist
of continuous joints. As a consequence of this a still lower peak
shear strength was obtained when completely separated surfaces were
sheared (Point marked 1) in Fig. 16 .

The strength of the intact rock was determined by two double shear
tests. It was found to be 12 MPa which is more than ten times the
strength of the fractured weakness planes.

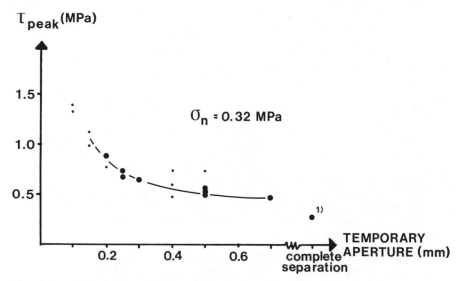

Fig. 16.   The peak shear strength as a function of temporary aperture
          1) Mean value from three experiments after complete
          separation. Corresponding peak angle of friction

$$\emptyset_{peak} = \arctan \frac{\tau_{peak}}{\sigma_n} = 39^{\circ}$$

The results in Fig. 16 were used as a basis for calculations.
Equations for the strength of weakness planes containing intact rock
bridges are given by Lajtai, 1969. The use of these equations indicate that weakness planes that had been given a temporary aperture
of 0.3 mm still had intact rock bridges covering 7% of the area of
the weakness plane. For increasing temporary aperture this area was
calculated to decrease.

Fig. 17 shows the peak angle of friction for continuous joints
evaluated from tests after different remaining apertures at the
start of shearing. Two sets of experimental data are shown. Each set
is a result of repeated shear tests on the same surfaces with repo-
sitioning of the surfaces and decreased remaining aperture at the
start of each new test.

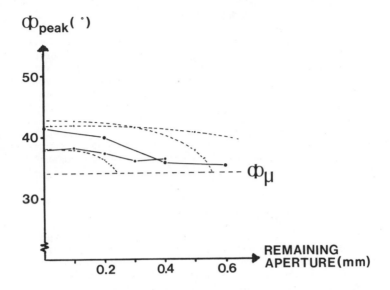

Fig. 17. The peak angle of friction as a function of remaining aperture.

The dotted horizontal line indicates the calculated basic angle of friction for the surfaces which is the angle of friction for a fictive flat surface of the material. The angle obtained agrees with the greatest measured residual angles of friction which were between 27° and 35°. The dotted curves show predicted peak angles of friction based on calculations.

The results in Fig. 17 show that the surfaces have a peak angle of friction which decreases if the joints are given a remaining aperture at the start of shearing. After a remaining aperture of more than 0.4 mm the peak angle of friction is close to the calculated basic angle of friction. By assuming that the peak strength is a result of contacts between surface roughnesses it was possible to predict the decrease of the peak angle of friction by studying the angles of inclination of the surface roughnesses. The calculations are reported in detail by Mäki, 1981.

The dotted curves in Fig. 17 show the results of the calculations based on two of the largest and one of the smaller roughnesses that were to be found on the surfaces. The positions of the experimental curves show that the peak angle of friction for these sets decreased due to remaining aperture in a way which is predicted by the change of the angle of inclination of the intermediate size roughnesses.

CONCLUSIONS

In Table 1 estimated average apertures are given for various distances behind the blasted round. For a fracture frequency of eight fractures per meter, the remaining apertures about 20 m from the round are estimated to less than 0.4 mm. The laboratory experiments indicate that for remaining apertures less than 0.4 mm the peak angle of friction could be used for stability calculations. The results presented here form a basis for such calculations. The rock damage model will be to great help for determination of the zone where peak shear strength values could be utilized. Laboratory experiments and field measurements will give the data necessary for evaluation of the peak strength of critical weakness planes.

ACKNOWLEDGEMENTS

The authors wish to express their appreciation to the Swedish Board for Technical Development (STU), to the Swedish Council for Building Research (BFR) and to the Swedish Industry who made this research project possible. The authors also greatfully acknowledge assistance given by the Boliden mine Aitik and the LKAB mine Leveäniemi where the field experiments were carried out and Paul Visca, University of Arizona and Per-Anders Persson, Nitro Nobel AB for valuable discussions.

REFERENCES

Call, R.D., et al., 1977, "Probalistic Approach to Slope Design for the Aitik Mine, Sweden", Rock Mechanics Day arranged by BeFo (Swedish Rock Mechanics Research Foundation), Stockholm, Sweden.

Frietsch, R., 1966, "Geology and Ores of the Svappavaara Area, Northern Sweden", Sveriges Geologiska Undersökning, Serie C Nr 604, Stockholm, Sweden.

Holmberg, R., 1981, "Field Studies of Rock Damage in the Leveäniemi Open Pit Mine", (In Swedish), SveDeFo, Report DS 1981:6, Stockholm, Sweden.

Holmberg, R. and Krauland, N., 1977, "Examination of Crack Frequencies Before and After a Blast with 250 mm Hole Diameter in Aitik", (In Swedish), SveDeFo, Report 1977:2, Stockholm, Sweden.

Holmberg, R. and Persson, P-A., 1979, "Design of Tunnel Perimeter Blasthole Patterns to Prevent Rock Damage", Proceedings, Tunnelling'79, Editor Jones, M.J., Institution of Mining and Metallurgy, London, U.K., March 12-16, 1979.

Lajtai, E.Z., 1969, "Strength of Discontinuous Rocks in Direct Shear", Geotechnique 19, No 2, pp 218-233.

Langefors, U. and Kihlström, B., 1963, "The Modern Technique of Rock Blasting", John Wiley, New York, USA, Almqvist and Wiksell, Stockholm, Sweden.

Mäki, K., 1981, "The Shear Strength of Planes of Weakness in Biotite Schist with Simulated Blast Damages", SveDeFo, Report DS 1981:4, Stockholm, Sweden.

Mäki, K. and Holmberg, R., 1980, "Rock Damage from Crater Blasting of a Raise", (In Swedish), SveDeFo, Report DS 1980:8, Stockholm, Sweden.

Mäki, K., Nord, G. and Persson, P-A., 1979, "The Influence of Blast Damages on the Shear Strength of Biotite Schist", (In Swedish), SveDeFo, Report DS 1979:11, Stockholm, Sweden.

Persson, P-A., Holmberg, R. and Persson, G., 1977, "Cautious Blasting Towards Pit Slopes in Open Pit Mines", (In Swedish), SveDeFo, Report DS 1977:4, Stockholm, Sweden.

Pincock, Allen & Holt, INC, 1979, "Slope Design Study for the Leveäniemi Open Pit", authors Visca, P.J., Jones, S.M. and Call, R.D., Tucson, USA.

Zweifel, H., 1972, "The Aitik Copper Mine Geology and Mineralization", Boliden Mineral AB, information sheet of the Aitik Mine, Boliden Mineral AB, Sweden.

Chapter 34

WASTE DUMP STABILITY AT FORDING COAL LIMITED IN B.C.

Robert S. Nichols

Senior Planning Engineer, Fording Coal Limited
Elkford, B.C.

ABSTRACT

Fording Coal Limited's mine in the Rocky Mountains near Elkford, B.C. has produced 21.8 million clean tonnes of metallurgical coal from 1971 to 1980, inclusive. This production has come from several pit areas and required the removal of 149.1 million bank cubic meters of waste rock and overburden by truck-shovel and dragline operations.

Waste dumps from 30m to 200m in height have been successfully constructed on natural slopes, generally between 10° and 26°. Design considerations for the dumps include foundation and soil conditions, natural slope angles and containment of weak overburden or rehandle materials.

Dump control is maintained by monitoring the crest and the face of an active dump area. Crest movement is measured by extensometers locally termed "Spoil Monitors". When crest movement rates progressively increase above normal, dumping operations are temporarily relocated. Some unstable conditions and mass failures have occurred as the result of natural foundation slopes being in excess of 26°, failure of weak water saturated foundation soils or weak material being placed on the dump.

INTRODUCTION

The Fording River Mine is located in the Rocky Mountains of southeastern British Columbia as shown in Figure 1.

All production is from multi-seam open pit mining at elevations ranging from 1600m to 2200m.

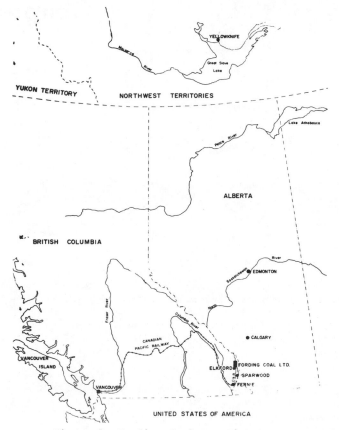

Figure 1. Fording Coal Location Map

The mine is located in a continental climatic zone with temperatures ranging from -40° C in January to +35° in July. Annual precipitation consists of rainfall between 220 to 350mm plus snowfall in the range of 240 to 680 cm.

Cleaned coal production at an annual rate of 3.0 million tonnes began in 1972. A total of 149.1 million bank cubic metres (bcm) of waste rock, overburden and rehandle has been moved by truck-shovel operations and a 46 m³ dragline.

In 1980, a major expansion program was initiated to increase annual production to 5.0 million tonnes. This will require 40.0 million bcm of waste to be removed per year.

Planning, constructing and maintaining stable waste piles along the valley slopes has been a significant factor in the success of

this operation. This paper describes the design considerations,
development and control used at Fording to maximize waste dump
stability.

GENERAL MINE LAYOUT

Multi-seam open pit mining is done in pit areas on the east and
west sides of the Fording River. The pits are designed to an overall average strip ratio of 7 bcm waste : 1 bcm raw coal. The Eagle
Mountain side east of the river with Clode, Turnbull, Taylor and
Blackwood truck-shovel pits has produced 53% of the waste from
1971-1980 inclusive. The Greenhills truck-shovel pits on the west
side of the river, have produced 13% of the waste. The remaining
34% waste, including rehandle, was moved by a 46 m^3 dragline.

Ten mineable coal seams occur in the lower Cretaceous Coal Bearing
Member of the Kootenay Formation. The 450m thick sequence consists
of interbedded sandstone, siltstone, mudstone, shale and sub-
bituminous coal.

DETERMINING WASTE DUMP LOCATIONS

Placement of the large quantities of waste material removed to
meet coal targets is a major concern to mine planning, environmental
and production departments. Figure 2 outlines the areas which
cannot be readily used for developing waste dumps because they are
within resource areas, plantsite and facilities areas, areas of
steep slopes and environmentally sensitive areas. Note that dumps
are numbered on the Eagle Mountain side and lettered on the
Greenhills side.

Coal Reserves

It is desireable to locate waste dumps outside areas of
potential economic open pit coal reserves. The exceptions to this
are relatively small dumps which are planned to be rehandled at a
later date. No. 1 Spoil, Turnbull Spoils and K Spoil were
developed during early stages of mining within reserve areas.

Pit development has not sufficiently advanced to begin backfilling except for the dragline pits.

Plant Site and Facilities

This area occupies a major portion of the valley floor.
Included in the facilities are the tailings pond areas, coarse
reject storage area and the railway and road access. The plant
site area includes the washplant complex, offices, warehouse,
shops and dry.

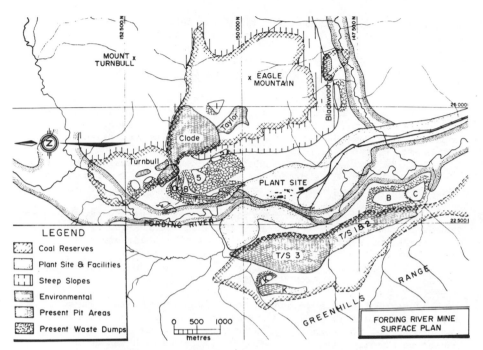

Figure 2. Plan Showing Pits and Dumps
Relative to Restricted Areas

Steep Slopes

   The valley walls in some areas, are too steep on which to develop
stable waste dumps. In general, a waste dump is not developed on a
foundation slope in excess of 24°. Some exceptions to this are in
areas where toe support is established first on shallow slopes or
where controlled development to reach shallower slopes is necessary
to minimize haulage distances.

Environmental

   The environmentally sensitive areas are associated with major
drainage patterns. Present planning allows for a 50 - 100m distance
between the 1,000 year flood plain limit and a 26° resloped dump
toe. This distance provides a long term wildlife corridor for
migrating animals in the valley.

Other Concerns

   Other factors considered when choosing a dump site are haulage
distances, haulroad grades, and drainage control.

Haulage distances are designed to be kept to a minimum. The hauls should be downhill or flat to maximize truck efficiency.

Seasonal drainage is presently diverted to larger creeks. However, future plans are to use coarse rock drains at crossing points. These will alleviate costly construction and maintenance of diversion ditches.

DESIGN CONSIDERATIONS

Parameters used in designing waste dumps include foundation conditions, strength characteristics of waste materials, height of the dump and the sequence of development. Pore water pressures within most piles are not considered. The piles are mostly free draining as the result of end dumping at the dump crest. This creates a naturally segregated, coarse, pervious layer of sandstone and siltstone boulders at the dump base. Long term stability is achieved by resloping the face of the piles to between 26° and 28° for reclamation purposes.

Foundation Conditions

Foundation conditions considered for a dump are natural slope angle, the slope profile and the strength characteristics of the foundation materials.

The natural slope angle and its' profile are determined using 1:1000 or 1:2000 scale sections spaced 50 or 100m apart. A foundation slope of 24° is now the accepted maximum on which dumps can be built. This is based on experience gained by dumping on steeper slopes. Controlled dumping on slopes greater than 24° was done in the early stages of Eagle Mountain mining. This was necessary to minimize haulage distances and to get the dump developed out to shallower slopes. Maintaining a stable dump was a problem during that stage of development.

The slope profile affects stability. Regular, concave slopes are the most desireable type. Irregular and convex slopes have resulted in unstable conditions being developed in the dumps.

Foundation material is checked by a site investigation which usually requires several test pits or sectional exposures. The sequence of material above bedrock is 0 - 10m of hard packed glacial till overlain by 0 - 7m of weak compressible organic soil and peat. The glacial till has a density of 2.18 kg/m^3. Samples of the till showed an average grain size content of 40% boulders and pebbles, 54% coarse to fine sand and 6% silt and clay.

Waste Dump Materials

Dump materials can be of three types: waste rock, overburden and rehandle.

Waste rock is estimated to comprise 80% of the total property waste (excluding dragline rehandle). The waste rock is a combination of 55% siltstone, 28% sandstone, 15% carbonaceous mudstone and shale and   2% non-recoverable coal. The quantities of each rock type delivered to a particular dump depend on what stratigraphic level a particular bench is at. This material has a dry density of 1.76 kg/m^3 and a friction angle of 37°.

The sandstone is typically coarse grained, durable, blocky and not significantly affected by weathering. The siltstone may grade to fine grained sandstone and can also be blocky depending on its sand content. Siltstone usually weathers well, although this is also a function of the sand content. The carbonaceous mudstone and shale is usually fine, friable and readily breaks down to fine particles when weathered.

Overburden is a mixture of glacial till and weathered rock. This material is not free draining and contains an average 15% moisture at a density of 1.72 kg/m^3. The undrained shear strength was determined to be 55.2 kPa with a 0° friction angle. Therefore, the height of a dump developed with this material is critical.

In Greenhills K-Spoil for example, this material, at a height of 24m, was found to slump from an initial face angle of 40° down to 26° over a period of several months. For planning purposes, the material must be kept to a minimum height and be contained.

Rehandle consists of a mixture estimated to be 85% waste rock and 15% overburden which has been subjected to weathering for 5 - 10 years. This material is not free draining but does drain significantly better than overburden. The friction angle is 33°. Samples of rehandle from Taylor Pit show a size distribution of 68% gravel, 27% sand and 5% silt and clay. It is interesting to note that after compaction testing, there is a shift in the size distribution between 15 and 20% to the finer side. This material has caused unstable conditions in No. 2 Spoil, which is 200m high, both at the time of dumping and many months later after it had been covered by waste rock. During initial dumping, the rehandle did not free-roll to the dump bottom. This created an oversteepened face of 40°. For planning purposes, this material should be contained and kept to a minimum height.

Dump Height

For waste rock, there is no restriction on the dump height. Where overburden is to be placed in an unconfined dump area, the maximum height designed is 25m. An estimated height of 60m is used for rehandle. This estimate is based on ½ of the free-rolling height of the rehandle observed on No. 2 Spoil.

Development of the Waste Dumps

Three main types of dumps developed at Fording are free, wrap-around and formed piles as illustrated in Figure 3. Dump areas are built using a combination of these types. The type of dump chosen depends on physical conditions of the site, the quantity of partic-ular materials expected from the pit and their delivery schedule.

Free dumping involves the placement of materials from a specific elevation in one lift only. These dumps are generally at least 90m high. The height of these dumps does not allow for the placement of significant quantities of overburden or rehandle. Free dumps on the Eagle Mountain side were established to accommodate short waste hauls from the higher elevations in Clode and Taylor Pits (see Figures 2 and 8).

Wrap-around dumps involve the construction of long dumps at successively lower elevations. They generally parallel the slope contours of the valley walls. This type of dump is used for road construction and containment of weak material. The lower portion of the wrap-around also provides toe support to the higher dumps. Re-sloping of these dumps for reclamation, requires the least amount of work. There are two ways in which the wrap-around method can be used to contain the weak materials as shown in Figure 4. The barrier type is used where there is sufficient quantity of waste rock available to precede delivery of the weak material. The waste rock toe type is used when the weak material is released before the waste rock is available. The waste rock wrap-around is built as soon as possible (within months) after the weak material is in place.

Formed dumps are developed by the deposition of waste in layers or lifts. This type of dump is used in the Greenhills pit areas to minimize uphill waste haulage (see Figure 9). This type of dump will accommodate overburden materials because of the relatively low lift height.

WASTE DUMP MONITORING AND CONTROL

Monitoring

The two types of monitoring are visual and measured. The visual examination is done by regular observation of the dump face for bulges and the dump surface for cracks. Significant bulges are those which interrupt the line of sight to the dump toe. Tension and shear cracks are common on all dumps. Large tension cracks parallel to the face, can be 30 to 60m long and up to 100m back from the face. These cracks do not appear to be related to dump

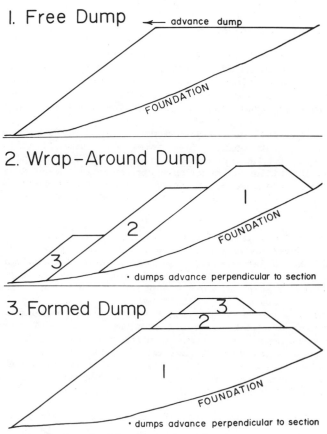

WASTE DUMP TYPES AT FORDING

I. Free Dump

← advance dump

FOUNDATION

2. Wrap-Around Dump

2

1

3

FOUNDATION

• dumps advance perpendicular to section

3. Formed Dump

3

2

1

FOUNDATION

• dumps advance perpendicular to section

Figure 3. Idealized Section Showing Dump Type
and Development Sequence

stability. They are believed to be a result of differential compaction due to increased dump height toward the crest. Shear cracks
are most common near the dump crest. These cracks indicate the
amount of crest movement which is directly related to dump stability.

The rate of movement at the dump crest is measured on a routine
basis using an extensometer, locally termed a "spoil-monitor." The
monitors consist of two stands with pulleys, a steel pin, a weight
and durable light wire. The set-up is illustrated in Figure 5.
Figure 6 shows a monitor set-up on No. 2 Spoil.

As the face of the pile settles, the relative displacement between the pin and weight is measured. The monitors are placed 30 to

CONTAINMENT METHODS AT FORDING

l. Waste Rock Barrier

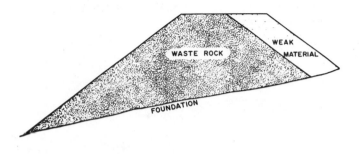

2. Waste Rock Toe Support

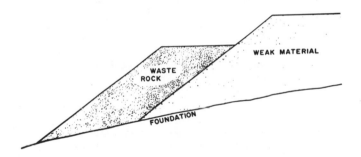

Figure 4. Sketch Showing Methods Used to
Contain Weak Material

100m beside the active face. Readings are taken once per shift
under normal circumstances. If the daily displacement increases
and exceeds 0.5m, then readings are taken more frequently.

The steepness of the foundation slope has an influence on
monitor rates. On No. 2 Spoil, crest movement decreased from 0.6 –
0.8m/day to less than 0.2m/day as the toe advanced over a 27° slope
to a 14° slope. Precipitation has not influenced the rate of
movement.

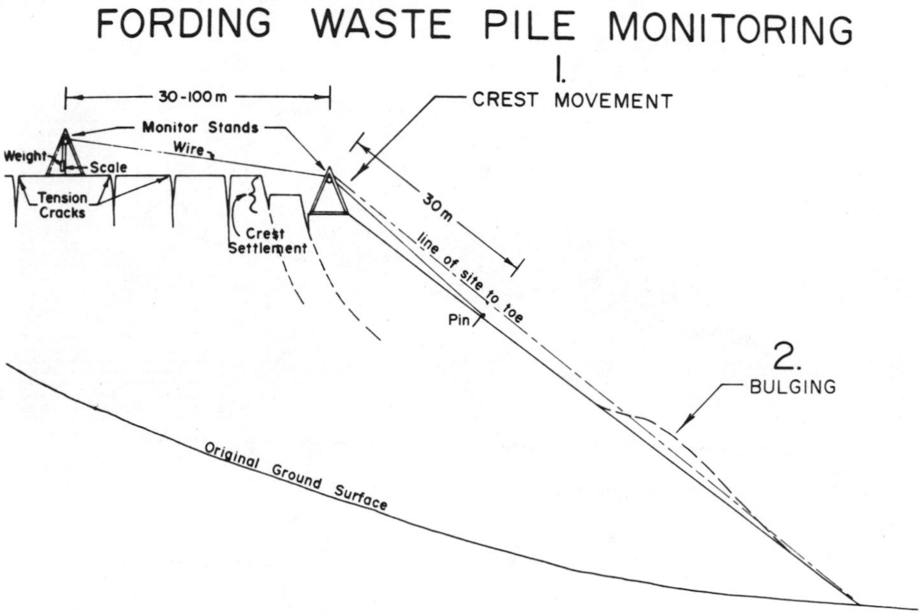

FORDING WASTE PILE MONITORING

Figure 5. Idealized Diagram Showing Monitoring
Done on Waste Dumps

Dump Control

Dump control is maintained by monitoring the crest movement and developing a long or alternate dump area. Most dumps are built with a 300m long face to enable development of several dump areas. When crest movement is accelerating and exceeds 0.5m/day, dumping in the area is temporarily stopped. This 0.5m/day limit is considered conservative. Rates of movement in excess of 1.5m/day have occurred on No. 2 Spoil without any subsequent mass failure.

Figure 7 shows a recent example of dump control on No. 2 Spoil. It is evident that shortly after times of "no dump", the crest movements decreased rapidly back to normal. The unstable conditions occurred directly as the result of rehandle being placed in the dump. Shortly after it had been built up, the rehandle area began to settle rapidly until it reached a stable condition. Once crest movements had decreased and stabilized, waste rock was dumped over it. The rehandle began to yield again after eight months. Rapidly increasing crest movements slowed back to normal rates after dumping was stopped.

Figure 7 also shows that precipitation does not significantly influence crest movement. There was also no effect when the average

Figure 6. Typical "Spoil Monitor" set-up. Note the
crack in the left foreground and the
crest settlement at the far stand.

rate of loading increased from 8,000 to 16,000 bcm/day.

REVIEW OF WASTE DUMPS AT FORDING

Generally, most dumps have been stable. Refer to Table 1 for
details on dump conditions and stability.

There are eleven distinct waste dumps presently developed. The
locations and number/letter of the dumps are shown in Figure 2.
Figure 8 shows the actual development of free and wrap-around dumps
on the west face of Eagle Mountain. Figure 9 shows the actual wrap-
around and formed combination of the largest Greenhills truck-
shovel dumps.

SUMMARY

The restrictions of building stable waste piles in mountainous
terrain can be successfully overcome. Careful planning is essential
to maximize designed stability during development of dumps. Monitoring and dump control has enabled Fording to maintain stable waste
piles on moderate to steeply dipping foundation slopes.

ACKNOWLEDGEMENTS

I would like to thank Fording Coal Limited for allowing me the
opportunity of presenting this paper. Many of the design features
and monitoring techniques were developed by Golder Associates and,

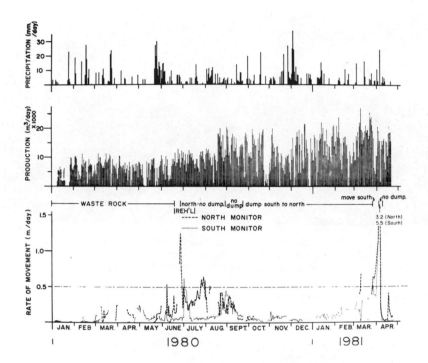

Figure 7. Chart showing rate of crest movement related to dump
activity, daily rate of waste production from Taylor
Pit and daily precipitation. Note that during
periods of "no dump", waste was hauled to an alternate
dump.

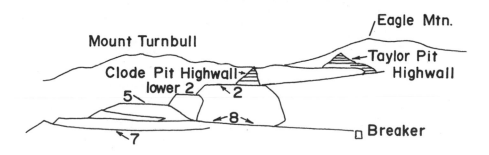

Figure 8. Panoramic view of Eagle Mountain pits and dump complex. This photograph was taken looking east from "K" Spoil. The height of No. 2 is approximately 200m.

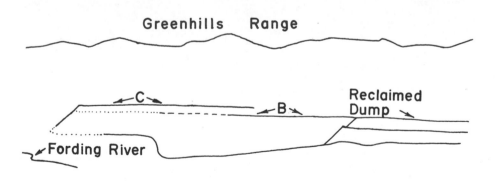

Figure 9. Photograph of "B" and "C" spoils in the Greenhills area.
Note the resloped and reclaimed dump in the right center.
At the left center is the wildlife corridor along the
Fording River.

Table 1: Waste Dumps at Fording

Dump	Features	Foundation Slopes	Dump Height	Dump Type	Review of Stability
EAGLE MOUNTAIN (see Figure 7)					
No. 1	1. Developed in a gulley 2. Contains initial Clode Pit waste which will be rehandled during Taylor Pit mining.	17° Max.	80M	Free	No Failures.
No. 2	1. Will ultimately contain 21.2 million bcm waste from Clode and Taylor Pits.	5°-31°	200M	Free	1. May, 1972. Two failures on foundation slopes of 25½° - 37½°. 2. July, 1974. A failure resulted from continued dumping despite abnormally high crest movement. 3. Nov. 1974. A failure resulted from change in slope profile from 14½° to 30°. 4. Nov. 1974. A failure resulted from yielding of a local area of weak organic foundation soils. 5. 1980-81. Dump control necessary to stabilize rehandle placed on the dump.
No. 5	1. Partially developed on slide debris from No. 2.	10°-20°	160m	Free	No failures.
No. 7	1. Lowest wrap in the Eagle Mtn. dump complex. 2. Used as coal haulage road for Turnbull.	5°	50m	Wrap-Around	No failures.
No. 8	1. Contains an estimated 15% overburden. 2. Used as coal haulage road for Clode and Taylor Pits.	5°	90m	Wrap-Around	No failures.
TURNBULL	1. These dumps have been resloped to 26 - 28° and reclaimed.	10°	50m	Free	No failures.
BLACKWOOD	1. Contains 2.3 million bcm waste rock.	7°-28°	60m	Free	1. Aug. 1980. A failure resulted from yielding of a local area of weak organic foundation soils.

Dump	Features	Foundation Slopes	Dump Height	Dump Type	Review of Stability
GREENHILLS RANGE (see Figure 8)					
"B"	1. Construction of a large waste rock "donut" around organic foundation material provided a place for the 20% overburden in this dump.	15° Max.	60m	Free + Wrap-Around	No failures.
	2. 40,700 bcm organic peat removed from the southeastern foundation area.	15° Max.	60m		
"C"	1. Overburden can be placed in the central portion of this dump.	0°	14m	Formed	No failures
"K"	1. Contains an estimated 60% overburden.	10°-12°	30m	Wrap-Around	No failures.
	2. The overburden is 24m high and is contained by a waste rock toe built 15m below the overburden elevation.				
DRAGLINE	1. There is no segregation of coarse rock to the bottom of the piles.	N/A	90m	N/A	1. Some failures occurred in the early stages of mining when saturated overburden was placed in piles on a 5° foundation slope.

Table 1 (cont.): Waste Dumps at Fording

in particular, D.B. Campbell. The engineering staff at Fording
were very helpful in critically reviewing this paper.

REFERENCES

Campbell, D.B., and Shaw, W.H., 1978 "Performance of a Waste Rock
    Dump on Moderate to Steeply Sloping Foundations", <u>Stability in
    Coal Mining</u>, proceedings of the first International Symposium
    on Stability in Coal Mining, Vancouver, B.C., April, 1978.

Question

Is dump movement occurring along the area of contact with the
foundation, in the foundation subsoils or confined to the waste
material.

Answer

The dump movement that we monitor occurs as a result of settlement
of the material within the piles near the crest. Movement of the
dump at the foundation only occurs in local areas where weak organic
soils greater than 0.3m thick have not been removed. Generally, there
is no dump movement along the foundation contact because the compacted till soil is stronger than the waste materials.

Question

In leaching dumps, time lags between water on-flow rates and
recovery rates at the dump toe may be several weeks. Have you considered this in your precipitation correlation.

Answer

No. The extreme case of this would be snowfall where the run-off or
"recovery" could be months after the precipitation. However, it has
been our experience that rainfall does not immediately (within one
week) affect the dump. It can affect the strength of organic foundation soils and cause them to fail.

Question

What possibility is there that with time, the readily degradable
shales will eventually settle to the lower portion of the dumps,
thusly reducing the permeability and increase dump weight due to
increased water.

Answer

The key points here are 'time' and 'slope'. We don't know how long it would take for this mechanism to occur if it will occur. Our experience is that No. 1 dump, which was developed over eight (8) years ago, still drains effectively at the toe. You may recall from my presentation, a slide which showed a large "bowl" above No. 1 dump. Snowfall accumulations in this bowl can be six metres per year. All the run-off from that area goes through No. 1 dump.

The second point is the foundation slope. Water flowing on a relatively steep slope will be able to carry a greater sediment load than on very shallow or flat slopes because of the hydraulic gradient. No. 1 dump has foundation slopes from 10° to 17°.

Question

Have you managed to correlate dump crest instability with rate of dump advance. At Bougainville, experience has indicated that 1 metre/ day is critical for this situation.

Answer

The rate of dump advance at Fording depends on the rate of loading and the dump height. We do not measure the rate of advance but use the "Spoil Monitors" to measure the rate of crest movement. The criteria that we consider representative of an unstable condition developing is when the monitors indicate an accelerating crest movement above a rate of 0.5 m/day.

As shown in Figure 7, the normal crest movement for our highest dump, No. 2 dump, does not change when the rate of loading is increased from 8,000 bcm/day to 16,000 bcm/day.

Question

Are the dumps located in a seismicly active area. If so, has this been incorporated in the design.

Answer

The property is not located in a seismicly active area and we do not incorporate this aspect into our design.

I should note that large blasts (approaching 500,000 Kg of explosives) in nearby Clode and Taylor Pits have not had any effect on dump stability.

Question

What preparations are made on natural ground surface prior to wasting on that surface.

Answer

Generally, where weak organic soils are in excess of 0.3m thick, they are removed. The glacial till below this is not removed because it is stronger than the waste materials placed on it.

Question

Did you establish a mathematical model to explain the type of failures you've got.

Answer

No. However, all failures have been a combination of circular arc type through the dump and plane failure along the foundation.

Question

It would appear that waste dumps at Fording are constructed by guidelines that are far from those that would be allowed by O.S.M. in the United States. Could you comment on the Provincial regulations that govern the stability of waste dumps in B.C.

Answer

Basically, four areas must be addressed when applying for Government approval to build dumps:

- a. A geotechnical assessment of the dump stability must be made for the development stage and final dump. At Fording, we use consultants to analyse the stability of a proposed dump and submit their report with the permit application.

- b. The effects on water quality are predicted. This work is usually done in-house at Fording.

- c. The land disturbance must be outlined. This involves assessing the ecological effect of building a waste dump in a certain location. This work is done in-house at Fording.

- d. A plan for reclamation is also included. The long term design must allow for resloping and revegetating the dumps.

Once the dump is completed, the following three things must be done as follow-up:

- a. The dump must be resloped to 26°.

- b. The 26° slopes must be revegetated to the satisfaction of the Minister of Energy, Mines & Petroleum Resources.

- c. Drainage in the area must be re-established to the satisfaction of the Minister of Environment

Chapter 35

EVALUATION OF SURFACE COAL MINE SPOIL PILE FAILURES

Peter M. Douglass and Michael J. Bailey

Hart-Crowser & Associates, Inc.
Seattle, Washington

INTRODUCTION

Spoil pile slope failures can have costly consequences. Instability in the form of a single major event or as a recurring
problem can mean lost production, lost resources and damaged
mining equipment. In extreme but not uncommon cases, a major
spoils slide can force the closure and abandonment of a pit.

Slope stability is affected by mine operating practices as
well as geologic conditions and material parameters. Mine
operators can enhance spoil pile stability through constructive
changes in their operations. Attention to the causes of instability can lead to development of operating practices
tailored to site specific problems. Action aimed at limiting
future instability is frequently more cost effective than continually "coping" with unstable spoils.

The effect of mine operations on material properties is
often overlooked because the "data" are not easily quantified
nor input directly into stability analysis equations. We have
found, however, that miners' experience with their own operation,
as well as data pooled from similar operations, yields relationships which can be utilized to solve stability problems. The
repetitive nature of area type surface coal mines with successive
spoil pile rows further enables the development of useful site
specific relationships. Stability problems can be mitigated
by applying these relationships .

815

STATEMENT OF THE PROBLEM

Cost control provides considerable incentive to improve spoil pile stability. Stability may sometimes be a concern because of the safety aspects and regulatory provisions, but costs are always a concern. Some reported mine experience suggests that over 95 percent of all slide related injuries result from highwall slope failures or rock falls, but almost 90 percent of all mine slope failures occur in the spoil piles.

Slope instability in spoil piles increases operating costs in several ways:

o   Slide stabilization and cleanup requires men and equipment that otherwise could be used on production or reclamation.

o   Mine production schedules are disrupted when slide debris blocks haul roads or damages power lines.

o   Exposed coal which is covered by a slide is frequently abandoned.

o   Coal left in place as a berm to provide stability between successive cuts is not recovered.

o   Development costs are lost if massive spoil failures force abandonment of part or all of a pit.

The frequency and severity of spoil pile failures, and resulting costs commonly control the level of effort expended in dealing with stability problems. Cost and effectiveness of mitigating actions are mine-site specific and no single approach is preferred. Minor changes in spoils handling may result in long-term cost benefits through avoidance of recurring stability problems.

Small or infrequent stability problems are commonly accepted as a reasonable risk in lieu of overconservative design or a costly program of preventative action. It should be recognized, however, that a minor slide could lead to larger consequences as when a small slide disrupts pit drainage and provokes massive instability. Similarly, in a tightly scheduled operation where "seconds per cycle" affect stripping costs, even elimination of nuisance slides may offer potential long-term cost savings.

In order to control spoil pile instability at an acceptable level of risk and cost, the following need to be understood:

o   What factors contribute to spoil instability?

o   How can the effects of these factors be isolated?

o   What is their relative importance?

The remainder of this paper addresses these concerns.

FACTORS THAT CONTRIBUTE TO SPOIL INSTABILITY

Forces tending to cause slope failure are a function of the weight and geometry of the spoil pile and the groundwater conditions. Factors tending to resist slope failure are shear strength of the spoil and underlying materials and, again, geometry.

Geometry of a spoil pile depends primarily on mine operations (equipment and procedures) and the characteristics of the spoil materials. Both can be affected by the mine operator to optimize spoil stability, within the constraints of existing site conditions and cost considerations.

Mine operations aspects cover both the method of stripping and the size of the excavation. Homogeneity of the spoils, that is uniformity of strength and drainage parameters, is influenced by whether the entire highwall is stripped in a single pass by a single piece of equipment or in layers with some combination of equipment. Draglines may provide more selectivity in mixing or segregating different overburden types than shovels, for instance, but much depends on the way in which the shovel or dragline is used.

The stripping equipment and the way it is used further influences the configuration of the spoil piles and the properties of the spoil material. For example, the bulking and the stacking angle (or angle of repose) of the spoil materials are partially controlled by the mine equipment and procedures employed.

Mine operations also include the ways in which groundwater and surface runoff are handled. Stability reducing forces result from water standing on the spoils, seepage, and excess pore water pressures associated with the underclay or the spoils.

Shear strength parameters affecting spoil stability depend first of all on the material itself, and secondly on the materials handling procedures. Material considerations are determined by the geology:

o Character and relative percentage of soil in the overburden

o Character of rock in the overburden

o Character of the underclay

In particular, the shear strength of spoil is related to its relative proportions of soil, slakable rock and sound rock. With this aspect in mind it is evident that the homogeneity or uniformity of spoils is of considerable interest in a slope stability analysis. Groundwater may cause changes in the nature of the spoil and underclay (eg, slaking, compactibility) which have an effect on shear

strength. Mine drainage thus affects both the forces causing instability as well as the ability of the spoil to resist shear.

The presence of particular materials in the overburden and the
materials handling operations should both be considered to exert a
significant influence on spoil pile stability. The relative importance of different contributing factors can be illustrated by examples.

CASE HISTORIES

Mine A: The first example is an area type surface coal mine in
southern Illinois. The mine has an annual production of almost
3,6000,000 tonnes (4,000,000 tons) from three pits.

The pit shown in Figure 1 is worked by a 107 cubic meter (140 yd^3)
stripping shovel in tandem with a 23 cubic meter (30 yd^3) pull-back
dragline. The overburden consists of about 21 meters (75 ft) of
clayey glacial till over about 12 meters (40 ft) of interbedded shales,
claystones and limestone. All production comes from a single seam,
the Illinois No. 6, which overlies a very slakable underclay.

During mining the shovel operator at Mine A tries to place blocky
rock near the base of the spoils to form a buckwall. Actually, quite
a bit of mixing of soil and rock occurs, resulting in a buckwall
ranging between 18 and 24 meters (60 and 80 ft) in height. More soil-
like or random spoils are placed in the upper part of the spoil piles
and then rehandled by the dragline which casts back two or three
spoil rows. Little water appears in the pit and what does collect is
pumped out over the highwall.

The spoils slope averaged about 39° with an oversteepened toe of
about 63° inclination. A large failure occurred with an apparent
circular slip surface, but it was not clear whether this surface was
confined to the spoils or passed through the underclay.

Spoil and underclay shear strength parameters were determined by
conventional laboratory tests and stability analyses performed to
determine the location of critical failure surfaces after the actual
failure. Analyses for both a toe failure and base failure resulted
in nearly coincident critical failure surfaces and a tension crack
located within about a meter and a half (5 ft) of the actual failure
scarp, as shown in Figure 1.

The analyses indicated that a base failure would likely have
occurred if the strength of the underclay was somewhat below its
maximum (peak) value. Failure through the underclay could have
resulted from disturbance or an increase in moisture content.

The lab tests indicate that the strength of the underclay declines from 27.5° at natural water content to about 23° when flooded, and further drops to about 16° when disturbed or remolded. Back-analysis of the actual failure, assuming a base failure through the underclay, would require a strength reduction to approximately 23°. For a failure surface to pass through the spoil pile toe but not the underclay, the cohesion of the spoil would have to be reduced about 12 percent below the measured data from 47.9 kPa (1000 psf) to 41.9 kPa (875 psf).

Since a difference of only 12 percent is within the expected natural variation of spoil shear strength values, it is not conclusive whether the underclay contributed materially to the slope failure. The analyses do indicate, however, the importance of reducing disturbance to the underclay and maintaining good pit drainage, especially where marginal conditions exist.

Remedial action that could be taken at Mine A to reduce the liklihood of a similar slope failure falls into two areas; maintenance of underclay strength and improvement of spoil strength. In the case of the underclay strength, it is clear that good drainage is essential. In general, such pit drainage was already being practiced at the mine, however, small surficial slides were common and frequently blocked drainage allowing flooding of the pit bottom.

A second factor, the effect of disturbance, is also important. Changes in the mine operation with respect to excavation and haulage of the coal could be made to minimize disturbance to the underclay by heavy equipment traffic. Coal trucks should be confined to maintained roadways on the underclay and special attention paid to the drainage of those roads. Changes might also be made in the excavation and loading of the coal to keep traffic to a minimum at the loading point. Blading off the surficial layer of softened or disturbed underclay immediately prior to placing new spoil could also be beneficial and accomplished at relatively minor cost.

Improvement of the spoil shear strength could be accomplished by more mixing of the highwall soil and rock to improve spoil uniformity. In this case it is not clear that the buckwall contributes materially to the stability of the spoil pile. Instead of constructing a buckwall with the rockier spoil, the shovel operator could make an effort to more uniformly stack the spoils, distributing the rock and improving the strength of the spoil mass throughout. As will be shown later, evidence at a number of surface coal mines in the central U.S. shows a good improvement in spoil strength as the relative percentage of sound rock increases.

Mine B: This is a two pit operation located in northern Missouri. The mine has an annual production of over one-half million tonnes (550,000 tons).

At the pit shown in Figure 2, a single fifty cubic meter (65 yd^3) shovel strips about eight meters (25 ft) of glacial till and about seventeen meters (55 ft) of shale interbedded with limestone and claystone. Two seams of coal, the Mulky and the Bevier, are re-covered. A third seam, only twenty centimeters (8 in) thick, is not recovered.

During mining the shovel travels on the upper Mulky Coal and ex-cavates about sixteen meters (55 ft) of overburden. This spoil is placed in the preceeding cut after the Bevier coal has been removed. The shovel then ramps down the parting to the lower Bevier Coal. The parting and some of the spoil placed next to it is then excavated and cast onto the top of the spoil pile.

Some seepage was observed through wet sand layers in the highwall, however most of the water in the pit results from precipitation and surface runoff. Pumping is required from a system of sumps cut into the underclay.

After a number of spoil slides, the operator attempted to stabilize an unstable spoil mass by leaving an in-place buckwall of undisturbed rock. This in-place buckwall was about fifty meters (160 ft) wide and included all of the parting as well as the unmined Bevier Coal. Spoil was placed above and behind this intact rock wall as shown in Figure 2.

When stripping of the Bevier Coal was resumed in the next pit, the underclay beneath the buckwall sheared through and the buckwall and spoil stacked on top of it moved into the excavated area. This resulted in a temporary closure of the pit.

This is an example of a sliding wedge or base type failure of the spoils. Similar wedge type failures occur through relatively weak underclay without an intact buckwall.

Spoil and underclay strength parameters were determined and wedge type stability analyses were carried out. The configuration of the failure wedge was varied during the analyses and the most critical slip surface was found to be nearly coincident with the surface of the previously placed spoil row behind the buckwall.

The analyses suggest that the formerly exposed surface of the spoil had a lower strength than the adjacent mass on either side. Strength reduction through an increase in absorbed moisture and slaking likely occurred, and is compatible with the observation of surface runoff into the pit.

A computerized limit equilibrium stability analysis, based on a method of slices, was used to conduct a sensitivity analysis. The sensitivity analysis was performed to determine the relative import-ance of the spoil and underclay shear strength parameters and the

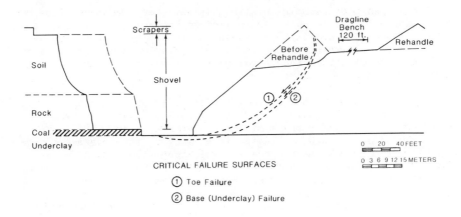

CRITICAL FAILURE SURFACES

① Toe Failure

② Base (Underclay) Failure

Figure 1. Profile of Highwall and Spoil Pile Showing Critical Failure Surface.

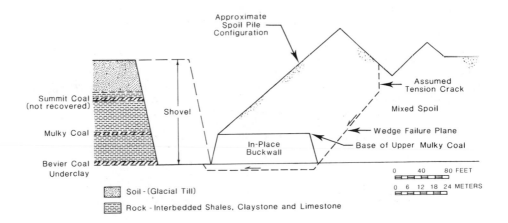

Figure 2. Profile of Highwall and Spoil Pile Showing Wedge Failure Under Buckwall.

spoil pile geometry associated with this failure. The conditions which were believed to exist just prior to the actual failure (i.e. factor of safety equals 1.0) were used as the initial starting point. Each of several parameters was individually varied over a reasonable range and the effect on the factor of safety was calculated. The results are shown in Figure 3.

It should be noted that a similar plot could be prepared for any type of slope failure using actual site conditions and observations to determine the likely range of the parameters at any particular mine. This form of plot is useful in evaluating the relative effect on stability of possible changes in spoil conditions. Such assessments of the relative importance of factors which control slope stability enables selection of the appropriate remedial action.

In the case shown, a unit change in the frictional strength of the underclay has a more pronounced affect on the factor of safety than a comparable change in the frictional strength of the spoils. Similarly, it can be seen that reasonable variation of the spoil pile height has a greater affect on stability than changes in the inclination of the spoil pile.

It should be noted that results of this particular analysis are not directly applicable to any other mine or to cases where two or more of the factors are changed simultaneously. For this mine and these conditions, however, the figure provides a clear picture of the relative importance of each parameter used in the stability analysis.

Results of analyses such as these enable a mine operator to focus remedial attention on those mine conditions which most directly influence the principal factors affecting spoil stability. In this instance, approximately a 10 percent increase in factor of safety would result from a change in any of the following:

o   Reduce the spoil pile height by about six meters (20 ft),

o   Flatten the average spoil slope from 47° to 39°,

o   Improve the frictional strength ($\emptyset$) of the underclay from 10° to 13°, or

o   Improve the frictional strength ($\emptyset$) of the spoil from 19° to 25°.

Presented in a manner such as this, a mine operator could compare the cost of another failure against the corresponding costs of changing his operations. Changes in stripping procedure, such as benching or a pull-back system could be used to reduce either the spoil pile height or inclination. A parameter such as frictional strength may be improved by changes in the handling of drainage or by some other

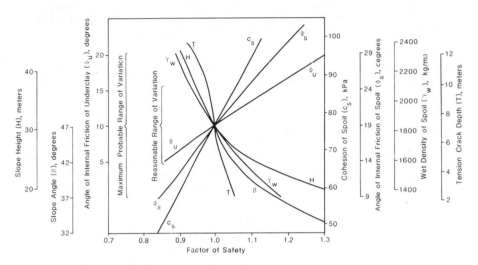

Note: To convert kg/m3 to pcf multiply by 0.0624, to convert kPa to psf
multiply by 20.9, and to convert m to ft multiply by 3.28.

Figure 3. Sensitivity Analysis of Wedge Type of Failure (Total
Strength Parameters).

indirect means. Observations of the relations between stability parameters and mine conditions can be used to evaluate the suitability of
attempting to change any particular parameter.

OBSERVED RELATIONS BETWEEN SPOIL PARAMETERS

Although conditions at every mine are somewhat different from every
other mine, general relations between stability parameters can
frequently be combined with on-site observations to improve understanding of a particular stability problem. These are not specifically cause and effect relations but are useful as empirical guidelines for stability improvement of surface coal mine spoils.

Observations at about two dozen pits in the central United States
indicate that both highwall height and the percentage of soil and
slakable rock in the highwall correlates reasonably well with spoil
instability. This is shown in Figure 4.

Miners frequently refer to a rule of thumb in this regard saying
that more than 40% soil in the overburden is a sign of probable spoil
instability. Disregarding minor sloughing, this is generally true in
our experience if we consider slakable and extensively weathered rock
in the same category as soil. Slakable rock is defined as rock with
a slake durability index of less than about 85, using the test method
developed by Franklin.

As Figure 4 shows, once the depth of excavation exceeds about
eighteen meters, (60 ft) pits with more than about 40% soil and
slakable rock frequently experience significant stability problems.
Between about 20% and 40%, the slope failures are less frequent or
less severe. Spoils that are predominantly sound rock (less than
about 20% soil and slakable rock) seem unlikely to experience spoil
pile instability, even in comparatively deep mines within the central
U.S.

The relative percentages of soil and rock in the overburden can
also be correlated with the results of shear strength tests on the
mine spoil, as shown in Figure 5. Although there is too much scatter
in the data for a good linear relationship, segregation of the test
results into three generic spoil groups (rock spoil, mixed spoil,
and soil spoil) results in clearly defined ranges in strength properties, as shown in Table 1.

Spoil shear strength parameters have also been found to roughly
correlate with the moisture and density of the spoils. Figure 6
shows the observed relation between the angle of internal friction
and water content of spoils at fifteen surface coal mines. An
apparent correlation also exists between cohesion of the spoils and
the in-situ unit weight. Figure 7 shows this relation using unit
weight generalized as a percent of the maximum unit weight of the

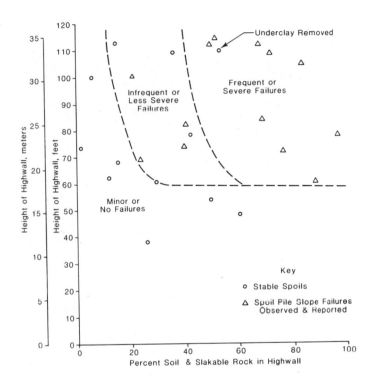

Figure 4. Effect of Highwall Height and Percentage of Soil and Slakable Rock in the Highwall on Spoil Pile Stability.

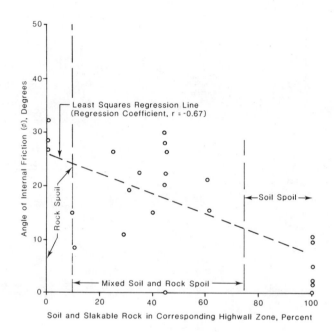

Figure 5. Angle of Internal Friction of Spoil Versus Soil and
Slakable Material in Corresponding Highwall Zone.

TABLE 1

ENGINEERING PROPERTIES OF GENERIC SPOIL TYPES

SPOIL TYPE	Natural Water Content w percent	Dry Density γ_D kg/m³	Angle of Internal Friction (Total) ϕ degrees	Cohesion (Total) c kPa	Plasticity Index PI
SOIL (Consists of 75% or more soil and slakable rock)	13 to 37 (21)	1380 to 1750 (1540)	0 to 10 (5)	9.6 to 95.8 (57.5)	12 to 30 (19)
No. of Tests	29	29	9	9	9
MIXED SOIL & ROCK	9 to 19 (13)	1310 to 2070 (1680)	4 to 30 (20)	19.1 to 134 (62.2)	14 to 32 (19)
No. of Tests	44	43	14	14	16
ROCK (Consists of 10% or less soil and slakable rock)	5 to 15 (9)	1440 to 1910 (1650)	27 to 32 (29)	28.7 to 47.9 (38.3)	14 to 20 (18)
No. of Tests	21	20	3	3	6

Note: Numbers in parenthese are median values for the range of data.
Median is defined as the value of a variable below and above which an equal number of variables fall.
To convert kg/m³ to pcf multiply by 0.0624, to convert kPa to psf multiply by 20.9.

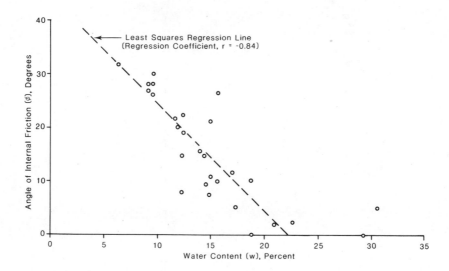

Figure 6. Angle of Internal Friction (Total Stress) Versus Water
Content for Spoil.

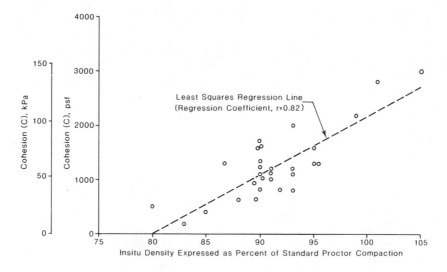

Figure 7. Cohesion Versus Percent Compaction for Spoil.

spoils achieved in the Standard Proctor compaction test.

Behavior of spoils can also be characterized by considering the relation between the Standard Proctor test optimum water content and the in-situ water content of the spoils. Figure 8 shows that predominantly soil spoils are characteristically wet of optimum in contrast to Figure 9 which shows that mixed soil and rock spoils are more nearly at their optimum water content or even on the dry side. This means that a higher degree of compaction is more readily obtained (with some constant degree of effort) for mixed soil and rock spoils than for predominantly soil spoils. Normal mine operations, thus generate a higher degree of compaction with mixed spoils and a higher shear resistance is the apparent result. In general, any attempt to increase the level of compactive effort during spoil placement is likely to be expensive and may involve a major change in stripping and placement methods. Therefore, it is worth noting that additional compactive effort will have little or no effect on wet soil spoils.

Another relation between overburden constituents and stability also pertains to the density of the spoils. Figure 10 shows the degree of bulking for spoils with different percentages of soil, slakable rock and sound rock. Such a relation would likely be further refined if it were exclusive to a particular type of stripping equipment.

While not specifically design or analysis tools, empirical relations such as these are quite useful in dealing with spoil instability problems. Understanding the characteristics of the spoil and how these might vary is essential to determining the relative importance of the various factors which affect stability.

SUMMARY

Observations at a number of area type surface coal mines in the United States indicate that relatively deep spoil pile failures are most problematic from a cost perspective. Identification of the principal factors contributing to a particular stability problem is necessary to develop the most cost effective remedial action.

Spoil pile failures are particularly a concern if any of the following results:

o Previously exposed coal is covered and unretrievable,

o An in-place buckwall, including unrecovered coal, is left to provide stability, or

o If spoil failures are a severe or frequently recurring problem.

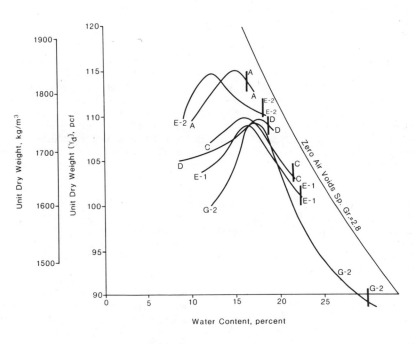

Figure 8. Compaction Curves for Soil Spoil.

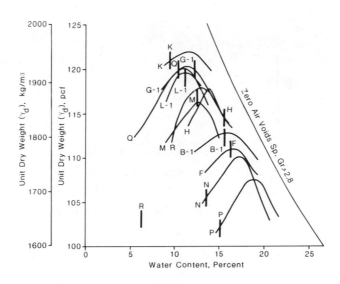

Figure 9. Compaction Curves for Mixed Soil and Rock Spoils.

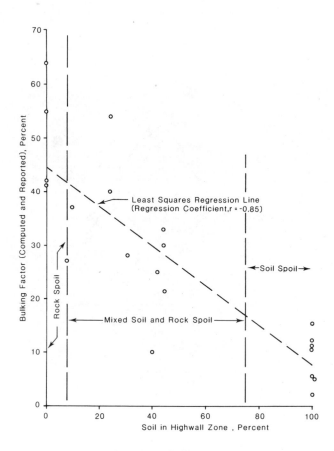

Figure 10. Bulking Factor Versus Percentage of Soil in Corresponding Highwall Zone.

In dealing with spoil pile slope failures the first objective is to isolate the factors which contribute to the problems and then determine their relative importance. To this end, empirical relations, back-analyses of previous slides, and sensitivity analyses can be used.

Stability investigations and remedial actions should be tailored to actual mine conditions and aimed at the factors which have the greatest apparent effect on the problem.

Use of the relations developed between overburden constituents and stability parameters may be applied to mine design as well. Planning pit and spoil pile dimensions and selection of stripping equipment should logically include consideration of spoil pile stability utilizing information from exploratory borings.

Slope stability analyses may or may not need to be elaborate, depending on the complexity of the actual problem. Various analysis techniques have been extensively described in the technical literature. Commonly, what is most needed is a link between mine conditions and the analysis rather than a particular analysis technique. Generally speaking, the chief constraint on selection of an analysis method is that the factor of safety should have an accuracy consistent with the available data and the level of risk considered acceptable. In this sense, for instance, a total stress analysis is likely to be more useful than effective stress analysis because of the difficulty and costs associated with determining reliable pore water pressure data for the latter.

An understanding of the relative effect of mine and geologic conditions on spoil stability enables a rational tradeoff to be made between alternative remedial actions. A partial solution to stability problems which can be economically achieved is likely more desirable than a costly solution to all of a mine's spoil stability problems. The most effective solutions are those tailored to the actual conditions at a particular mine.

ACKNOWLEDGEMENTS

Much of the work described herein was financially supported by the U.S. Bureau of Mines under contract No. J0275013. Mr. Peter Douglass was Technical Director on the project and employed by Shannon & Wilson, Inc. at the time. Mr. Michael Bailey was Technical Project Officer under the employ of the U.S. Bureau of Mines. The work was performed between September, 1977 and August, 1979 and the final report released in June, 1980. The authors wish to gratefully acknowledge the cooperation received from all the participating mining companies and other organizations.

Question

Do you have an explanation for the apparent sensitivity of the
friction angle to water content and how did you determine the
relationship between water content and friction angle.

Answer

The relationship between water content and friction angle was obtained by comparing a number of spoil parameters measured at different
mines and testing for correlation using a least squares regression
technique. The friction angle was determined from unconsolidated,
undrained triaxial strength tests. The tests included both undisturbed samples and samples recompacted at their natural moisture content
to the unit weight observed in the field. ASTM Test procedures were
used.

The effects of moisture on the shear strength of soil and rock
materials are complex and imperfectly understood. Concerning spoil
materials, the main factor appears likely to be the relative percentages of soil and rock in the spoil samples. Almost all the spoils
which were principally soil had water contents above fifteen percent
(consistently above optimum for compaction), rock spoils had water
contents below about nine percent, (at or below optimum moisture
content) and mixed soil and rock spoils had water contents between
nine percent and about eighteen percent, (generally within about 3
percent of optimum moisture content). We found that as the percentage
of rock in the spoil increases, so does the shear strength.

Other aspects of the effect of water content on shear strength
include the effect of excess pore pressures in saturated materials and
capillary tension in unsaturated materials, the effect of moisture
on density (considering a corresponding relationship between density
and shear strength) and other phenomena such as slaking, softening
and lubrication.

Question

How often is compaction of spoil performed. Is it economical.

Answer

Actually, it was not our intent to imply deliberate compaction of
spoils per se, but rather to variations in the unit weight of spoils,
and percentages of compaction relative to the maximum density.
Considerable variation has been observed in mine spoils, with dry
unit weights ranging from about 1040 kg/m^3 (65 lb/ft^3) to 2160 kg/m^3
(135 lb/ft^3) and compaction ranging from 80 to 105 percent.

These ranges depend on several characteristics which may or may not
be controllable by the mine operator, such as:
    1) Composition of the spoil (principally the relative amounts of
       soil and rock, and secondarily the type(s) of soil and rock).
    2) Moisture content.
    3) Method of spoil placement.

"Compaction" can be controlled by changes in equipment and/or for spoil handling procedures which influence such factors as fragmentation of the overburden rock, mixing of soil and rock spoil components, segregation and selective handling of particular strata, drainage affecting spoil water content, etc.

The authors are aware of some surface mining operations where the principal method of spoils handling is by scraper or truck and the equipment routing is controlled to maximize spoil compaction. At another mine, dragline operators are encouraged to dump spoils from the maximum possible height so as to increase the compactive effort and resulting spoil density.

The authors are not familiar with any cases where spoils in area-type surface mines are placed in lifts and compacted in the manner used for earth dam construction. Such an approach is described in spoil disposal regulations (i.e. for valley fills) adopted by the Office of Surface Mining, however these are considered more applicable to contour-type strip mining and to mountain-top removal mining than to area-type stripping.

Regarding the economics of "compacting" spoils, this depends on a number of mine-site specific factors. Frequently minor change(s) in mine operations which serve to increase the unit weight of the spoils can be accomplished cost effectively. The magnitude of the resulting unit weight changes and shear strength improvements and cost of achieving such changes, are questions which must be addressed on a case by case basis.

Question

Were your case studies based on spoil piles designed before current OSM regulations came into affect and how will the regulations affect your empirical techniques in terms of their usefulness.

Answer

Unlike spoil disposal for other types of surface coal mines (such as contour mining) and other aspects of area-type mining, regulations of the Office of Surface Mining do not prescribe the design or construction of spoil piles in area-type surface mines. The type of spoil piles described in the paper are constructed the same now as they were prior to PL 95-87.

Regardless of the limits of surface mine regulations, the approach described in the paper is one of observation and development of regional and site specific relationships based on observed behavior. The manner in which mining is carried out is fundamental, in that a change in mining practice affects the resulting spoil pile and the relationships which may be developed. The usefulness of the approach is unaffected by such changes but the specific relationships could change considerably.

Question

Of the 16 mines looked at how many were area strip types.

Answer

All of the mines referred to in the paper are "area-type" surface mines. The mines were distributed across six states and both over-burden conditions and mining practices varied somewhat. Principal overburden stripping was usually accomplished by dragline or shovel. In about a third of the cases, this primary equipment was working in tandem with a bucket-wheel excavator or a second dragline. Trucks and scrapers were used to move a substantial portion of overburden in three of the mines.

Chapter 36

SLOPE STABILITY IN RECLAIMED CONTOUR STRIPPING

G. Faulkner, C. Haycocks, M. Karmis and E. Topuz

Department of Mining and Minerals Engineering
Virginia Polytechnic Institute and State University
Blacksburg, Virginia

INTRODUCTION

The Appalachian coal region of Virginia, southern West Virginia and Kentucky and the extreme eastern portion of Tennessee constitutes a unique area of surface coal mining activities. The area contains a number of contiguously placed coal seams, up to a total of fourteen in some places, which vary in thickness from a few centimeters to 2,5 meters. Much of this coal is of high quality, some being the highest grade metallurgical coal found anywhere in the world. The region may be broadly characterized physiologically as a dissected plateau with a relief above the valley floors of up to 500 meters and a dendritic drainage pattern. Up to six seams may outcrop in the hillsides above drainage, many of which have been subjected to contour stripping operations.

Prior to the passage of the surface mining reclamation act, little effort was made to reclaim stripped areas. The passage of the strip mining act, including the controversial requirement for returning the land to its approximate original contour, has forced intensive reclamation efforts from the mining industry. The wisdom of this original contour requirement has been the subject of considerable debate and litigation, mainly because of its cost, inconvenience, and value of the land after it has been reclaimed. Regardless of whether reclamation is carried out, under current practices slope failures do occur. This in effect circumvents the intent of the law and can expose large unvegetated spoil areas, and contribute to hillside erosion and stream silting. Such failures must be eliminated if further costly reclamation efforts are to be avoided.

GEOLOGIC AND PHYSIOLOGIC ENVIRONMENT

The region which is the focus of this study is depicted in Figure
1. The topography in this area is relatively steep with some mining
operations conducted on slopes in excess of 30°. Annual rainfall is
between 85 and 130 cm. per year, and the high erosion rates have
limited soil thickness to less than 30 cm. in many locations (Skelly
and Loy, 1979). The highly dissected nature of the plateau in this
area means that most of the stripping operations are located on curved
benches either on the inside of valleys or outside of curves. This
phenomenon complicates both the mining and reclamation operations.
Geologically the rocks above and below the coal seams consist of
shales, sandstones and, occasionally, limestones in varying combinations and thicknesses. The area is very gently folded with dips
never exceeding 4°, and a series of synclinal and anticlinal structures may be recognized. Significant faulting is rare; however,
satellite imagery shows numerous linears existing throughout the region which exhibit no predominant directional trends. Such linear
zones with their concentrated joint patterns often comprise the areas
of lower relief. Jointing is predominantly at very high angle, much
being discontinuous and limited to individual strata layers only.
Localized low angle jointing and slickensiding may occur in the vicinity of sandstone channels, particularly where these intersect the
coalbeds. The sandstones and limestones of the region are reasonably
stable and where these are the predominant rock types excavated the
slopes stabilize well. Compared to the limestones and sandstones,
the shales characteristically degrade fairly easily forming amorphous
clays. These clays are the principal source of problems in the reclaimed slopes. Pyrite in the shales is not a particular problem in
most of this region, with the exception of the northern areas where
unreclaimed mines have given rise to some acid mine water run-off.

MINING METHODS AND LEGAL REQUIREMENTS

The initial phase of site preparation prior to commencing excavation is removal of timber and other vegetation from the site. At
this point erosion is prevented through a series of simple but effective methods such as ditches, water bars, straw dikes and riprap.
Drill benches are then cut and the overburden drilled using rotary
units. Blasting is done with ammonium nitrate and the broken rock
loaded into trucks with front-end loaders. Bulldozers may be used to
push the blasted material to the loader. This is conventional truck
haul back mining and the material is taken a short distance from the
face area and dumped. Coal is removed using a small front-end loader
and may also need to be drilled and blasted prior to loading. Augering the outcrop seam may also take place at this point, then as the
coal face is advanced the fill material is brought back along the
bench and replaced with surplus swell material being removed to
another site. Roadways are maintained along the bench elevation for
truck and other equipment access. Topsoil is placed over the returned
fill and the site is ready for seeding and final reclamation.

The post-1977 legal requirements on stripping are complex; briefly, however, the major requirements are stabilization of the whole mined area to control erosion and restore the land to its approximate contour and use that existed prior to mining. Minimum hydrologic damage must be done and permanent vegetation must be installed. The fill material must cover the highwall and the ground above the highwall must not be disturbed. On slopes greater than 20° no spoil must be dumped over the downslope side of the bench.

Multi-seam strip mining is possible in some areas, in which case the uppermost coal seam is normally excavated first and then successive beds are excavated progressively downward. However, the opposite sequence may be used on occasions (Curry, 1977).

FAILURE MECHANISMS

In examining slope failure mechanisms it is appropriate to divide the mining areas into those which have been reclaimed with backfilling to original contour and those which have not. Unfilled slopes were mined prior to the 1977 Surface Mining Act, and constitute the majority condition in the area under consideration at the present time.

Unfilled Benches

These benches are subject to three primary types of slope failure in the highwall (Fig. 2), all of which place failed spoil material onto the bench. Due to the original slope angles and bench width they do not directly place material onto the downslope fill of the bench and thereby endanger areas below the mining levels. These three types of failure are:

Ravelling. Ravelling constitutes a major problem in shale rocks and occurs excessively when shales are the predominant materials in the highwalls. Uninterrupted, the material will break down fairly rapidly until it reaches an angle of repose, which can vary from 32 to 37° depending upon the amount of sandstone present in the original shale layer. Large continuous calcite coated joint surfaces may inhibit degradation, but this is only temporary.

Toppling. Severe toppling can occur, particularly in isolated headlands, where a high density of jointing is encountered. This may occur in limestone, sandstone or shale; however, it is often accelerated in the shale materials due to degradation and weathering. Joint dips commonly vary from 90 to 77° and may be continuous across an entire lithologic sequence in the highwall.

Plane failures. The presence of plane failures in the highwall tends to be rather rare due to the normally vertical or near vertical nature of the joints. However, some localized high angle plane failures have occurred, resulting in the accumulation of considerable material on

the bench. Such failures normally follow periods of high rainfall
and are probably due to joint surface degradation and uplift
pressures.

The major effect of these failures is a cutting back on the highwall angle with significant degradation of the slope above the highwall, which is strictly prohibited by law. In addition, the dumping
of considerable material onto the bench can seriously restrict the
usage of the bench for agricultural or other purposes. Rotational
failures such as reported by Jaworski and Zook (1979) for tertiary
sandstone highwalls have not been observed.

Reclaimed (Original Contour) Benches

Reclaiming benches using fill material is now standard procedure
and final layouts with access roads are illustrated in Figure 3.
Once the bedrock material has been replaced, topsoil is laid over
this and forms a foundation for subsequent seeding and revegetating.
These reclaimed banks are subject to two types of failure mechanisms:

Face erosion and soil creep. Face erosion and soil creep cause a
progressive cutback of the reclaimed spoil bank and loss of topsoil.

Rotational failures. Rotational failures have occurred at a number
of sites through the fill material (Fig. 4). These comprise a serious
and very damaging phenomenon since they normally result in spoil material being dumped over the downslope side of the bench where it can
seriously damage structures below and perhaps find its way into the
local water system. The occurrence of these rotational failures is
due primarily to super-saturation of the bench material which normally results from poor drainage provisions. Water may enter the slope
from the highwall, from rainfall, or directly down from the upslope
side of the bench. The predominantly shale fill materials are particularly vulnerable to these type failures.

REMEDIAL METHODS

The maintenance of stable slopes in contour mining is basic to
both the mining and reclamation concepts. Since the strip benches
extend for many miles, site specific slope support or failure prevention methods are not practical unless they are low cost and can be
integrated with the mining or reclamation efforts. Proposed and
current practices for slope stabilization are as follows:

(1) To prevent face erosion and soil creep in back-to-contour
spoil banks various combinations of grasses are sown using hydro-
seeding immediately after reclamation. Trees, such as black locusts,
may also be planted to fully stabilize the area.

(2) Rotational failures in this fill are best prevented by superior drainage of the slope. Water which originates from direct

rainfall onto the slope as well as drainage from the upslope can be controlled and channeled off the slope using ditches. However, water may progress into the fill from joints in the old highwall, particularly in the sandstones, and some routine method of draining this fill such as French drains may be necessary under selected geologic conditions. Chimney drains, as shown in Figure 5, extending through to the bench elevation have proved very effective.

(3) Stabilizing the highwall in strip operations mined prior to the 1977 act, or for future operations when exceptions to the act are granted, presents a problem best taken care of during the mining operation. Cutting back of the highwall to a stable angle depending upon the lithologic sequence would be the most economical procedure. This would require further disturbing the upslope beyond the requirements for mining, but the cut back slope would probably more closely represent the final eroded surface. Surface stabilization using such methods as rock bolts and mine mesh might be used on a limited site by site basis.

The sandstone and limestone highwalls show good stabilization of dips less than 75°. However the shale rocks, due to the tendency to decompose into clays, must be cut back to less than 40° and provisions made for drainage as these slopes themselves can be ultimately subject to rotational failures.

CONCLUSIONS

Slope failures can and do occur along contour mining stripping operations. Rotational failure in the fill can be prevented with adequate drainage including, where necessary, in the fill itself. Highwall control is best made by cutting back during mining. Cut back angles will depend upon the lithologic sequence comprising the highwall. Progressive degradation of shales in the highwalls can provide the most serious long-term threat to stability due to their long-term tendency to degrade into clays. Fill materials that contain a high percentage of sandstone or limestone are particularly stable and are most desirable.

REFERENCES

Curry, J.A., 1977, "Surface Mining Coal on Steep Slopes: Back-to-Contour Demonstration," Fifth Symp. on Surface Mining and Reclamation, Louisville, KY.

Jaworski, W.E., and Zook, R.L., 1979, "Considerations in the Stability Analyses of Highwalls in Tertiary Rocks," Stability in Coal Mining, C.O. Brawner and I.P.F. Dorling, eds., Chap. 4, Miller Freeman Publications, Inc., San Francisco.

Nielsen, G.F., 1980, Keystone Coal Industry Manual, McGraw-Hill, New York.

Skelly and Loy, 1979, <u>Illustrated Surface Mining Methods</u>, McGraw-Hill, New York.

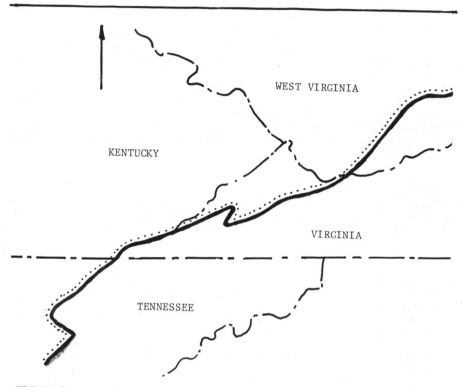

FIGURE 1. Appalachian coal region considered in this study. The area on the north, or dotted, side of the thick solid line bisecting the region contains coal, while the area to the south contains no coal.

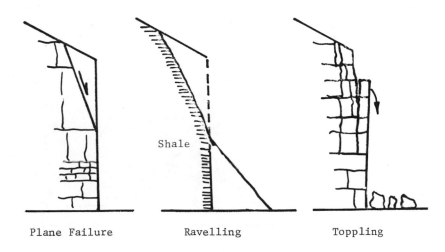

FIGURE 2.   Three possible types of highwall failure.

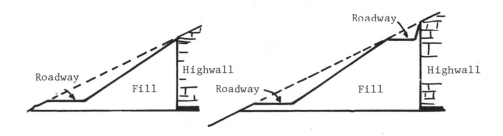

FIGURE 3.   Reclaimed strip benches showing optional road layouts.

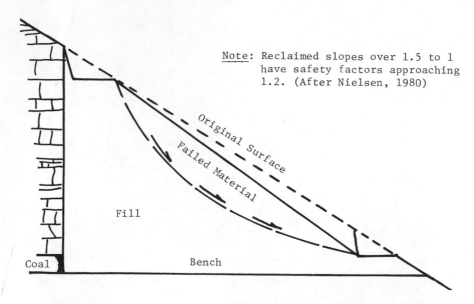

Note: Reclaimed slopes over 1.5 to 1
have safety factors approaching
1.2. (After Nielsen, 1980)

FIGURE 4.   Example of rotational failure in reclaimed fill material.

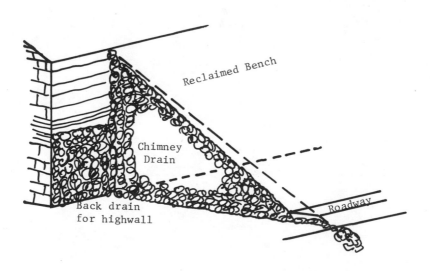

FIGURE 5.   Chimney drains as used for draining reclaimed strip
benches.

Question

You indicated rotational failures are a common problem for A.O.C. reclaimed areas, have you observed wedged type failures in the same areas.

Answer

No wedge type failures have been observed in the reclaimed material. This is not surprising since this material constitutes a relatively homogenous soil/rock mass. No wedge type failures originating in the highwall material itself have been observed after A.O.C. reclamation. This is probably due to the presence of the reclaimed material, which provides sufficient toe support to inhibit any tendency towards such a failure.

Question

Did rotational failures pass through or above the under-clay.

Answer

The observed failures passed only through the reclaimed material itself and above the under-clay which occasionally underlies the coal.

Question

What effects of augering on highwall stability have been observed.

Answer

No highwall failures have been observed that could be directly related to augering. This appears to be mainly due to the spacing selected for the auger holes which has insured stability of the pillars between holes. Some localized rock loosening may occur during the actural augering operation due to vibrations, but significant slope failures are apparently very rare and were not observed.

Question

How much highwall failure is attributable to over-steepening slopes as opposed to actual mining of coal by augering or other excavation techniques which under-mine or under-cut slopes.

Answer

The only highwall failures observed were probably due to over-steepening slopes as opposed to mining activities. This is hard to prove as such, but no particular concentration of highwall failures was observed in areas that had been augered. It may wll be that further studies will bring this point out.

Question

What are typical slope angles for filled slopes in A.O.C. reclaimed areas.

Answer

Actual slope angles vary with the pre-mining or virgin slope angle
which may be as high as 30 degrees. A reclaimed slope angle may be
slightly in excess of pre-mining slope angles due to cut for
roadway.

Question

What is a reasonable cut-off for stability and/or successful
vegetation.

Answer

Again, it is difficult to give a simple figure. Natural slopes in the
area in excess of 45 degrees are successfully vegetated. However,
as noted at the beginning of the paper, the average soil thickness in
the region is less than 12 inches, indicating a high degree of
superficial erosion. At the higher reclaimed slope angles, of say
33 degrees, vegetation can be successfully established, particularly
with the use of trees in addition to grasses and other vegetation.
Likewise, stability can be assured at the higher angles provided
adequate care is taken to insure proper slope drainage and to
avoid the buildup of ground water pressures.

Chapter 37

THE IMPACT OF THE FEDERAL SURFACE MINING CONTROL AND

RECLAMATION ACT UPON STABILITY DESIGNS FOR SURFACE COAL MINES

Robert W. Thompson and Darry A. Ferguson

President, CTL/Thompson, Inc.
Denver, Colorado

Vice President, Mineral Resources and Engineering, Inc.
Golden, Colorado

ABSTRACT

In August, 1977 the United States Congress passed Public Law
95-87 which has been commonly referred to as the Surface Mining
Control and Reclamation Act (or SMCRA). This title is somewhat
misleading in that the act creates as many regulatory requirements for underground coal mines as for surface coal mines. However, this paper will concentrate on the effects of SMCRA on
designs and stability for various activities of surface coal
mines, specifically those concerning "Disposal of Excess Spoil".

In the past, the objective of most laws and regulations applicable to the mining industry has been to achieve goals of safety
or environmental concerns for all types of mining. However, SMCRA
is unique in that it sought to regulate all activities of coal
mining from initial data gathering, exploration, mine design and
facilities design to reclamation activities and design.

More specific to the stability of the activities of surface
coal mining, special criteria for design and construction performance standards have been promulgated for the construction and maintenance of roads, dams and embankments, coal processing waste, excess spoil piles, backfilled reclamation and pit slides. The criteria for design for the previously mentioned facilities has been
enumerated by regulations and in many ways restrict the use of site
specific design parameters. The restrictions imposed by regulations
significantly inhibit the engineer's ability to develop the
best practical and economical design and in some cases may re-

sult in a design less stable than a site-specific engineered
design. This paper addresses some of the problems created by the
SMCRA regulations for achieving stability in certain structures
in surface coal mines.

INTRODUCTION

The U.S. Congress passed the Surface Mining Control and
Reclamation Act (SMCRA) in August, 1977 which was a law designed
to set standards of reclamation for coal mining activities, both
surface and underground.

Most laws and regulations passed by Congress have been developed in response to a particular protection which permeates most industries such as water quality (The Federal Water Pollution Control
Act, As Amended), air quality (The Clean Air Act of 1970, As
Amended), endangered species, (The Endangered Species Act of 1973),
and noise (The Noise Control Act of 1972). Other laws have been
enacted to cover hazardous wastes, safety, historic preservation
and safe water. This SMCRA legislation was unique in that it
sought to regulate all activities of surface coal mining utilizing
"cradle-to-grave" regulatory concepts to ensure proper reclamation
was achieved.

SMCRA's cradle-to-grave objectives started with baseline environmental data gathering, geology, hydrology, premining plans of
operations, details of operations, and reclamation, all of which
were combined into a permit application. The permit application
must also demonstrate how the operator planned to comply with
the "performance standards", which are regulatory standards for
the day to day operations.

In its efforts to totally regulate all phases and activities
of the mine operation, the newly created Office of Surface Mining
(OSM) promulgated regulations which set forth design criteria for
certain activities, such as road design, spoil embankments, refuse
piles and dams. OSM established rigid approaches to design
which in many cases have proven to be less effective, less safe
and/or less desirable than site-specific engineered designs. This
paper will examine the general design criteria established by OSM
for disposal of excess spoil and examine alternative considerations for site-specific design.

It should be noted that the Office of Surface Mining is
currently repromulgating many of its regulations in response to
a change in administration. The original regulations were
developed under the Carter administration. The present Reagan
administration believes prudent environmental objectives and
safeguards can be accomplished with less detail in the regulations,
thus allowing more flexibility for operator designs and state

programs. The reader should be aware that, as of this date, revised regulations may modify OSM's design criteria discussed herein. However, the everyday administrator may continue to use the now established design criteria as a guideline. Therefore, the mine planner or permitter should be conscious of the use of the site-specific design methods.

OSM DESIGN CRITERIA

As indicated in the title, this paper sets forth discussions of the impact of the Federal Surface Mining Control and Reclamation Act on slope stability investigations and specifically, with the requirements imposed by various regulations on disposal of excess spoil waste and how these regulations impact the design.

OSM has established the general requirements for analysis under 30 CFR, Part 780 which presents the requirements for permit application. Under Part 780.35, titled "Disposal of Excess Spoil", various items are required in the permit. The spoil disposal area must be designed to meet the specific performance requirements presented in Parts 816.71 through 816.74. Under Part 780.35 the permit application is required to contain a geotechnical investigation which describes the character of the bedrock, geologic conditions, potential seepage areas, any areas of known subsidence or potential subsidence due to future mining, a technical description of the materials to be incorporated in the structure, and specifically a stability analysis which includes the strength parameters, pore pressures and long term safety conditions anticipated in the design. The engineer is required to present all design assumptions, calculations, alternatives considered, and specific design specifications.

From the general requirements cited above, specific design criteria are presented under Paragraph 816.71 through 816.74. These sections deal with general requirements and specific design requirements for fills defined as "Valley Fills", "Head of Hollow Fills", and "Durable Rock Fills". Fills that are not classified fall under the "general requirements". All fills, irrespective of classification, must have a static factor of safety of 1.5. Fills that are classified as either "Valley Fills" or "Head of Hollow Fills" have additional requirements regarding drainage and internal drain systems. Classification of a spoil disposal fill as a Durable Rock Fill results in some relaxation of the rigid design requirements.

In order to be considered a Durable Rock Fill, 80 percent of the materials contained within the disposal area must be sandstone, limestone or other rocks that do not breakdown with time as determined by the Slake Durability Test. There are many specific requirements within the regulations which essentially define the configuration of any spoil disposal facility, such as terraces,

6 m   (20 ft.) wide, spaced at 15 m   (50 ft.) vertically to meet
the conditions imposed to limit erosion over the face of the embankments. Furthermore, the slopes between benches are limited
to a maximum of two horizontal to one vertical.

The slope of the top of a spoil disposal area is specified to
not exceed five percent and must slope away from the face of the
disposal site. With the regulations, as they presently exist, the
geotechnical engineer is forced into a basic configuration on the
face of the disposal site. The only variable factor is the slope
between benches. The maximum allowed slope is 2:1 and variations
from this configuration require considerable theoretical analysis
and negotiations at various levels in order to obtain approval of
the design.

The stability analysis usually results in an evaluation of the
foundation conditions to determine if the maximum slope permitted
by the regulations can be constructed, and secondly, involves
evaluation of the configuration for a specific site. Factors
which influence any stability analysis include; 1) the strength
of the materials incorporated within the disposal facility, 2)
the strength of foundation materials, 3) ground water conditions
within the foundations, 4) possible future seepage conditions
within the disposal embankment, 5) geometry of the embankment and
foundation, 6) potential seismic hazard, and 7) possible site
conditions such as existing or future underground mining considerations.

The general configuration or geometry of the slope, in most
cases, starts out as a maximum slope permitted by the regulations.
Effectively, this results in a 2:1 slope between horizontal benches
located approximately at each 15 m   (50 ft.) of vertical height of
the embankment. The existing ground surface geometry is normally
defined initially by site surveys and topography, and the geotechnical investigation will normally be conducted to evaluate the
various strength parameters of the foundation materials. An
essential part of the stability analysis is evaluation of design
parameters to be used to represent the strength, settlement, and
other physical characteristics of the materials in the spoil pile.

To illustrate these problems, this paper will discuss CTL/
Thompson's evaluation of these parameters for two spoil disposal
facilities located in Colorado. One of the facilities is an unclassified fill, constructed as part of the Kerr Coal Company's
mine near Walden, Colorado. The other facility is an excess spoil
pile designed as a Valley Fill located in the Eckman Park portion
of the Energy Fuel's mine near Steamboat Springs, Colorado. Both
fills occurs in areas which were developed as extensions of existing
mines and evaluation of the physical characteristics of the spoil
to be placed in these fills was considerably easier than an

evaluation based on drill cores from an investigation for a new
site.

CASE I - ENERGY FUELS' ECKMAN PARK SITE

At the Eckman Park spoil disposal site, the mine is located
in the Williams Fork Formation, which is a part of the Upper Cretaceous Mesa Verde Group. The Williams Fork Formation includes a
fine to very fine-grained sandstone interbedded with thin layers
of siltstone and shale. In addition, there are massive marine
shale members within the formation. The shale series are predominantly claystone with some siltstones and are usually a dark
gray to gray in color.

Some of the sandstones within this formation are sufficiently
durable to be considered as a Durable Rock Fill. The predominant
rock within the Williams Fork Group however, degrades sufficiently
upon weathering to not permit design as a Durable Rock Fill. The
upper surfaces of the formation materials have weathered in-place
to essentially a very stiff clay, the thickness of which ranged
from 0 to 3 m (0 to 10 ft.) in the areas of the site where the
proposed disposal facility was to be located. At the base of the
disposal area the valley had as much as 6 m (20 ft.) of soft to
medium stiff residual clays and topsoils.

In order to evaluate the materials that would be placed in the
overburden disposal site, samples of the overburden materials from
the active mine were obtained. These samples were selected based
on a review of cored borings drilled in the new mine area for reserve evaluation. Sandstones and shales which overlie the coal
were exposed in existing cuts and were available for evaluation of
physical characteristics from the existing mining operation. For
the site evaluation, these materials were divided into a series of
four typical materials. For an evaluation of a bulking factor to
estimate volume change from the in-place formation to the disposal
site, bulk specific gravity tests were performed. The typical
samples are described below.

Sample	Description	Specific Gravity	Unit Weight kg/m³	(pcf)
I	Dark gray, massive shale	2.39	2390	(149)
II	Light tan to gray sandstone	2.13	2130	(133)
III	Banded gray shale	2.29	2290	(143)
IV	Dark gray shale	2.35	2350	(147)

To provide an estimate of the amount of bulking and to evaluate
the approximate in-place density, a number of field density tests
in the active mining area were made with two types of tests. A
large sand cone density test was made using a plate with a

305 mm (12 in.) diameter. At all locations where sand tests were made, tests were performed with a Troxler moisture density gauge which had a probe extending 305 mm (12 in.) below the gauge. Multiple readings were made with the nuclear test gauge and the gauge was rotated at the test locations to develop average densities at each test location. Other tests were attempted in loose spoil piles which had been deposited by a dragline operation. Where the material had been deposited by the dragline, the moist, unit weight measured using large diameter sand cones ranged from 1490 to 1710 kg/m^3 (93 to 107 lb. per cu. ft.). After reviewing the conditions in the field, it was decided that the nuclear density tests may be a more reliable measurement.

In addition to tests located in the loose spoil areas, density tests were made in areas containing lightly compacted spoil. In these areas, the moist unit weight ranged from 1810 to 1940 kg/m^3 (113 to 121 lbs. per cu. ft.). Generally, the lower moist weights were associated with more clayey type rocks and higher densities were measured where the sandstones were exposed. For comparative purposes, several density tests were made in haulroads. In these areas, the moist density was generally in the range of 220 kg/m^3 (126 lbs. per cu. ft.) with a moisture content ranging from 10 to 13 percent.

Based on these observations, it was estimated that the average moist density of the spoil would be on the order of 1860 kg/m^3 (116 lbs. per cu. ft.). This would result in a bulking factor of approximately 23 percent. The experience at the mine indicated 26 to 30 percent bulking, however, these factors were estimated on a basis of the dragline operation. A bulking factor on the order of 25 percent appears to be reasonable where the materials are laid in the lifts of four feet as specified by the regulations.

In addition to the field density tests, the average angle of repose was measured at numerous locations at the existing mine. The angle of repose ranged from 34 to 40 degrees for the loose, dumped spoil. For purposes of the design, a value of 37 degrees was used. The Canadian Pit Slope Manual reports typical values for the angle of internal friction for shale to run on the order of 34 degrees and sandstones to range from 35 to 45 degrees.

In order to model the strength characteristics of the material two large samples were obtained at the mine and a gradation test was performed on each of these samples, each of which weighed about 181 kg (400 lbs.). Based on the observations and data from fill utilized for haulroads and other uses at the active mine area, CTL/Thompson, Inc. estimated that about 40 percent of the materials would be 305 mm (12 in.) or larger, after compaction. Thus, the performance of the fill would be influenced primarily by the minus 305 mm (12 in.) fraction of the spoil. Therefore,

gradation curves were made on minus 305 mm (12 in.) material. Remolded samples were constructed with gradation curves similar to
the field gradation with coarse particles passing the number 4
sieve in an attempt to model the field condition. Shear tests on
these samples indicated peak shear strengths ranging from 30 to 37
degrees.

The slope was analyzed using three methods. A computer search
routine using Spencer's method as developed by Steve Wright at the
University of Texas was used to locate critical circles for various
strength assumptions. Critical circles were checked using a graphical method of slices with the hand checking method. In addition
to these methods a sliding wedge analysis was made and a specified
shear surface analysis was made using one of the options in
Wright's program.

The compacted spoil was assumed to have an angle of internal
friction of 37 degrees and a moist unit weight of 1840 kg/m^3 (115
lbs. per cu. ft.). All failures occurred in the residual clay
soil until artificial water conditions were imposed to sufficient
elevations to force failure within the spoil. The spoil was assumed to have unit weight of 1840 kg/m^3 (115 lbs. per cu. ft.), no
cohesion and an angle of internal friction of 37 degrees. The
upper, weathered materials were modeled with a strength of $\emptyset$ = 15
degrees, and C = 95.8 kPa (2,000 psf) based on the results of laboratory testing. The underlying near-surface bedrock was assigned
a $\emptyset$ = 10 degrees and C = 144 kPa (3,000 psf). These values correspond to the more weathered phases which normally occurred at the
surface of the shale. For drained conditions, the minimum factor
of safety was computed to be 1.8. With full saturation the factor
of safety was computed to be 1.4. The likelihood of saturation
was considered to be low because of the moisture regime and the
drainage to be proivded within the fill.

Subsequent to analysis by CTL/Thompson, the Energy Fuels' disposal facility was reviewed and approved by OSM for construction.
Construction of the facility commenced in the summer of 1980.
During the construction of this fill, density tests were taken on-
site to check the in-place densities of the fill "as constructed"
as compared with the assumptions made during design. Fill was
constructed in approximately 1.2 m (4.0 ft.) thick layers, with
the exception of a buttress area located near the toe of the
slope. In the buttress area the fill was laid down in maximum
lift thicknesses of 0.6 m (2 ft.) and compacted to very high densities. For the buttress fill area, sandstone was used to construct the buttress and the densities were higher than indicated
by previous calculations. In the areas of fill constructed using
1.2+m (4.0+ ft.) thick lifts several field density tests were
made and indicated values ranging from 1760 to 1920 kg/m^3 (110 to
120 lbs. per cu. ft.), moist. This is the range that was

anticipated by the design. The Energy Fuels' disposal site and structure, appears to be performing satisfactorily and in accordance with the design predictions.

The general design criteria for the Eckman Park disposal site conformed with the rigid approach established by the SMCRA regulations, although some minimal variations in design planning were accepted by the regulatory agency. It should be noted that the same facility could have been constructed under different engineered configurations providing an equivalent factor of safety and the same environmental protection.

CASE II - KERR COAL COMPANY'S DISPOSAL STRUCTURE

The unclassified three million cubic yard fill constructed at the Kerr Coal Mine, near Walden, Colorado, is unusual because of its method of construction which was totally by Caterpillar 657-pp scrapers. This particular mine is located in a relatively flat area with very steep dipping beds and the mining sequence follows along the strike of a syncline. The mining operation is such that a substantial volume of material was generated during the initial cut. During the steady state phase of operation, material is removed from the active mine area and used to place in the previous mined area. The entire mine is located in shales, which are relatively soft when compared with the highwall formed by a relatively hard sandstone. All of the materials are late-Cretaceous in age. The overburden pile was constructed from the fall of 1978 to June of 1980.

The pile was constructed in a relatively flat area which slopes slightly to the east. The area was located a few hundred feet from the active mine area. Scrapers from the mine area were routed from the pit onto the stockpile and the overburden was laid down in relatively thin lifts. Based on field observation, the layer thickness ranged from 0.3 m to 0.9 m (1 to 3 ft.). The material was compacted by the haul traffic over the individual layers as additional layers were placed. These layers were periodically leveled using large Caterpillar D-9H dozers and motor graders. CTL/Thompson was contacted to evaluate the density and stability of this structure after it had been constructed. This structure was not designed and constructed under the rigid OSM specifications. The pile was slightly over 30.5 m (100 ft.) high and in some areas there were no benches. In other areas there were haulroads (See Figure 1). This structure offered the opportunity of evaluating the strength of the slope in the materials after a spoil structure had been completed.

The company was interested in the density of the pile in order to evaluate how much swell was occurring from the pit to the pile to allow for estimates of the need for additional spoil disposal

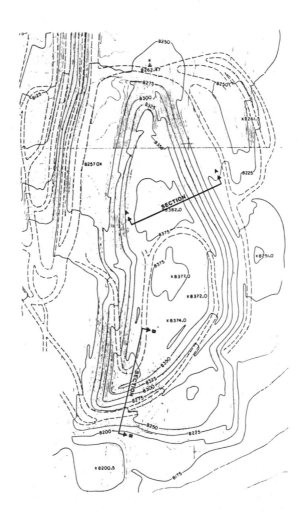

FIGURE 1. STOCKPILE TOPOGRAPHY

facilities. A large number of field density tests were made at
this site. The tests were conducted at 36 locations and at 18 of
the locations with both sand and nuclear density tests being made.
Three different materials were identified as the most typical material in any given test. The materials were identified as brown
claystone, gray claystone, and a friable sandstone. Results of
the density tests are summarized in Tables 1, 2 and 3. This data
is of particular interest because of the scarcity of published and
documented data.

Density tests were performed using a large diameter plate in
the sand tests and the nuclear tests were made by rotating the
gauge at least three times at each test location to get an average
result. Procedure for the nuclear tests consisted of inserting
the probe to a depth of approximately six inches, making a reading,
rotating the gauge 90 degrees, making additional readings, and
rotating the gauge again 90 degrees to get a minimum of three readings at each depth at each location. Test locations were selected
at various locations on the spoil pile in order to get an average
of the density of the spoil pile. Additional analyses were performed on all samples from the density tests, then with this data
an adjusted laboratory gradation was prepared for remolded samples
for shear strength tests. The shear tests were made in a three-
inch diameter direct shear box on samples which had an adjusted
gradation to resemble the field material. Direct shear tests
were conducted at two moisture conditions. The tests were conducted at field moisture content and the second series of tests
was conducted under a flooded condition. A summary of the results of these tests is presented on Figure 2.

The stability for this pile was analyzed using the Modified
Bishop method of analysis. A computer search routine with a circular search procedure was used to conduct the analysis. The computer analysis was checked using a graphic method of slices procedure. For the portions of pile where no benches were constructed the lowest calculated factor of safety was 1.2. In areas
where the benches were constructed for haulroads, the indicated
factor of safety was 1.6 for dry conditions and 1.5 for the assumed moist condition.

CONCLUSIONS

1) The Federal Surface Mining Action and consequent regulations impose many limitations on the planner when pre-
paring designs for mine operations requiring disposal of
excess spoil. The geometry of the slope is, to a large
degree, limited by the various bench requirements and maximum slopes which are specified by the regulations. Wherever
possible, it appears most advantageous to construct the disposal facilities as unclassified fills. At most mine

TABLE I. ANALYSIS OF DENSITY TEST RESULTS

GROUP I - BROWN CLAYSTONE

		TEST METHOD						
		NUCLEAR			SAND (TOTAL)			
		WET DENSITY	WATER CONTENT	DRY DENSITY	WET DENSITY	WATER CONTENT	DRY DENSITY	
TEST NO.	DEPTH	kg/m³ (pcf)		kg/m³ (pcf)	kg/m³ (pcf)		kg/m³ (pcf)	
6	SURF	1910 (119)	23.5	1540 (96)	1840 (115)	16.6	1590 (99)
8	0.9m (3')	2000 (125)	14.9	1750 (109)	1810 (113)	10.1	1650 (103)
10	SURF	1950 (122)	23.2	1590 (99)	1950 (122)	14.5	1710 (107)
14	1.2m (4')	1840 (115)	21.5	1520 (95)	1680 (105)	16.4	1440 (90)
31	SURF	1870 (117)	14.8	1630 (102)	-- --	--	-- --
32	SURF	1890 (118)	19.0	1590 (99)	-- --	--	-- --
33	SURF	1990 (124)	13.1	1780 (111)	-- --	--	-- --
34	SURF	1920 (120)	15.0	1680 (105)	-- --	--	-- --
AVERAGE		1920 (120)	18.1	1630 (102)	1830 (114)	14.4	1600 (100)

AVERAGE USING BOTH TEST METHODS

WET DENSITY kg/m³ (pcf)	WATER CONTENT	DRY DENSITY kg/m³ (pcf)
1870 (117)	16.3	1620 (101)

TABLE 2. ANALYSIS OF DENSITY TEST RESULTS

GROUP II - GRAY CLAYSTONE

		TEST METHOD						
		NUCLEAR			SAND (TOTAL			
		WET DENSITY	WATER CONTENT	DRY DENSITY	WET DENSITY	WATER CONTENT	DRY DENSITY	
TEST NO.	DEPTH	kg/m³ (pcf)		kg/m³ (pcf)	kg/m³ (pcf)		kg/m³ (pcf)	
2	1.2m (4')	1840 (115)	12.1	1630 (102)	1760 (110)	12.0 (est)	1570 (98)
3	1.4m (4.5')	2020 (126)	12.3	1790 (112)	1810 (113)	9.3	1650 (103)
4	1.4m (4.5')	2000 (125)	11.7	1780 (111)	1950 (122)	9.5	1780 (111)
5	SURF	2070 (129)	19.9	1730 (108)	1870 (117)	18.8	1590 (99)
7	0.9m (3')	1870 (117)	15.6	1620 (101)	1860 (116)	9.2	1700 (106)
9	1.2m (4')	1990 (124)	16.4	1710 (107)	1790 (112)	11.8	1600 (100)
17	1.2m (4')	1890 (118)	18.5	1600 (100)	1790 (112)	11.7	1600 (100)
18	1.2m (4')	1860 (116)	13.6	1630 (102)	1680 (105)	8.0	1550 (97)
19	SURF	1970 (123)	14.9	1710 (107)	--- ---	---	--- ---
20	SURF	1990 (124)	16.9	1700 (106)	--- ---	---	--- ---
22	SURF	2000 (125)	14.6	1750 (109)	--- ---	---	--- ---
23	SURF	2100 (131)	11.1	1890 (118)	--- ---	---	--- ---
24	2.4m (8')	1760 (110)	18.7	1490 (93)	--- ---	---	--- ---
25	4.7m (15.5')	1750 (109)	13.4	1540 (96)	--- ---	---	--- ---
26	SURF	1940 (121)	12.9	1710 (107)	--- ---	---	--- ---
27	SURF	1910 (119)	13.0	1680 (105)	--- ---	---	--- ---
29	SURF	1920 (120)	11.5	1730 (108)	--- ---	---	--- ---
35	SURF	1910 (119)	22.5	1550 (97)	--- ---	---	--- ---
AVERAGE		1940 (121)	15.0	1680 (105)	1810 (113)	11.3	1630 (102)

AVERAGE USING BOTH TEST METHODS

WET DENSITY kg/m³ (pcf)	WATER CONTENT	DRY DENSITY kg/m³ (pcf)
1870 (117)	13.2	1670 (104)

TABLE 3. ANALYSIS OF DENSITY TEST RESULTS

GROUP III - FRIABLE SANDSTONE

		TEST METHOD								
		NUCLEAR					SAND (TOTAL)			
TEST NO.	DEPTH	WET DENSITY kg/m³ (pcf)		WATER CONTENT	DRY DENSITY kg/m³ (pcf)		WET DENSITY kg/m³ (pcf)		WATER CONTENT	DRY DENSITY kg/m³ (pcf)	
1	SURF	2030	(127)	7.9	1890	(118)	1990	(124)	7.8	1840	(115)
15	1.2m (4')	1940	(121)	16.5	1670	(104)	1620	(101)	9.2	1490	(93)
21	SURF	2100	(131)	13.0	1710	(107)	---	---	---	---	---
28	SURF	2100	(131)	9.8	1910	(119)	---	---	---	---	---
30	SURF	1920	(120)	10.5	1750	(109)	---	---	---	---	---
AVERAGE		1990	(124)	11.5	1780	(111)	1810	(113)	8.5	1670	(104)

AVERAGE USING BOTH TEST METHODS

WET DENSITY kg/m³ (pcf)		WATER CONTENT	DRY DENSITY kg/m³ (pcf)	
1910	(119)	10.0	1730	(108)

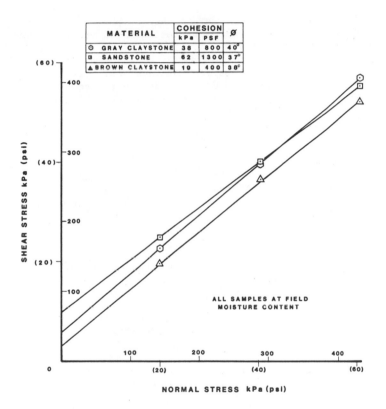

FIGURE 2. DIRECT SHEAR TEST RESULTS

locations, the surrounding topography is sufficiently steep to fall within the categories that require the fill to be classified either as a "Valley Fill" or "Head of Hollow Fill". Stability analyses of these structures are most significantly influenced by the Engineer's estimate of the physical characteristics of the material to be incorporated in the fill structure and the foundation soils or bedrock. These parameters are the only ones not rigidly specified by the existing regulations.

2) There are very little data presently published regarding the measured physical characteristics of materials incorporated in these piles. It appears from CTL/Thompson's experience, that numerous economies could be achieved by allowing site-specific designs for each spoil structure. The rigid design approached by OSM does not necessarily create more stable and environmetally acceptable structures than site-specific designs.

ACKNOWLEDGEMENT

Considerable information in this paper was derived from studies conducted for Energy Fuels Corporation at their Eckman Park Mine No. 1 and the Kerr Coal Company's mine near Walden, Colorado. The authors wish to thank these companies for their permission to make site-specific data available.

REFERENCES

_____, 1977, "Engineering and Design Manual - Coal Refuse Disposal Facilities", Prepared for Mine Safety and Health Administration by D'Appolonia Consulting Engineers.

_____, 1977, "Surface Mining Reclamation and Enforcement Provisions, Department of Interior, IN Federal Register, December 13, 1977.

_____, 1979, "Surface Coal Mining and Reclamation Operations - Permanent Regulatory Program", Department of Interior, IN Federal Register, March 13, 1979.

Registrants List

John P. Ashby
Golder Associates
10628 N.E. 38th Place
Kirkland, WA 98033

Luis Vicente Alfaro
Planning Engineer
CIA. Minera Disputada
  de las Condes
Las Urbinas 53 - 10th Floor
Santiago, Chile

Lars Alm
Mine Superintendent
LKAB
S-980 20 Svappavaara, Sweden

John D. Austin
Klohn, Leonoff Ltd.
10180 Shellbridge Way
Richmond, BC V6X 2W7 Canada

Michael J. Bailey
Hart Crowser & Assoc. Inc.
1910 Fairview Ave. East
Seattle, WA 98102

T. Balakrishnan
Staff Engineer
Iron Ore Co. of Canada
P. O. Box 1000
Labrador City, Nfld., A2V 2L8,
  Canada

Jim Balmer
Geotechnical Engineer
Gibraltar Mines Ltd.
Box 130
McLeese Lake, BC VOL 1PO
  Canada

George Barker
Geological Technician
Gibraltar Mines Ltd.
Box 130
McLeese Lake BC VOL 1PO,
  Canada

N. Barton
Terra Tek Inc.
420 Wakara Way
Salt Lake City, UT 84108

Driss Benhima
Engineer
Khouribga Open Pit
Mining Division
OCP - Khouribga, Morocco

Brad Bjornson
B. C. Coal Ltd.
Box 2000
Sparwood, BC VOB 2G0, Canada

G. H. Blackwell
Brenda Mines Ltd.
Box 420
Peachland, BC VOH 1XO, Canada

W. D. Blenkhorn
Sr. Min Engineer
Utah Mines Ltd.
Box 370
Port Hardy, BC VON 2PO, Canada

Paul Bluekamp
Geological Engineer
Cleveland-Cliffs
504 Spruce Street
Ishpeming, MI 49849

Larry Bolivar
Senior Geologist
International Minerals
  & Chemicals Corp.
P. O. Box 867
Bartow, FL 33830

R. Bone
B. C. Min. of Energy, Mines
  & Petrol. Res.
Rm. 105 - 525 Superior St.
Victoria, BC V8V 1T7, Canada

C. O. Brawner
P. O. Box 91651
West Vancouver, BC V7V 3P3,
  Canada

A. Brown
Golder Associates Inc.
12345 W. Alameda Parkway
Lakewood, CO 80228

Paul Buckley
Senior Geologist
Placer Developments Ltd.
Endako Mines Div.
Endako, BC VOJ 1L0, Canada

Jim Burt
Staff Mining Engineer
Cyprus Industrial Minerals
P. O. Box 516
Ennis, MT 59729

P. N. Calder
Queen's University
Dept. of Geol. Science
Kingston, Ont. K7L 3N6, Canada

R. Call
Call & Nichols Inc.
6420 East Broadway
Tucson, AZ 85710

John Catchpole
Exxon Minerals Co.
P. O. Box 2180
Houston, TX 77001

Samuel S. M. Chan
Professor of Mining Engineering
College of Mines
University of Idaho
Moscow, ID 83843

Gerald H. Clark
Chief Geologist
Empresa De. Cobre Cerro
   Colorado S. A.
Apartado 6 3497 El Dorado
Panama City, Rep. of Panama

Peter Collins
Manager - Mine Engineering
Iron Ore Co. of Canada
P. O. Box 1000
Labrador City, Nfld, A2V 2L8,
   Canada

Andy Cooney
Geologist
Cleveland-Cliffs
504 Spruce St.
Ishpeming, MI 49849

Martial Cote
Supervisor of Mine Engineering
Quebec Cartier Mining Co.
Fermont
Duplessis, Que. G0G 1J0, Canada

V. E. Dawson
Ministry of Energy, Mines
   & Petrol. Res. (B.C.)
Rm. 105 525 Superior St.
Victoria, BC V8V 1T7, Canada

Eric A. De Boor
Shell South Africa Ltd.
P. O. Box 31769
Johannesburg 2000, Rep.
   So. Africa

C. Dinis da Gama
DMGA - I.P.T.
P. O. Box 7141
Sao Paulo, Brazil

Pierre Desautels
Assistant Pit Geologist
Placer Developments Ltd.
Endako Mines Div.
Endako, BC V0J 1L0, Canada

Adriano E. De Souza
Av. Graca Aranha, 26 - 18 Floor
20005 Rio De Janeiro, Brazil

H. J. Dennis
B. C. Min. of Energy, Mines
   & Petrol. Res.
Rm. 105 - 525 Superior St.
Victoria, BC V8V 1T7, Canada

Ernesto Diaz
Planning Engineer
CIA. Minera Disputada De
   Las Condes, S. A.
Las Urbina 53, 10th Floor
Santiago, Chile

Real Doucet
Qit-Fer Et Titane Inc.
P. O. Box 160
Havre St.-Pierre, Que. G0G 1PO,
   Canada

Peter Douglass
Hart, Crowser & Associates
1910 Fairview Ave. East
Seattle, WA 98102

B. M. Dudas
B. C. Ministry of Energy, Mines
   & Petrol. Res.
Rm. 103, 2747 E. Hastings St.
Vancouver, BC V5K 1Z8 Canada

Martin Dzidrums
Manager of Engineering
Westroc Industries, Ltd.
2650 Lakeshore Highway
Mississauga, Ont. L5J 1K4,
　Canada

Wade N. Ellett
Mine Manager
Freeport Gold Co.
Mountain City Star Route
Elko, NV 89801

S. Erickson
Mining Engineer
Anamax Mining Co.
P. O. Box 127
Sahuarita, AZ 84629

Fosco V. Facca
Geologist
Griffith Mine
Pickands Mather & Co.
Red Lake, Ont. POV 2MO, Canada

Alan Fair
Senior Geotechnical Engineer
Syncrude Canada Ltd.
Bag 4009
Fort McMurray, Alta. T9H 3L1,
　Canada

Robert Falletta
Mine Engineer
Phelphs Dodge Corp.
Morenci, AZ 85540

Dave Farnsworth
Mine Production Supervisor
Monsanto Co.
P. O. Box 816
Soda Springs, ID 83276

G. Faulkner
Mining & Minerals Engineering
Virginia Tech
213 Holden Hall
Blacksburg, VA 24061

David Fawcett
Senior Project Engineer
Coal Mining Research Centre
11111 - 87 Ave. Suite #202
Edmonton, Alta. T6G 0X9,
　Canada

Frank Fenwick
Production Mgr.
Rossing Uranium Ltd.
P. O. Box 22391
Windhoek, South-West Africa 9100

Ron Ferrel
Mining Technical Supervisor
Monsanto Co.
P. O. Box 816
Soda Springs, ID 83276

T. Fright
Sr. Mining Engineer
Utah Mines Ltd.
Box 370
Port Hardy, BC V0N 2P0, Canada

Ray Frost
BHP Civil Engineer Mgr.
Broken Hill Prop. Co. Ltd.
North Sydney, NSW,
　2060 Australia

D. M. Galbraith
B. C. Min. of Energy, Mines
　& Petrol. Res.
Rm 105 - 525 Superior St.
Victoria, BC V8V 1T7, Canada

Om P. Garg
Chief, Long Range Planning
Iron Ore Co. Canada
P. O. Box 1000
Labrador City, Nfld. A2V 2L8,
　Canada

Charles Glass
Dept. of Mining & Geological
　Engineering
University of Arizona
Tucson, AZ 85723

Richard Goodman
Dept. of Civil Engineering
University of California
Berkeley, CA 94720

H. S. Gopinath
Addl. Chief Engineer
Western Coalfields Ltd.
Coal Estate
Nagpur, India

Garry B. Gould
Mine Engineer
Gregg River Resources Ltd.
P. O. Box 2880
Calgary, Alta. T2P 2M7, Canada

H. Graham
South African Iron & Steel
    Industrial Corp Ltd.
Private Bag
Sishen, Cape Prov.,
    South Africa 8445

Gordon Grams
Mine Engineer
Gibraltor Mines Ltd.
Box 130
McLeese Lake, BC V0L 1PO,
    Canada

J. Grimes
Manager, Mining
South African Iron & Steel
    Industrial Corp. Ltd.
Private Bag
Sishen, Cape Prov.,
    South Africa 8445

Tim Hagan
Principal Blasting Consultant
Golder Associates Pty. Ltd.
466 Malvern Rd.
Prahran, Vict. 3181, Australia

Louis W. Hamm
Geological Engineer
Fred C. Hart Assoc. Ltd.
1320 17th St.
Denver, CO 80202

Jerry Hanford
Senior Geotechnical Eng.
Syncrude Canada Ltd.
Mildred Lake, Bag 4009
Fort McMurray, Alta. T9H 3L1,
    Canada

Ted Hannah
Senior Geologist
Crows Nest Resources Ltd.
P. O. Box 2003
Sparwood, BC V0B 2G0, Canada

G. Harries
ICI Australia Operations Pty. Ltd.
1 Nicholson St.
Melbourne, Australia

P. Mark Hawley
D. R. Piteau & Associates Ltd.
408 Kapilano 100
S. Park Royal
West Vancouver, BC V7T 1A2,
    Canada

D. I. R. Henderson
B. C. Min. of Energy, Mines
    & Petrol Res.
Rm. 105 - 525 Superior St.
Victoria, BC V8V 1T7, Canada

Dr. Herget
Mining Research Laboratories
P. O. Box 100
Elliot Lake, Ont. P5A 2J6, Canada

Francois E. Heuze
Lawrence Livermore National
    Laboratory
L201
Livermore, CA 94550

James Hodos
Placer Service Corp.
12050 Charles Dr., Ste A
Grass Valley, CA 95945

Evert Hoek
Golder Associates
224 West 8th Ave.
Vancouver, BC, Canada

Roger Holmberg
SveDeFo
Box 32058
12611 Stockholm, Sweden

D. S. Holmes
Development Engineer
State Electric Commission of Vict.
15 William St.
P. O. 2765 Y GPO
Melbourne, Australia

I. Holubec
Vice President
Geocon (1975) Ltd.
909 - 5th Ave. S. W.
Calgary, Alta. T2P 3G5, Canada

Henry Hong
Slope Stability Engineer
Lornex Mining Co. Ltd.
P. O. 1500
Logan Lake, BC V0K 1W0 Canada

Stan Hoyle
Safety Engineering Consultant
Government of Alberta
2nd Floor, Oxbridge Place
9820 - 106 St.
Edmonton, Alta. P5K 2J6, Canada

S. J. Hunter
B. C. Min. of Energy, Mines &
    Petrol. Res.
Rm. 105 - 525 Superior St.
Victoria, BC V8V 1T7, Canada

Sall Ibrahima
Snim Sem
Service 820
Zouerate, Mauritania

Abdeslam El Ibrahimi
Engineer
Khouribga Open Pit
Mining Division
OCP-Khouribga, Morocco

Haruo Inoue
Geologist
Idemitsu Kosan Co. Ltd.
1200 - 840 - 7th Ave. S. W.
Calgary, Alta., Canada

Roosevelt Isaac
Senior Geotechnical Engineer
Syncrude Canada Ltd.
Bag 4009
Fort McMurray, Alta. T9H 3L1,
    Canada

Wayne D. Jackson
Mining Engineer
Amax Exploration Inc.
1707 Cole Blvd.
Golden, CO 80401

Jacobs
Rock Mechanics Engineering
Gecamines
56 Rue Royale
Bruxelles, Belgium

Fernando E. Jaramillo
Mining Engineer - Cerrejon
    Intercor
Carrera 7a No. 37069 Piso 8
Bogota, Columbia

Vincente M. Jayme Jr.
Resident Manager
CDCP Mining Corp.
2nd Floor, Alco Bldg.
391 Buendia Ave. Ext.
Makati, Metro Manila, Phillipines

Klaus W. John
Professor
Ruhr University Bochum
Im Haarmannsbusch 114A
D-4630 Bochum, West Germany

Gordon Jorgenson
Manager Consulting Services
Cil. Inc., Explosives Div.
P. O. Box 10
Montreal, Que. H3C 2R3, Canada

Tahkaoja Kalevi
Mine Engineer
Outokumpu Oy
P1 8
94101 Kem 10, Finland

S. S. Kantak,
Director Technical
M/s. Fomento
P. B. No 31
Margao, Goa, India 403601

Victor L. Kastner
Geological Engineer
Inspiration Consolidated
    Copper Co.
P. O. Box 4444
Claypool, AZ 85532

Hiroshi Kawakami
Manager, Mining Section
Overseas Mineral Resources
    Development
Sabah Bhd Mamut Mine
P. O. Box No. 5
Ranau, Sabah, Malaysia

Francis S. Kendorski
Engineers International Inc.
5107 Chase Ave.
Downers Grove, IL 60515

Ross W. Kenway
Director of Mining
The UMA Group Ltd.
#1700 -1600 West Hastings St.
Vancouver, BC V6E 3X2, Canada

A. Komar
P. T. Timah (Persero)
J. L. Gatot Subroto
Jakarta, Indonesia

Rinus Kotter
Mine Manager
Gecamines Kakanda
56 Rue Royalle,
1000 Bruxelles, Belgium

Gerry Kulbieda
Operating Engineer
Cleveland-Cliffs Iron Co.
Empire Iron Mining Partnership
Palmer, MI 49871

Lutz Kunze
Manager, Geoteck Eng. Dept.
Pincock, Allen & Holt Inc.
4370 S. Fremont Ave.
Tucson, AZ 85714

Fernand Lambert
Mine Engineer
Amok/Cluff Mining
825 - 45th St. West
Saskatoon, Sask. S7K 3X5,
   Canada

Pekka Lappalainen
Mining Engineer
Outokumpu Oy
Mining Technology Group
83500 Outokumpu, Finland

Ron Lau
Project Engineer
EBA Engineering Consultants Ltd.
14535 - 118 Ave.
Edmonton, Alta. T5L 2M7,
   Canada

John Leahy
Asst. Supt. Open Pit
Climax Molybdenum Co.
Amax Inc.
Climax, CO 80429

John Leighton
Graduate Student University of
   British Columbia
1594 E. Lorne St.
Kamloops, BC V2C 1X6, Canada

Yannick Le Mailloux
SOGEREM
Rue Frederic Joliot B. P. 24
13762 Les Milles Cedex, France

R. W. Lewis
B. C. Ministry of Mines, Energy
   & Petrol. Resources
Rm 105 - 525 Superior St.
Victoria, BC V8V 1T7, Canada

Tim Liutman
Soil Sampling Service Inc.
5815 North Meridian
Puyallup, WA 98371

Burke Lokey
Project Engineer
Fred C. Hart Assoc. Inc.
1320 17th St.
Denver, CO 80202

Trevor Lumb
Crippen Consultants Ltd.
1605 Hamilton Ave.
North Vancouver, BC, Canada

Tracy Lyman
Stone & Webster Engineering
   Corp.
Box 5406
Denver, CO 80217

J. P. MacCulloch
B. C. Ministry of Energy, Mines
   & Petrol. Resources
Rm 105 - 525 Superior St.
Victoria, BC V8V 1T7, Canada

Greg MacMaster
Mine Engineer
Cassiar Resources
General Delivery
Cassiar, BC, Canada

Eric Magni
Mgr., Engineering Geol. &
   Rock Eng.
Trow Ltd.
P. O. Box 430, Station B
Hamilton, Ont. L8L 7W2, Canada

Michael M. Maier
Exxon Minerals Co.
P. O. Box 2180
Houston, TX 77001

David Malcolm
Chief Engineer
McIntyre Mines Ltd.
Box 2000
Gran Cache, Alta. T0E 0X0,
   Canada

C. W. Mallett
CSIRO
P. O. Box 54
Mount Waverley, Vic., 2149
   Australia

Dennis C. Martin
Associate
D. R. Piteau & Associates Ltd.
408 - 100 S. Park Royal
West Vancouver, BC V7T 1A2,
   Canada

R. T. Martin
B. C. Ministry of Energy, Mines
   & Petrol. Resources
Rm 105 - 525 Superior St.
Victoria, BC V8V 1T7, Canada

Rick McCosh
Group Leader Earth Sciences
Techman Engineering Ltd.
840 - 7th Ave. S. W.,
Calgary, Alta. T2P 2M7, Canada

B. K. McMahon
McMahon Burgess & Yeates
Level 4, Chatswood Plaza
1 Railway St.
P. O. Box 648
Chatswood, 2067 NSW, Australia

Richard D. McNeely
Operations Supt.
Freeport Gold Co.
Mountain City Star Route
Elko, NV 89801

Greg Melton
Mining Engineer
Reynolds Metals Co. - Bauxite
P. O. Box 398
Bauxite, AR 72011

Jack Miller,
Asst. Mine Geol.,
Equity Silver Mines Ltd.,
P. O. Box 1450
Houston, BC V0J 1Z0, Canada

Stanley M. Miller
Dept. of Civil Engineering
The University of Wyoming
Laramie, WY 82071

Terrell W. Miller
Research Associate
Exxon Production Research Co.
P. O. Box 2189
Houston, TX 77001

Victor Miller
Senior Mining Engineer,
Dravo Corp.
1250 - 14th St.
Denver, CO 80202

Roger Missavage
Southern Illinois University
R. R. #1
West Frankfort, IL 62896

David J. W. Mitchell
Project Geologist
B. P. Canada Inc.
333 - 5th Ave. S. W.
Calgary, Alta. T1Y 1J7, Canada

Bibhu Mohanty
Research Physicist
C-I-L Inc.
Explosive Research Lab.
McMasterville, Que. J3H 1R3,
   Canada

Arturo Montano
General Mine Supt.
CIA Minera De Cananea
P. O. Box 1139
Douglas, AZ 85607

Jose Moreno-Campoy
Chief of Mine Planning
CIA Minera De Cananea
P. O. Box 1139
Douglas, AZ 85607

J. R. Morgan
Principal Geotechnical Eng.
Golder Associates Pty. Ltd.
466 Malvern Rd.
Prahran, Vic. 3181, Australia

Michael W. Mosley
Mine Engineering Supt.
Estech Gen. Chem. Corp.
P. O. Box 208,
Bartow, FL 33830

Allan Moss
Steffen Robertson & Kirsten
#500 - 1281 W. Georgia St.
Vancouver, BC V6E 3J7, Canada

Pascal Muzard
Sr. Planning Engineer
Petro-Canada
665 8 St. S.W.
Calgary, Alta. T2P 3E3, Canada

Frank Mylrea
Crippen Consultants Ltd.
1605 Hamilton Ave.
North Vancouver, BC, Canada

Honorio M. Narciso
Esso Minerals Canada
Coal Dept., Rm 1011
Esso Plaza
237 4th Avenue S.W.
Calgary, Alta. T2P 0H6, Canada

Cid E. Neves
MBR
Av. Graca Aranha
26 - 18 Floor
20005 Rio De Janeiro, Brazil

Rob S. Nichols
Senior Planning Engineer
Fording Coal Ltd.
P. O. Box 100
Elkford, BC V0B 1H0, Canada

Ralph O'Grady
Mine Planning Engineer
Syncrude Canada Ltd.
BAG 4009,
Fort McMurray, Alta. T9H 3L1,
   Canada

Terry L. Olmsted
Principal Engineering Geologist
Hart Crowser & Associates Inc.
Design Service Bldg.
1910 Fairview Ave. East
Seattle, WA 98102

Lou Oriard
3502 Sagamore Dr.
Huntington Beach, CA 92649

R. Pakalnis
Engineer
C/o Dept. of Mining Engineering
University of British Columbia
Vancouver, BC, Canada

Roger A. Paul
Engineering Geologist
WIDCO
1015 Big Hanaford Rd.
Centralia, WA 98531

D. L. Pentz
Golder Associates
10628 N.E. 38th Place
Kirkland, WA 98033

W. Gordon Peters
Mine Manager - Cerrejon Intercor
Carrera 7a No. 37069 Piso 8
Bogota, Columbia

D. R. Piteau
D. R. Piteau & Associates Ltd.
408 - 100 S. Park Royal
West Vancouver, BC, Canada

Shri M.G.R. Rao
Sr. Executive Engineer
Western Coalfields Ltd.
Coal Estate
Nagpur, India

John J. Reed
Consultant
P. O. Box 126
Anacortes, WA 98221

Ray Reipas
Mining Engineer
Crows Nest Resources Ltd.
P. O. Box 2003
Sparwood, BC V0B 2G0, Canada

Jovenal A. Relativo
Mine Supt., Open Pit
CDCP Mining Corp.
2nd Floor, Alco Bldg.
391 Buendia Ave. Ext.
Makati, Metro Manila, Philippines

R. J. Rennie
Monenco Ltd.
900 - 1 Palliser Square
125, 9th Ave. S.E.
Calgary, Alta. T2G 0P6, Canada

A. J. Richardson
B.C. Ministry of Energy, Mines
   & Petrol. Resources
Rm 105 - 525 Superior St.
Victoria, BC V8V 1T7, Canada

Michael Richings
Golder Associates Inc.
12345 W. Alameda Parkway
Lakewood, CO 80228

John C. Roberts
Project Engineer
Anaconda Copper Co.
P. O. Box 1268
Tonopah, NV 89049

J. W. Robinson
B.C. Ministry of Energy, Mines
& Petrol. Resources
2569 Kenworth Rd.
Nanaimo, BC V9T 4P7, Canada

Keith E. Robinson
Executive Manager
Robinson Dames & Moore
445 Mountain Highway
North Vancouver, BC V7J 2L1,
Canada

Robert Robinson
Engineering Geologist
Shannon & Wilson Inc.
1105 N. 38th St.
Seattle, WA 98103

W. C. Robinson
Ministry of Energy, Mines &
Petrol. Resources
Rm 105 - 525 Superior St.
Victoria, BC V8V 1T7, Canada

Phil Rosen
Engineer of Mines
Dept. of Labour, Occupational
Health & Safety
1150 Rose St.
Regina, Sask. S4R 1Z6, Canada

Mr. Rosengren
Golder Associates Pty. Ltd.,
466 Malvern Rd.
Prahran, Vict. 3181, Australia

Ian Rozier
Golder Associates
224 West 8th Ave.
Vancouver, BC, Canada

E. Sadar
B.C. Ministry of Mines, Energy &
Petrol. Resources
Rm 105 - 525 Superior St.
Victoria, BC V8V 1T7, Canada

Gilbert Sauriol
Chief Mines Engineer
Noranda Mines Ltd.
Division Mines Gaspe
Murdochville, Que. G0E 1W0,
Canada

Javier Sauza Sodinez
Minera Del Norte S.A.
Zaragoza 1000 S
D-208, Monterrey NL, Mexico

James P. Savely
Chief Geological Engineer
Inspiration Consolidated Copper
Co.
P. O. Box 4444
Claypool, AZ 85532

Udo F. Schmeling
MBR Mineracoes Brasileiras
Reunidas S.A.
Ave. Graca Aranha, 26 - 18 Floor
20005 Rio De Janeiro, Brazil

Robert Scott
Amax of Canada
P. O. Box 1000
Kitsault, BC V0V 1J0, Canada

Ben Seegmiller
Seegmiller Associates
447 East 200 South
Salt Lake City, UT 84111

A. A. Serrano
Professor
Laboratorio Carreteras &
Geotecnia
Alfonso XII - 3
Madrid 7, Spain

Rohini P. Sharma
Advanced Mining Engineer
Duval Corp.
4715 E. Fort Lowell Rd.
Tucson, AZ 85712

Robert Sharon
Geologist
Seegmiller International
143 South 400 East
Salt Lake City, UT 84111

Kunihiko Shibata
Geologist
Idemitsu Kosan Co. Ltd.
1200 - 840, 7th Ave. S. W.,
Calgary, Alta., Canada

Yong Fook Shin
General Manager,
Rahman Hydraulic Tin Berhad
1280, Jalan Bahru
Chai Leng Park
Prai, Malaysia

Don H. Shultz
Mine Superintendent
Dawn Mining Co.
P. O. Box 25,
Ford, WA 99013

Frank Sibbel
Operations Manager,
Greenbushes Tin Mine
P. O. Box 30
Greenbushes, WA 6254, Australia

Joe Siefke
Geologist
U.S. Borax
Boron, CA 93516

Casey Simms
Geophysicist
Cooksley Geophysics Inc.
P. O. Box 1602
Redding, CA 96099

Dave Smith
Manager
Thurber Consultants Ltd.
305 - 1550 Alberni St.
Vancouver, BC V6G 1A5, Canada

D. Smith
B.C. Ministry of Energy, Mines &
    Petrol. Resources
Rm 105 - 525 Superior St.
Victoria, BC V8V 1T7, Canada

Gary Smith
Mine Geologist
Amax of Canada
General Delivery
Kitsault, BC V0V 1J0, Canada

Joe Smith
Development Supervisor
Iron Ore Co. of Canada,
P. O. Box 1000
Labrador City, Nfld. A2V 2L8,
    Canada

Rod Smith
Klohn Leonoff Consultants
10180 Shellbridge
Richmond, BC, Canada

Vince Solano
Senior Mines Engineer
Syncrude Canada Ltd.
Mildred Lake
Bag 4009,
Fort McMurray, Alta. T9H 3L1,
    Canada

Peter F. Stacey
Golder Associates
224 West 8th Ave.
Vancouver, BC V5Y 1N5, Canada

Alan F. Stewart
Associate
D. R. Pitteau & Associates Ltd.
408 - 100 S. Park Royal
West Vancouver, BC V7T 1A2,
    Canada

Doug Stewart
Afton Mines,
P. O. Box 937
Kamloops, BC V2C 5N4, Canada

Brian Stimpson
Associate Professor of Mining
University of Alberta
Edmonton, Alta. T6G 2G6,
    Canada

Gerald T. Sweeney, AIPG
21869 56th Ave.
Langley, BC V3A 7N6, Canada

Simon T. Tarbutt
Chief Engineer at Toquepala,
    Peru
Southern Peru Copper Corp.
120 Broadway
New York, NY 10271

John Taylor
Manalta Coal Ltd.
Box 100
Halkirk, Alta. T0C 1M0, Canada

C. Scott Thomas
Geotechnical Engineer
Amax Coal Co.
105 S. Meridian St.
Indianapolis, IN 46225

R. W. Thompson
CTL/Thompson, Inc.
1971 W. 12th Ave.
Denver, CO 80204

Jack A. Thomson
Inspector of Mines, Technician
Province of British Columbia
Ministry of Mines
101, 2985 Airport Dr.
Kamloops, BC V2B 7W8, Canada

Bill Tijman
Slope Indicator Co.
3668 Albion Place North
Seattle, WA 98103

Joe Tomica
Senior Mining Engineer
Energy Resources Conservation
   Board
640, 5th Ave. S.W.
Calgary, Alta. T2P 3G4, Canada

W. A. Trythall
Chief Engineer
Placer Development Ltd.
700 Burrard Bldg.
1030 W. Georgia St.
Vancouver, BC V6E 3A8, Canada

Louis Tsang
Highmont Operating Corp.
Logan Lake, BC V0K 1W0,
   Canada

Mark G. Utting
Sr. Staff Hydrogeologist
Hart Crowser & Associates Inc.
Design Service Bldg.
1910 Fairview Ave. East
Seattle, WA 98102

Dominador C. Uy
Asst. Vice President
Atlas Cons. Mining &
   Development Corp.
P. O. Box 223
Cebu, Philippines

Doug VanDine
Professor
Queen's University
Dept. of Geol. Sciences
Kingston, Ont. K1L 3N6, Canada

T. Vaughan-Thomas
B. C. Ministry of Mines, Energy
   & Petrol. Resources
Rm 105 - 525 Superior St.
Victoria, BC V8V 1T7, Canada

Tom Valdut
Techman Engineering Ltd.
840 - 7th Ave. S. W.
Calgary, Alta. T2P 2M7, Canada

J. C. Wells
Chief Mine Engineer
Anamax Mining Co.
P. O. Box 127
Sahuarita, AZ 84629

Richard Willette
Superintendent, Hector Mine
N. L. Chemicals
P. O. Box 219
Newberry Springs, CA 92365

Kevin Williams
Project Engineer
Can Geol, Eng. Inc.
4381 Gallant Ave.
North Vancouver, BC, Canada

Roger Williams
Amax of Canada
P. O. Box 1000
Kitsault, BC V0V 1J0, Canada

Clive Wilson
7 Carmichael Place
Woodland, Durban, South Africa

Buck Wong
Mine Engineer
Denison Mines Ltd.
Box 11575 - 650 West Georgia St.
Vancouver, BC V6B 4N7, Canada

Ray H. Yamamoto
Geotechnical Engineer
Syncrude of Canada Ltd.
705 - 132, 21 MacDonald Dr.
Fort McMurray, Alta., Canada

Larry Yano
Geotechnical Engineer
Syncrude Canada Ltd.
Mildred Lake, Bag 4009
Fort McMurray, Alta, T9H 3L1,
   Canada

C. Zanbak
Department of Geology
Kent State University
Kent, OH 44242

Zavis Zavodni
Kennecott Minerals Co.
Engineering Center
P. O. Box 11248
Salt Lake City, UT 84147